AF821014

# Translational Medicine Research

**Series editors**

Zhu Chen, Shanghai Jiaotong University, Shanghai, China

Xiaoming Shen, Shanghai Jiaotong University, Shanghai, China

Saijuan Chen, Shanghai Jiaotong University, Shanghai, China

Kerong Dai, Shanghai Jiaotong University, Shanghai, China

**Aims and Scope**

In collaboration with National Infrastructures for Translational Medicine (Shanghai), the largest translational medicine research center in China, the book series "Translational Medicine Research" offers a state-of-the-art resource for physicians and researchers alike who are interested in the rapidly evolving field of translational medicine. It features original and observational investigations in the broad fields of laboratory, clinical and public health research, providing practical and up-to-date information on significant research from all subspecialties of medicine and broadening readers' horizons, from bench to bed and bed to bench.

With a focus on global interdisciplinary academic collaboration, the series aims to expedite the translation of scientific discovery into new or improved standards of management and health outcomes practice.

**Series Description**

Translational medicine converts promising laboratory discoveries into clinical applications and elucidates clinical questions with the use of bench work, aiming to facilitate the prediction, prevention, diagnosis and treatment of diseases. The development of translational medicine will accelerate disease control and the process of finding solutions to key health problems. It is a multidisciplinary endeavor that integrates research from the medical sciences, basic sciences and social sciences, with the aim of optimizing patient care and preventive measures that may extend beyond health care services. Therefore, close and international collaboration between all parties involved is essential to the advancement of translational medicine.

To enhance the aforementioned international collaboration as well as to provide a forum for communication and cross-pollenation between basic, translational and clinical research practitioners from all relevant established and emerging disciplines, the book series "Translational Medicine Research" features original and observational investigations in the broad fields of laboratory, clinical and public health research, aiming to provide practical and up-to-date information on significant research from all subspecialties of medicine and to broaden readers' vision horizons, from bench to bed and bed to bench.

Produced in close collaboration with National Infrastructures for Translational Medicine (Shanghai), the largest translational medicine research center in China, the book series offers a state-of-the-art resource for physicians and researchers alike who are interested in the rapidly evolving field of translational medicine. Prof. Zhu Chen, the Editor-in-Chief of the series, is a hematologist at Shanghai Jiao Tong University, China's former Minister of Health, and chairman of the center's scientific advisory board.

More information about this series at http://www.springer.com/series/13024

Daxiang Cui
Editor

# Gastric Cancer Prewarning and Early Diagnosis System

上海交通大學出版社
SHANGHAI JIAO TONG UNIVERSITY PRESS

Springer

*Editor*
Daxiang Cui
Institute of Nano Biomedicine and Engineering
Shanghai Jiao Tong University
Shanghai
China

ISSN 2451-991X ISSN 2451-9928 (electronic)
Translational Medicine Research
ISBN 978-94-024-0949-9 ISBN 978-94-024-0951-2 (eBook)
DOI 10.1007/978-94-024-0951-2

The print edition is not for sale in China Mainalnd. Customers from China Mainland please order the print book from: Shanghai Jiao Tong University Press.

This co-published book was advertised with a copyright holder "Springer Science+Business Media Dordrecht" in error, whereas "Springer Science+Business Media B.V. and Shanghai Jiao Tong University Press, Shanghai" are holding the copyright.

Library of Congress Control Number: 2016961959

© Springer Science+Business Media B.V. and Shanghai Jiao Tong University Press, Shanghai 2017
This work is subject to copyright. All rights are solely and exclusively licensed by the Publisher, whether the whole or part of the material is concerned, specifically the rights of translation, reprinting, reuse of illustrations, recitation, broadcasting, reproduction on microfilms or in any other physical way, and transmission or information storage and retrieval, electronic adaptation, computer software, or by similar or dissimilar methodology now known or hereafter developed.
The use of general descriptive names, registered names, trademarks, service marks, etc. in this publication does not imply, even in the absence of a specific statement, that such names are exempt from the relevant protective laws and regulations and therefore free for general use.
The publisher, the authors and the editors are safe to assume that the advice and information in this book are believed to be true and accurate at the date of publication. Neither the publisher nor the authors or the editors give a warranty, express or implied, with respect to the material contained herein or for any errors or omissions that may have been made. The publisher remains neutral with regard to jurisdictional claims in published maps and institutional affiliations.

Printed on acid-free paper

This Springer imprint is published by Springer Nature
The registered company is Springer Science+Business Media B.V.
The registered company address is: Van Godewijckstraat 30, 3311 GX Dordrecht, The Netherlands

# Preface

Gastric cancer is the fourth most common cancer and the second leading cause of cancer-related death worldwide. It ranks number two among all malignant tumors in China. The gastric cancer prognosis is very poor with 5-year survivals below 24 %. How to realize gastric cancer prewarning and early diagnosis has become a key scientific problem. In 1998, my tutor Professor Chenzhi Su and Professor Xiaojun Yan in the Fourth Military Medical University put forward the concept of gastric cancer prewarning and early diagnosis system, designed its frame, obtained a key budget from the Department of Public Health of Shanxi Province, and then initiated its development.

Nanotechnology advances very fast. Nanomaterials and nanotechnology have integrated into gastric cancer theranostics, as a new emerging multidisciplinary frontier, exhibiting great potential in applications such as the prewarning and early diagnosis of gastric cancer. Under the support of the Chinese Nano Fundamental Research Project (2010CB933900), wrapping four key scientific problems such as gastric cancer biomarkers, ultrasensitive detection methods, nanoprobe-based qualitative and quantitative visualization, and biosafety of nanoprobes, our team finished some important works; it is very necessary to publish the book to introduce the advances associated with gastric cancer prewarning and early diagnosis system, which will be helpful to clinical doctors, undergraduate students, PhD students, and postgraduates.

This book contains a collection of major research accomplishments in the past decade or so in the area of specifically gastric cancer prewarning and early diagnosis system based on molecular biology, biological information, and nanotechnology in my team, mainly including screening gastric cancer biomarkers, developing ultrasensitive detection methods of gastric cancer biomarkers, designing and synthesizing molecular imaging probes for in vivo gastric cancer-targeted imaging and simultaneous therapy, gastric cancer operation boundary identification, metastasis lymph node tracking, gastric cancer stem cells identification and killing, and enhanced immuno-protection methods of gastric cancer patients, as well as clinical translation study and bioinformatic database.

This book includes 15 chapters. Chapter 1 exhibits the background of gastric cancer prewarning and early diagnosis system; Chaps. 2, 3, 4, 5, 6, and 7 exhibit gastric cancer-associated biomarker screening and ultrasensitive detection methods; Chaps. 8, 9, and 10 exhibit established detection methods for gastric cancer biomarkers; Chaps. 11, 12, 13 and 14 exhibit multifunctional nanoprobes for theranostics of gastric cancer and biosafety assessment; and Chap. 15 exhibits part of the gastric cancer prewarning database and bioinformatic analysis method.

This book aims to improve the clinical translation of gastric cancer prewarning and early diagnosis system and to be the definitive reference book for scientists in the field of nanomedicine, theranostics, molecular imaging, and therapy. It is my hope that it can stimulate the interest of researchers and clinical doctors associated with these fields.

Finally, I would like to thank all members of my team and the staff of Springer for the support in making this book into reality.

Shanghai, China Daxiang Cui

# Contents

# Chapter 1
# Background of Gastric Cancer Prewarning and Early Diagnosis System

**Daxiang Cui**

Gastric cancer is the fourth most common cancer and the second leading cause of cancer-related death worldwide [1, 2]. In China, according to the latest cancer disease statistics, gastric cancer ranks as no. two in incidence and no. three in mortality rate of all malignant tumors [3]. The prognosis of gastric cancer is very poor, with 5-year survival less than 24 % [3, 4]. Early gastric cancer can be cured with surgery, while advanced gastric cancer often needs combined multidisciplinary therapy. Gastric cancer seems to be insensitive to current chemotherapy agents, a feature that may be closely related to the characteristics of stomach cancer stem cells [4, 5]. "Early gastric cancer" refers to in situ gastric cancer, which is located only in the gastric mucous membrane, and does not infiltrate into the thin submucosa. In China, the diagnosis rate of early gastric cancer is less than 20 %, whereas the diagnosis rates of early gastric cancer in Japan and South Korea have reached 30–50 % [6, 7]. Solving the problem of prewarning and early diagnosis of gastric cancer could not only save many patients with gastric cancer but also reduce the cost of treatment and even cure or significantly prolong the lifespan of patients.

## 1.1 Current Clinical Status of Gastric Cancer

The stomach is the body's most important digestive organ. The human stomach is located in the left hypochondrium, largely on the left side of the body's midline, with only a small portion on the right. The stomach is divided into four layers (from

D. Cui
Institute of Nano Biomedicine and Engineering, Shanghai Engineering Research Center for Intelligent Diagnosis and Treatment Instrument, National Center for Translational Medicine, Collaborative Innovational Center for System Biology, Shanghai Jiao Tong University, 800 Dongchuan Road, Shanghai, 200240, P. R. China
e-mail: dxcui@sjtu.edu.cn

© Springer Science+Business Media B.V. and Shanghai Jiao Tong University Press, Shanghai 2017

D. Cui (ed.), *Gastric Cancer Prewarning and Early Diagnosis System*, Translational Medicine Research, DOI 10.1007/978-94-024-0951-2_1

the innermost to the outermost): mucosa, submucosa, muscularis, and serosa. Gastric cancer originates in the epithelial cells of the outermost layer, but it can occur in various parts of the stomach, such as the pyloric antrum (the most common area), the gastric cardia region, and, less likely, the gastric body. It may invade the stomach wall at different depths and breadths. With the naked eye or by endoscopy, gastric cancer can be observed to have a variety of forms, such as superficial, ulcerative, infiltrative, ulcerative carcinoma (cancer arising from chronic gastric ulcer), or mass. There are many histological classifications of cancer, such as adenocarcinoma (accounting for about 90 %; including papillary adenocarcinoma, tubular adenocarcinoma, mucinous adenocarcinoma, and signet-ring-cell carcinoma), adeno-squamous cell carcinoma, squamous cell carcinoma, undifferentiated carcinoma, and carcinoid tumor [8, 9].

Gastric cancer is a malignant tumor; new cases of gastric cancer number more than 1 million each year globally, and worldwide more than 0.6 million deaths each year are due to the disease, with Japan, China, and South Korea having the highest incidence worldwide [10, 11, 31]. According to the 2012 China Cancer Registration Report, the incidence and mortality of gastric cancer rank as nos. two and three, respectively, among all malignant tumors in China [3]. In this report, the ratio of morbidity to mortality in both men and women was 2:1, and the incidence increased with age, being significantly higher among those aged 50 to 80 years; however, a younger trend has recently been noted, with the proportion of patients with gastric cancer aged 19–to 40 years increased from the originally reported 1.7 % to the currently reported 3.3 % [12]. An annual growth rate of 2.3 % is predicted in the number of patients with gastric cancer in China [13, 14].

Gastric cancer rarely arises directly from the normal gastric mucosa; cancer formation shows a long process of evolution: from normal gastric mucosa to non-atrophic gastritis to atrophic gastritis to intestinal metaplasia to dysplasia to cancer. In gastric cancer science, this sequence of changes is universally recognized as the gastric cancer development model [15, 16]. The occurrence and progression of gastric cancer is a multifactorial process involving multiple gene interactions. It involves the evolution of a number of external and internal change factors. External environmental factors include *Helicobacter pylori* infection, dietary factors causing chronic inflammation, and other environmental factors; internal factors include the activation of oncogenes and inactivation of tumor suppressor genes, as well as the presence of some growth factors involved in DNA microsatellite instability caused by unlimited gastric epithelial cell proliferation [17].

As early as 1978, the World Health Organization (WHO) London Conference unified the concept of the pathological changes occurring in the body before gastric carcinoma occurs. Concepts of a precancerous state and precancerous lesions were defined. A precancerous state means a state associated with gastric diseases that pose a risk of gastric cancer, such as chronic atrophic gastritis, gastric polyps, gastric ulcers, gastritis, and other conditions. Precancerous lesions are a pathological concept, referring to pathological changes that are easily transformed into cancerous tissue, including atypical hyperplasia or dysplasia and intestinal metaplasia. Chronic atrophic gastritis is considered to be a common precancerous condition, and gastric glandular atrophy and chronic inflammation are its main pathological features, accompanied by gastric dysplasia and intestinal metaplasia [18, 19]. Numerous studies indicate that there is a certain relationship between chronic atrophic gastritis and gastric cancer [20]. According to statistics, 67.8–92.3 % of gastric carcinoma specimens included chronic atrophic gastritis with intestinal metaplasia, while about 50 % of gastric cancer specimens included precancerous diseases such as atrophic gastritis [21, 34]. In Japan, 4665 cases of healthy asymptomatic individuals were observed for a period of 7.7 years, and it was found that gastric cancer risk showed a gradual upward trend from chronic superficial gastritis, to chronic atrophic gastritis, and to severe atrophic gastritis with intestinal metaplasia; the probability of gastric cancer being caused by a low degree of dysplasia was 9 %, and the probability of gastric cancer being caused by a high degree of dysplasia was 74 % [22]. Inoue et al. [23] studied 5373 cases of chronic atrophic gastritis; the median follow-up time was 10 years, and 117 cases of cancer occurred. Moderate atrophic gastritis exhibited the highest risk of gastric cancer, with a risk factor of 2.22, and after a 4- to 6-year follow-up, the peak value for cancer risk was 5.0 [24]. Shigeto et al. [25] reported on 2859 healthy people (1011 males and 1848 females) followed-up at health examinations for 11 years, of whom ultimately 61 persons (33 men and 28 women) were found to have gastric cancer. The report also revealed that patients suffering from both chronic *H. pylori* infection and atrophic gastritis had a high risk of gastric cancer, while those patients suffering from chronic atrophic gastritis without *H. pylori* infection had the highest risk of gastric cancer [16, 26]. Chinese clinical studies also found that 1.2–7.1 % of patients with atrophic gastritis finally developed gastric cancer [27]. Diagnosis and treatment at an early stage is the most critical factor in improving the survival rate of gastric cancer. For patients with advanced gastric cancer and advanced metastasis, in whom repeated lymph node, blood, and peritoneal relapses occur, the prognosis is very poor; even in those who underwent radical surgery, the 5-year survival rate was only 5–20 % [3, 4]. After surgical resection and adjuvant therapy, patients with early gastric cancer have a 5-year survival rate of 90–95 % [3, 4].

In China, the current status of the diagnosis and treatment of gastric cancer presents "one high, three low" characteristics; namely, a high incidence of mortality, and low rates of early diagnosis, radical surgery, and 5-year survival [4]. In China, the early gastric cancer detection rate is only 10–20 % [16, 27], while in South Korea the rate is about 30 % [28, 29] and in Japan the rate is up to 30–50 % [23, 30]. There is a considerable gap in the prevention and treatment of gastric cancer in China compared with the international advanced levels in Japan and South Korea, etc. It has been shown that nearly 80 % of early gastric cancers have no symptoms, and the remaining 20 % show only a few symptoms, such as mild ulcer-like dyspepsia, or similar nonspecific symptoms, which means that finding early gastric cancer in an asymptomatic population entails certain difficulties [16, 27, 29].

## 1.2 Main Diagnostic Methods for Gastric Cancer in the Clinical Setting

Currently, four methods are ordinarily used for the diagnosis of early gastric cancer: gastric endoscopy, imaging tools such as computed tomography (CT), pathological examination, and serological examination.

### *1.2.1 Endoscopy*

Endoscopy (gastroscopy) is the most effective preferred method for early gastric cancer examination in the clinical setting. It can provide much information about gastric cancer lesions, such as their position, shape, and size; it has a wide field of vision and strong resolution; and it can be used to collect many samples for pathology examination, with high accuracy. The use of biopsy samples sprayed with pigment; brush cell slices; and smears can obviously improve the detection rate [35]. Without endoscopic examination, it is difficult to assess the level of tumor invasion and more difficult to assess the surrounding organs that have been invaded. However, gastroscopy is expensive and requires the operator to have certain experience; it may also cause pain and secondary damage. Patients with high blood pressure, blood diseases, and heart diseases are unfavorable candidates for gastroscopy.

### *1.2.2 Imaging Diagnosis*

X-ray and CT are conventional methods for gastric cancer imaging examinations. CT imaging shows the normal stomach outline very clearly, in which the internal and external status of the stomach can be observed, as well as organs with remote metastasis; CT has a unique diagnostic effect for gastric cancers that grow inside the stomach wall or between the stomach layers, and it is superior to endoscopy and gastrointestinal angiography examination. CT is used for the assessment of lymph nodes; lymph nodes with a size of more than 1 cm can be diagnosed as lymph node metastases. Therefore, CT imaging improves the detection rate of gastric cancer, especially early gastric cancer, and it allows precise tumor staging by observing abnormal thickening of the gastric wall and changes in the gastric mucosa. Relevant data show that the sensitivity of CT detection of T1 and T2 gastric cancers was 75 %, and the concurrence rate of CT and clinical diagnosis was 68 %, while the sensitivity of CT detection of T3 and T4 gastric cancers was 98 %, and the concurrence rate of CT and clinical diagnoses was 88 %; these findings strongly suggest that CT has a high detection rate for the diagnosis of advanced gastric cancer, but for the diagnosis of early gastric cancer there is a low concurrence rate between CT detection and clinical diagnosis. Therefore, CT imaging is mainly used to determine whether distant lymph node metastasis (N phase) and liver metastasis (M phase) exist [32, 36].

Early gastric cancer and precancerous lesions are not usually found by gastroscopy, X-ray, CT, or magnetic resonance imaging (MRI) and are easily missed. For those patients with early lesions who often self-medicate and do not see a doctor, the symptoms are ignored, and the patients are often found to have distant metastasis; thus, the best treatment opportunity is lost. It is reported that in 84.5 % of gastric carcinoma patients the disease is first diagnosed when it is medium or late gastric cancer. How to use these imaging tools to increase the diagnostic ratio of early gastric cancer is a key scientific problem.

### *1.2.3 Histopathological Examination*

Histopathology is the gold standard for the diagnosis of gastric cancer. Many samples, collected from multiple gastric areas by using endoscopy, can be used for pathology examination, and gastric cancer patients can be clearly diagnosed and precise pathological classification can be realized. Postoperative pathology examination can lead to the comprehensive understanding of the classification and staging of the tumor, and lymph node metastasis, and the relationship of the tumor with the adjacent organs, and can provide the basis for postoperative adjuvant therapy and the determination of prognosis. However, because the sample-collecting depth is limited, pathology examination often cannot determine the depth of tumor invasion and the relationship between the tumor and adjacent organs, and comprehensive analysis combining imaging data and other information is needed [32, 33].

### *1.2.4 Serological Biomarker Examination*

In order to identify patients with early gastric cancer in a large asymptomatic population, the best way is to carry out a natural population census of mass serological biomarkers. According to advice from the American Gastroenterological Association, patients who are more than 45 years of age or those who have atypical hyperplasia should be examined regularly by using gastroscopy. For persons above 40 years of age who have signs such as repeated abdominal discomfort, bloating, pain, and unexplained indigestion, the possibility of gastric cancer should be considered, and relevant inspections in a timely manner are necessary. In those areas with high rates of *H. pylori* infection, individuals less than 40 years of age should be checked regularly with endoscopy. Japan's early gastric cancer screening is the most successful of that in any country. In a recent 10–year period in Japan, gastric air bubble and barium double-contrast imaging, and gastroscopy examination were used to screen around 5 million people per year, and 3000–6000 cases of gastric cancer per year were found; the discovery rate of gastric cancer was about 0.12 %, and 50 % of cases were early gastric cancer [37].

China has a population of 1.4 billion, and economic conditions and medical resources are not sufficient for the large-scale screening of gastric cancer patients as is done in Japan; therefore, it is necessary to develop low-cost, simple, and rapid methods suitable for early gastric cancer screening based on new technology principles. Tumor biomarker screening methods, compared with gastroscopy and imaging examinations such as CT and MRI, have the following obvious advantages: they are economical, fast, accurate, and are able to find early tumors, thus gaining enough time for early clinical treatment. At present, tumor biomarker detection technology could be a unique and effective method to screen out those patients with asymptomatic gastric cancer, and could be a valuable index for the detection of early gastric cancer. In China, the development of simple and rapid biomarker detection methods and devices could be one of the best pathways to solve the problem of early gastric cancer screening.

## 1.3 Concept of Gastric Cancer Prewarning and Early Diagnosis System

To date, there has been no effective gastric cancer prewarning and early diagnosis system that could be used clinically. Molecular biology studies show that the molecular mechanism of gastric cancer is associated with many genes and proteins. The process of gastric cancer formation has three stages: the normal gastric mucosa stage, the precancerous lesion stage, and the gastric cancer stage [38]. In each stage, there are many differentially expressed genes and proteins. Understanding the biological processes of cancer initiation at the gene expression level is very important for early cancer detection. The study of gene expression levels at different stages of growth and disease, at different cell cycle timings, and on response to stimulation helps to answer why different stages of cancerous development occur. In the course of the development of normal gastric mucosa to precancerous lesions and the development of precancerous lesions to gastric cancer, characteristic genes and proteins are differently expressed. Real-time examination of the changes in the transcript and translation levels of these genes and proteins could lead to the realization of a gastric cancer prewarning and early diagnosis system. How to find the key differentially expressed genes and proteins involved in the initiation and progression of gastric cancer has become a key scientific problem.

Our team first proposed the concept of a gastric cancer prewarning and early diagnosis system in 1998 [39, 40]. This system employs a high-tech information-processing system and a biochip technology platform for support. Changes in the expression levels of the genes or proteins differentially expressed in normal gastric mucosa, precancerous lesions, and gastric cancer were investigated in gastric biopsy specimens or blood samples, and a bioinformatics analysis tool was used to compare the examined results with those in a prewarning database, and thus to determine whether the patient was in the gastric cancer, precancerous lesion, or normal mucosa stage. The system consists of three parts; that is, the gastric cancer prewarning chip, the gastric cancer prewarning database, and the information-processing system. In order to establish the system, we completed the screening of the genes differentially expressed in normal gastric mucosa, precancerous lesions, and gastric cancer; designed and fabricated the gastric cancer gene expression profile chip; developed the gastric cancer information-processing software; and established a primary gastric cancer prewarning data library.

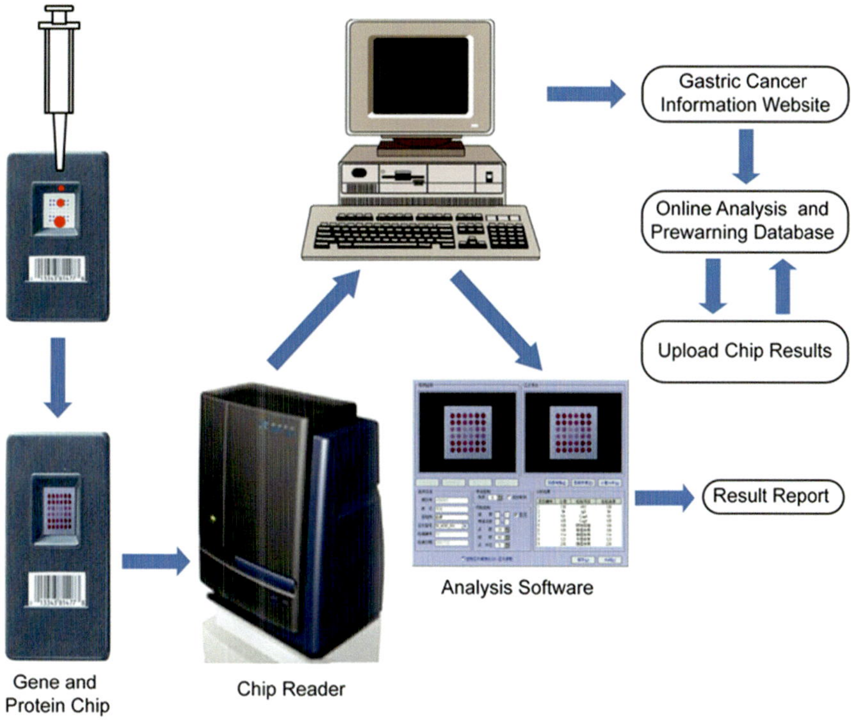


A total of 412 genes associated with the three stages of gastric cancer development were identified. There were 216 genes displaying higher expression in gastric cancer, 85 genes displaying higher expression in precancerous lesions, and 88 genes displaying higher expression in normal gastric mucosa. Also, 15 genes associated with gastric cancer metastasis and 8 genes associated with risk factors were included from classification as target genes in the early gastric cancer diagnosis chip. As shown in Table 1.1, the threshold values of the 412 genes selected to distinguish gastric cancer from precancerous lesions and normal gastric mucosa were defined as $6.01 \pm 2.4$, $4.84 \pm 1.94$, and $5.42 \pm 2.17$, respectively. These 412 selected genes and their critical threshold values were compiled into an analytical software, which could automatically provide reports by analyzing the results of the 412 genes obtained by the examination of gastric tissues. All data were compiled into a prewarning database for gastric cancer by using CGO is a compiling software name, it is very general, CGO software [41].

**Table 1.1** Gene expression threshold for distinguishing three kinds of gastric mucosa

| Gene classification | Gastric cancer tissue (GC/N) | Precancerous lesion (PC/N) | Normal gastric mucosa (N*/GC or N*/PC) |
|---|---|---|---|
| 216 genes associated with gastric cancer | 6.01±2.40 | 1.18±0.47 | <0.75 |
| 85 genes associated with precancerous lesions | 1.32±0.53 | 4.86±1.94 | 2.54±0.41 |
| 88 genes associated with normal mucosa | 1.31±0.54 | 2.50±0.75 | 5.42±2.17 |
| 15 genes associated with metastasis of gastric cancer | 5.81±2.32 (M) | 1.13±0.58 | 0.65±0.35 |
| | 2.32±1.19 ($N^1$) | | |
| 8 genes associated with risk factors | | >2.0 | |

Specification: The above data indicate the relative expression levels between GC/N, PC/N, N/PC, and N/GC mean ratio and minimum values
*M* metastasis, $N^1$ no metastasis, *GC* gastric cancer, *PC* precancerous lesion, *N* normal mucosa, *N** selected gene expression levels in normal gastric mucosa

The gastric cancer prewarning and early diagnosis system that we developed can be used for the diagnosis of advanced gastric cancer, analysis of prognosis, and evaluation of therapeutic effectiveness. However, the system cannot solve all the problems associated with gastric cancer diagnosis; for example, problems such as gastric cancer prewarning and gastric cancer genotyping. The gastric cancer prewarning and early diagnosis database system lacks a database for the serological examination of gastric cancer and lacks an imaging detection database. In order to make this gastric cancer prewarning and early diagnosis system suitable for clinical application, it is very important to improve the system. In order to optimize and perfect the system, three key scientific problems must be solved; that is, the problem of biomarkers for early gastric cancer, the problem of the ultrasensitive simultaneous detection of biomarkers, and the in-vivo quantitative and qualitative visual problems with gastric cancer biomarkers.

With the rapid development of nanotechnology, emerging nanomaterials and nanotechnology offer bright new opportunities to solve the three key scientific problems in the early warning and early diagnosis system for gastric cancer. Nanomaterials have four unique effects such as small size effects, quantum size effects, surface effects and grand quantum tunnel effects. In particular, their sound, light, electricity, heat, and magnetic properties not only enhance the sensitivity and specificity of biomarker detection, but can also provide qualitative and quantitative visualization of gastric cancer biomarkers [42]. With the support of the Nano Key Fundamental Research Project (2010CB933900), investigating key scientific problems associated with the early diagnosis of gastric cancer, we started the following studies: gastric cancer-related-biomarker screening and identification, nanoparticle-controlled synthesis and characterization, the development of methods to enhance the sensitivity and specificity of biomarkers for gastric cancer detection, the preparation of multifunctional nanoprobes and their application for multimodal imaging and in-vivo genotyping, the development of integrated theranostic technology, the evaluation of nanoprobe biosafety, and the performance of clinical trials.

In this book, I review and summarize recent progress in the development of our gastric cancer prewarning and early diagnosis system, with the main purposes being to optimize this system with the use of nanotechnology, actively promote the clinical validation of the system, and speed up the commercial development and clinical translation of this system.

## References

1. Cancer IARC (International Agency for Research on Cancer). Globocan 2012: estimated cancer incidence, mortality and prevalence worldwide in 2012. World Health Organization. http://globocan.iarc.fr/Pages/fact_sheets_cancer aspx. Accessed 2014; 9.
2. Siegel R, Naishadham D, Jemal A. Cancer statistics, 2013. CA Cancer J Clin. 2013;63:11–30.
3. Chen WQ, Zheng RS, Zhang SW, Zeng HM, Zou XN. The incidences and mortalities of major cancers in China, 2010. Chin J Cancer. 2014;33:402–5.
4. Takahashi T, Saikawa Y, Kitagawa Y. Gastric cancer: current status of diagnosis and treatment. Cancers (Basel). 2013;5:48–63.
5. Vinogradov S, Wei X. Cancer stem cells and drug resistance: the potential of nanomedicine. Nanomedicine (Lond). 2012;7:597–615.
6. Kong SH, Park DJ, Lee HJ, et al. Clinicopathologic features of asymptomatic gastric adenocarcinoma patients in Korea. Jpn J Clin Oncol. 2004;34(1):1–7.
7. Hamashima C, Shibuya D, Yamazaki H, et al. The Japanese guidelines for gastric cancer screening. Jpn J Clin Oncol. 2008;38(4):259–67.
8. Allum WH, Blazeby JM, Griffin SM, et al. Guidelines for the management of oesophageal and gastric cancer. Gut. 2011;60(11):1449–72.
9. Okines A, Verheij M, Allum W, et al. Gastric cancer: ESMO clinical practice guidelines for diagnosis, treatment and follow-up. Ann Oncol. 2010;21 suppl 5:v50–4.
10. Lee JH, Kim JJ. Endoscopic mucosal resection of early gastric cancer: experiences in Korea. World J Gastroenterol. 2007;13(27):3657–61.
11. Matsuda T, Marugame T, K-i K, et al. Cancer incidence and incidence rates in Japan in 2005: based on data from 12 population-based cancer registries in the Monitoring of Cancer Incidence in Japan (MCIJ) project. Jpn J Clin Oncol. 2011;41(1):139–47.
12. Zhao P, Dai M, Chen W, et al. Cancer trends in China. Jpn J Clin Oncol. 2010;40(4):281–5.
13. Sun X-D, Mu R, et al. Analysis of mortality rate of stomach cancer and its trend in twenty years in China. Chinese J Cancer Prevention Treatment. 2004;26(1):6–11.
14. Pinglu Z, et al. Comparative analysis of gastric cancer in young and pathological features and biological behavior of gastric cancer in elderly. Zhejiang Clin Med J. 2007;9(11):1518–9.
15. Sipponen P, Kimura K. Intestinal metaplasia, atrophic gastritis and stomach cancer: trends over time. Eur J Gastroenterol Hepatol. 1994;6 Suppl 1:S79–83.
16. Leng WK, Lin SR, Ching JY. Factors predicting progression of gastric intestinal metaplasia: results of a randomized trial on *Helicobacter pylori* eradication. Gut. 2004;53(9):1244–9.
17. Jemal A, Center MM, DeSantis C, et al. Global patterns of cancer incidence and mortality rates and trends. Cancer Epidemiol Biomarkers Prev. 2010;19(8):1893–907.
18. Mizuno S, Miki I, Ishida T, et al. Prescreening of a high-risk group for gastric cancer by serologically determined Helicobacter pylori infection and atrophic gastritis. Dig Dis Sci. 2010;55(11):3132–7.
19. Okamura T, Tsujitani S, Korenaga D, et al. Lymphadenectomy for cure in patients with early gastric cancer and lymph node metastasis. Am J Surg. 1988;155(3):476–80.
20. Everett S, Axon A. Early gastric cancer in Europe. Gut. 1997;41(2):142–50.
21. Nakamura K, Ueyama T, Yao T, et al. Pathology and prognosis of gastric carcinoma. Findings in 10,000 patients who underwent primary gastrectomy. Cancer. 1992;70(5):1030–7.

22. Tan YK, Fielding JW. Early diagnosis of early gastric cancer. Eur J Gastroenterol Hepatol. 2006;18(8):821–9.
23. Inoue M, Tajima K, Matsuura A, et al. Severity of chronic atrophic gastritis and subsequent gastric cancer occurrence: a 10-year prospective cohort study in Japan. Cancer Lett. 2000;161(1):105–12.
24. Lee HJ, Yang HK, Ahn YO. Gastric cancer in Korea. Gastric Cancer. 2002;5(3):177–82.
25. Shigeto N, Shimizu S. Eradication of Helicobacter pylori for idiopathic thrombocytopenic purpura. Nihon Shokakibyo Gakkai Zasshi. 2004;101(6):598–608.
26. Sipponen P, Kosunen T, Valle J, et al. Helicobacter pylori infection and chronic gastritis in gastric cancer. J Clin Pathol. 1992;45(4):319–23.
27. Sun Jie, et al. Correlation analysis. Gastric endoscopic diagnosis and biopsy results. Modern J Integrated Traditional Chinese Western Med. 2009;18(26):3175–6.
28. Choi IJ. Screening and surveillance of gastric cancer. Korean J Gastroenterol. 2007;48(supp 1):15–22.
29. Yang L. Incidence and mortality of gastric cancer in China. World J Gastroenterol. 2006; 12(1):17–20.
30. Jian S, et al. Early diagnosis and treatment of gastric cancer. Chinese J Pract Surg. 2011; 31(8):717–9.
31. Levy L, Gralnek IM. Complications of diagnostic colonoscopy, upper endoscopy, and enteroscopy. Best Pract Res Clin Gastroenterol. 2016;30(5):705–18.
32. Pan Z, et al. Evaluation of multislice helical CT scanning in TNM staging of gastric cancer. J Clin Radiol. 2002;21(8):614–7.
33. Vukobrat-Bijedic Z, Husic-Selimovic A, Sofic A, et al. Comparability of diagnostic methods: proximal endoscopy, CT and EUS in determining stomach tumor localization and their importance in the preoperative analysis of process progression. Acta Inform Med. 2013;21(4): 246–9.
34. Nie L, Li M, He X, Feng A, Wu H, Fan X. Gastric mixed adenoneuroendocrine carcinoma: correlation of histologic characteristics with prognosis. Ann Diagn Pathol. 2016;25:48–53.
35. Fukao A, Tsubono Y, Tsuji I, et al. The evaluation of screening for gastric cancer in Miyagi prefecture, Japan: a population-based case-control study. Int J Cancer. 1995;60(1):45–8.
36. Leung WK, Wu MS, Kakugawa Y, et al. Screening for gastric cancer in Asia: current evidence and practice. Lancet Oncol. 2008;9(3):279–87.
37. Zhu Y, et al. Clinicopathological and prognostic significance of multiple molecular genetic abnormalities in gastric cancer. Acta Universitatis Medicinalis Secondae Shanghai. 1996;16(4): 233–6.
38. Wang J, et al. Situation screening and diagnosis of early gastric cancer progression. Chongqing Med J. 2009;38(20):2634–6.
39. Xu B, Wang J. Epidemiological studies of gastric cancer. Chinese J Cancer Prevention Treatment. 2006;13(1):81–7.
40. Cui DX, Zhang L, Yan XJ, et al. A microarray-based gastric carcinoma prewarning system. World J Gastroenterol. 2005;11(9):1273–82.
41. Xiaojun Y. Early diagnosis of gastric cancer and the establishment of early warning systems. World Chinese J Digestol. 1999;7(2):96–7.
42. Ruan J, Ji JJ, Song H, et al. Fluorescent magnetic nanoparticle-labeled mesenchymal stem cells for targeted imaging and hyperthermia therapy of in vivo gastric cancer. Nanoscale Res Lett. 2012;7:12.

# Chapter 2
# Human Serum Protein Markers for Gastric Cancer Detection

**Hualin Fu and Daxiang Cui**

## 2.1 Introduction

By incidence rate, gastric cancer is the no. 2 cancer in China, after lung cancer; by mortality rate, gastric cancer is the no. 3 cancer in China, second only to lung cancer and liver cancer. Gastric cancer patients accounted for about 12.6 % of all cancer patients, and the number of deaths due to gastric cancer accounted for about 14 % of the number of all cancer-related deaths in the country. According to cancer statistics in 2010, 273,000 people died of gastric cancer in China [1]. In terms of mortality, gastric cancer is one of the most malignant cancers in all cancers. Its mortality rate is second only to pancreatic cancer, mesothelioma, liver cancer, lung cancer, and esophageal cancer. A very important reason for the relatively high mortality of gastric cancer patients is that early gastric cancer is often difficult to detect. Patients with gastric cancer are often found at the advanced stage of cancer. According to the relevant study, the 5-year survival rate of early gastric cancer patients after treatment can reach 60–90 %, and the 5-year survival rate of patients with advanced gastric cancer is only 5–30 %. The early diagnosis of gastric cancer often leads to better gastric cancer prognosis [2–4]. Screening of gastric cancer has been done nationwide in Japan and South Korea. These two countries and China are having the highest gastric cancer incidence in Asia. For 40 years, gastric cancer had been the most deadly cancer in Japan. The Japanese government advocated the barium meal X-ray method for early gastric cancer screening in 1957 [5]. By 1975, Japan's

H. Fu (✉) • D. Cui
Institute of Nano Biomedicine and Engineering,
Shanghai Engineering Research Center for Intelligent Diagnosis and Treatment Instrument,
National Center for Translational Medicine, Collaborative Innovational Center
for System Biology, Shanghai Jiao Tong University, 800 Dongchuan Road,
Shanghai 200240, P. R. China
e-mail: hfu@sjtu.edu.cn; dxcui@sjtu.edu.cn

© Springer Science+Business Media B.V. and
Shanghai Jiao Tong University Press, Shanghai 2017

D. Cui (ed.), *Gastric Cancer Prewarning and Early Diagnosis System*,
Translational Medicine Research, DOI 10.1007/978-94-024-0951-2_2

stomach cancer mortality rate began to decline with the help of the large-scale gastric cancer screening. In 1983, Japan adopted the Elderly Health Care Act, providing cancer screening services for the old people. According to the plan, the government provides a voluntary base screening program for gastric cancer, cervical cancer, breast cancer, and lung cancer [6]. However, the participation rate of gastric cancer screening program hovered at around 13 % since the acceptance rate of barium meal X-ray screening in the elderly is not high. South Korea began a national cancer screening program in 2003 to provide free stomach cancer, breast cancer, and cervical cancer screening services. A liver cancer screening service was added in 1999, and a colorectal cancer screening service was added in 2004 [7]. South Korea cancer screening plans provide gastric cancer screening for people over the age of 40 with either barium meal X-ray screening or endoscopic screening methods. In South Korea, the participating rate of target population of gastric cancer screening is 70 %. There are still nearly 30 % of the people who did not participate in the gastric cancer screening program. The top 3 reasons why people do not participate in the gastric cancer screening program are as follows: (1) since there is no symptom, so there is no need to participate, (2) they have no time to attend, and (3) they are afraid of the screening process. In gastric cancer screening program in South Korea, endoscopic screening acceptance rate (70 %) is higher than barium meal X-ray screening (30 %) [8]. It appears in both Japan and Korea that there are people afraid of gastric cancer screening with barium meal X-ray or endoscopy method. It is necessary to develop some simpler and easier methods for gastric cancer screening, such as serum marker cancer screening, which is the main topic of this chapter.

Since 1991, Japan began to use pepsinogen I (PGI) and 2 (PGII) serum levels to screen high-risk populations of gastric cancer in atrophy gastritis patients. Pepsinogen 1 is mainly made by gastric chief cells and neck cells. Pepsinogen 2 is synthesized by many parts of the stomach including the gastric cardia and pylorus regions. During atrophy gastritis, PGI secretion was reduced. As the main gastric glands are converted into the pyloric glands, PG II expression is not unaffected. With the progression of gastric atrophy, PGI levels will decrease and the ratio of PG I and PG II will fall. A Japanese study has shown that, with PGI < 70 ng/ml and PGI:PGII < 3.0 as the cutoff value, one could detect gastric cancer with the sensitivity of 84.6 % and specificity of 73.6 % [9]. However, because of the specific expression patterns of PGI and PGII, this method is difficult to detect gastric cancer in the pyloric region. Since 2008, South Korea began to use pepsinogen 1 and 2 index in the screening of early gastric cancer. When using PGI:PGII < 3.0 as an indicator for gastric cancer, a detection sensitivity of 59.2 % and specificity of 61 % were achieved [10]. The sensitivity and specificity of gastric cancer detection with PG serum concentration and ratio indexes were also affected by *Helicobacter pylori* infection, gender, gastric cancer type (intestinal type of gastric carcinoma and poorly differentiated carcinoma), and age of patients. In addition, the relation between pepsinogen levels and gastric cancer risk might be affected by race since a Singaporean research pointed out that Indians have high *H. pylori* infection rate and low concentration of PGI, but their gastric cancer rates were much lower than the Chinese and Malays

[11]. In summary, early screening of gastric cancer is mainly performed by barium meal X-ray and endoscope imaging screening methods. Although there are many serum markers of gastric cancer such as the pepsinogens, the serum screening results are still not very promising. It is necessary to develop some new methods for serum screening of gastric cancer.

The concept of tumor serum markers is now extended from protein markers and carbohydrate antigens to other molecular species, such as metabolic markers, miRNA markers, and noncoding RNA markers. Because of the limitation of space, this article will primarily focus on the studies of serum protein markers in gastric cancer research. At the same time, the expression pattern of these proteins in the stomach and gastric cancer tissues in the Human Protein Atlas archives is also discussed [12, 13] since the expression change of protein markers in the tissues was closely related to the change of protein serum levels. Besides PGI and PGII, a lot of protein markers of gastric cancer are now being investigated. Literature search found that, in the study of gastric cancer, at least 30 different proteins were considered. These proteins could be classified as eight groups based on different characters as the following (Table 2.1):

Group 1, broad-spectrum serum tumor markers such as CEA, CA125, CA19-9, AFP, and M2-pyruvate kinase.
Group 2, gastric-specific secreting proteins such as PGI, PGII, gastrin 17, and TFF3.
Group 3, gastric disease-associated antibodies such as HP, CagA, p53, and anti-parietal cell antibodies.
Group 4, cell signaling or inflammatory signaling molecules such as SAA, MIF, leptin, dickkopf, olfactomedin 4, and VAP-1.
Group 5, proteases and protease inhibitors, such as UPA, cathepsin B, and HMW kininogen.
Group 6, apoptosis-related proteins such as cytokeratin 18.
Group 7, cell-cycle-related proteins such as RegIV, IPO-38, and S100A6.
Group 8, blood coagulation-related proteins such as thrombin light chain and fibrinopeptide A.

In this chapter, we reviewed the application of these different types of serum protein markers in gastric cancer detection.

## 2.2 Broad-Spectrum Tumor Serum Markers in the Detection of Gastric Cancer

### *2.2.1 CEA*

CEA (carcinoembryonic antigen) is a cell adhesion molecule with an IgG structural domain. CEA is encoded by a gene named CEACAM5 (carcinoembryonic antigen-related cell adhesion molecule 5) in the human genome [70]. CEACM gene family has 30 members;17 of them are actively transcribed genes. According to the

**Table 2.1** Serum protein markers

| Protein species | Protein names | Related cancers | Reference |
|---|---|---|---|
| Broad-spectrum tumor serum marker | CEA | Colon cancer, gastric cancer, breast cancer, prostate cancer | [14–16] |
| | CA19-9 | Colon cancer, pancreatic cancer, gastric cancer | [14, 17–19] |
| | CA125 | Ovarian cancer, gastric cancer | [14, 20] |
| | AFP | Liver cancer, gastric cancer | [21, 22] |
| | M2-pyruvate kinase | Lung cancer, pancreatic cancer, liver cancer, breast cancer, gastric cancer | [23–25] |
| Gastric-specific secreting protein | PGI | Gastric cancer | [10, 11, 26] |
| | PGII | Gastric cancer | [10, 11, 26] |
| | Gastrin 17 | Gastric cancer | [26] |
| | TFF3 | Gastric cancer | [27, 28] |
| Gastric cancer-related antibodies | HP | Gastric cancer | [29] |
| | CagA | Gastric cancer | [30, 31] |
| | P53 | Many cancers including gastric cancer | [32–34] |
| | Anti-parietal cell antibody | Gastric cancer | [35] |
| Inflammatory signaling molecules | SAA | Kidney cancer, colon cancer, prostate cancer, gastric cancer | [36, 37] |
| | MIF | Melanoma, prostate cancer, liver cancer, gastric cancer | [38, 39] |
| | Leptin | Endometrial cancer, breast cancer, gastric cancer | [40–42] |
| | Dickkopf | Lung cancer, prostate cancer, gastric cancer | [43–45] |
| | Olfactomedin 4 | Prostate cancer, gastric cancer | [46, 47] |
| | VAP-1 | Colon cancer, gastric cancer | [48, 49] |
| Proteases and protease inhibitors | UPA | Breast cancer, gastric cancer | [50, 51] |
| | Cathepsin B | Cervical cancer, endometrial cancer, gastric cancer | [50, 52, 53] |
| | HMW kininogen | Colon cancer, gastric cancer, prostate cancer | [54–56] |
| Apoptosis-related proteins | Cytokeratin 18 | Bladder cancer, lung cancer, gastric cancer | [57–59] |
| Cell-cycle-related proteins | RegIV | Pancreatic cancer, colon cancer, gastric cancer | [60–62] |
| | S100A6 | Pancreatic cancer, gastric cancer | [63, 64] |
| | IPO-38 | Gastric cancer | [65] |
| Blood coagulation-related proteins | Thrombin light chain | Gastric cancer | [66] |
| | Fibrinopeptide A | Cervical cancer, breast cancer, gastric cancer | [67–69] |

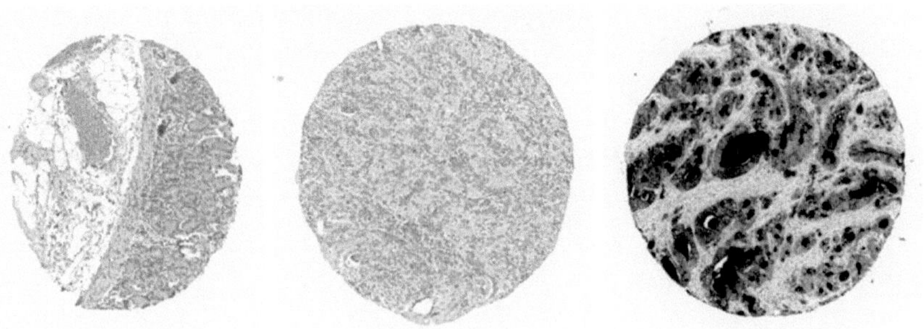


**Fig. 2.1** CEACAM5 expression in normal gastric tissues and gastric cancer tissues (Data obtained from the Human Protein Atlas)

homology of the genes, this gene family is further divided into CEA-related genes and pregnancy-specific glycoprotein (PSG)-related genes. CEA was found to be highly expressed in serum of patients with rectal cancer, and high serum CEA concentrations were also found in breast cancer, gastric cancer, and prostate cancer patients [15, 16, 71]. High preoperative CEA levels often indicate a high probability of cancer recurrence and poor prognosis. The cutoff value for CEA concentration that is commonly used in clinical practice is 5 ng/ml. The reported positive rate of CEA in the serum of gastric cancer varied from 9.6 to 57.6 % between different studies. CEA was used together with other serum protein markers to detect gastric cancer [14]. The mechanism why CEA is overexpressed in gastric cancer is not clear. However, a recent study showed that gastric cancer cells contained bacteria-related DNA sequences from *Pseudomonas*. These bacterial sequences were frequently inserted into the UTR of specific human genes such as CD74, thymosin B10, CEACM5, and CEACM6. CEACM5 and CEACM6 are two genes from the CECAM gene family, encoding CEA and CEA-like proteins. It is not clear if the insertion of foreign bacteria DNA sequence into the UTR of CEACM5 and CEACM6 genes induces CEA overexpression [72]. According to the data from the Human Protein Atlas, the expression of CEA in gastric tissues is very low, while CEA is highly expressed in more than 50 % of the gastric cancer tissues, suggesting that the high serum CEA concentration might be due to the high expression of CEA in the gastric cancer tissues [73] (Fig. 2.1).

### 2.2.2 *CA19-9*

CA19-9 (carbohydrate antigen 19–9), also known as Sialyl-Lewis$^A$, was found in the serum of patients with colorectal cancer for the first time in 1981 [74]. CA19-9 is a ligand molecule that can bind with E-selection, which mediates the interaction of cancer cells and endothelial cells to promote cancer metastasis. CA19-9 is the most frequently used serum marker in pancreatic cancer, which is also used in the screening of gastric cancer when combined with other serum markers. The study showed that the positive rate of CA19-9 antibody detection of gastric cancer was about 30 % [19]. However, CA19-9 is expressed in some of the normal pancreatic duct cells, and also CA19-9 expression can be increased in case of inflammation, so CA19-9 test sometimes shows false-positive results.

### 2.2.3 *CA125*

CA125, a cancer antigen found to be highly expressed in ovarian cancer in 1981, is actually encoded by a mucin gene, muc16. CA125 is one of the most commonly used serum markers in ovarian cancer [20]. The sensitivity of CA125 as a serum marker for gastric cancer was about 40 % [14, 75, 76]. From the Human Protein Atlas database, CA125/Muc16 expression was very low in normal tissues, but it was highly expressed in about 1/3 of gastric cancer tissues, which might be the reason of increased level of CA125 in the serum of gastric cancer patients (Fig. 2.2).

### 2.2.4 *AFP*

AFP (alpha-fetoprotein) was first found in mouse and human liver cancers in 1963, which is one of the earliest discovered tumor markers. Until today, it is still widely used in serological detection of hepatocellular carcinoma (HCC) [21]. However, the sensitivity of AFP to gastric cancer is much lower than that of liver cancer. The sensitivity of AFP for gastric cancer detection is about 4.7 %. However, AFP is often highly expressed in a specific type of gastric cancer, hepatoid gastric adenocarcinoma [22, 77–79]. Based on the data from the Human Protein Atlas database, AFP was weakly expressed in the normal stomach, but highly expressed in about 10 % of gastric cancer tissues, which was consistent with the relatively low sensitivity of serum AFP detection of gastric cancer (Fig. 2.3).

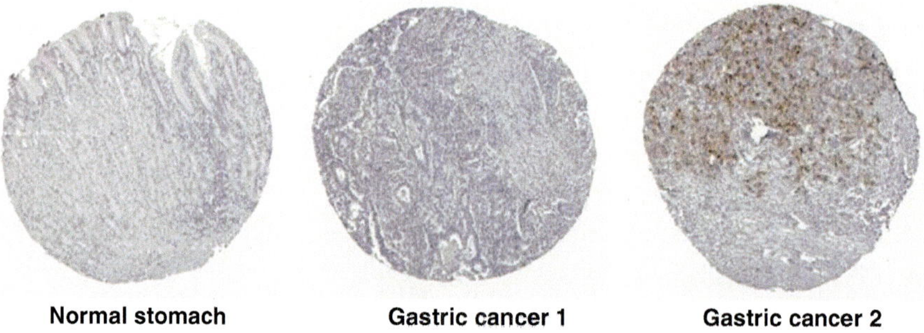


**Fig. 2.2** CA125/Muc16 expression in the normal gastric tissue and gastric cancer tissues (Data obtained from the Human Protein Atlas)

### 2.2.5 *M2-Pyruvate Kinase*

M2-pyruvate kinase is a low-activity type of pyruvate kinase. It is an enzyme related to glycolysis, which was first described as a tumor-associated metabolic marker in 1985 [23]. M2-pyruvate kinase activity increased in multiple cancers such as lung cancer, pancreatic cancer, liver cancer, kidney cancer, and breast cancer [25]. M2-pyruvate kinase activity can be measured from the blood and can be also detected from the feces. When 15 U/mL value was used as the cutoff value of M2-pyruvate kinase in the blood, the sensitivity of M2-pyruvate kinase activity to detect gastric cancer was 64 %, and the specificity was 89 %, which is better for gastric cancer detection than CEA and CA19-9 [24, 80]. However, it is worth noting that the serum level of M2-pyruvate kinase is affected by some non-cancer diseases, such as chronic heart failure, diabetic nephropathy, rheumatism, pancreatitis, and inflammatory bowel disease. When using M2-pyruvate kinase serum levels for the detection of cancer, these inflammatory conditions should be prescreened and excluded.

## 2.3 Gastric Secreted Protein as Serum Markers of Gastric Cancer

### 2.3.1 *PGI*

PGI (pepsinogen group I) was first studied in 1964. PGI expression was localized to the gastric mucosa in 1971 [81]. In the latest nomenclature, PGI is renamed as PGA (pepsinogen A). Human PGA in fact is encoded by three very similar genes, namely, PGA3, PGA4, and PGA5. Their protein sequences are only different from each other with a few amino acids. Pepsinogen is the precursor of pepsin, a main protease produced by the stomach, which is very important for stomach digestive function. PGI is a stomach-specific protein, most of which are directly secreted into the

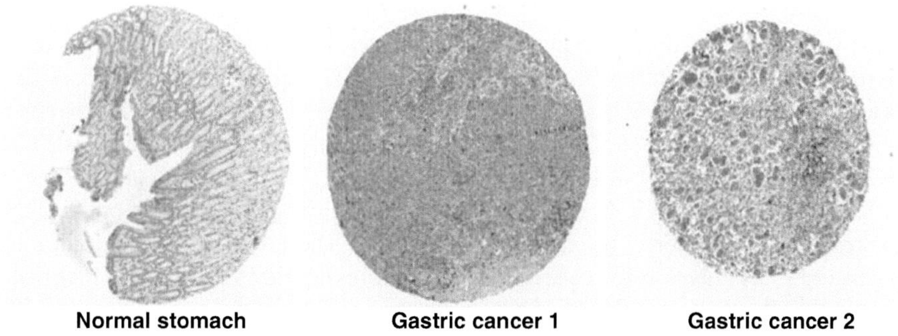


**Fig. 2.3** AFP expression in normal stomach tissue and in gastric cancer tissues (Data obtained from the Human Protein Atlas)

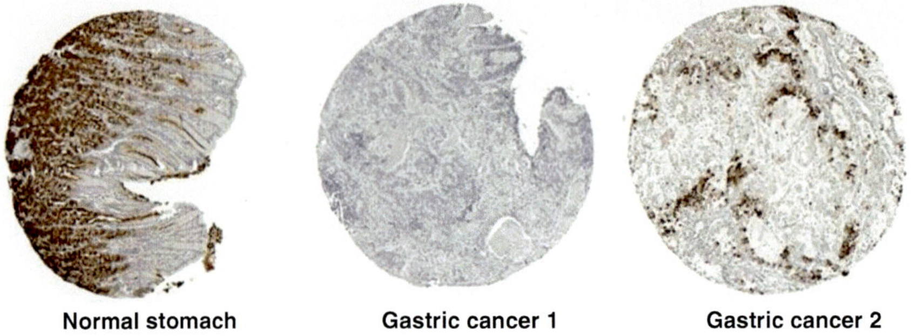


**Fig. 2.4** PGI gene family member PGA3 expression in normal stomach and gastric cancer tissues (Data from the Human Protein Atlas)

stomach, with about 1 % PGI leaking into the peripheral blood. Since PGI is rarely expressed in other organs, pepsinogen level can reflect the function of the stomach and the gastric disease progression, which is an important reason to use pepsinogen serum level for gastric cancer detection. In the course of gastric cancer development from atrophic gastritis to intestinal metaplasia and to gastric cancer, the level of PGI decreased gradually, which is an important indicator for gastric cancer progression. Although PGI levels can indicate the risk of stomach cancer development, however, PGI index cannot directly pinpoint gastric cancer. In clinical practice, PGI measurement often needs to be used in combination with other indicators. The final diagnostic results need to be confirmed with X-ray or endoscopy to verify whether the patients are suffering from gastric cancer [10, 11, 26]. From the Human Protein Atlas database, PGI was strongly expressed in the gastric tissues but weakly expressed in the vast majority of gastric cancer tissues. Low PGI serum concentration is a reflection of its low expression in the gastric cancer tissues (Fig. 2.4).

### 2.3.2 *PGII*

PGII (pepsinogen group II) expression was located in the stomach and duodenum tissues in 1973 [82]. In the latest nomenclature, PGII was renamed as PGC, which is encoded by a single PGC gene in the human genome. PGII was often used together with PGI for gastric cancer detection. The commonly used cutoff value is PGI/PGII $< 3.0$ as an indicator of gastric cancer risk [10, 11, 26]. From the Human Protein Atlas database, PGII/PGC was strongly expressed in the gastric tissues but weakly expressed in the majority of gastric cancer tissues. The low PGII concentration in gastric cancer patients is the reflection of low expression of PGII in the gastric cancer tissues. We think that the PGI/PGII indicator for gastric cancer was not specific, because PGII and PGI were both decreased in gastric cancer tissues (Fig. 2.5).

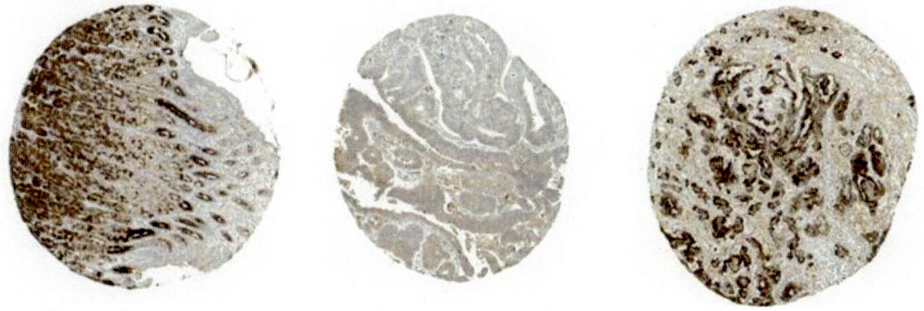


**Fig. 2.5** PGII/PGC expression in normal stomach and gastric cancer tissues (Data from the Human Protein Atlas)

### 2.3.3 *Gastrin 17*

Gastrin 17, first studied in 1905, is a hormone secreted by the gastric antrum, which is important for stimulating gastric acid secretion [83]. Gastrin 17 is a truncated peptide of the gastrin protein, also known as "little gastrin," but is considered with full biological gastrin activity. Since pepsinogen was mainly secreted by the upper part of the gastric body of the stomach and gastrin 17 mainly secreted by the bottom half of the gastric body, the serological detection of PG and gastrin 17 can detect the functional status of the stomach and specific position of gastric atrophy gastritis. Recent studies have indicated that if PG and gastrin 17 are both low, gastritis may happen in both gastric body and gastric antrum, with so-called multifocal atrophy gastritis. Patients with multifocal atrophy gastritis have the highest risk for gastric cancers comparing to other types of gastritis patients. When using PGI < 70 ng/ml, PGI:PGII < 3.0, and gastrin 17 < 1 pm/l as the criteria to detect gastric cancers, the sensitivity is 12.3 %, while the specificity is 98.9 % [26]. From the Human Protein Atlas database, gastrin has a strong expression in the gastric glandular cells, and the expression of gastrin in gastric cancer tissues is low. However, research shows that the level of gastrin increases with age, so the age factor should be considered when using gastrin levels to detect gastric cancer (Fig. 2.6).

### 2.3.4 *TFF3*

TFF3 (trefoil factor 3) is a member of the TFF gene family, which also includes the homologous genes TFF1 and TFF2. The TFF family genes encode secreted proteins that are thermodynamically stable and resistant to protease digestion. They are thought to play an important role in the injury response of gastrointestinal mucosa. Among them, TFF1 and TFF2 were expressed in the stomach, while TFF3 was

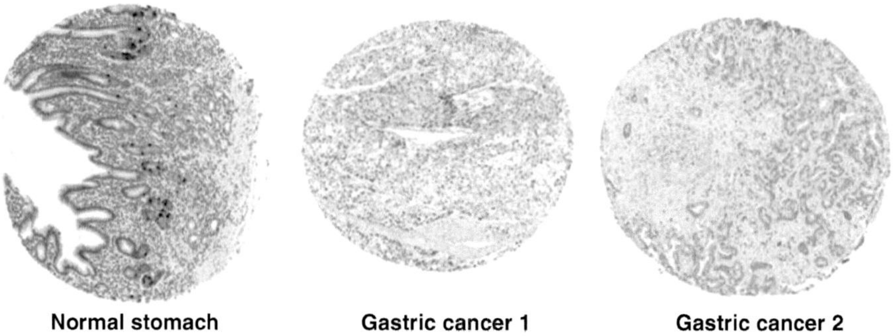


**Fig. 2.6** Gastrin expression in normal stomach and gastric cancer tissues (Data from the Human Protein Atlas)

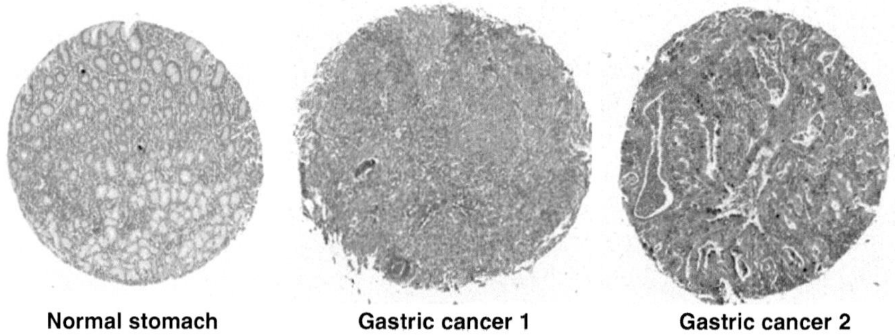


**Fig. 2.7** TFF3 expression in normal stomach and gastric cancer tissues (Data from the Human Protein Atlas)

mainly expressed in the goblet cells of the small intestine. During the course of gastric cancer development, there is an intermediate stage called gastric intestinal metaplasia with the appearance of small intestine goblet-like cells in gastric mucosa. As a result, the expression of TFF3 likely increases in the course of intestinal metaplasia and gastric cancer development. Preliminary studies have shown that serum TFF3 increased the detection sensitivity of gastric cancer to nearly 66–80 %, and specificity reached to about 83 %. The serum concentration of TFF3 was not affected by the treatment of *Helicobacter pylori*, which showed high stability [27, 28, 84, 85]. From the Human Protein Atlas database, TFF3 was not expressed in normal tissues of the stomach, but TFF3 expression was observed in epithelial cells of gastric intestinal metaplasia and increased significantly in gastric cancer tissues. The increased serum level of TFF3 in gastric cancer reflected the increase of gastric cancer tissue expression of TFF3 (Fig. 2.7).

## 2.4 Gastric Infection and Gastric Cancer-Associated Antibodies

### *2.4.1* H. pylori *Antibody*

*H. pylori* antibody detects the infection by *H. pylori*, which is one of the most important causes for gastric cancer. Gastric cancer is induced by a series of gastric tissue changes in the form of superficial gastritis, atrophic gastritis, intestinal metaplasia, and atypical hyperplasia to gastric cancer [86]. In the early stage, the most important infection source is *H. pylori*. Because of the important contribution of *H. pylori* discovery to gastric diseases and gastric cancer research, Marshall Barry and Robin Warren were awarded the 2005 Nobel Prize for physiology. *H. pylori* can survive in the strongly acidified environment of the stomach. *H. pylori* is a very successful human pathogen that infects 50 % of the people on earth. Average rate of *H. pylori* infection in Chinese is 70 %. In gastric ulcer patients, *H. pylori* infection rate is 60–80 %. In duodenal ulcer patients, *Helicobacter pylori* infection rate is 70–100 %. Antibiotics for the treatment of gastric *H. pylori* have a good effect on gastric cancer prevention. Since *H. pylori* infection is very common, *H. pylori* antibody positivity is not directly related to gastric cancer. However, *H. pylori* infection indicates the risk of gastric cancer. *H. pylori* antibody is often used together with pepsinogen level measurements for gastric cancer detection, which was named as the ABC method in Japan. Another variation of this detection method is by including gastrin 17 measurement [29].

### *2.4.2 CagA Antibody*

CagA is an important cytotoxic protein carried by the major pathogenic *H. pylori* strains, which can induce epithelial-to-mesenchymal transformation of gastric epithelial cells. CagA transgenic animal model showed that the high expression of CagA in the stomach of transgenic mice could induce gastric cancer. CagA-positive *H. pylori* had a higher risk of inducing gastric cancer [30]. The sensitivity of CagA antibody detection of gastric cancer can reach 70 %, but the specificity is poor, about 40 %. The statistical analysis showed that CagA antibody is not an ideal serum marker for gastric cancer, but it still has some clinical application value [31, 87].

### *2.4.3 p53 Antibody*

P53 is an important cancer suppressor gene. p53 gene mutation is very frequent in many cancers. Some p53 mutations changed the structure and stability of p53 protein, which could be detected as a cancer cell antigen by

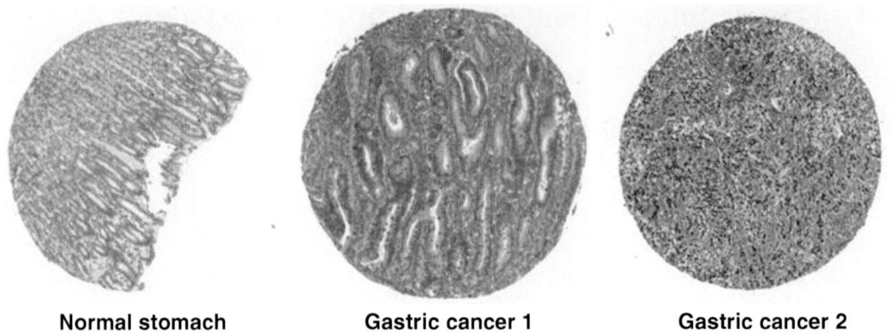


**Fig. 2.8** p53 expression in normal stomach and gastric cancer tissues (Data from the Human Protein Atlas)

the immune cells. p53 mutant protein antibodies can be detected in the patient serum of many types of cancers, including breast cancer, blood cancer, esophageal cancer, ovarian cancer, lung cancer, pancreatic cancer, and gastric cancer [33, 34]. The study showed that the detection rate by serum p53 antibody for gastric cancer was about 15 %. p53 antibody seems to detect a different group of gastric cancer patients that are not overlapped by CA19-9 or CEA detection. P53 antibody detects early phase of gastric cancer, and early gastric cancer is usually not detected by CEA or CA19-9 antibodies [32]. From the Human Protein Atlas database, p53 expression in the gastric gland cells is weak, but p53 (or p53 mutant protein) expression is very strong in about 40 % of gastric cancer tissues. p53 antibody serum level as an indicator of gastric cancer detection has a certain value for early-stage gastric cancer detection (Fig. 2.8).

### 2.4.4 Anti-parietal Cell Antibody

Anti-parietal cell antibody stands for anti-parietal cell self antibody. In the process of *H. pylori* infection of the stomach and atrophy gastritis development, sometimes autoantibodies are produced, especially antibodies against gastric parietal cells significantly increased in gastritis patients. To some extent, the parietal cell antibody titer may reflect the degree of atrophy gastritis development and predict the risk for gastric cancer. It was found that the anti-parietal cell antibody titers increased during active *H. pylori* infection or at early stage of atrophic gastritis, while decreased in the late-stage severe atrophic gastritis. When combined with pepsinogen measurements, gastritis patients with both low pepsinogen levels and anti-parietal cell antibody levels have higher gastric cancer risks [35].

## 2.5 Inflammation Signaling Molecules in the Diagnosis of Gastric Cancer

### 2.5.1 *SAA (Serum Amyloid A)*

SAA (serum amyloid A) was identified by proteomic analysis as a protein related to gastric cancer. SAA is one of the acute proteins during inflammation reaction, which has a low level in the serum of healthy people. SAA serum levels were found to increase in many cancers including kidney cancer, colorectal cancer, prostate cancer, and breast cancer. Studies have shown that SAA levels were significantly higher in patients with gastric cancer than in healthy controls and that high SAA levels were also predicted to have a worse prognosis and higher cancer recurrence rate [37]. However, the researchers found that, from histology studies, SAA was not expressed in gastric cancer cells but expressed in the blood vessels. The mechanism of SAA association with the malignant tumors is not clear. It is speculated that increasing SAA levels may be caused by the inflammatory reaction and the rise of IL-6 levels [36].

### 2.5.2 *MIF (Macrophage Migration Inhibitory Factor)*

MIF (macrophage migration inhibitory factor) is a key molecule in the inflammatory reaction and immune system diseases. Macrophage migration inhibitory factor level increased in a variety of cancers, including melanoma, prostate cancer, liver cancer, and breast cancer, which is a possible target for cancer immunotherapy [39]. In the course of *Helicobacter pylori* infection, MIF levels increased in both gastric epithelial cells and inflammatory cells [88]. Further study found that the MIF serum level with a cutoff value of 3.23 ng/ml can detect gastric cancer with a sensitivity of 83.5 %, higher than the detection efficiency of classic cancer marker CEA. Extremely high MIF levels also correlated to a lower 5-year survival rates after surgery [38]. However, because MIF level changes in many diseases, the specificity of detection of gastric cancer by MIF serum levels is not high.

### 2.5.3 *Leptin*

Leptin is a small protein that regulates appetite and energy metabolism, mainly synthesized by fat cells. However, studies have found that gastric chief cells also secrete leptin [89, 90]. As atrophic gastritis and gastric cancer affect the normal function of the stomach, the serum level of leptin is also affected by these diseases.

Further studies show that the serum levels of leptin and PGI show similar increase in the early stages of gastritis and early intestinal metaplasia, while decreased in the late stage of intestinal metaplasia and atypical hyperplasia. Leptin level could be a good marker for the detection of early intestinal metaplasia, but might not be a suitable marker for gastric cancer detection. Leptin level in patients is affected by many factors, such as obesity, BMI, and diet habits, which could affect the specificity of leptin as a marker for gastric cancer detection [40].

### 2.5.4 *Dickkopf*

Dickkopf (DKK1) is a secreted protein, which is an inhibitor of WNT/beta-catenin signaling pathway. The activation of Wnt/beta-catenin signaling pathway is a common phenomenon in many cancers including gastric cancer. However, the WNT inhibitor DKK1 has demonstrated a high expression in multiple cancers, including liver cancer, lung cancer, cervical cancer, breast cancer, and gastric cancer [44, 45]. Immunohistochemical analysis showed that DKK1 was highly expressed in 50 % of gastric cancer, and the expression of DKK1 was highly correlated with the invasion and metastasis of cancer cells. Patients with high DKK1 expression had much lower rates of 5-year survival after surgery than those patients with low DKK1 expression [43]. Another study showed, when using 31.915 pg/ml as the cutoff value for DKK1 serum concentration, DKK1 level can detect gastric cancer with 87.6 % sensitivity and 87.9 % specificity [91].

### 2.5.5 *Olfactomedin 4*

Olfactomedin 4 is a secreted glycoprotein of 72 KDa, which may function to regulate cell adhesion, cell migration, and cell division. Olfactomedin 4 is expressed in the stem cell niche of the small intestine. Olfactomedin 4 knockout in gastric cells will inhibit cell division and enhance the apoptosis induced by TNF-alpha and H2O2 [92]. Further studies showed that olfactomedin 4 was highly expressed in 50 % of gastric cancer tissues. Olfactomedin 4 expression in early gastric cancers was higher than late-stage cancers. Its expression in intestinal type of gastric cancers is also higher than its expression in diffuse-type gastric cancers. Olfactomedin 4 expression decreases in metastatic gastric cancers suggesting that olfactomedin 4 is inhibiting gastric cancer metastasis [46, 93]. In prostate cancer, olfactomedin 4 was also found to inhibit cancer metastasis [47].

### 2.5.6 *VAP-1 (Vascular Adhesion Protein-1)*

VAP-1 (vascular adhesion protein-1) (also known as AOC3, amine oxidase, copper containing 3) is a glycoprotein that mediates the adhesion of lymphocytes to the blood vessel wall, which may regulate leukocyte infiltration into the tumors through

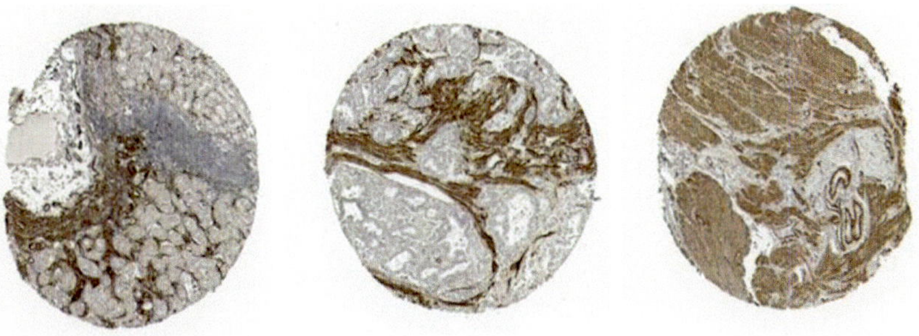


**Fig. 2.9** VAP-1 expression in normal stomach and gastric cancer tissues (Data from the Human Protein Atlas)

blood vessels. When infiltrating the tumors, lymphocytes have to go through the following four steps: (1) rolling and adhesion to the blood vessel walls, (2) leukocyte cell activation, (3) firm adhesion, and (4) migration across the blood vessel wall. In the first step, VAP-1 plays a role in regulating the binding of lymphocytes and blood vessel walls. VAP-1 will transform into sVAP-1 (soluble VAP-1) and enter the bloodstream by the degradation of immune cell MMPs. Further study showed that the serum levels of sVAP-1 in patients with gastric cancer (317.8 ± 196.5 pg/ml) were significantly higher than those in normal subjects (118.04 ± 37.1 pg/ml). sVAP-1 levels in serum of patients with different stages of gastric cancer showed that the sVAP-1 level gradually decreased with the advancement of gastric cancer. The 5-year survival rate of patients with high sVAP-1 was lower than that of patients with low sVAP-1 levels [48, 94]. In colorectal cancer, it also found that the level of VAP-1 in patients with advanced cancer decreased [49]. From the Human Protein Atlas database, VAP-1 expression in the gastric epithelial tissues is weak, but it is highly expressed in gastric blood vessels. In gastric cancer tissues, VAP-1 expression in the blood vessels is still very strong. Occasionally, some gastric cancer cells also expressed VAP-1. It can be seen that the increase of VAP-1 in serum of gastric cancer patients is a reflection of the increased expression in the cancer tissues (Fig. 2.9).

## 2.6 Protease and Protease Inhibitors in the Diagnosis of Gastric Cancer

### *2.6.1 UPA*

UPA (urokinase plasminogen activator) is a serine protease, which can promote the invasion of cancer cells by activating the fibrinolytic enzymes to degrade the extracellular matrix. Recent studies have shown that the serum levels of UPA in patients with gastric cancer are significantly increased compared to normal healthy people controls and also patients with atypical hyperplasia. When using 0.21 ng/ml as the threshold value, the sensitivity of UPA level to detect gastric cancer was 70 %. In

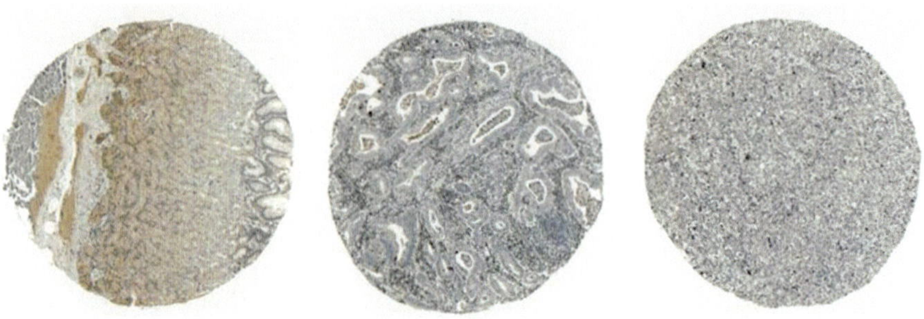


**Fig. 2.10** UPA expression in normal stomach and gastric cancer tissues (Data from the Human Protein Atlas)

addition, the serum level of UPA in patients with gastric cancer metastasis was significantly higher than that of nonmetastatic cancer patients [50, 95, 96]. From the Human Protein Atlas database, UPA expression in the normal gastric gland tissues was weak. UPA was occasionally strongly expressed in immune cells of gastric cancer tissues. Histology studies do not support UPA as a marker for gastric cancer (Fig. 2.10).

### 2.6.2 *Cathepsin B*

Cathepsin B is a cysteine protease that can activate collagenase zymogens and UPA, which plays an important role in the intracellular and extracellular protein degradation pathways [52, 53]. Cathepsin B is highly expressed in many cancers, such as breast cancer, lung cancer, head and neck cancer, prostate cancer, and ovarian cancer. Studies showed that the serum levels of cathepsin B in patients with gastric cancer were significantly higher than those of healthy controls and patients with gastric dysplasia. Using 4.5 ng/ml as the threshold of detection, cathepsin B assay can detect gastric cancer with 86 % sensitivity. The serum levels of cathepsin B were significantly associated with the progression of cancer and increased with the increasing TNM stages. The serum level of cathepsin B was also higher in metastatic cancer patients than in nonmetastatic cancer patients [50, 96–98].

### 2.6.3 *HMW Kininogen*

HMW kininogen, which refers to the high-molecular-weight kininogen, is an auxiliary factor of a factor XII, a cysteine protease inhibitor. HMW kininogen was digested by the protease kallikrein to produce bradykinin and Hka. Hka and the D5 domain of kininogen (kininostatin) have antiangiogenesis and cancer inhibition

**Fig. 2.11** Kininogen expression in normal stomach and gastric cancer tissues (Data from the Human Protein Atlas)

effects. Mass spectrometry analysis found significantly higher level of a 2209 Da degradation peptide corresponding to the D5 kininogen region in the serum of gastric cancer patients compared to healthy controls. The underlying reason could be due to the degradation of kininostatin and the loss of the original cancer growth inhibition activity of kininostatin. The mass spectrum 2209 Da peak signals in the patients of gastric cancer are much stronger than signals in healthy controls. Using the signal of kininogen HMW 2209 Da peak can detect gastric cancer with 39 % sensitivity and 95 % specificity. One drawback of using this signal to diagnose gastric cancer is that it is difficult to detect the degradation products of HMW kininogen by simple detection method, such as ELISA, because kininogen's multiple degradation products are likely to contain this 2209 Da peptide segment, which can only be detected by HPLC and mass spectrum [54, 99]. From the Human Protein Atlas database, kininogen was expressed in vascular tissues in the normal gastric tissues, but the expression of kininogen in gastric cancer tissues was more abundant. The increased level of kininogen in gastric cancer tissues maybe derived from the increase of blood vessel density (Fig. 2.11).

## 2.7 Apoptosis-Related Molecules

### *2.7.1 Cytokeratin 18*

Cytokeratin 18 is an abundant intermediate filament protein that accounts for 5 % of all cellular proteins, which has a high expression in epithelial cells and vascular endothelial cells. In the course of cancer development, apoptosis, necrosis, and autophagy are all very frequent, which leads to various forms of cancer cell death, so that the detection of cell death can be used as a means of detecting cancer. Cytokeratin 18 is a marker of cell death since activated caspases 3, 9, and 7 in the process of cell apoptosis are able to cleave cytokeratin 18 into small peptide fragment, which causes the collapse of the cytoskeleton and the formation of apoptotic bodies. A cleaved

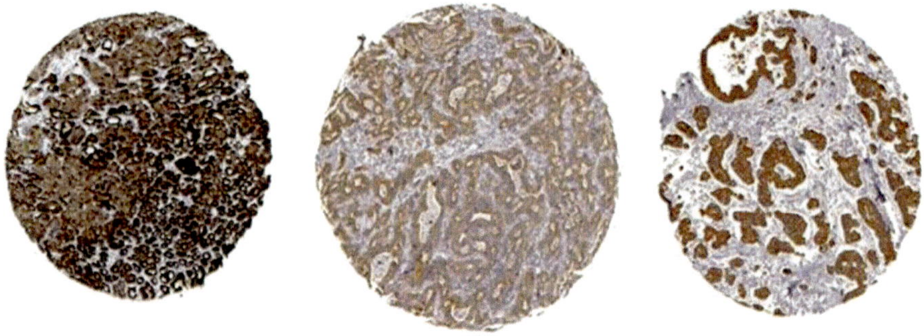


**Fig. 2.12** Cytokeratin 18 expression in normal stomach and gastric cancer tissues (Data from the Human Protein Atlas)

cytokeratin 18 fragment such as M30 will often be released and detected in the blood when apoptotic events happen [100]. Under the situation of necrotic whole cell death, it is possible that the full-length cytokeratin 18 protein M65 will be released into the blood. A study found that patients with gastric cancer can produce monoclonal antibodies against cytokeratin 18 [101]. Further ELISA test showed that the serum levels of M30 and M65 in gastric cancer patients were significantly higher than those of healthy controls. The sensitivity of using M30 concentration (with a cutoff value of 155 u/L) to detect gastric cancer was 67.5 %, and the specificity was 90.9 %. The sensitivity using M65 concentration (with a cutoff value of 142.2 U/L) to detect gastric cancer was 70.1 %, and the specificity was 90.9 %. In addition, the serum levels of M30 and M65 have positive correlation with gastric cancer lymph node metastasis and distant metastasis. From the perspective of prognosis, when using 199.2 U/L as a cutoff value of M65, the 5-year survival rate of gastric cancer patients with high M65 concentration was significantly lower than that of patients with low M65 concentration. The detection of serum level of cytokeratin 18 is valuable to gastric cancer diagnosis, staging, and prognosis [57]. Based on the Human Protein Atlas database, cytokeratin18 has a strong expression in the normal gastric tissues, and the expression of cytokeratin 18 in gastric cancer tissues showed a downward trend. The increase of serum cytokeratin 18 level in gastric cancer patients may be the result of the increased degradation of cytokeratin 18 by related proteases (Fig. 2.12).

## 2.8 Cell Division-Related Molecules

### *2.8.1 RegIV*

RegIV is a secreted protein belonging to the regenerating gene family of proteins. It is highly expressed in gastric cancer and colorectal cancer tissues. RegIV may be associated with cancer development because it activates EGFR/Akt/AP-1 signaling pathway [61]. Since RegIV is a secreted protein, it may become a promising cancer

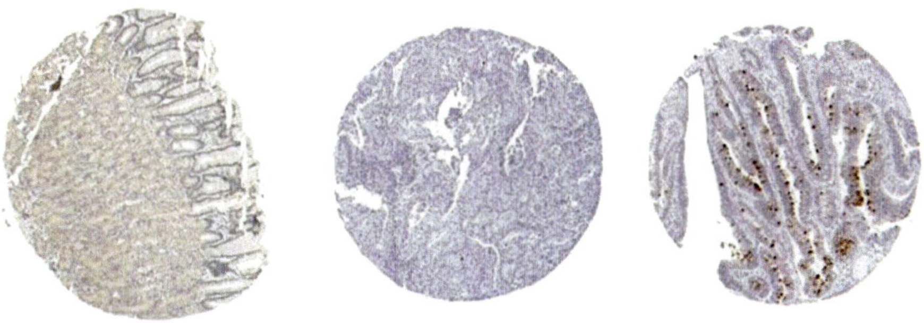


**Fig. 2.13** RegIV expression in normal stomach and gastric cancer tissues (Data from the Human Protein Atlas)

marker. RegIV levels in serum of patients with gastric cancer were significantly higher than those in healthy controls and patients with gastritis. With a cutoff value of 2 ng/ml, the sensitivity of RegIV serum level to detect gastric cancer was 36.1 %, and the specificity was 99 %. The sensitivity of serum RegIV on the detection of gastric cancer was higher than that of CEA (5.6 %) and CA19-9 (8.3 %) [60]. Based on the Human Protein Atlas database, RegIV was not expressed in the normal gastric tissues, and its expression in some gastric cancer tissues was increased. The increase of RegIV level in the serum of gastric cancer patients may be caused by the increased expression of RegIV in gastric cancer tissues (Fig. 2.13).

### *2.8.2 S100A6*

S100A6 (calcyclin) is a molecule of S100 gene family. S100A6 is a calcium-regulated protein that exerts a variety of cell functions, such as cell cycle, calcium balance and signal regulation, ion transport, cytoskeleton remodeling, etc. [64]. There was a significant correlation between S100A6 expression levels with several cancers such as pancreatic cancer, gastric cancer, melanoma, liver cancer, and colorectal cancer. Further studies showed that the serum levels of s100A6 in patients with gastric cancer were significantly higher than that of healthy controls, and the serum levels of s100A6 were positively correlated increasing tumor stages, lymph node metastasis, and distant metastasis. High serum levels of S100A6 also indicated a decreased survival rate of 5 years after surgery for gastric cancer [63].

### *2.8.3 IPO-38*

IPO-38 is a highly expressed protein marker in the serum of gastric cancer patients identified with protein antibody microarray. The sensitivity of gastric cancer detection by IPO-38 with a cutoff value of 140.399 pg/mL is 57.4 % and the specificity is

90.2 %, better than some traditional tumor markers such as CEA, CA72-4, and CA19-9. The serum level of IPO-38 was positively correlated with the size of gastric cancers. There is also a negative correlation between the high IPO-38 serum level and the prognosis of the 5-year survival rate of the tumor patients. These studies show that IPO-38 serum level can be used as an index for the diagnosis and prognosis of gastric cancer. By immunoprecipitation and mass spectrometry analysis, IPO-38 was identified as histone 2B, which might be related to the histone turnover of the gastric cancer cells [65].

## 2.9 Plasma Coagulation Factors

### *2.9.1 Thrombin Light Chain A*

Thrombin light chain A was identified as a serum marker for gastric cancer by proteomic analysis. The sensitivity of thrombin light chain A 3946 Da peak to detect gastric cancer was 85.4 %, and the specificity was 91.8 %66. Based on the Human Protein Atlas database, thrombin was expressed in blood vessels in the normal gastric tissues. Thrombin expression of gastric cancer tissues was increased because of the overgrowth of blood vessels. The increase of thrombin A chain levels in serum of gastric cancer patients may be caused by the increase of thrombin expression and tissue damage in gastric cancer tissues (Fig. 2.14).

### *2.9.2 Fibrinopeptide A*

Fibrinopeptide A was identified as a marker of gastric cancer by MALDI-MS analysis of the serum sample of gastric cancer patients [67]. Further studies showed that fibrinopeptide A was significantly higher in patients with lymph node metastasis.

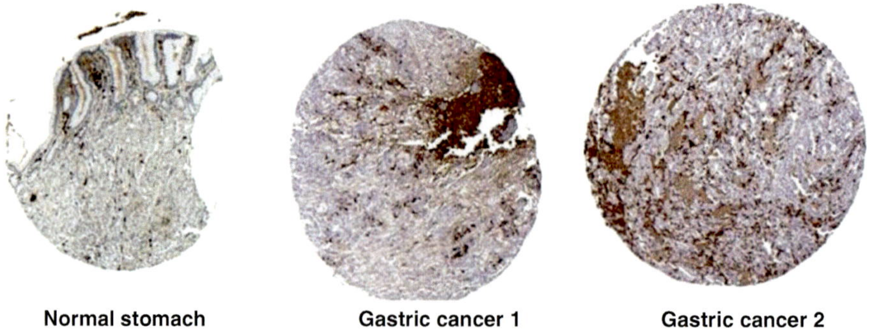


**Fig. 2.14** Thrombin expression in normal stomach and gastric cancer tissues (Data from the Human Protein Atlas)

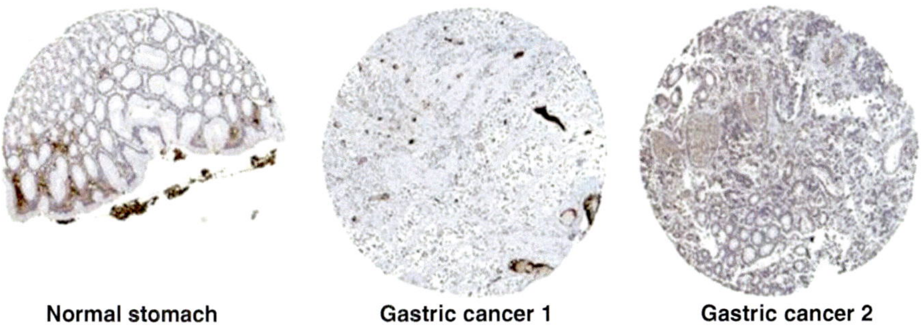


**Fig. 2.15** Fibrinopeptide A expression in normal stomach and gastric cancer tissues (Data from the Human Protein Atlas)

The sensitivity of fibrinopeptide A detection of gastric cancer lymph node metastasis was 85.4 % [102]. In the process of gastric cancer growth or metastasis, tissue damage may happen and lead to the degradation of fibrinopeptide A and B by thrombin, which leads to the increase of serum levels of fibrinopeptide A in the serum. Based on the data from the Human Protein Atlas database, fibrinopeptide A was expressed in blood vessels in the normal stomach. The increased level of fibrinopeptide A in the serum of gastric cancer patients may be caused by tissue damage and coagulation reaction, and the fibrinopeptide A staining of blood coagulation is observed in some gastric cancer samples (Fig. 2.15).

## 2.10 Conclusions

The current development of serum markers of gastric cancer has been helpful for the gastric cancer risk assessment, gastric cancer early diagnosis, finding of gastric cancer locations, gastric cancer typing and grading, and cancer metastasis and prognosis assessment. For example, the combinations of gastrin 17 and pepsinogen detection could be used to locate gastritis. Anti-parietal cell antibody and *H. pylori* antibody could be used to monitor *H. pylori* infection status in gastritis patients. CagA antibody, *H. pylori* antibody, and PGI and PGII measurements could be used in screening high-risk gastritis patients and early gastric cancer screening. TFF3 measurement could be helpful for identifying gastric intestinal metaplasia and early gastric cancers. AFP measurement is useful for identifying hepatoid gastric adenocarcinoma. CEA marker is useful to identify late-stage gastric cancers. M2-pyruvate kinase, p53-auto antibody, leptin, RegIV, olfactomedin, and VAP-1 are useful markers for detecting early-stage gastric cancers. Dickkopf, fibrinopeptide A, s100A6, UPA, cathepsin B, and MIF are valuable markers for gastric cancer metastasis. Cytokeratin 18 serum detection is useful for gastric cancer staging and cancer metastasis monitoring and prognosis after surgical treatments. At present, the large-scale screening of gastric cancer is still based on the combination of PG and HP antibodies, and other serum markers still need to be further developed. Up to date, the detection ability of a single serum marker for gastric cancer is generally low in sensitivity and specificity. In addition, many serum markers are affected by inflammation or other kinds of organ lesions, so that the combined detection of multiple serum markers is still the mainstream for gastric cancer detection in the present and foreseeable future.

## References

1. Chen WQ, Zheng RS, Zhang SW, Zeng HM, Zou XN. The incidences and mortalities of major cancers in china, 2010. Chin J Cancer. 2014;33:402–5.
2. Onodera H, et al. Surgical outcome of 483 patients with early gastric cancer: prognosis, postoperative morbidity and mortality, and gastric remnant cancer. Hepatogastroenterology. 2004;51:82–5.
3. Yuasa N, Nimura Y. Survival after surgical treatment of early gastric cancer, surgical techniques, and long-term survival. Langenbecks Arch Surg (Deutsche Gesellschaft fur Chirurgie). 2005;390:286–93. doi:10.1007/s00423-004-0482-y.
4. Howlader N, Noone AM, Krapcho M, Garshell J, Miller D, Altekruse SF, Kosary CL, Yu M, Ruhl J, Tatalovich Z, Mariotto A, Lewis DR, Chen HS, Feuer EJ, Cronin KA, (eds). SEER cancer statistics review, 1975–2011, National Cancer Institute. Bethesda, MD, http://seer.cancer.gov/csr/1975_2011/, based on November 2013 SEER data submission, posted to the SEER web site. 2014.
5. Wakai S, Tanabe N, Suzuki H. The impact of high-density barium use in double contrast radiographic methods for gastric cancer screening in Niigata, Japan. Tohoku J Exp Med. 2005;205:343–9.
6. Wang B, Yanagawa H, Sakata K. Gastric cancer screening programme in Japan: how to improve its implementation in the community. J Epidemiol Community Health. 1994;48:182–7.
7. Yoo KY. Cancer control activities in the Republic of Korea. Jpn J Clin Oncol. 2008;38:327–33. doi:10.1093/jjco/hyn026.
8. Park B, Choi KS, Lee YY, Jun JK, Seo HG. Cancer screening status in Korea, 2011: results from the Korean National Cancer Screening Survey. Asian Pac J Cancer Prev (APJCP). 2012;13:1187–91.
9. Kitahara F, et al. Accuracy of screening for gastric cancer using serum pepsinogen concentrations. Gut. 1999;44:693–7.
10. Kang JM, et al. The role of serum pepsinogen and gastrin test for the detection of gastric cancer in Korea. Helicobacter. 2008;13:146–56. doi:10.1111/j.1523-5378.2008.00592.x.
11. Kim N, Jung HC. The role of serum pepsinogen in the detection of gastric cancer. Gut Liver. 2010;4:307–19. doi:10.5009/gnl.2010.4.3.307.
12. Uhlen M, et al. Towards a knowledge-based Human Protein Atlas. Nat Biotechnol. 2010;28:1248–50. doi:10.1038/nbt1210-1248.
13. Uhlen M, et al. Proteomics. Tissue-based map of the human proteome. Science. 2015;347:1260419. doi:10.1126/science.1260419.
14. Hwang GI, et al. Predictive value of preoperative serum CEA, CA19-9 and CA125 levels for peritoneal metastasis in patients with gastric carcinoma. Cancer Res Treat (Official Journal of Korean Cancer Association). 2004;36:178–81. doi:10.4143/crt.2004.36.3.178.
15. Fakih MG, Padmanabhan,A. CEA monitoring in colorectal cancer. What you should know. Oncology (Williston Park). 2006;20:579–87; discussion 588, 594, 596 passim.
16. Rubach M, Szymendera JJ, Kaminska J, Kowalska M. Serum CA 15.3, CEA and ESR patterns in breast cancer. Int J Biol Markers. 1997;12:168–73.
17. Poruk KE, et al. The clinical utility of CA 19–9 in pancreatic adenocarcinoma: diagnostic and prognostic updates. Curr Mol Med. 2013;13:340–51.

18. Shibutani M, et al. Significance of CEA and CA19-9 combination as a prognostic indicator and for recurrence monitoring in patients with stage II colorectal cancer. Anticancer Res. 2014;34:3753–8.
19. Shimada H, Noie T, Ohashi M, Oba K, Takahashi Y. Clinical significance of serum tumor markers for gastric cancer: a systematic review of literature by the Task Force of the Japanese Gastric Cancer Association. Gastric Cancer (Official Journal of the International Gastric Cancer Association and the Japanese Gastric Cancer Association). 2014;17:26–33. doi:10.1007/s10120-013-0259-5.
20. Karam AK, Karlan BY. Ovarian cancer: the duplicity of CA125 measurement. Nat Rev Clin Oncol. 2010;7:335–9. doi:10.1038/nrclinonc.2010.44.
21. Bei R, Mizejewski GJ. Alpha fetoprotein is more than a hepatocellular cancer biomarker: from spontaneous immune response in cancer patients to the development of an AFP-based cancer vaccine. Curr Mol Med. 2011;11:564–81.
22. Lin HJ, Hsieh YH, Fang WL, Huang KH, Li AF. Clinical manifestations in patients with alpha-fetoprotein-producing gastric cancer. Curr Oncol. 2014;21:e394–9. doi:10.3747/co.21.1768.
23. Eigenbrodt E, Reinacher M. Carbohydrate metabolism in neoplastic tissue. Infusionsther Klin Ernahr. 1986;13:85–90.
24. Schneider J, Schulze G. Comparison of tumor M2-pyruvate kinase (tumor M2-PK), carcino-embryonic antigen (CEA), carbohydrate antigens CA 19–9 and CA 72–4 in the diagnosis of gastrointestinal cancer. Anticancer Res. 2003;23:5089–93.
25. Iqbal MA, Gupta V, Gopinath P, Mazurek S, Bamezai RN. Pyruvate kinase M2 and cancer: an updated assessment. FEBS Lett. 2014;588:2685–92. doi:10.1016/j.febslet.2014.04.011.
26. Kikuchi R, et al. Low serum levels of pepsinogen and gastrin 17 are predictive of extensive gastric atrophy with high-risk of early gastric cancer. Tohoku J Exp Med. 2011;223:35–44.
27. Kaise M, et al. The combination of serum trefoil factor 3 and pepsinogen testing is a valid non-endoscopic biomarker for predicting the presence of gastric cancer: a new marker for gastric cancer risk. J Gastroenterol. 2011;46:736–45. doi:10.1007/s00535-011-0396-8.
28. Huang Z, et al. Serum trefoil factor 3 is a promising non-invasive biomarker for gastric cancer screening: a monocentric cohort study in China. BMC Gastroenterol. 2014;14:74. doi:10.1186/1471-230X-14-74.
29. Miki K. Gastric cancer screening by combined assay for serum anti-Helicobacter pylori IgG antibody and serum pepsinogen levels – "ABC method". Proc Jpn Acad Ser B Phys Biol Sci. 2011;87:405–14.
30. Parsonnet J, Friedman GD, Orentreich N, Vogelman H. Risk for gastric cancer in people with CagA positive or CagA negative Helicobacter pylori infection. Gut. 1997;40:297–301.
31. Shiota S, Matsunari O, Watada M, Yamaoka Y. Serum Helicobacter pylori CagA antibody as a biomarker for gastric cancer in east-Asian countries. Future Microbiol. 2010;5:1885–93. doi:10.2217/fmb.10.135.
32. Shimizu K, Ueda Y, Yamagishi H. Titration of serum p53 antibodies in patients with gastric cancer: a single-institute study of 40 patients. Gastric Cancer (Official Journal of the International Gastric Cancer Association and the Japanese Gastric Cancer Association). 2005;8:214–9. doi:10.1007/s10120-005-0337-4.
33. Pedersen JW, et al. Early detection of cancer in the general population: a blinded case-control study of p53 autoantibodies in colorectal cancer. Br J Cancer. 2013;108:107–14. doi:10.1038/bjc.2012.517.
34. Suppiah A, Greenman J. Clinical utility of anti-p53 auto-antibody: systematic review and focus on colorectal cancer. World J Gastroenterol (WJG). 2013;19:4651–70. doi:10.3748/wjg.v19.i29.4651.
35. Sugiu K, et al. Anti-parietal cell antibody and serum pepsinogen assessment in screening for gastric carcinoma. Dig Liver Dis (Official Journal of the Italian Society of Gastroenterology and the Italian Association for the Study of the Liver). 2006;38:303–7. doi:10.1016/j.dld.2005.10.021.
36. Chan DC, et al. Evaluation of serum amyloid A as a biomarker for gastric cancer. Ann Surg Oncol. 2007;14:84–93. doi:10.1245/s10434-006-9091-z.

37. Moshkovskii SA. Why do cancer cells produce serum amyloid A acute-phase protein? Biochemistry (Biokhimiia). 2012;77:339–41. doi:10.1134/S0006297912040037.
38. Xia HH, et al. Serum macrophage migration-inhibitory factor as a diagnostic and prognostic biomarker for gastric cancer. Cancer. 2009;115:5441–9. doi:10.1002/cncr.24609.
39. Babu SN, Chetal G, Kumar S. Macrophage migration inhibitory factor: a potential marker for cancer diagnosis and therapy. Asian Pac J Cancer Prev (APJCP). 2012;13:1737–44.
40. Capelle LG, et al. Serum levels of leptin as marker for patients at high risk of gastric cancer. Helicobacter. 2009;14:596–604. doi:10.1111/j.1523-5378.2009.00728.x.
41. Artac M, Altundag K. Leptin and breast cancer: an overview. Med Oncol. 2012;29:1510–4. doi:10.1007/s12032-011-0056-0.
42. Wang PP, He XY, Wang R, Wang Z, Wang YG. High Leptin level is an independent risk factor of endometrial cancer: a meta-analysis. Cell Physiol Biochem (International Journal of Experimental Cellular Physiology, Biochemistry, and Pharmacology). 2014;34:1477–84. doi:10.1159/000366352.
43. Gao C, Xie R, Ren C, Yang X. Dickkopf-1 expression is a novel prognostic marker for gastric cancer. J Biomed Biotechnol. 2012;2012:804592. doi:10.1155/2012/804592.
44. Rachner TD, et al. High serum levels of Dickkopf-1 are associated with a poor prognosis in prostate cancer patients. BMC Cancer. 2014;14:649. doi:10.1186/1471-2407-14-649.
45. Li S, et al. Dickkopf-1 is oncogenic and involved in invasive growth in non small cell lung cancer. PLoS One. 2013;8:e84944. doi:10.1371/journal.pone.0084944.
46. Yu L, Wang L, Chen S. Olfactomedin 4, a novel marker for the differentiation and progression of gastrointestinal cancers. Neoplasma. 2011;58:9–13.
47. Chen L, et al. Olfactomedin 4 suppresses prostate cancer cell growth and metastasis via negative interaction with cathepsin D and SDF-1. Carcinogenesis. 2011;32:986–94. doi:10.1093/carcin/bgr065.
48. Kaplan MA, et al. Relationship between serum soluble vascular adhesion protein-1 level and gastric cancer prognosis. Oncol Res Treat. 2014;37:340–4. doi:10.1159/000362626.
49. Toiyama Y, Miki C, Inoue Y, Kawamoto A, Kusunoki M. Circulating form of human vascular adhesion protein-1 (VAP-1): decreased serum levels in progression of colorectal cancer and predictive marker of lymphatic and hepatic metastasis. J Surg Oncol. 2009;99:368–72. doi:10.1002/jso.21246.
50. Herszenyi L, et al. Serum cathepsin B and plasma urokinase-type plasminogen activator levels in gastrointestinal tract cancers. Eur J Cancer Prev. 2008;17:438–45.
51. Harbeck N, et al. Prospective Biomarker Trials Chemo N0 and NNBC-3 Europe Validate the Clinical Utility of Invasion Markers uPA and PAI-1 in Node-Negative Breast Cancer. Breast Care (Basel). 2008;3:11–5. doi:10.1159/000151734.
52. Wu D, et al. Cathepsin B may be a potential biomarker in cervical cancer. Histol Histopathol. 2012;27:79–87.
53. Devetzi M, et al. Cathepsin B protein levels in endometrial cancer: potential value as a tumour biomarker. Gynecol Oncol. 2009;112:531–6. doi:10.1016/j.ygyno.2008.10.030.
54. Umemura H, et al. Identification of a high molecular weight kininogen fragment as a marker for early gastric cancer by serum proteome analysis. J Gastroenterol. 2011;46:577–85. doi:10.1007/s00535-010-0369-3.
55. Wang J, et al. Identification of kininogen-1 as a serum biomarker for the early detection of advanced colorectal adenoma and colorectal cancer. PLoS One. 2013;8:e70519. doi:10.1371/journal.pone.0070519.
56. Liu Y, et al. Cleaved high-molecular-weight kininogen and its domain 5 inhibit migration and invasion of human prostate cancer cells through the epidermal growth factor receptor pathway. Oncogene. 2009;28:2756–65. doi:10.1038/onc.2009.132.
57. Oyama K, et al. Serum cytokeratin 18 as a biomarker for gastric cancer. Clin Exp Med. 2013;13:289–95. doi:10.1007/s10238-012-0202-9.
58. Ramazan Sekeroglu M, et al. Diagnostic value of cytokeratin-18 as a tumor marker in bladder cancer. Clin Biochem. 2002;35:327–31.
59. De Petris L, et al. Diagnostic and prognostic role of plasma levels of two forms of cytokeratin 18 in patients with non-small-cell lung cancer. Eur J Cancer (Oxford, England: 1990). 2011;47:131–7. doi:10.1016/j.ejca.2010.08.006.

60. Mitani Y, et al. Reg IV is a serum biomarker for gastric cancer patients and predicts response to 5-fluorouracil-based chemotherapy. Oncogene. 2007;26:4383–93. doi:10.1038/sj.onc.1210215.
61. Bishnupuri KS, et al. Reg IV activates the epidermal growth factor receptor/Akt/AP-1 signaling pathway in colon adenocarcinomas. Gastroenterology. 2006;130:137–49. doi:10.1053/j.gastro.2005.10.001.
62. Wang F, et al. Identification of RegIV as a novel GLI1 target gene in human pancreatic cancer. PLoS One. 2011;6:e18434. doi:10.1371/journal.pone.0018434.
63. Zhang J, Zhang K, Jiang X, Zhang J. S100A6 as a potential serum prognostic biomarker and therapeutic target in gastric cancer. Dig Dis Sci. 2014;59:2136–44. doi:10.1007/s10620-014-3137-z.
64. Ohuchida K, et al. The role of S100A6 in pancreatic cancer development and its clinical implication as a diagnostic marker and therapeutic target. Clin Cancer Res (An Official Journal of the American Association for Cancer Research). 2005;11:7785–93. doi:10.1158/1078-0432.ccr-05-0714.
65. Hao Y, et al. IPO-38 is identified as a novel serum biomarker of gastric cancer based on clinical proteomics technology. J Proteome Res. 2008;7:3668–77. doi:10.1021/pr700638k.
66. Ebert MP, et al. Identification of the thrombin light chain a as the single best mass for differentiation of gastric cancer patients from individuals with dyspepsia by proteome analysis. J Proteome Res. 2005;4:586–90. doi:10.1021/pr049771i.
67. Ebert MP, et al. Identification and confirmation of increased fibrinopeptide a serum protein levels in gastric cancer sera by magnet bead assisted MALDI-TOF mass spectrometry. J Proteome Res. 2006;5:2152–8. doi:10.1021/pr060011c.
68. Gadducci A, et al. Pretreatment plasma levels of fibrinopeptide-A (FPA), D-dimer (DD), and von Willebrand factor (vWF) in patients with operable cervical cancer: influence of surgical-pathological stage, tumor size, histologic type, and lymph node status. Gynecol Oncol. 1993;49:354–8. doi:10.1006/gyno.1993.1139.
69. Sagripanti A, Carpi A, Ferdeghini M, Pinori E, Nicolini A. The measurement of plasma fibrinopeptide A in breast cancer patients. Rays. 1987;12(65–69):107–9.
70. Kuespert K, Pils S, Hauck CR. CEACAMs: their role in physiology and pathophysiology. Curr Opin Cell Biol. 2006;18:565–71. doi:10.1016/j.ceb.2006.08.008.
71. Gold P, Freedman SO. Specific carcinoembryonic antigens of the human digestive system. J Exp Med. 1965;122:467–81.
72. Riley DR, et al. Bacteria-human somatic cell lateral gene transfer is enriched in cancer samples. PLoS Comput Biol. 2013;9:e1003107.
73. Liu JN, Wang HB, Zhou CC, Hu SY. CEACAM5 has different expression patterns in gastric non-neoplastic and neoplastic lesions and cytoplasmic staining is a marker for evaluation of tumor progression in gastric adenocarcinoma. Pathol Res Pract. 2014;210:686–93. doi:10.1016/j.prp.2014.06.024.
74. Chang TH, Steplewski Z, Sears HF, Koprowski H. Detection of monoclonal antibody-defined colorectal carcinoma antigen by solid-phase binding inhibition radioimmunoassay. Hybridoma. 1981;1:37–45.
75. Bast Jr RC, et al. Reactivity of a monoclonal antibody with human ovarian carcinoma. J Clin Invest. 1981;68:1331–7.
76. Yin BW, Lloyd KO. Molecular cloning of the CA125 ovarian cancer antigen: identification as a new mucin, MUC16. J Biol Chem. 2001;276:27371–5. doi:10.1074/jbc.M103554200.
77. Liu X, Sheng W, Wang Y. An analysis of clinicopathological features and prognosis by comparing hepatoid adenocarcinoma of the stomach with AFP-producing gastric cancer. J Surg Oncol. 2012;106:299–303. doi:10.1002/jso.23073.
78. Khramkova NI, Guelstein VI. Antigenic Structure of Mouse Hepatomas. V. Organospecific Liver Antigens and Embryonic Alpha-Globulin in Hepatomas of Mice Induced with Orthoaminoazotoluene (Aat). Neoplasma. 1965;12:239–50.
79. IuS T. Detection of embryo-specific alpha-globulin in the blood serum of a patient with primary liver cancer. Vopr Med Khim. 1964;10:90–1.
80. Kumar Y, Tapuria N, Kirmani N, Davidson BR. Tumour M2-pyruvate kinase: a gastrointestinal cancer marker. Eur J Gastroenterol Hepatol. 2007;19:265–76. doi:10.1097/MEG.0b013e3280102f78.

81. Samloff IM. Cellular localization of group I pepsinogens in human gastric mucosa by immunofluorescence. Gastroenterology. 1971;61:185–8.
82. Samloff IM, Liebman WM. Cellular localization of the group II pepsinogens in human stomach and duodenum by immunofluorescence. Gastroenterology. 1973;65:36–42.
83. Edkins JS. The chemical mechanism of gastric secretion. J Physiol. 1906;34:133–44.
84. Kaise M, et al. Influence of Helicobacter pylori status and eradication on the serum levels of trefoil factors and pepsinogen test: serum trefoil factor 3 is a stable biomarker. Gastric Cancer (Official Journal of the International Gastric Cancer Association and the Japanese Gastric Cancer Association). 2013;16:329–37. doi:10.1007/s10120-012-0185-y.
85. Aikou S, et al. Tests for serum levels of trefoil factor family proteins can improve gastric cancer screening. Gastroenterology. 2011;141(837–845):e831–7. doi:10.1053/j.gastro.2011.05.040.
86. Correa P. A human model of gastric carcinogenesis. Cancer Res. 1988;48:3554–60.
87. Zhao Z, Li Y, Liu S, Fu W. Serum Helicobacter pylori CagA antibody may not be used as a tumor marker for diagnosing gastric cancer in east Asian countries. Tumour Biol (The Journal of the International Society for Oncodevelopmental Biology and Medicine). 2014;35:12217–24. doi:10.1007/s13277-014-2530-8.
88. Xia HH, et al. Helicobacter pylori infection is associated with increased expression of macrophage migratory inhibitory factor–by epithelial cells, T cells, and macrophages–in gastric mucosa. J Infect Dis. 2004;190:293–302. doi:10.1086/421915.
89. Zhang Y, et al. Positional cloning of the mouse obese gene and its human homologue. Nature. 1994;372:425–32. doi:10.1038/372425a0.
90. Bado A, et al. The stomach is a source of leptin. Nature. 1998;394:790–3. doi:10.1038/29547.
91. Lee HS, et al. Clinical significance of serum and tissue Dickkopf-1 levels in patients with gastric cancer. Clin Chim Acta (International Journal of Clinical Chemistry). 2012;413:1753–60. doi:10.1016/j.cca.2012.07.003.
92. Liu RH, et al. Depletion of OLFM4 gene inhibits cell growth and increases sensitization to hydrogen peroxide and tumor necrosis factor-alpha induced-apoptosis in gastric cancer cells. J Biomed Sci. 2012;19:38. doi:10.1186/1423-0127-19-38.
93. Luo Z, et al. OLFM4 is associated with lymph node metastasis and poor prognosis in patients with gastric cancer. J Cancer Res Clin Oncol. 2011;137:1713–20. doi:10.1007/s00432-011-1042-9.
94. Yasuda H, et al. Serum soluble vascular adhesion protein-1 is a valuable prognostic marker in gastric cancer. J Surg Oncol. 2011;103:695–9. doi:10.1002/jso.21877.
95. Koshikawa N, Yasumitsu H, Umeda M, Miyazaki K. Multiple secretion of matrix serine proteinases by human gastric carcinoma cell lines. Cancer Res. 1992;52:5046–53.
96. Herszenyi L, et al. Proteases in gastrointestinal neoplastic diseases. Clin Chim Acta (International Journal of Clinical Chemistry). 2000;291:171–87.
97. Hirano T, Yoshioka H. Serum cathepsin B levels, urinary excretion of cathepsin B and tissue cathepsin B content in the patients with gastric cancer. Nihon geka hokan Archiv fur japanische Chirurgie. 1993;62:217–21.
98. Ebert MP, et al. Overexpression of cathepsin B in gastric cancer identified by proteome analysis. Proteomics. 2005;5:1693–704. doi:10.1002/pmic.200401030.
99. Liu W, et al. Proteomic identification of serum biomarkers for gastric cancer using multidimensional liquid chromatography and 2D differential gel electrophoresis. Clin Chim Acta (International Journal of Clinical Chemistry). 2012;413:1098–106. doi:10.1016/j.cca.2012.03.003.
100. Caulin C, Salvesen GS, Oshima RG. Caspase cleavage of keratin 18 and reorganization of intermediate filaments during epithelial cell apoptosis. J Cell Biol. 1997;138:1379–94.
101. Abe T, et al. Human monoclonal antibodies against cytokeratin 18 generated from patients with gastric cancer. Jpn J Cancer Res (Gann). 1989;80:271–6.
102. Zhang MH, et al. A prognostic biomarker for gastric cancer with lymph node metastases. Anat Rec. 2013;296:590–4. doi:10.1002/ar.22642.

# Chapter 3
# Screening and Detection of Gastric Cancer Circulating MicroRNA Biomarkers

**Zhang Jingpu and Daxiang Cui**

## 3.1 MicroRNA and Gastric Cancer

### *3.1.1 MicroRNA*

MicroRNAs (miRNAs) are endogenous noncoding RNAs with a length of ~22 nt [1]. It was first found in *C. elegans* by Victor Ambros, Rosalind Lee, and Rhonda Feinbaum in 1993 [1]. Until the year 2010, the registered miRNAs in the miR-Sanger base reached up to more than 10,500 from 115 species [2]. miRNAs are involved in the control of development, differentiation, proliferation, apoptosis, the stress response, and other cell processes through the cleavage of target mRNA or repressing its expression. One mRNA is regulated by multiple miRNAs; in turn, one miRNA is involved in regulations of various mRNAs, thus forming a complicated miRNA functional network [3]. More importantly, most of human miRNAs are closely correlated with human diseases, including cancers. miRNAs are usually located in cancer-associated genomic regions (CAGRs) [3], such as minimal regions of loss of heterozygosity (LOH) and amplification, common breakpoint regions, and fragile sites (FRA), and thus frequently dysregulated in cancers [4]. miRNAs have been proposed to contribute to oncogenesis because they can function either as tumor suppressors or as oncogenes [3]. Compared to mRNA expression profiles, miRNA expression profiles may be more accurate in cancer classification [5]. Therefore, miRNAs are potent to be valuable biomarkers for the diagnosis and prognosis of cancers.

---

Z. Jingpu (✉) • D. Cui
Institute of Nano Biomedicine and Engineering, Shanghai Engineering Research Center for Intelligent Diagnosis and Treatment Instrument, National Center for Translational Medicine, Collaborative Innovational Center for System Biology, Shanghai Jiao Tong University, 800 Dongchuan Road, Shanghai 200240, Peoples Republic of China
e-mail: zhangjp@sjtu.edu.cn; dxcui@sjtu.edu.cn

© Springer Science+Business Media B.V. and Shanghai Jiao Tong University Press, Shanghai 2017

D. Cui (ed.), *Gastric Cancer Prewarning and Early Diagnosis System*, Translational Medicine Research, DOI 10.1007/978-94-024-0951-2_3

### *3.1.2 MicroRNA and Gastric Cancer*

Gastric cancer (GC) is currently the fourth most common cancer and the second leading cause for cancer-related deaths worldwide [6]. According to the annual report on the status of cancer in China in 2010 [7], GC was ranked as the second among the most common cancers and the third leading cause of death in China. No typical symptoms suggestive of GC exist until the cancer is advanced [8], leading to a great difficulty in early GC diagnosis. And the limited treatment modalities for current clinical applications make GC a very poor prognosis with 5-year survivals below 24 % [9]. In general, it is imperative to develop rational approaches for the diagnosis and prognosis of GC to improve its detection and treatment.

Dysregulation of miRNA occurs in GC and other diseases. The mechanisms by which miRNAs take part in tumor promotion and progression are complex and numerous. However, most of them converge on common signaling mechanisms that govern cell proliferation, apoptosis, and invasiveness [10]. Wu and Lee et al. summarized that a substantial number of miRNAs associated with cell cycle, apoptosis, metastasis, and transcriptional regulation show differential expression in GC tissues [10]. Some miRNAs play oncogenic roles in GC. miR-21, one well-known oncogenic miRNA, has been frequently observed to be aberrantly overexpressed in various tumors, as well as GC. Forced expression of miR-21 significantly enhanced cell proliferation and invasiveness in GC cells. The oncogenic effect of miR-21 may be mediated by regulation of RECK, a tumor suppressor gene [11]. In addition, tissue miR-21 is an antiapoptotic and pro-survival factor for GC and other tumors by targeting phosphatase and tensin homolog (PTEN, deleted on chromosome ten) [3]. Some other miRNAs acted as suppressors. miR-451 could be considered as a new suppressor gene by downregulating macrophage migration inhibitory factor (MIF), a proinflammatory cytokine involved in oncogenic transformation and tumor progression [12].

Furthermore, the miRNA clusters including miR-17-5p, miR-18a, miR-19a, miR-20a, miR-19b-1, and miR-92, which were recognized as oncogenic miRNAs, were also dysregulated in GC. miR-19b, miR-20a, and miR-92a could sustain the self-renewal function of GC stem cells and promote the proliferation of GC cells by targeting the E2F1 and HIPK1 proteins, which suppressed Wnt–b-catenin signaling. A real-time PCR analysis of miR-19b, miR-20a, and miR-92a expression in 97 GC specimens suggested that miR-92a could be used as an independent prognostic factor in GC [13]. Additionally, miR-17-92 cluster members miR-19a/miR-19b were found to be overexpressed in GC tissues and significantly associated with the patients' metastasis of GC. Reporter gene assay and Western blot showed that MXD1 is a direct target of miR-19a/miR-19b. Thus, miR-19a/miR-19b facilitated GC cell migration, invasion, and metastasis through targeting the antagonist of c-Myc-MXD1, implicating a novel mechanism for the malignant phenotypes of gastric cancer [14].

## 3.2 Circulating MicroRNA Biomarkers for Gastric Cancer

### *3.2.1 Background*

MicroRNAs from both the solid tissues and the body fluids are capable of being predictive biomarkers. miRNAs can be reliably extracted and detected from frozen and paraffin-embedded tissues [4], cultured cell lines [15], and even circulating cells in peripheral blood [16]. Moreover, miRNAs are shown to present in human blood (including whole blood, plasma, and serum) in a remarkably stable form that is protected from endogenous RNase activity [17]. miRNAs are also found in other body fluids, such as saliva, human breast milk [18], circulating exosomes, and even urine [4]. Excitingly, Mitchell and his colleagues employed TaqMan RT-PCR detecting the serum levels of miR-141 and demonstrating miR-141 from serum as a biomarker to distinguish patients with prostate cancer from healthy controls [17]. Nearly at the same time, sera miR-25 and miR-223 were found as non-small cell lung cancer-specific biomarker by Solexa sequencing, and this fact was further validated by quantitative reverse transcription polymerase chain reaction assays [19].

Recently multidisciplinary evaluation, complete history and physical, and basic laboratory studies are suggested as the standard workup for gastric cancer diagnosis according to the National Comprehensive Cancer Network (NCCN) guidelines, which include complete blood count and chemistry profile, abdominal computed tomography (CT) with contrast, CT/ultrasound of the pelvis (females), chest imaging, esophagogastroduodenoscopy, optional positron emission tomography or positron emission tomography/CT, optional endoscopic ultrasound, and testing for *H. pylori* [20]. Each of the single examination has its advantages and disadvantages. Tumor circulation biomarkers including serological markers undoubtedly provide a noninvasive and highly practical method to assess GC [18]. However, the currently known tumor biomarkers either in the sera (CEA, CA19.9, CA72.4, CA50) or in the gastric juice (CEA, CA19.9, fetal sulfoglycoprotein) have been found to exhibit low sensitivity and specificity in gastric cancer diagnosis [8, 21]. Therefore, more studies need to be conducted to screen out other alternative tumor biomarkers [22], and circulating miRNA biomarkers have recently emerged as novel diagnostic and prognostic biomarkers in oncology [23, 24].

### *3.2.2 Advantages on Circulating MicroRNA Biomarkers*

miRNAs have the following advantages to serve as biomarkers: (a) miRNAs are involved in tumorigenesis; (b) miRNAs are tissue, tumor, or even pathology specific; and (c) some miRNAs are related to treatment response or patients' survival [11]. However, ideal biomarkers of tumors must fit into two criteria: (a) specificity, i.e., their presence is associated with the occurrence of only a specific type of tumor, and (b) proportionality, i.e., their levels in serum, plasma, urine, or other body fluids should be correlated with the extent of tumor development [25]. Therefore, great efforts should be devoted to screen miRNA biomarkers characterized by specificity and proportionality for GC detection.

### 3.2.3 *Correlation Between Tissue miRNAs and Circulating miRNAs*

As shown in Fig. 3.1, three major potential pathways have been hypothesized for miRNAs to enter the circulation: (i) free miRNAs directly secreted by cells, in a similar fashion to cytokines and hormones, and miRNAs selectively packed into microparticles (MPs) (ii) and exosomes (iii) and then released by cells via shedding of microvesicles (MVs) [25]. On the contrast, a wide range of miRNAs correlated with physiologic and pathologic events would be selectively loaded into the vesicles, making the relations of the miRNA levels more complex between the tissues and circulation system [26].

### 3.2.4 *Challenges on Screening of Circulating MicroRNA Biomarkers*

Although a large number of circulating miRNAs are being identified as biomarkers for cancer and diseases, various problems challenge the screening results for clinical application, such as the lack of standardized collection of clinical specimen, the platform with rigorous performance for detection, and the conventional reference genes for data normalization [27]. Firstly, biological variances including age, sex,

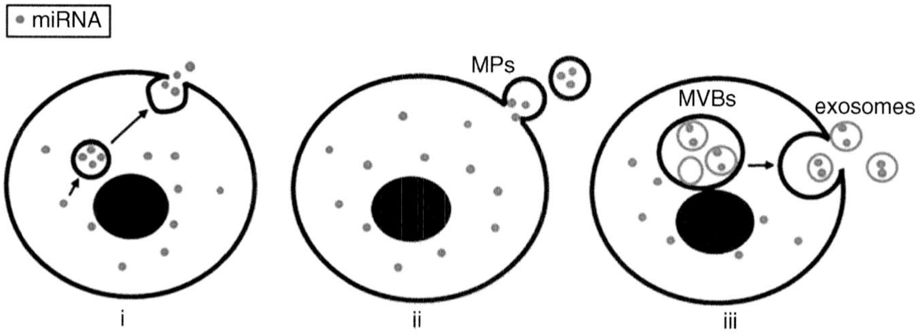


**Fig. 3.1** Schematic view of the possible sources of serum or plasma miRNAs. (**i**) miRNAs are secreted via exocytosis of secretory vesicles and granules; (**ii**) miRNAs are released via cell plasma "pinch-off" microparticles (MPs); and (**iii**) miRNAs are released via cell-derived exosomes. *MVB* multivesicular body (Copyright permission from Zen and Zhang [25])

and race acted as confounding factors which might lead to false-positive or false-negative results and, thus, are required to be better characterized [28]. Secondly, microarrays, qPCR microfluidic arrays, and next-generation sequencing (NGS) are the principal strategies available for large-scale miRNA profiling. However, different sensitivity and specificity in platform as critical variables cause discrepancies on the expression of miRNAs between studies [28].

In addition, there is still neither convention regarding the quantification of miRNAs nor consensus in using the reference miRNAs in plasma or serum [29]. miRNA was detected by quantitative reverse transcription polymerase chain reaction (qRT-PCR) employing either the absolute quantification method [30, 31] or the relative one [32–34], the former of which quantifies miRNA based on the standard curve, thus avoiding the option of the appropriate reference miRNA, and the latter one was comparatively easier to implement especially for high-throughput screening of large quantities of samples. For the relative quantification method, miR-16 has been found expressed at similar levels in most tissues [26] and was commonly used as an endogenous control for many studies [35–38]. In addition, there are alternative reference microRNAs including small RNAs (5S, U6), total RNA, or 18S rRNA [29]; however, it was claimed that 5S and U6 are not present in the plasma and serum [19, 36] and, thus, not applicable in plasma or serum miRNA analysis. Nevertheless, three known synthetic nonmammalian miRNAs cel-miR-39, cel-miR-54, and cel-miR-238 as exogenous controls, mainly derived from *Caenorhabditis elegans*, were initially introduced by Mitchell's group to plasma or serum samples after the addition of the denaturing solution during the RNA extraction process [17]. Since then, the exogene normalization method was widely promoted to give quality control of the RNA extracted from the source material. It can't be denied that the single exogenous miRNA may not be enough to reflect the landscape within the source material. Therefore, the integrating use of different normalization methods was strongly recommended, preferably the use of both endogenous and exogenous controls [39].

### *3.2.5 Circulating MicroRNA Biomarkers for Gastric Cancer*

As for GC, a number of circulation miRNAs have been found to be dysregulated in GC patients by Solexa sequencing or microarray profiling. And some of them have been identified by quantitative reverse transcription polymerase chain reaction (qRT-PCR) as biomarkers to detect GC, such as miR-106b, let-7a [30], miR-451, and miR-486 [40] in plasma, five miRNA signatures (miR-1, miR-20a, miR-27a, miR-34, and miR-423-5p) [31], and miR-378 [32] in serum. However, most of these studies failed to determine at which point in a cancer's evolution these miRNA biomarkers can be detected in the bloodstream, whereas an optimal circulating miRNA biomarker should be able to differentiate different courses in cancer's development [41]. In addition, since different data normalization methods for qRT-PCR were adopted in different assays, which affect the final selection of differentially expressed miRNAs [42], a reliable data normalization method for circulating miRNA study should be explored.

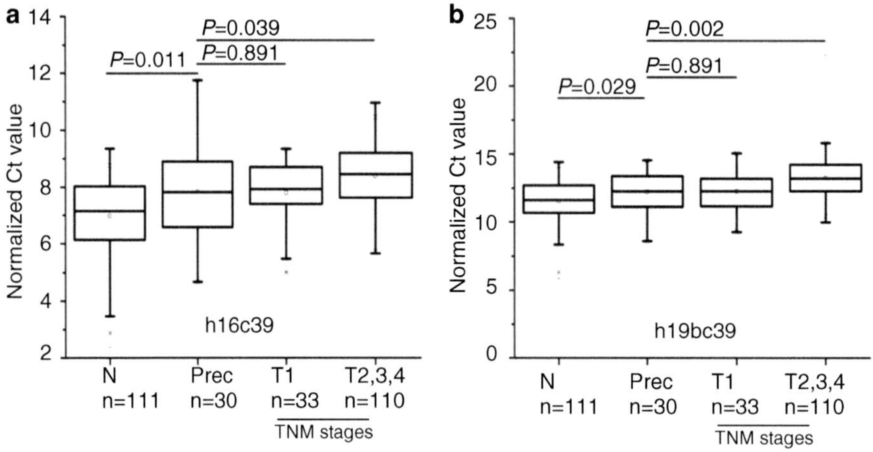


**Fig. 3.2** The expression of miR-16-5p (**a**), miR-19b-3p (**b**) among GC patients and normal controls. *N* normal control, *Prec* precancerous patients, *T1–4* GC patients with different TNM stages, respectively, from I(T1), II(T2), and III(T3) to IV(T4)

Daxiang Cui and his group were for the first time to report that miR-16-5p and miR-19b-3p were identified to be the novel potential plasma biomarkers to detect gastric cancer [43]. Differentially expressed miRNAs were initially screened out by genome-wide miRNA profiling microarrays between 16 plasma samples of gastric cancer and 18 matched normal controls and then were quantified and validated by quantitative reverse transcription-PCR method between 155 gastric cancer cases and 111 normal controls. Additionally, 30 plasma samples from precancerous lesions and 18 paired samples from gastric cancer cases with gastrectomy were further detected. Results showed that based on two normalization methods, miR-16-5p and miR-19b-3p in plasma were found to be capable of distinguishing normal population from GC cases with different TNM stages and differentiation grades, particularly including the early cancer cases ($P<0.05$). And the two miRNAs were downregulated in GC cases (FC<0.5). Especially, the downregulation degree was correlated with the progression of the GC cases from the early stage to the advanced stage ($rs>0.2$, $P<0.01$, Fig. 3.2). And the same weak downregulation of the two biomarkers as the early GC occurred initially in the precancerous diseases ($P<0.05$, Fig. 3.2). The corresponding performance of the two miRNAs to detect GC in ROC analysis gradually performed better with the disease progressing from the earlier stages or lower grades to the advanced stages (TNM IV stage, AUC=0.832 for miR-16-5p; TNM III stage, AUC=0.822 for miR-19b-3p) or high grade (poorly differentiated, AUC=0.801, 0.791, respectively, for miR-16-5p and miR-19b-3p). Additionally, miR-19b-3p remained downregulated in the patient's plasma within 9 days after gastrectomy. In conclusion, miR-19b-3p and miR-16-5p may be prospective biomarkers to detect gastric cancer and indicate its progression and thus may own great potential in applications such as early screening and progression evaluation of gastric cancer in the near future.

### 3.2.6 Conclusions and Prospects

The broad range of screening of miRNA biomarkers from clinical specimen is impressive, and the fast pace of evolution of miRNA-based technologies is quite exciting. The enormous potential of circulating miRNAs as a class of ideal cancer biomarkers is based on the following facts: (i) they are remarkably stable molecules, well preserved in harsh conditions, and resistant to RNase activity; (ii) their expression profiles are specifically correlated with certain type of cancer or pathognomonic condition; and (iii) they are easily accessible and can be sampled in a relatively noninvasive manner and readily detected by various techniques [25]. However, all the pre-analytical and analytical issues should be addressed by establishing a standard process of biomarker identification and characterization in order to render miRNAs an effective diagnostic strategy and even for clinical application in the near future [28].

## 3.3 Nanoparticle-Based Optical Detection of MicroRNA

### 3.3.1 Background

For the purpose of developing miRNAs as reliable diagnostic and prognostic biomarkers, it is primarily important to develop or optimize the efficient, sensitive, and reproducible detection methods [4]. Different methods and technologies for miRNA detection have different developmental levels from a single miRNA detection to mass population or multiplexed miRNA profiling. Currently widely used methods are Northern blotting, quantitative reverse transcription polymerase chain reaction (qRT-PCR), microarray for miRNAs with the already known sequences, and next-generation sequencing for miRNAs with unknown sequences. As the conventional standard method, Northern blotting bears the following defects, the huge time requirement, necessity for large RNA samples, and potentially hazardous radioactive labels, making it somewhat unsuitable for expression level studies [44]. miRNA detection by RT-PCR not only needing the prior reverse transcription with low efficiency easily makes underestimation or even false-negative PCR reactions but also needing dedicated design of primers, specified instruments, difficult normalization methods, and complex data analysis makes it time-consuming and costly per sample [45]. The microarray methods—which hybridize all the target miRNAs or cDNAs to the probes under the same condition—are prone to cross-hybridization, thus resulting in false positives or negatives [46]. Next-generation sequencing could discover novel miRNAs with unknown sequences with high sensitivity, specificity, and throughput, but the large quantity of data to analysis and the high-cost reagents and instruments limited its wide use [46]. Microarray and deep sequencing with higher throughput and RT-PCR with higher sensitivity and specificity are usually combined to screen the miRNA expression differences between normal and abnormal (such as samples with cancer and other diseases) samples and confirm microRNA biomarker for disease diagnosis, prognosis, and even therapy [31, 32].

Recently, with the development of nano-science and technology, emerging miRNA detection methods—which exploit properties including electrics [47], electrochemistry [48], mechanics [49], and optics [50] of nanomaterials such as nanopores [51], nanowires [52], and nanoparticles [53]—are booming. The electrical, electrochemical, and mechanical detections of miRNAs are limited for further in vivo or in situ detection because of the prior miRNA isolation step and the high susceptibility to the environment resulting in too many background signals and other noise present in the detection system [47]. Optical detections of microRNAs based on nanoparticles translate the miRNA hybridization, adsorption, or binding events into measurable optical signals including fluorescence, scattering, colorimetric signals, and so on. Compared to other properties, miRNA optical detection has greater potential for further biological sample detection, in vivo detection, or even in situ imaging, as well as achieves high sensitivity and specificity within less time by less labor. In the present paper, emphasis is laid on employing quantum dots (QDs) and silver nanocluster (AgNC)-based fluorescence and silver nanorod-based surface-enhanced Raman scattering (SERS) to detect miRNAs.

### *3.3.2 Characteristics of MicroRNA Detection*

Both the single-molecule detection and multiplexed detection of miRNAs encounter the following challenges due to the intrinsic characteristics of miRNAs [54, 55]. Firstly, miRNAs with ~22 nts are too short to label and amplify and thus hard for probe design. Secondly, miRNAs from different families show great differences in GC content, expression levels in biological samples, and hybridization efficiency with complementary DNAs, which leads to difficulty in normalization of the hybridization efficiency for each microarray element. Moreover, miRNAs within the same family may differ by a single nucleotide (e.g., let-7 family). The lack of a common sequence among miRNAs also makes them difficult to selectively purify. Finally, the primary transcripts (pri-miRNAs), precursors (pre-miRNAs), and other noncoding short RNAs may affect the detection signals from mature miRNAs.

To solve the above problems in miRNA detection, various improvements are made for the specific detection methods. In miRNA profiling with probes, LNA and PNA are usually introduced into the probes to obtain excellent detection results. LNAs comprise a class of bicyclic high-affinity RNA analogs in which the furanose ring of LNA monomers is conformationally locked in an RNA-mimicking C3′-endo/N-type conformation [56]. LNA exhibits the following advantages as hybridization probes. On the one hand, LNA mediates high-affinity hybridization without compromising base-pairing selectivity and excellently distinguishes single-base mismatching. On the other hand, LNA incorporations display high stability and low toxicity in biological systems, efficient transfection into mammalian cells, good aqueous solubility, and potent antisense activity in vivo [56]. LocMeCytidine-5′-triphosphate

(LNA-mCTP) was synthesized by Kore and his colleagues and further used to transcribe into the short $^{33}$P-LNA RNA probes, which exhibit strong binding affinity to target oligonucleotides in experiments to detect miR-124 within different tissue regions from mouse and let-7C in human cancer tissues by in situ hybridization [57]. The PNA is a DNA analog in which the negatively charged deoxyribose phosphate backbone is replaced by an electroneutral peptide-like backbone consisting of N-(2-aminoethyl)-glycine scaffold [29]. The PNA probe exhibits superior hybridization characteristics, improved chemical and enzymatic stability, and unprecedented thermal stability relative to natural nucleic acids [58]. On the one hand, the uncharged backbone of PNA exerts no internal repulsion to the complementary target oligonucleotide, enabling highly specific, rapid binding at low ionic strength. On the other hand, PNA increases melting temperatures by approximately 1.0 °C per base pair for PNA–DNA duplex and thus contributes to distinguish single-base mismatch [58–60].

As to miRNA profiling without probes, a variety of chemical and enzymatic methods to directly modify miRNAs through attaching labels or tags allow sensitive detection [54]. More importantly, label-free miRNA detection with the help of wonderful optical technologies emerges as the novel miRNA detection methods characterized by no need for sample preparation and microRNA amplification, small amount of crude sample in demand, simple detection methods, less time, high specificity and sensitivity, and stringent discrimination between microRNAs and other RNAs [44].

According to the environment for the binding or hybridization between the miRNA targets and signal molecules, the optical detection of miRNAs can be classified as solid-phase or surface-based detection and solution-phase or solution-based detection [54, 61]. Solid-phase detection can be further developed into in vitro noninvasive detection, such as detecting miRNAs in blood, plasma, serum [62], or other kinds of body fluids despite needing post-hybridization wash procedure. Solution-phase detection has the potential for further in vivo or in situ detection [61] as well as enables directly detecting miRNAs in complex matrices [44].

### 3.3.3 Fluorescence-Based MicroRNA Detection

Recently in addition to conventional fluoroscopic technologies based on fluorescence emission and quenching, various improved fluorometric analysis methods are employed to achieve sensitive detection of miRNAs, such as total inner reflection fluorescence [63], time-resolved fluorescence [64, 65], fluorescence correlation spectroscopy [66], and fluorescence resonance energy transfer (FRET), bioluminescence resonance energy transfer (BRET) [67], or chemiluminescence resonance energy transfer (CRET) [68]. Correspondingly, the fluorescent materials for miRNA detection cover from organic dyes (e.g., 6-carboxyfluorescein (FAM), Texas Red fluorophore (TEX615) [69], Cy3, and Cy5 [70]) and fluorescent proteins (e.g., Renilla luciferase (Rluc) [71]) to inorganic nanostructures (e.g., quantum dots (QDs) [72] and silver nanoclusters (AgNCs)) and other nonfluorescent nanomaterials conjugated or encapsuled with the fluorescent dyes [73, 74].

### 3.3.3.1 Quantum Dots

Quantum dots (QDs) are superior to organic fluorophores as follows: (1) size-tune fluorescent emission; (2) the broad excitation and narrow emission spectra, which allow excitation of mixed QD populations at a single wavelength far removed (>100 nm) from their respective emissions [75]; and (3) high quantum yield and good photo stability [76]. All of these advantages of QDs play important roles in miRNA detection.

Song and his cowokers established a ligase-free sensor with one-step hybridization based on quantum dot-functionalized reporter DNA for the specific detection of miRNA mimics (the dual short RNA) and demonstrated that quantum dot-modified reporter probes could increase the melting temperature [72]. In the present paper, two sizes of quantum dots (QDs) with emission wavelengths at 605 nm and 655 nm functionalized, respectively, by SA and anti-FAM (denoted as SA-QDs and anti-FAM QDs) were bound to biotinylated oligonucleotides reporter A and FAM reporter B, in which one end was complementary to the capture probe immobilized on 96 well plate and the other end was complementary to the short RNA target (Fig. 3.3). The reporter-functionalized QDs and appropriate amounts of target RNAs were mixed in a reaction volume and then added into the capture probe-modified microwells. After full hybridization and washing procedures, the fluorescence signals excited at 355 nm can be obtained on a fluorescence spectrophotometer.

The present sensor was shown to detect as low as 10 pM in 100 μL of solution and span three orders of magnitude. Because the applied QDs with modulated sizes exert advantages on wide excitation and narrow emission wavelength ranges, the

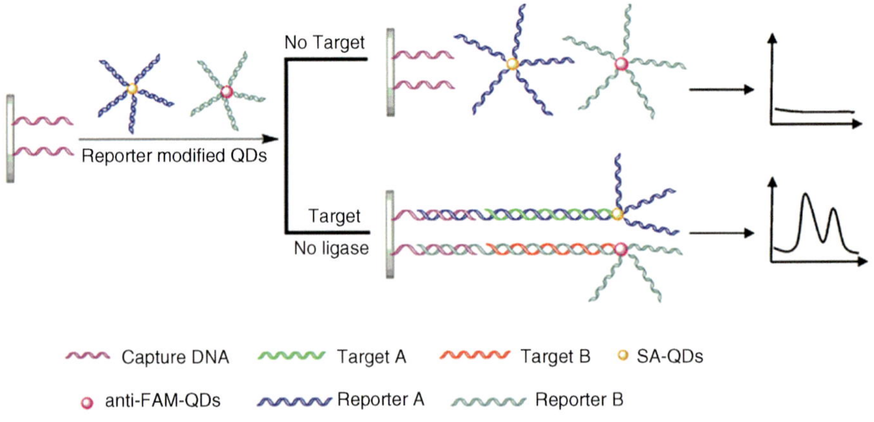


**Fig. 3.3** Scheme of the QD-enhanced detection of multiple short RNA targets via one-step template-dependent surface hybridization (Copyright permission from Song et al. [72])

proposed method is capable to detect the dual RNAs simultaneously in a single sample at one excitation wavelength without interference and is prospective for multiplexed miRNA detection. Moreover, the QD-modulated reporter probes enhance the binding affinity among the ternary oligonucleotides, eliminating the use of enzymatic ligation and thus reducing the costs. However, it should be noted that the length of capture probes differently affected the different targets, making the design of the capture probe and reporter probes more sophisticated. In addition, the present method was no further applied to real samples, reducing its robustness validation.

In addition, the particular properties enable QDs as good energy donor or acceptor in QD-based FRET [77], CRET [76], and BRET [78]. Recently, a rapid, sensitive, and specific miRNA detection method was developed by combining the two-stage isothermal exponential amplification reaction (EXPAR) with the single-QD-based nanosensor [77]. The principle of the present miRNA assay is shown in Fig. 3.4. The target miRNA is first exponentially amplified through the first-stage EXPAR and then linearly converted into reporter oligonucleotide Y in the second-stage EXPAR. The reporter oligonucleotide Y then hybridized with the biotinylated capture probe and Cy5-labeled reporter probe to form the sandwich hybrid which were then caught and assembled on the surface of 605QDs through specific

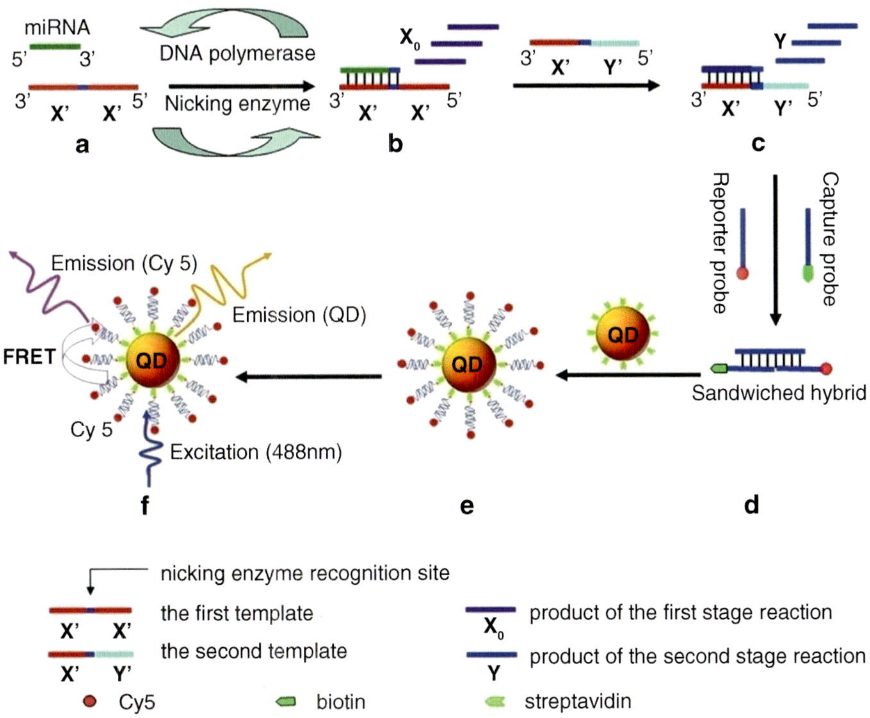


**Fig. 3.4** Scheme of the miRNA assay based on the two-stage EXPAR and single-QD-based nanosensor. (**a**) first-stage amplified; (**b**) amplified products; (**c**) second-stage amplified products; (**d**) sandwiched hybrid; (**e**) self-assembled on the surface of QDs; (**f**) detection of light signal. (Copyright permission from Zhang, et al. [77])

streptavidin–biotin binding. The consequent 605QD/reporter oligonucleotide Y/Cy5 complexes were detected by nanosensor in a microfluidic flow at an excitation wavelength of 488 nm.

The promising assay in the paper was applied to detect let-7a miRNA with the limit of detection (LOD) as low as 0.1 aM. More excitingly, the single-QD-based nanosensor can discriminate single-nucleotide differences between let-7 miRNA family members including let-7a, let-7b, and let-7c. What is more important is that miR-21 and let-7a were detected simultaneously by designing two sets of corresponding templates with the same capture and reporter probes. Undoubtedly, attributed to the combination of high amplification efficiency of EXPAR, high sensitivity of single-particle detection, and near-zero background noise of the single-QD-based nanosensor, the present method exhibits highly improved sensitivity, specificity, and bright promise for further multiplexed miRNA detection. However, before the clinical application, the method there should seek further improvement on miRNA detection in real samples without interference from the complex environment. And it is worthwhile to develop a microfluidic setup capable of simultaneously manipulating massive sample detection.

#### 3.3.3.2 Silver Nanoclusters

DNA strands with specific sequences were used as templates to create the time-dependent and size-specific silver nanoclusters (AgNCs) that follow the contour of the DNA templates [79]. As the promising substitutes for organic dyes and QDs, these silver nanoclusters encapsulated in single-stranded DNA (ssDNA) do well in high emission rates throughout the visible and near IR, photostability [80], biocompatibility, small size (<2 nm) [81], and water solubility [82], thus exhibiting great potential for nontoxic, sensitive high-resolution biological imaging (Patel, Richards et al. 2008) and chemical/biological detection [82] by enhancement or quenching of the emission intensity and/or by shifting the emission wavelength [83].

Yang and his coworkers developed a DNA–silver nanocluster (DNA/AgNC) probe to detect target miRNA using the fluorescence dropping of the probe (Fig. 3.5)

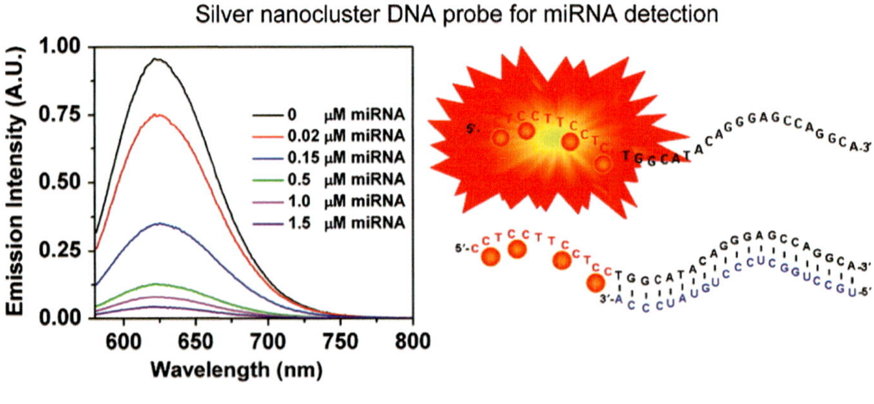


**Fig. 3.5** Scheme of silver nanocluster DNA probe for miRNA detection (Copyright permission from Yang and Vosch [84])

[84]. In the present paper, the DNA-12 nt-RED, a DNA sequence capable of creating red-emitting AgNC, was attached with DNA-160, a DNA sequence complementary to miR-160, to obtain the final DNA template DNA-12 nt-RED-160. After that, the DNA-12 nt-RED-160 probe and miRNA target were mixed together, followed by the addition of $AgNO_3$ and reduction by $NaBH_4$ to form AgNC and monitoring the changes on its red emission after 1 h. The AgNC-based fluorescence method has great prospective for picomole level detection of miR-160 and the capacity to detect miR-160 in RNA extracts from plants.

In the later work by the above team, they performed a series of experiments to illustrate the formation of bright red emissive silver nanoclusters (AgNCs) in a DNA sequence (DNA-12 nt-RED-160), demonstrating that the mismatch self-dimer formation of the DNA probes contributes to the bright red emission. And based on the mechanism, a novel DNA/AgNC probe for miR-172 detection was successfully developed [83].

Hairpin DNA-templated AgNCs (AgNCs/HpDNA) were prepared and introduced into strand displacement amplification (SDA) by Daxiang Cui's group as a novel probe for miRNA detection. The light-up platform was achieved by guanine (G)-rich fluorescence enhancement which converted the excitation/emission pair of AgNCs/HpDNAs from a shorter wavelength to a longer wavelength. Based on this mechanism, single and duplex detections were conducted on two plasma biomarkers including miR-16-5p and miR-19b-3p for the diagnosis of gastric cancer. The compatibility of AgNCs/HpDNA probes with SDA reaction was also studied. The results showed that the probe (AgNCs/RED 16(7 s)C) for miR-16-5p adopted a better conformation with high specificity to recognize the single-base mismatch, whereas the probe (AgNCs/GRE 19b(5 s)C) for miR-19b-3p generated high fluorescence signal during the amplification, but it would be partially digested by AgNCs. The method for the first time integrated the generation of enhanced fluorescent AgNCs into the target recognition process, which offers great opportunity for rapid miRNA detection.

As shown in the scheme of reaction mechanism (Fig. 3.6), a light-up system to detect miRNAs was developed by integrating AgNCs/HpDNA with SDA. Different from the traditional HpDNA structures for MB, the HpDNAs in our study include a 3′ overhang (HpO) as well as the stem (HpS) and loop (HpL) regions. The cytosine-rich strand at 3′ overhang mainly contributes to the generation of AgNCs, which are expected to emit red (HpRO, 5′-CCCTTAATCCCC-3′, 12 nt) or green fluorescence (HpGO, 5′-CCCCCCCCCCCCCCC GCCCGCC-3′, 22 nt), respectively, upon hybridization with their complementary strands with guanine-rich overhang at 5′ end [16, 18]. To stabilize the hairpin structure with such a long overhang, 11-nt-long stem was adopted. And to make the hairpin better be opened by the target miRNAs, we introduced the shared stem proposed by Tsourkas [19], of which some bases close to 5′ end of the stem were shared with the loop region together binding to the miRNAs. Accordingly, the miRNAs can be divided into the long part complementary with the loop (MLc) and the short part complementary with the stem (MSc). Another key point is the two region-containing primers we used, the PSc region at 3′ end complementary with the stem sequences of HpDNAs and the 5′ overhang of the primer (PO) full of guanines as a fluorescence enhancer to specifically light up AgNCs.

Generally, in the presence of miRNAs, they hybridized to the AgNCs/HpDNAs with little fluorescence by pairing to the whole loop region and part of the stem sequences, the latter of which acted as toeholds to unfold the HpDNAs. Subsequently, with the primers annealing to the 3′ stem, the SDA reaction was triggered together with dNTP and polymerase, in which the complementary strand with G-rich overhang (HpDNAc) was extended to generate duplex (AgNCs/HpDNA–HpDNAc) and the target miRNAs were simultaneously displaced. Upon the formation of duplex, the fluorescence enhancement at special wavelength can be obtained. Consequently, the displaced miRNAs go into the next cycling. By contrast, in the absence of miRNAs, the HpDNAs are folded, and the polymerization reaction can't be triggered, leading to the AgNCs on HpDNAs in the dark at special wavelength.

Undoubtedly, the present method paves the way for the application of fluorescence properties of DNA/AgNC to detect miRNAs, which integrates the probe design and target detection into one step. However, lack of full understanding of the

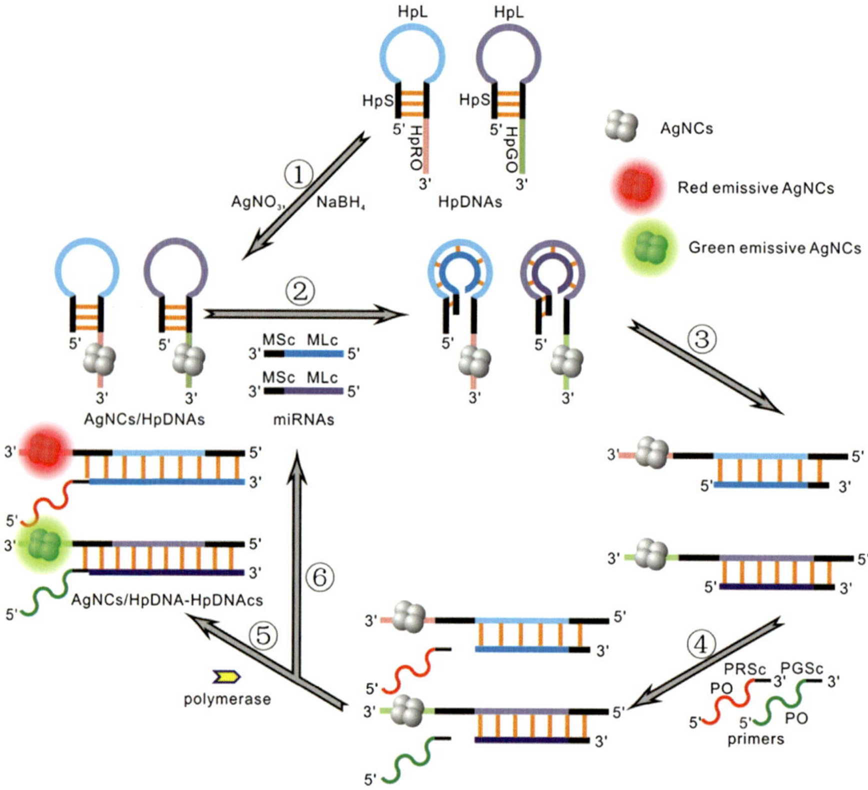


**Fig. 3.6** Schematic illustration of the design and detection strategy. (1) Formation of AgNCs on the HpDNA template. (2) Hybridization of target miRNAs to the loop and part of 5′ stem in AgNCs/HpDNAs. (3) Unfolding of HpDNAs upon the hybridization. (4) Primers annealing to 3′ end of the stem with an overhang left at 5′end. (5) Extension of HpDNAcs to form AgNCs/HpDNA–HpDNAcs. (6) Displacement of the target miRNAs for the next cycling (Copyright permission from Zhang, et al. [43])

mechanisms underlying the AgNC-based fluorescence changes, it is difficult to eliminate the effects of non-miRNAs. Additionally, the synthesis of AgNC by DNA template containing the relevant target sequences makes it necessary to design the specific DNA/AgNC probe to detect one target miRNA sequence. Therefore, the method encounters great challenges on the design of DNA templates for multiplexed miRNA detection.

#### 3.3.3.3 Magnetic Nanoparticles and Others

Recently, magnetic beads conjugated with molecular beacons [85] and magnetic fluorescent (MF) nanoparticle-based molecular beacon (MB) [74, 86] have been developed as multimodal miRNA detecting or imaging probes for miRNA-based diagnosis and even therapy. In the specific probe, magnetic particles are exploited to facilitate the separation or delivery. And to introduce stable fluorescence generating sources, such as organic dyes, silica and MBs are widely used as the scaffold or linker.

Utilizing the array surface provided by magnetic beads conjugated with molecular beacons, the cation exchange-based fluorescence amplification (CXFluoAmp) method to detect miRNA was developed [85]. As shown in Fig. 3.7, the molecular beacon bound on magnetic beads worked as capture probe to hybridize with the target miRNA and link with detection probe containing nonfluorescent ionic nanocrystals (NCs) CdSe after ligation, thus triggering the release of $Cd^{2+}$ by adding $Ag^{+}$ to the above complex and finally turning on the fluorescence response of the dye Rhod-5 N. The promising assay enabled detecting synthesized miR-21 as low as 100 fM with the LOD at 35 fM and dynamic range spanning seven orders of magnitude and discriminated the different miR-21 levels in the total RNA extracts from the breast cancer tissue and normal breast tissue. In addition, the method showed

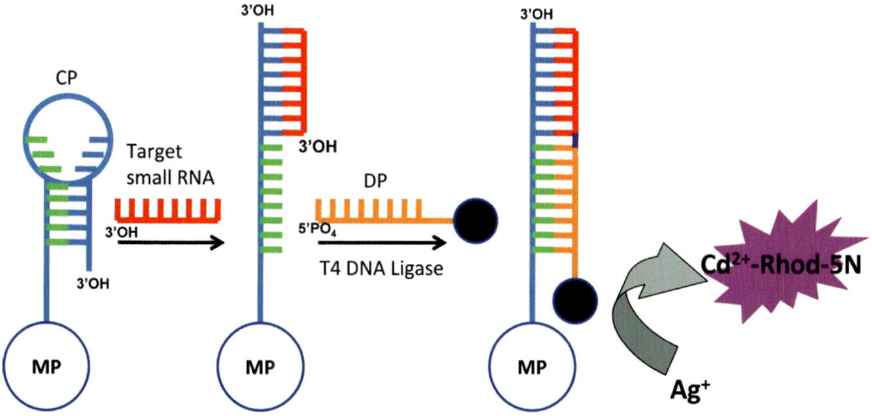


**Fig. 3.7** Scheme of the small RNA detection using CXFluoAmp. *MP* magnetic particle, *CP* capture probe, *DP* detection probe (Copyright permission from Li et al. [85])

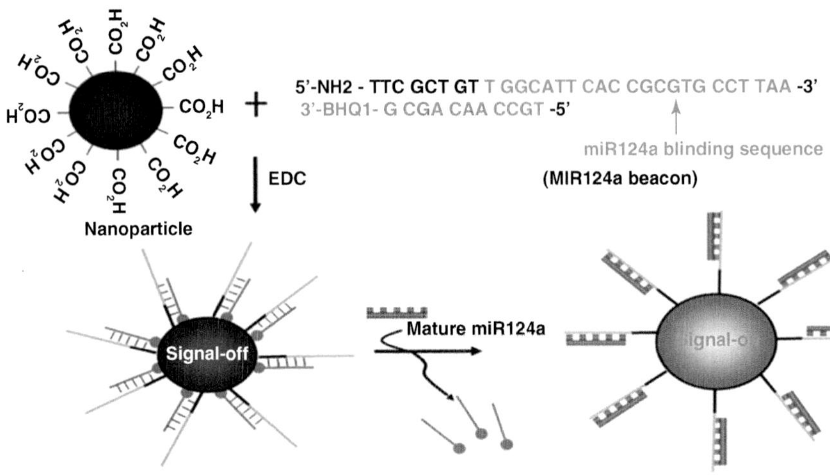


**Fig. 3.8** Scheme of the fluorescent nanoparticle miR-124a beacon (MF-miR-124a beacon) for targeting miR-124a (Copyright permission from Do et al. [74])

high specificity to distinguish one or few nucleotide mutations. The ligation reaction by T4 DNA ligase contributed to the prominent sensitivity and specificity. However, it should be noted that the self-release of $Cd^{2+}$ in nanocrystals may adversely affect the assay.

Do Won et al. first reported the application of magnetic fluorescent (MF) nanoparticle-conjugated molecular beacon to image intracellular miRNAs [74]. As seen in Fig. 3.8, the MF nanoparticle which was composed of cobalt ferrite in the central core and a rhodamine fluorescent dye coated with a silica shell was linked with miR-124a-specific molecular beacon with a partially double-stranded DNA including aptamers that contain nonfluorescent dye, a black hole-quenching (BHQ) molecule. In the absence of the target miR-124a, the quencher close to the MF nanoparticles makes the fluorescence signal off, while in the presence of the target miR-124a, the stronger complementary pairing between miR-124a and its binding sequence in MB leads to the quencher detached from the MF nanoparticles and consequently the signal is on. Based on the MF particle-conjugated miR-124a beacon, the present method achieved specific miR-124a detection in microtube, in vitro intracellular miR-124a imaging in P19 cells induced and cultured as neuronal differentiation of cells, and in vivo miR-124a imaging in engineered mouse.

In addition to magnetic particle-based fluorophores, metal nanoparticle fluorophores have emerged as attractive probes in miRNA detection and imaging in the past decades [73]. Take silver shell/silica core fluorophore, for example. It was employed to detect miR-486 in lung cancer cell by fluorescence in situ hybridization (FISH). As shown in Fig. 3.9, the probe was composed of a silica core (50 nm) encapsulated with the dye $Ru(bpy)_3^{2+}$ and the silver nanoshell (10 nm) with the single-strand oligonucleotide complementary to the target miRNA. Based on the metal–fluorophore interaction resulting in the enhanced intensity and shortened

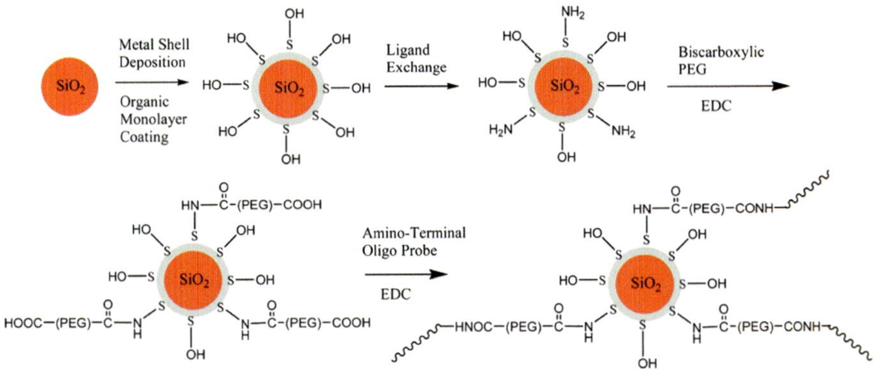


**Fig. 3.9** Scheme of the preparation of metal shell/silica core structure bound with the single-strand oligonucleotide probe (Copyright permission from Zhang et al. [73])

lifetime of fluorescence, the probes indicated the location of single miR-486 in lung cancer cells by counting the numbers of spots with certain fluorescence lifetime. The metal nanoparticle fluorophore-based FISH assay provided prospect for cell-to-cell variations in miRNA expression including quantity and location. However, it is strange that the assay selected Cy3 covered by silver nanoshell with the already shortened fluorescence lifetime to image the miRNAs in cells with shorter fluorescence lifetime than the former one, thus resulting in the question on the use of the silver nanoshell.

## 3.3.4 *Scattering-Based MicroRNA Detection*

### 3.3.4.1 SERS

Surface-enhanced Raman scattering (SERS) technique, with huge enhancements of adsorbate Raman signals and a drastic quenching of fluorescence [87], shows great strengths in sensitivity to detect extremely low levels of analyte and specificity to provide the molecular fingerprint of the analyte in near real time [88, 89], thus providing a prominent pathway for miRNA detection. However, the practical application of SERS-based sensing suffers from the difficulty in the fabrication of large area, uniform, and high enhancement substrates [90]. In the current reports on SERS-based miRNA detection, the substrates with aligned silver nanorod arrays (AgNRs) prepared by oblique angle deposition (OAD) method were proposed to have prospect for routine miRNA expression profiling in future clinic laboratories [89].

Exploiting OAD-fabricated silver nanorod arrays (AgNRs) and affinity differences between samples and probes on the substrates, the SERS-based miRNA detection method was developed by Driskell and his coworkers [91]. In the detection (Fig. 3.10), after the hybridization of small RNAs including the target RNAs with the thiolated complementary probes, oligo-samples were incubated with the

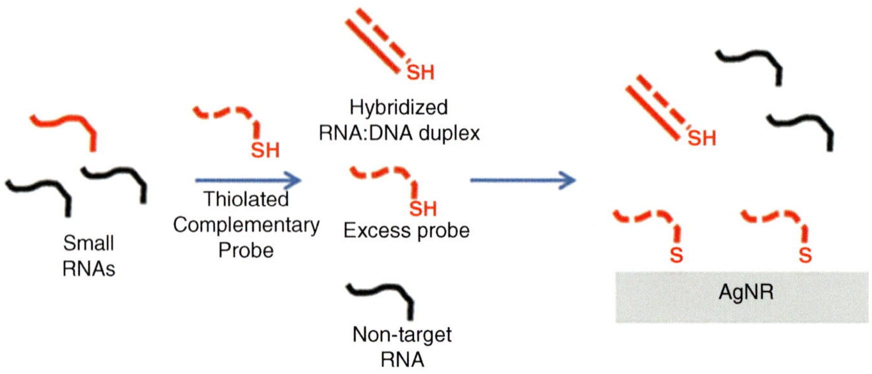


**Fig. 3.10** Scheme of label-free SERS detection of miRNA based on affinity (Copyright permission from Driskell and Tripp [91])

AgNR substrate, rinsed, and dried, resulting in the mere adsorption of the excess thiolated probes and the corresponding production of SERS signals because only thiolated ssDNA showed strong adsorption on the substrate, while the dsRNA–DNA and nontarget RNA exhibited negligible adsorption. The present method opens possibility for multiplexed label-free miRNA detection. However, with limit of detection of 28 nM, the sensitivity should be further enhanced to compete with other optical miRNA detection methods, as well as the specificity must be validated by detecting different miRNAs synthesized or even extracted from real samples.

In addition to solid–gas interface-based SERS, other interfaces including solid–solid, solid–liquid, and solid–vacuum have been observed to have SERS effect. Wang et al. developed the miRNA detection method based on SERS effect on solid–liquid interface using plasmonic coupling phenomenon combined with the LNA-based nucleic acid hybridization process, in which plasmonic coupling enhancement in the gap regions between nanoparticles contributes to the high SERS enhancement [92]. As seen in Fig. 3.11, silver nanoparticles (AgNPs) functionalized with the thiolated oligonucleotides with the same sequences as the miRNA targets and labeled with Cy3 Raman dye worked as reporter NP, while AgNPs functionalized with the complementary LNAs worked as capture NP. In the absence of the miRNA targets, the two probes assembled into a three-dimensional (3D) nanonetwork through hybridization, inducing plasmonic coupling between adjacent NPs to produce increased SERS signal of the Raman levels upon laser excitation. In the presence of the miRNA targets, the plasmonic coupling is interfered by the formation of LNA–miRNA duplex between capture NPs and miRNA, thus leading to the reduction in SERS intensity. The method was successfully applied to 100 nM miR-21 detection. However, the initial experiments should be further validated by detecting a series of miRNAs with different concentrations and sequences. Therefore, there is still a long way to go for multiplexed simultaneous assays of miRNAs in a single sample solution.

In conclusion, the SERS technique for miRNA detection is in its fledging period, facing the problem on sophisticated readout system, complicated analysis skill, and

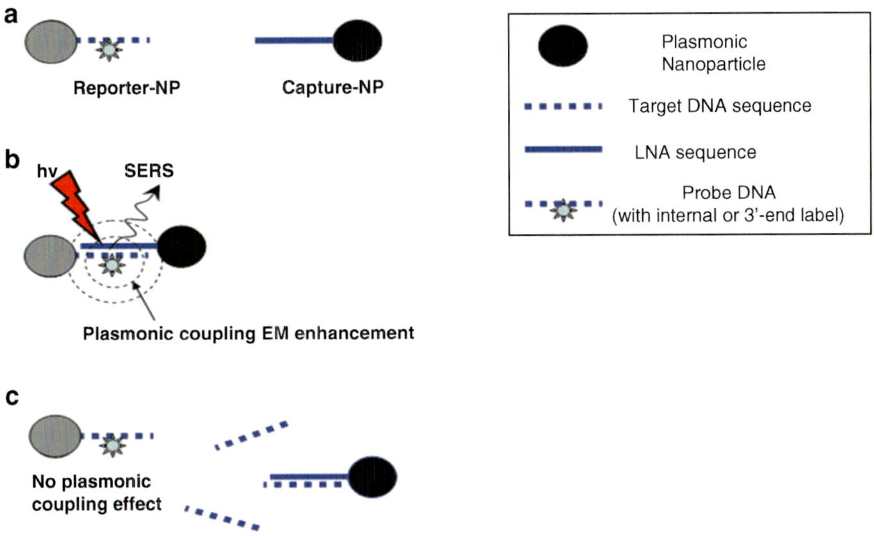


**Fig. 3.11** Scheme of plasmonic coupling interference (PCI) nanoprobes for nucleic acid detection. (**a**) reporter-NP and capture-NP; (**b**) plasmonic coupling EM enhancement; (**c**) no plasmonic coupling effect (Copyright permission from Wang and Vo-Dinh [92])

convoluted data interpretation and verification [88]. However, as for label-free and multiplexed detection of miRNA, SERS is undoubtedly a primary selection and has potential for the development of novel miRNA detection tools.

#### 3.3.4.2 Other Scattering

Besides SERS, other light scattering signals including Rayleigh scattering are also capable of miRNA detection with the help of advanced optical detection technology.

Exploiting the Brownian motion and the stronger resonance Rayleigh scattering of silver nanoparticles (AgNPs) [93], the single-silver-nanoparticle counting system (SSNPC) by photon-burst counts with ultrahigh sensitivity, high spatial resolution (0.5 fL), and good reproducibility was established to detect miRNA in homogeneous solution. The principle of the method is shown in Fig. 3.12, in which two kinds of AgNPs conjugated with oligonucleotides, respectively, complementary to the two ends of the target sequence worked as probes, producing high signals without miRNA targets while low signals with miRNA targets in SSNPC detection. The method was able to detect synthetic miRNA as low as 0.5 fM with high reproducibility. Though with higher sensitivity by further reducing the sample volume using microfluidic droplet technique, the method is hard to simultaneously detect multiplexed miRNAs.

Based on the enhancement of light scattering by gold nanoparticles (AuNPs), a high-density Scanometric miRNA (Scano-miR) microarray was constructed [94]. As seen in Fig. 3.13, miRNA targets in a sample mixture were firstly attached with

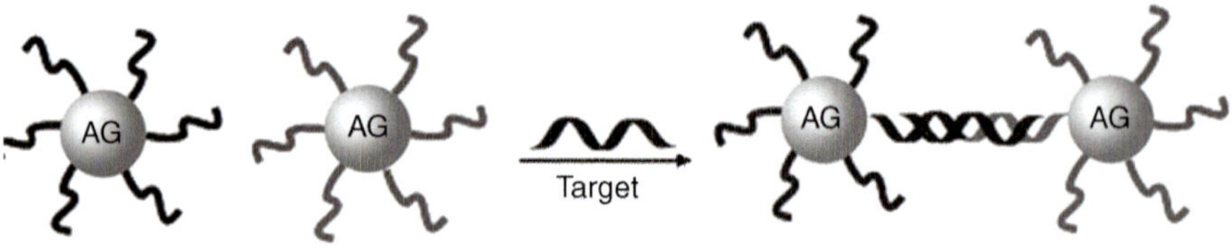


**Fig. 3.12** Scheme of homogeneous detection of miRNA based on SSNPC system (Copyright permission from Xu et al. [93])

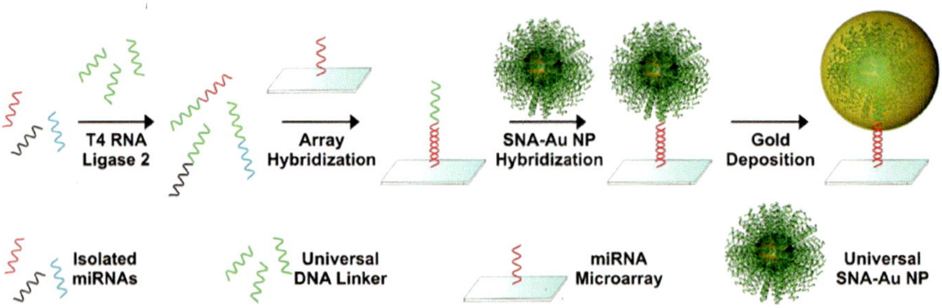


**Fig. 3.13** Scheme of scanometric array-based multiplexed detection of miRNA species (Scano-miR platform) (Copyright permission from Alhasan et al. [94])

universal DNA linker by T4 RNA ligase and then hybridized with the DNA complementary to the miRNA target on the microarray. After washing the unbound miRNA, spherical nucleic acid–gold nanoparticles (SNA–AuNPs) conjugated with oligonucleotides complementary to the universal linker sequences were added and hybridized onto the microarray followed by a second washing. The final microarray after further gold enhancement was detected by scanometric system. The Scano-miR system was first applied to detection of miR-16 with detection limit at 1 fM (~600 copies/μ L serum), as well as human serum miRNAs with good performances in specificity, reliability, and reproducibility. More importantly, the array was useful for miRNA profiling in cell lines and even tissue samples, thus showing as robust high-throughput tools for screening miRNA biomarkers.

### *3.3.5 Colorimetry-Based MicroRNA Detection*

Colorimetric method for miRNA detection is generally superior to other optical detection methods in terms of low-cost equipment. And the widely used nanomaterial in this method is microarray-based gold nanoparticles combined with silver enhancement to realize the high-throughput detection [53].

With the help of gold nanoparticle and silver enhancement, a simple, sensitive, and specific colorimetric method for miRNA detection was developed [95]. The principle of the method is shown in Fig. 3.14, in which the biotinylated oligonucleotides as capture probes complementary to the 3′ end of the target miRNA sequence and gold nanoparticles (AuNPs) as reporter probes conjugated with oligonucleotides

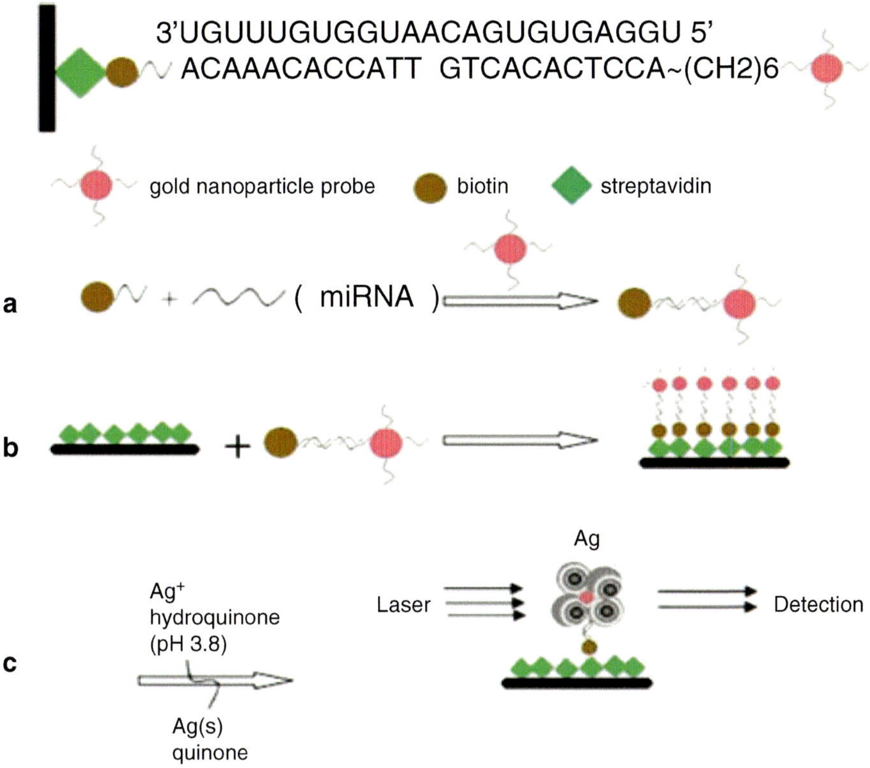


**Fig. 3.14** Scheme of quantification of miRNA by functionalized gold nanoparticles. (**a**) gold nanoparticle probes combine with biotin-labeled miRNA; (**b**) streptavidin-conjugated capture probes combine with biotin-labeled miRNA with gold nanoprobes; (**c**) quantitative detection (Copyright permission from Yang et al. [95])

complementary to the 5′ end sequences were hybridized with the miRNA target, followed by immobilizing onto the streptavidin-coated microplate and the final colorimetric detection by a microplate reader after washing and silver enhancement steps. The present method was able to detect miR-122a/miR-128 in isolated RNA from only 2-ng sample of mouse brain or liver tissue and detect synthesized miR-122a ranging from 10 fM to 10 pM with the capability to discriminate one single-oligonucleotide mismatch. However, it is limited for multiplexed miRNA detection due to complex probe design. Anyway, the colorimetric method does not require expensive equipment and an advanced readout system and can be performed in any standard laboratory [96].

### *3.3.6 Conclusions and Outlook*

In this part, the current nanoparticle-based miRNA detection by optical technologies was summarized from the following aspects about fluorescence, scattering, and colorimetry (shown in Table 3.1). Among three of them, more attention has been drawn by the former two-based miRNA detection in the last few years.

**Table 3.1** Summarization and comparison of the current nanoparticle-based optical miRNA detection methods

| Solution-/surface -based or FISH | miRNA modification (amplification, labeling, ligation, etc.) | Probes | Detection signal | Samples | Performance (sensitivity, specificity, dynamic range, etc.) | Ref. |
|---|---|---|---|---|---|---|
| Surface based | No modification | Capture probes, quantum dot-modified reporter probes | Fluorescence emission intensity | Synthetic dual short RNAs | LOD at 10 pM in 100 μL with dynamic range of 10 pM–15 nM | [72] |
| Solution based | Isothermal amplification | 605QD/reporter Oligonucleotide Y/Cy5 complexes | Burst counts in FRET | let-7a, let-7b, and let-7c; miR-21 | Near zero Background noise<br>LOD at 0.1 aM<br>Clearly discriminate let-7a, let-7b, let-7c | [77] |
| Solution based | No modification | DNA/AgNC probe | Fluorescence dropping value | miR-160 | Picomole level detection of miRNA | [84] |
| -[a] | No modification | MF-miR-124a beacon | Fluorescence emission intensity. | miR-124a | Capable of in vitro intracellular imaging and in vivo imaging | [74] |
| Solution based | No modification | Capture probe: molecular beacon<br>Detection probe: aptamer with NCs | Fluorescence emission intensity in CXFluoAmp | Synthetic miR-21<br>RNA extracts from tissues | LOD at 35 fM with dynamic range of seven orders of magnitude | [85] |
| FISH | No modification | Metal shell/silica core structure bound with the single-strand oligonucleotide probe | Fluorescence emission intensity and lifetime in FISH | miR-486 in cell lines positive to lung cancer | Capable of detecting and locating low levels of miR-486 in cells | [73] |
| Surface based | No modification | Thiolated ssDNA | SERS | let-7f | LOD at 28 nM | [91] |
| Solution based | No modification | Capture NPs: AgNPs with thio-LNAs<br>Reporter NPs: AgNPs with Cy3-labeled DNA | SERS in PCI | miR-21 | 100 nM | [92] |

| | | | | | | |
|---|---|---|---|---|---|---|
| Solution based | No modification | AgNPs conjugated with oligonucleotides | Photon counts in SSNPC system | miR-21, miR-23a, P | LOD at 0.5 fM | [93] |
| Surface based | miRNA ligation by T4 RNA ligase | Universal DNA linker, miRNA microarray, and universal SNA–Au NPs | Light scattering on Scano-miR arrays | Synthetic miR-16 RNA extracts from cell line and serum | LOD at 1 fM with single-nucleotide polymorphism (SNP) selectivity | [94] |
| Surface based | No modification | Biotinylated probe (capture probe) and gold nanoparticle probe | Colorimetric absorbance | Synthetic miR-122a, RNA extracts from tissues | LOD at 10 fM ranging from 10 fM to 10 pM with specificity to discriminate one single-base mismatch | [95] |

[a]Denotes no classification as solution- or surface-based detection

Fluorescence-based miRNA detection is peculiar to in vivo and in situ imaging, as well as high sensitivity in single miRNA quantification in solution. And a myriad of nanomaterials are able to be applied to fluorescence detection, including self-fluorescent structures such as quantum dots (QDs), silver nanoclusters (AgNCs), and conjugated fluorophores such as fluorescent magnetic nanoparticles and metal nanoparticle fluorophores, thus providing sufficient material elements for selecting to develop the optimal miRNA detection method with high sensitivity or imaging method with identified safety. Moreover, the development of multimodal probes for fluorescent imaging emerges as the promising tool for miRNA-based cancer diagnosis and therapy.

Scattering-based methods to detect miRNA have great advantages in label-free multiplexed detection. The widely used nanomaterials for this method are silver or gold nanoparticle-based substrate or solution. The limited scattering substrates create bottleneck for SERS-based miRNA detection with high sensitivity. However, with the deep development of the study, it is prospective to construct miRNA microarray with high throughput based on scattering detection.

Compared to the above methods, colorimetric miRNA detection with simple and low-cost instrument or even naked eyes makes it a potential portable sensor for public use out of the laboratories.

In a word, the emerging nanoparticle-based optical technologies pave the way for detection of miRNA by developing novel nanosensors or probes. Based on the current study, there is still a lot of work to do in order to put it into clinical practice in the next few years. Firstly, environment-friendly and biosafe nanomaterial should be further synthesized to meet the application to in vivo imaging or detection of miRNA in homogeneous solution or on microarray. For example, silver nanoclusters produced by DNA templates showing good biocompatibility have attracted great attention for direct miRNA detection. On the other hand, we should pay attention to the novel properties of the nanomaterials when applied to miRNA detection. Secondly, sufficient experiments should be performed to validate the robustness of the designed method including sensitivity, specificity, reliability, reproducibility, and so on. The combination of conventional miRNA detection methods with the emerging nanotechnologies is a promising way to obtain ultrahigh sensitivity, such as the isothermal amplification of miRNA combined with QD-based nanosensor [77]. Last but not the least, based on the above elements, the advanced optical detection technology such as the microfluidic setup or surface plasmon resonance imaging (SPRi) [97] is becoming the hot study for miRNA detection with ultrahigh sensitivity.

Anyway, employing nanoparticle-based optical technology, multiplexed miRNA detection, or in vivo imaging will be widely used in biomarker identification for cancer diagnosis, prognosis, and even therapy in the near future.

## References

1. Bartel DP. MicroRNAs: genomics, biogenesis, mechanism, and function. Cell. 2004;116:281–97.
2. Meyer SU, Pfaffl MW, Ulbrich SE. Normalization strategies for microRNA profiling experiments: a 'normal' way to a hidden layer of complexity? Biotechnol Lett. 2010;32:1777–88.

3. Calin GA, Croce CM. MicroRNA signatures in human cancers. Nat Rev Cancer. 2006;6:857–66.
4. Iorio MV, Croce CM. MicroRNA dysregulation in cancer: diagnostics, monitoring and therapeutics. A comprehensive review. Embo Mol Med. 2012;4:143–59.
5. Lu J, Getz G, Miska EA, Alvarez-Saavedra E, Lamb J, Peck D, Sweet-Cordero A, Ebet BL, Mak RH, Ferrando AA, et al. MicroRNA expression profiles classify human cancers. Nature. 2005;435:834–8.
6. Ferlay J, Shin HR, Bray F, Forman D, Mathers C, Parkin DM. Estimates of worldwide burden of cancer in 2008: GLOBOCAN 2008. Int J Cancer. 2010;127:2893–917.
7. Chen WQ, Zheng RS, Zhang SW, Zhao P, Zeng HM, Zou XN, He J. Annual report on status of cancer in China, 2010. Chinese J Cancer Res. 2014;26:48–58.
8. Hartgrink HH, Jansen EPM, van Grieken NCT, van de Velde CJH. Gastric cancer. The Lancet. 2009;374:477–90.
9. Wang J, Yu JC, Kang WM, Ma ZQ. Treatment strategy for early gastric cancer. Surg Oncol. 2012;21:119–23.
10. Wu WKK, Lee CW, Cho CH, Fan D, Wu K, Yu J, Sung JJY. MicroRNA dysregulation in gastric cancer: a new player enters the game. Oncogene. 2010;29:5761–71.
11. Wang J, Wang Q, Liu H, Hu B, Zhou W, Cheng Y. MicroRNA expression and its implication for the diagnosis and therapeutic strategies of gastric cancer. Cancer Lett. 2010;297:137–43.
12. Bandres E, Bitarte N, Arias F, Agorreta J, Fortes P, Agirre X, Zarate R, Diaz-Gonzalez JA, Ramirez N, Sola JJ, et al. microRNA-451 Regulates Macrophage Migration Inhibitory Factor Production and Proliferation of Gastrointestinal Cancer Cells. Clin Cancer Res. 15;2009:2281–90.
13. Wu Q, Yang Z, Wang F, Hu S, Yang L, Shi Y, Fan D. MiR-19b/20a/92a regulates the self-renewal and proliferation of gastric cancer stem cells. Journal of cell science, 2013;126:4220–9
14. Wu Q, Yang Z, An Y, Hu H, Yin J, Zhang P, Nie Y, Wu K, Shi Y, Fan D. MiR-19a/b modulate the metastasis of gastric cancer cells by targeting the tumor suppressor MXD1. Cell Death Dis. 2014;5:e1144.
15. Kim DN, Chae H-S, Oh ST, Kang J-H, Park CH, Park WS, Takada K, Lee JM, Lee W-K, Lee SK. Expression of viral microRNAs in Epstein-Barr virus-associated gastric carcinoma. J Virol. 2007;81:1033–6.
16. Zhou H, Guo J-M, Lou Y-R, Zhang X-J, Zhong F-D, Jiang Z, Cheng J, Xiao B-X. Detection of circulating tumor cells in peripheral blood from patients with gastric cancer using microRNA as a marker. J Mol Med-JMM. 2010;88:709–17.
17. Mitchell PS. Circulating microRNAs as stable blood-based markers for cancer detection. P Natl Acad Sci U S A. 2008;105:10513.
18. Kosaka N, Iguchi H, Ochiya T. Circulating microRNA in body fluid: a new potential biomarker for cancer diagnosis and prognosis. Cancer Sci. 2010;101:2087–92.
19. Chen X, Ba Y, Ma L, Cai X, Yin Y, Wang K, Guo J, Zhang Y, Chen J, Guo X, et al. Characterization of microRNAs in serum: a novel class of biomarkers for diagnosis of cancer and other diseases. Cell Res. 2008;18:997–1006.
20. Lawson JD, Sicklick JK, Fanta PT. Gastric Cancer. Curr Probl Cancer. 2011;35:97–127.
21. Catalano V, Labianca R, Beretta GD, Gatta G, de Braud F, Van Cutsem E. Gastric cancer. Crit Rev Oncol Hemat. 2009;71:127–64.
22. Taback B, Hoon DSB. Circulating nucleic acids in plasma and serum: Past, present and future. Curr Opin Mol Ther. 2004;6:273–8.
23. Thiel A, Ristimaki A. Gastric Cancer: Basic Aspects. Helicobacter. 2012;17:26–9.
24. Berger F, Reiser MF. Micro-RNAs as potential new molecular biomarkers in oncology: have they reached relevance for the clinical imaging sciences? Theranostics. 2013;3:943–52.
25. Zen K, Zhang C-Y. Circulating MicroRNAs: a novel class of biomarkers to diagnose and monitor human cancers. Med Res Rev. 2012;32:326–48.
26. Reid G, Kirschner MB, van Zandwijk N. Circulating microRNAs: Association with disease and potential use as biomarkers. Crit Rev Oncol Hemat. 2011;80:193–208.
27. Waldman SA, Terzic A. Translating microRNA discovery into clinical biomarkers in cancer. Jama-J Am Med Assoc. 2007;297:1923–5.
28. De Guire V, Robitaille R, Tetreault N, Guerin R, Menard C, Bambace N, Sapieha P. Circulating miRNAs as sensitive and specific biomarkers for the diagnosis and monitoring of human diseases: Promises and challenges. Clin Biochem. 2013;46:846–60.

29. Mo M-H, Chen L, Fu Y, Wang W, Fu SW. Cell-free Circulating miRNA Biomarkers in Cancer. J Cancer. 2012;3:432–48.
30. Tsujiura M, Ichikawa D, Komatsu S, Shiozaki A, Takeshita H, Kosuga T, Konishi H, Morimura R, Deguchi K, Fujiwara H, et al. Circulating microRNAs in plasma of patients with gastric cancers. Brit J Cancer. 2010;102:1174–9.
31. Liu R, Zhang CN, Hu ZB, Li G, Wang C, Yang CH, Huang DZ, Chen X, Zhang HY, Zhuang R, et al. A five-microRNA signature identified from genome-wide serum microRNA expression profiling serves as a fingerprint for gastric cancer diagnosis. Eur J Cancer. 2011;47:784–91.
32. Liu HS, Zhu L, Liu BY, Yang L, Meng XX, Zhang W, Ma YY, Xiao HS. Genome-wide microRNA profiles identify miR-378 as a serum biomarker for early detection of gastric cancer. Cancer Lett. 2012;316:196–203.
33. Zhou J, Yu L, Gao X, Hu J, Wang J, Dai Z, Wang J-F, Zhang Z, Lu S, Huang X, et al. Plasma MicroRNA Panel to Diagnose Hepatitis B Virus-Related Hepatocellular Carcinoma. J Clin Oncol. 2011;29:4781–8.
34. Schultz NA, Dehlendorff C, Jensen BV, Bjerregaard JK, Nielsen KR, Bojesen SE, Calatayud D, Nielsen SE, Yilmaz M, Hollander NH, et al. MicroRNA Biomarkers in Whole Blood for Detection of Pancreatic Cancer. Jama-J Am Med Assoc. 2014;311:392–404.
35. Heneghan HM, Miller N, Lowery AJ, Sweeney KJ, Newell J, Kerin MJ. Circulating microRNAs as novel minimally invasive biomarkers for breast cancer. Ann Surg. 2010;251:499–505.
36. Lawrie CH, Gal S, Dunlop HM, Pushkaran B, Liggins AP, Pulford K, Banham AH, Pezzella F, Boultwood J, Wainscoat JS, et al. Detection of elevated levels of tumour-associated microRNAs in serum of patients with diffuse large B-cell lymphoma. Brit J Haematol. 2008;141:672–5.
37. Liu CJ, Kao SY, Tu HF, Tsai MM, Chang KW, Lin SC. Increase of microRNA miR-31 level in plasma could be a potential marker of oral cancer. Oral Dis. 2010;16:360–4.
38. Huang ZH, Huang D, Ni SJA, Peng ZL, Sheng WQ, Du X. Plasma microRNAs are promising novel biomarkers for early detection of colorectal cancer. Int J Cancer. 2010;127:118–26.
39. Zampetaki A, Mayr M. Analytical challenges and technical limitations in assessing circulating MiRNAs. Thromb Haemostasis. 2012;108:592–8.
40. Konishi H, Ichikawa D, Komatsu S, Shiozaki A, Tsujiura M, Takeshita H, Morimura R, Nagata H, Arita T, Kawaguchi T, et al. Detection of gastric cancer-associated microRNAs on microRNA microarray comparing pre- and post-operative plasma. Brit J Cancer. 2012;106: 740–7.
41. Cortez MA, Calin GA. MicroRNA identification in plasma and serum: a new tool to diagnose and monitor diseases. Expert Opin Biol Th. 2009;9:703–11.
42. Mestdagh P, Van Vlierberghe P, De Weer A, Muth D, Westermann F, Speleman F, Vandesompele J. A novel and universal method for microRNA RT-qPCR data normalization. Genome Biol. 2009;10:R64.
43. Zhang J, Song Y, Zhang C, Zhi X, Fu H, Ma Y, Chen Y, Pan F, Wang K, Ni J, et al. Circulating MiR-16-5p and MiR-19b-3p as Two Novel Potential Biomarkers to Indicate Progression of Gastric Cancer. Theranostics. 2015;5:733–45.
44. Hunt EA, Goulding AM, Deo SK. Direct detection and quantification of microRNAs. Anal Biochem. 2009;387:1–12.
45. Dijkstra JR, Mekenkamp LJM, Teerenstra S, De Krijger I, Nagtegaal ID. MicroRNA expression in formalin-fixed paraffin embedded tissue using real time quantitative PCR: the strengths and pitfalls. J Cell Mol Med. 2012;16:683–90.
46. Koshiol J, Wang E, Zhao Y, Marincola F, Landi MT. Strengths and Limitations of Laboratory Procedures for MicroRNA Detection. Cancer Epidem Biomar. 2010;19:907–11.
47. Zhang G-J, Chua JH, Chee R-E, Agarwal A, Wong SM. Label-free direct detection of MiRNAs with silicon nanowire biosensors. Biosens Bioelectron. 2009;24:2504–8.
48. Yin H, Zhou Y, Zhang H, Meng X, Ai S. Electrochemical determination of microRNA-21 based on graphene, LNA integrated molecular beacon, AuNPs and biotin multifunctional bio bar codes and enzymatic assay system. Biosens Bioelectron. 2012;33:247–53.
49. Husale S, Persson HHJ, Sahin O. DNA nanomechanics allows direct digital detection of complementary DNA and microRNA targets. Nature. 2009;462:1075–8.

50. Wang L, Cheng Y, Wang H, Li Z. A homogeneous fluorescence sensing platform with water-soluble carbon nanoparticles for detection of microRNA and nuclease activity. Analyst. 2012;137:3667–72.
51. Gu L-Q, Wanunu M, Wang MX, McReynolds L, Wang Y. Detection of miRNAs with a nanopore single-molecule counter. Expert Rev Mol Diagn. 2012;12:573–84.
52. Fan Y, Chen X, Trigg AD, Tung C-h, Kong J, Gao Z. Detection of microRNAs using target-guided formation of conducting polymer nanowires in nanogaps. J Amer Chem Soc. 2007;129:5437–43.
53. Liang RQ, Li W, Li Y, Tan CY, Li JX, Jin YX, Ruan KC. An oligonucleotide microarray for microRNA expression analysis based on labeling RNA with quantum dot and nanogold probe. Nucleic Acids Res. 2005;33(2):e17.
54. Wark AW, Lee HJ, Corn RM. Multiplexed detection methods for profiling microRNA expression in biological samples. Angew Chem Int Ed. 2008;47:644–52.
55. Benes V, Castoldi M. Expression profiling of microRNA using real-time quantitative PCR, how to use it and what is available. Methods. 2010;50:244–9.
56. Stenvang J, Silahtaroglu AN, Lindow M, Elmen J, Kauppinen S. The utility of LNA in microRNA-based cancer diagnostics and therapeutics. Semin Cancer Biol. 2008;18:89–102.
57. Kore AR, Hodeib M, Hu Z. Chemical synthesis of LNA-mCTP and its application for microRNA detection. Nucleos Nucleot Nucl. 2008;27:1–17.
58. Fabani MM, Abreu-Goodger C, Williams D, Lyons PA, Torres AG, Smith KGC, Enright AJ, Gait MJ, Vigorito E. Efficient inhibition of miR-155 function in vivo by peptide nucleic acids. Nucleic Acids Res. 2010;38:4466–75.
59. Kim H, Choi J-j, Cho M, Park H. A PNA microarray platform for miRNA expression profiling using on-chip labeling technology. Biochip J. 2012;6:25–33.
60. Fabani MM, Gait MJ. miR-122 targeting with LNA/2 '-O-methyl oligonucleotide mixmers, peptide nucleic acids (PNA), and PNA-peptide conjugates. RNA-Publ RNA Soc. 2008;14:336–46.
61. Cissell KA, Deo SK. Trends in microRNA detection. Anal Bioanal Chem. 2009;394:1109–16.
62. Lodes MJ, Caraballo M, Suciu D, Munro S, Kumar A, Anderson B. Detection of cancer with serum miRNAs on an oligonucleotide microarray. PLoS One. 2009;4:e6229.
63. Matveeva EG, Gryczynski Z, Stewart DR, Gryczynski I. Ratiometric FRET-based detection of DNA and micro-RNA on the surface using TIRF detection. J Lumin. 2010;130:698–702.
64. Dodgson BJ, Mazouchi A, Wegman DW, Gradinaru CC, Krylov SN. Detection of a Thousand Copies of miRNA without Enrichment or Modification. Anal Chem. 2012;84:5470–4.
65. Jiang L, Duan D, Shen Y, Li J. Direct microRNA detection with universal tagged probe and time-resolved fluorescence technology. Biosens Bioelectron. 2012;34:291–5.
66. Neely LA, Patel S, Garver J, Gallo M, Hackett M, McLaughlin S, Nadel M, Harris J, Gullans S, Rooke J. A single-molecule method for the quantitation of microRNA gene expression. Nat Methods. 2006;3:41–6.
67. Cissell KA, Hunt EA, Deo SK. Resonance energy transfer methods of RNA detection. Anal Bioanal Chem. 2009;393:125–35.
68. Bi S, Zhang J, Hao S, Ding C, Zhang S. Exponential amplification for chemiluminescence resonance energy transfer detection of microRNA in real samples based on a cross-catalyst strand-displacement network. Anal Chem. 2011;83:3696–702.
69. Broyles D, Cissell K, Kumar M, Deo S. Solution-phase detection of dual microRNA biomarkers in serum. Anal Bioanal Chem. 2012;402:543–50.
70. Thomson JM, Parker J, Perou CM, Hammond SM. A custom microarray platform for analysis of microRNA gene expression. Nat Methods. 2004;1:47–53.
71. Cissell KA, Rahimi Y, Shrestha S, Hunt EA, Deo SK. Bioluminescence-based detection of MicroRNA, miR21 in breast cancer cells. Anal Chem. 2008;80:2319–25.
72. Song W, Qiu X, Lau C, Lu J. Quantum dot-enhanced detection of dual short RNA sequences via one-step template-dependent surface hybridization. Anal Chim Acta. 2012;735:114–20.
73. Zhang J, Fu Y, Mei YP, Jiang F, Lakowicz JR. Fluorescent metal nanoshell probe to detect single miRNA in lung cancer cell. Anal Chem. 2010;82:4464–71.

74. Do Won H, In Chan S, Dong Soo L, Soonhag K. Smart Magnetic Fluorescent Nanoparticle Imaging Probes to Monitor MicroRNAs. Small. 2010;6:81–8.
75. Medintz IL, Uyeda HT, Goldman ER, Mattoussi H. Quantum dot bioconjugates for imaging, labelling and sensing. Nat Mater. 2005;4:435–46.
76. Huang X, Ren J. Nanomaterial-based chemiluminescence resonance energy transfer: A strategy to develop new analytical methods. Trac-Trend Anal Chem. 2012;40:77–89.
77. Zhang Y, Zhang C-y. Sensitive detection of microRNA with isothermal amplification and a single-quantum-dot-based nanosensor. Anal Chem. 2012;84:224–31.
78. Cissell KA, Campbell S, Deo SK. Rapid, single-step nucleic acid detection. Anal Bioanal Chem. 2008;391:2577–81.
79. Petty JT, Zheng J, Hud NV, Dickson RM. DNA-templated Ag nanocluster formation. J Am Chem Soc. 2004;126:5207–12.
80. Richards CI, Choi S, Hsiang J-C, Antoku Y, Vosch T, Bongiorno A, Tzeng Y-L, Dickson RM. Oligonucleotide-stabilized Ag nanocluster fluorophores. J Am Chem Soc. 2008;130:5038–9.
81. Patel SA, Richards CI, Hsiang J-C, Dickson RM. Water-soluble Ag nanoclusters exhibit strong two-photon-induced fluorescence. J Am Chem Soc. 2008;130:11602–3.
82. Zhou Z, Du Y, Dong S. DNA-Ag nanoclusters as fluorescence probe for turn-on aptamer sensor of small molecules. Biosens Bioelectron. 2011;28:33–7.
83. Shah P, Rorvig-Lund A, Ben Chaabane S, Thulstrup PW, Kjaergaard HG, Fron E, Hofkens J, Yang SW, Vosch T. Design aspects of bright red emissive silver nanoclusters/DNA probes for microRNA detection. ACS Nano. 2012;6:8803–14.
84. Yang SW, Vosch T. Rapid detection of microRNA by a silver nanocluster DNA probe. Anal Chem. 2011;83:6935–9.
85. Li J, Schachermeyer S, Wang Y, Yin Y, Zhong W. Detection of microRNA by fluorescence amplification based on cation-exchange in nanocrystals. Anal Chem. 2009;81:9723–9.
86. Kim JK, Choi KJ, Lee M, Jo MH, Kim S. Molecular imaging of a cancer-targeting theragnostics probe using a nucleolin aptamer- and microRNA-221 molecular beacon-conjugated nanoparticle. Biomaterials. 2012;33:207–17.
87. Muniz-Miranda M, Gellini C, Pagliai M, Innocenti M, Salvi PR, Schettino V. SERS and computational studies on microRNA chains adsorbed on silver surfaces. J Phys Chem C. 2010;114:13730–5.
88. Wang Z, Yang B. MicroRNA Expression Detection Methods. Canada: Springer-Verlag; 2010.
89. Driskell JD, Seto AG, Jones LP, Jokela S, Dluhy RA, Zhao YP, Tripp RA. Rapid microRNA (miRNA) detection and classification via surface-enhanced Raman spectroscopy (SERS). Biosens Bioelectron. 2008;24:917–22.
90. Zhao Y.-P. and Liu Y.-J. The Silver Nanorod Array SERS Substrates. AIP Conf Proc. 2010;1267:277–8.
91. Driskell JD, Tripp RA. Label-free SERS detection of microRNA based on affinity for an unmodified silver nanorod array substrate. Chem Commun. 2010;46:3298–300.
92. Wang H-N, Vo-Dinh T. Plasmonic coupling interference (PCI) nanoprobes for nucleic acid detection. Small. 2011;7:3067–74.
93. Xu FG, Dong CQ, Xie C, Ren JC. Ultrahighly sensitive homogeneous detection of DNA and microRNA by using single-silver-nanoparticle counting. Chem-Eur J. 2010;16:1010–6.
94. Alhasan AH, Kim DY, Daniel WL, Watson E, Meeks JJ, Thaxton CS, Mirkin CA. Scanometric microRNA array profiling of prostate cancer markers using spherical nucleic acid-gold nanoparticle conjugates. Anal Chem. 2012;84:4153–60.
95. Yang W-J, Li X-B, Li Y-Y, Zhao L-F, He W-L, Gao Y-Q, Wan Y-J, Xia W, Chen T, Zheng H, et al. Quantification of microRNA by gold nanoparticle probes. Anal Biochem. 2008;376:183–8.
96. Wang Z, Yang B. MicroRNA Expression Detection Methods. Canada: Springer-Verlag; 2010.
97. Zhou WJ, Chen YL, Corn RM. Ultrasensitive microarray detection of short RNA sequences with enzymatically modified nanoparticles and surface plasmon resonance imaging measurements. Anal Chem. 2011;83:3897–902.

# Chapter 4
# Screening and Identification of Biomarkers from Gastric Cancer Saliva Metabolites

**Shangli Cheng and Daxaing Cui**

## 4.1 Introduction

Gastric cancer is the second leading cause of death worldwide. Based on recent statistics, the number of men is twice than women [1]. The latest statistics of gastric cancer all over the world showed that gastric cancer had caused 723,100 deaths, and there were 951,600 new cases in 2012 [2]. The mortality of gastric cancer had regional differences [2, 3]. The mortality rate of gastric cancer in Eastern Asia was higher than other regions [4, 5]. The causing factors associated with gastric cancer included genetic factors, environmental risk factors, infection of *Helicobacter pylori*, intake of salt, intake of alcohol, and smoking of tobacco [6–9]. The gastric cancer patients in the late stage had a low five-year survival rate of less than 24 % [10]. However, the five-year survival rate in Japan was 96.5 %, which was much better than European countries of 23 % and the USA of 29 %. This owned to the early gastric cancer diagnosis [3, 11, 12]. In this case, how to improve the five-year survival of gastric cancer patients by early diagnosis has become our concern.

In diagnosis of gastric cancer, the traditional methods were usually using imaging methods, such as CT and PET/CT [13–16]. In early gastric cancer diagnosis, endoscopy was more successful than X-ray-based methods [17, 18]. From the statistics in Korea, the sensitivity and specificity in gastric cancer diagnosis using endoscopy was, respectively, 59.0 % and 96.3 %, higher than the sensitivity of 42.1 % and the specificity of 89.8 % using X-ray-based methods [19]. As the vague symptoms of the early gastric cancer patients, it was hard for the diagnosis of gastric

S. Cheng (✉) • D. Cui
Institute of Nano Biomedicine and Engineering, Shanghai Engineering Research Center for Intelligent Diagnosis and Treatment Instrument, National Center for Translational Medicine, Collaborative Innovational Center for System Biology, Shanghai Jiao Tong University, 800 Dongchuan Road, Shanghai 200240, P. R. China
e-mail: chengshangli@sjtu.edu.cn; dxcui@sjtu.edu.cn

© Springer Science+Business Media B.V. and
Shanghai Jiao Tong University Press, Shanghai 2017
D. Cui (ed.), *Gastric Cancer Prewarning and Early Diagnosis System*,
Translational Medicine Research, DOI 10.1007/978-94-024-0951-2_4

cancer in early stage [13, 20]. Most of the gastric cancer patients were diagnosed in the late stage. In China, early gastric cancer diagnosis ratio was less than 20 %. Therefore, how to find early gastric cancer is still a great challenge [20].

Many studies proposed that the biomarkers have the potential in early gastric cancer diagnosis [21, 22]. Up to date, the biomarkers such as SNP sites, plasma miRNAs, proteins, etc. have been found in GC research [23–25]. For example, the germline mutation and hypermethylation of E-cadherin was found as a hereditary cause of GC [24, 26]. The miRNA expressions of miR-125b, miR-199a, and miR-100 were shown to be upregulated in GC patients, and circulating miR-16-5p and miR-19b-3p were identified to be downregulated in GC patients, which can be used to indicate progress of gastric cancer [27, 28]. The protein of apoC-I, apoC-III, CA19-9, and CRP were used for the prediction of GC and received the accuracy of ~74 % [29]. Furthermore, the biomarkers of small metabolic molecules were considered equally important as DNA mutation, miRNA, and protein in diagnosis of GC. For example, some volatile biomarkers from breath metabolism of gastric cancer patients were screened out [30]. Up to date, more and more studies focused on small metabolites, such as amino acids, nucleic acids, carbohydrates, lipids, and so on [31–34]. And the metabolites were used in diagnosis of gastric cancer patients [35]. The method had become a hot spot [31, 36]. As blood was one of the means for disease diagnosis, and saliva is the ultrafiltration of blood, saliva can reflect the condition of tissues to a certain degree [37].

In this study, 38 types of amino acids in saliva were screened out as metabolites, and the molecular concentration in saliva was obtained. Eight molecules were identified as biomarkers in gastric cancer diagnosis, which had higher concentrations in gastric cancer patients than that in healthy subjects. The eight metabolites were mapped into eight metabolic pathways, in which three pathways were associated with central carbon metabolism in cancer. The results showed that the eight metabolites, especially glutamate, 4-hydroxyproline, and 4-aminobutanoate, extracted from saliva were highly correlated with cancer metabolism. Furthermore, the diagnosis accuracy using the eight metabolites was calculated with the 23 independent saliva samples. The glutamate had the highest accuracy of 0.838 in diagnosis of gastric cancer. The results suggested that the eight identified biomarkers had the potential in gastric cancer diagnosis and had significant clinical value for the noninvasive diagnosis of gastric cancer.

## 4.2 Molecular Identification in Saliva of Gastric Cancer Patients

The tumor stage information followed the AJCC Cancer Staging Manual. In all, there were 17 saliva samples from gastric cancer patients and 30 plasma samples from healthy subjects. The average age was 44 in gastric cancer patients and 41 in healthy subjects. The range of age and sex distribution is shown in Table 4.1. The age and sex distribution were not significantly different among gastric cancer patients and healthy subjects, and it would lead the results without bias (Table 4.2).

**Table 4.1** The information of the samples

| | | Case | Control | *p*-value |
|---|---|---|---|---|
| Age | Average | 44 | 41 | $p^a$=0.721 |
| | Range | 54~31 | 52~31 | |
| Sex | Male | 9 | 8 | $p^b$=0.263 |
| | Female | 8 | 5 | |
| Stage | I | 1 | | |
| | II | 1 | | |
| | III | 9 | | |
| | IV | 4 | | |
| | Unknown | 2 | | |

The case and control samples represent gastric cancer patients and healthy subjects, respectively
[a]Independent *T*-test
[b]Pearson chi-square

**Table 4.2** The information of the samples taken as testing set

| | | GC patients | Control |
|---|---|---|---|
| Age | Average | 63 | 25 |
| | Range | 73~54 | 29~23 |
| Sex | Male | 13 | 4 |
| | Female | 2 | 4 |
| Stage | I | 0 | |
| | II | 4 | |
| | III | 5 | |
| | IV | 2 | |
| | Unknown | 4 | |

The molecules in saliva were screened out using HPLC-MS. The analyses of OPLS-DA between gastric cancer patients and healthy subjects were shown (Fig. 4.1). At last, 38 types of amino acids were screened out as metabolites, and the molecular concentration in saliva was obtained (Table 4.3).

## 4.3 Biomarkers in Saliva of Gastric Cancer Patients and Metabolic Characteristics

All the biomarkers were validated with standard sample. Firstly, the datum obtained was qualified. Five metabolites, including 1-methylhistidine, 3-methylhistidine, homocitrulline, cysteine, and homocysteine, were excluded because of low concentration in many samples (Table 4.4). The 33 metabolites were further analyzed (Table 4.3).

In order to identify the regulation of the metabolites, for each metabolite, the p-value of Mann–Whitney *U* test was calculated between the 17 gastric cancer patients and the 13 healthy subjects. Eight metabolites, including 4-hydroxyproline,

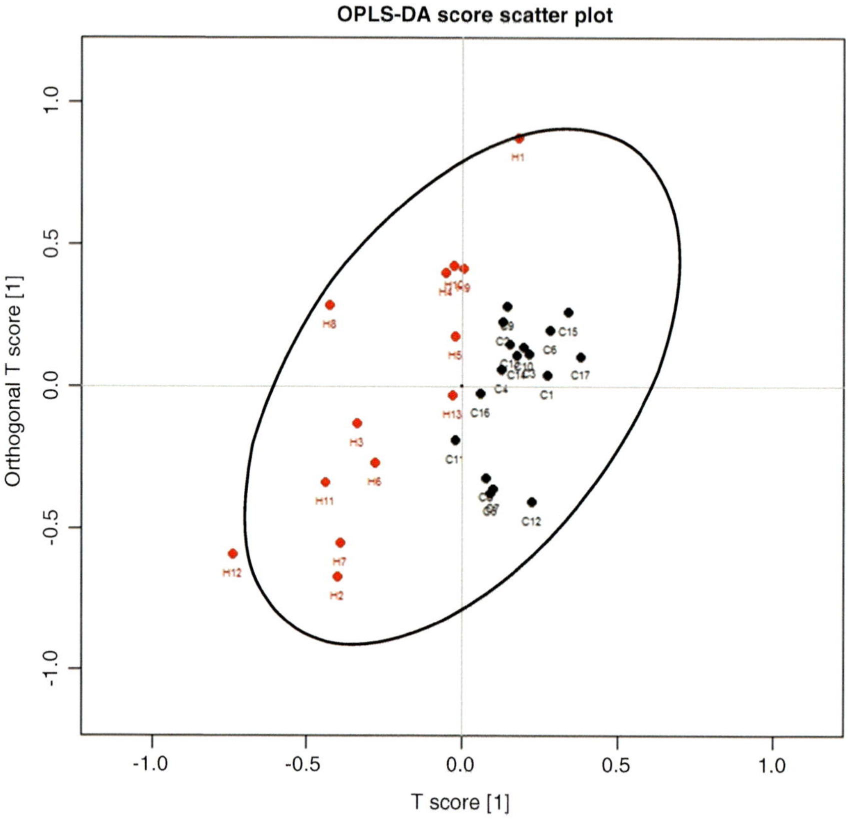


**Fig. 4.1** The OPLS-DA of gastric cancer patients and healthy subjects. The *red* and *black* points represent healthy subjects and gastric cancer patients, respectively

aspartate, histidine, glutamate, carnosine, α-aminoadipate, 4-aminobutanoate, and 3-aminoisobutanoate, were identified and taken as a potential biomarker in early diagnosis of gastric cancer. Furthermore, the fold change of the eight metabolites was analyzed. The results showed that they had higher concentration in saliva of gastric cancer patients than that in healthy subjects (Table 4.5).

Referenced to central carbon metabolism in cancer from KEGG database, five pathways are associated with amino acid metabolism, which are glycine, serine, and threonine metabolism; alanine, aspartate, and glutamate metabolism; arginine and proline metabolism; and D-glutamine metabolism. In this study, the eight molecules were associated with eight metabolic pathways (Fig. 4.2a). Glutamate, 4-hydroxyproline, and 4-aminobutanoate were associated with arginine and proline metabolism. Aspartate, glutamate, and 4-aminobutanoate were associated with alanine, aspartate, and glutamate metabolism. Aspartate was associated with glycine, serine, and threonine metabolism. Furthermore, the Pearson correlation coefficient between the metabolic levels of the metabolites was calculated (Table 4.6 and Fig. 4.2b). The high correlated metabolic regulations among the eight molecules are exhibited.

**Table 4.3** The Mann–Whitney U test and FC of the molecules

| Molecules | *p* | FC |
|---|---|---|
| PSer | 0.201 | 0.728 |
| PEtN | 0.155 | 1.401 |
| Tau | 0.202 | 1.230 |
| Asn | 0.392 | 1.239 |
| Ser | 0.615 | 0.957 |
| Hyp | 0.006 | 2.530 |
| Gly | 0.233 | 1.337 |
| Gln | 0.517 | 1.506 |
| Asp | 0.038 | 1.872 |
| EtN | 0.052 | 1.962 |
| His | 0.036 | 2.116 |
| Thr | 0.691 | 0.820 |
| Cit | 0.325 | 1.292 |
| Ala | 0.691 | 1.630 |
| Glu | 0.049 | 2.115 |
| MHis1 | 0.885 | 2.289 |
| MHis3 | 0.778 | 0.129 |
| Car | 0.032 | 1.428 |
| Ans | 0.849 | 1.519 |
| Hcit | 0.248 | 1.439 |
| Arg | 0.867 | 0.714 |
| Aad | 0.037 | 1.814 |
| GABA | 0.031 | 2.401 |
| bAib | 0.003 | 3.118 |
| Abu | 0.753 | 0.595 |
| Hyl | 0.900 | 1.021 |
| Pro | 0.250 | 1.241 |
| Orn | 0.572 | 1.090 |
| Cth | 0.931 | 1.348 |
| Cys | 0.342 | 0.802 |
| Lys | 0.738 | 1.228 |
| Val | 0.202 | 0.642 |
| Tyr | 0.950 | 1.088 |
| Hcy | 0.516 | 0.657 |
| Ile | 0.031 | 0.382 |
| Leu | 0.069 | 0.414 |
| Phe | 0.258 | 0.727 |
| Trp | 0.285 | 0.712 |

*FC* is the fold change between GC and HC samples

**Table 4.4** The concentration of the molecules in saliva (μmol/L)

| **Sample No.** | **GC1** | **GC2** | **GC3** | **GC4** | **GC5** | **GC6** |
|---|---|---|---|---|---|---|
| PSer | 0.0 | 1.1 | 0.0 | 3.5 | 1.9 | 1.5 |
| PEtN | 15.3 | 9.4 | 31.1 | 25.6 | 1.9 | 109.4 |
| Tau | 29.4 | 11.5 | 55.5 | 56.6 | 36.1 | 85.6 |
| Asn | 0.3 | 0.1 | 0.1 | 0.1 | 0.1 | 0.0 |
| Ser | 4.7 | 2.8 | 5.9 | 9.8 | 35.8 | 11.4 |
| Hyp | 0.8 | 0.6 | 1.7 | 0.8 | 0.2 | 2.5 |
| Gly | 61.0 | 23.2 | 62.4 | 84.9 | 162.2 | 142.3 |
| Gln | 5.4 | 4.1 | 7.2 | 23.7 | 27.9 | 32.5 |
| Asp | 7.3 | 6.1 | 4.4 | 19.1 | 45.5 | 22.0 |
| EtN | 21.8 | 6.2 | 25.3 | 52.5 | 14.5 | 104.9 |
| His | 12.2 | 7.7 | 21.4 | 21.3 | 9.6 | 36.9 |
| Thr | 1.4 | 0.6 | 2.1 | 5.0 | 17.5 | 5.4 |
| Cit | 11.4 | 1.7 | 7.2 | 18.1 | 40.0 | 18.2 |
| Ala | 26.6 | 8.8 | 18.7 | 33.1 | 216.4 | 48.7 |
| Glu | 12.7 | 4.1 | 23.8 | 52.5 | 106.4 | 46.8 |
| 1MHis | 0.5 | 0.0 | 0.0 | 0.0 | 0.0 | 0.0 |
| 3MHis | 0.2 | 0.0 | 0.0 | 0.0 | 0.0 | 0.0 |
| Car | 5.0 | 2.2 | 2.5 | 1.3 | 1.8 | 1.8 |
| Ans | 1.0 | 0.2 | 1.6 | 1.0 | 0.2 | 0.1 |
| Hcit | 0.0 | 0.7 | 0.0 | 0.2 | 0.0 | 0.0 |
| Arg | 9.7 | 4.3 | 8.7 | 8.3 | 2.9 | 11.6 |
| Aad | 0.6 | 0.3 | 0.6 | 1.5 | 2.8 | 1.9 |
| GABA | 8.7 | 3.1 | 4.7 | 2.9 | 2.7 | 8.8 |
| bAib | 0.5 | 0.6 | 1.2 | 1.4 | 8.0 | 3.8 |
| Abu | 0.2 | 0.1 | 0.6 | 2.6 | 8.6 | 1.3 |
| Hyl | 16.8 | 20.7 | 27.7 | 46.0 | 32.2 | 23.8 |
| Pro | 70.0 | 10.3 | 67.9 | 24.3 | 73.8 | 64.0 |
| Orn | 14.3 | 5.2 | 13.4 | 22.8 | 61.0 | 34.9 |
| Cth | 0.3 | 0.0 | 0.2 | 1.0 | 0.5 | 0.7 |
| Cys | 1.0 | 0.0 | 0.0 | 0.0 | 0.3 | 0.0 |
| Lys | 16.5 | 3.8 | 11.1 | 23.3 | 142.2 | 52.4 |
| Val | 2.6 | 0.7 | 1.1 | 16.1 | 44.4 | 13.3 |
| Tyr | 10.2 | 7.0 | 29.4 | 21.1 | 22.2 | 31.3 |
| Hcy | 0.0 | 0.0 | 0.3 | 0.0 | 0.0 | 0.3 |
| Ile | 0.8 | 1.0 | 0.2 | 3.7 | 13.5 | 5.2 |
| Leu | 1.9 | 1.3 | 1.2 | 18.4 | 30.3 | 10.9 |
| Phe | 3.5 | 2.9 | 16.0 | 18.0 | 17.4 | 16.2 |
| Trp | 0.4 | 0.2 | 0.4 | 3.0 | 5.7 | 0.4 |
| **Sample No.** | **GC7** | **GC8** | **GC9** | **GC10** | **GC11** | **GC12** |
| PSer | 2.5 | 0.0 | 1.6 | 1.4 | 1.0 | 2.1 |
| PEtN | 4.4 | 1.2 | 25.2 | 15.8 | 0.7 | 1.0 |

**Table 4.4** (continued)

| | | | | | | |
|---|---|---|---|---|---|---|
| Tau | 120.7 | 82.9 | 13.6 | 60.5 | 144.5 | 81.6 |
| Asn | 0.0 | 0.1 | 0.1 | 0.2 | 0.1 | 0.0 |
| Ser | 18.5 | 9.9 | 4.4 | 6.8 | 9.0 | 8.7 |
| Hyp | 1.6 | 1.0 | 0.5 | 1.4 | 1.5 | 0.5 |
| Gly | 259.2 | 144.0 | 30.4 | 49.4 | 146.1 | 193.9 |
| Gln | 98.8 | 23.0 | 4.9 | 19.3 | 17.1 | 43.0 |
| Asp | 36.8 | 16.8 | 5.8 | 12.1 | 16.5 | 23.5 |
| EtN | 63.9 | 13.5 | 8.2 | 17.1 | 22.4 | 31.9 |
| His | 76.6 | 16.9 | 3.7 | 9.6 | 17.1 | 65.3 |
| Thr | 11.1 | 6.2 | 2.0 | 2.2 | 3.3 | 5.3 |
| Cit | 31.2 | 16.2 | 3.2 | 7.2 | 12.8 | 26.2 |
| Ala | 91.6 | 48.5 | 11.2 | 24.0 | 36.2 | 45.0 |
| Glu | 101.3 | 31.6 | 28.0 | 28.1 | 28.6 | 57.9 |
| 1MHis | 0.0 | 0.0 | 0.0 | 0.0 | 0.0 | 0.0 |
| 3MHis | 0.0 | 0.0 | 0.0 | 0.0 | 0.0 | 0.0 |
| Car | 4.0 | 0.2 | 0.9 | 1.8 | 0.9 | 2.3 |
| Ans | 3.9 | 0.0 | 0.3 | 0.3 | 0.0 | 0.1 |
| Hcit | 0.8 | 0.4 | 0.0 | 0.0 | 0.0 | 0.0 |
| Arg | 7.9 | 3.3 | 7.4 | 10.1 | 6.4 | 6.5 |
| Aad | 1.1 | 1.0 | 0.9 | 1.0 | 1.4 | 0.6 |
| GABA | 10.0 | 4.0 | 1.9 | 7.2 | 7.1 | 5.2 |
| bAib | 1.4 | 2.0 | 1.1 | 3.6 | 4.3 | 0.4 |
| Abu | 3.1 | 1.5 | 0.3 | 0.4 | 2.7 | 3.0 |
| Hyl | 23.2 | 16.8 | 25.5 | 20.3 | 20.1 | 26.9 |
| Pro | 168.7 | 93.6 | 10.8 | 14.9 | 46.9 | 58.6 |
| Orn | 42.8 | 25.6 | 3.0 | 11.4 | 41.8 | 36.4 |
| Cth | 1.4 | 0.3 | 0.0 | 0.2 | 0.9 | 0.0 |
| Cys | 0.0 | 0.0 | 0.0 | 0.6 | 0.0 | 0.0 |
| Lys | 96.3 | 24.9 | 5.0 | 14.2 | 46.0 | 56.1 |
| Val | 45.9 | 16.6 | 3.2 | 5.0 | 26.1 | 15.9 |
| Tyr | 105.6 | 23.5 | 5.2 | 18.0 | 33.5 | 55.9 |
| Hcy | 0.0 | 0.0 | 0.0 | 0.3 | 0.0 | 0.3 |
| Ile | 13.4 | 7.2 | 1.4 | 1.5 | 9.7 | 7.1 |
| Leu | 47.0 | 16.1 | 3.7 | 5.1 | 25.7 | 14.3 |
| Phe | 57.1 | 18.7 | 3.1 | 9.8 | 16.1 | 20.4 |
| Trp | 2.0 | 1.5 | 0.5 | 0.3 | 2.3 | 1.6 |
| **Sample No.** | **GC13** | **GC14** | **GC15** | **GC16** | **GC17** | **HC1** |
| PSer | 1.2 | 2.8 | 2.5 | 2.6 | 1.8 | 2.0 |
| PEtN | 9.6 | 3.1 | 26.7 | 22.4 | 41.4 | 60.6 |
| Tau | 110.4 | 73.2 | 66.7 | 66.4 | 173.1 | 65.5 |
| Asn | 0.1 | 0.1 | 0.1 | 0.1 | 0.1 | 0.1 |
| Ser | 13.0 | 7.7 | 7.1 | 19.5 | 33.0 | 5.9 |

(continued)

**Table 4.4** (continued)

| | | | | | | |
|---|---|---|---|---|---|---|
| Hyp | 1.0 | 0.9 | 4.1 | 0.3 | 2.7 | 1.2 |
| Gly | 91.1 | 75.9 | 50.7 | 128.4 | 123.3 | 25.3 |
| Gln | 13.9 | 11.7 | 10.2 | 12.1 | 46.3 | 10.4 |
| Asp | 18.6 | 8.2 | 10.1 | 13.3 | 40.7 | 5.2 |
| EtN | 63.3 | 36.8 | 41.7 | 47.6 | 113.1 | 47.5 |
| His | 26.5 | 11.0 | 27.4 | 16.3 | 17.0 | 2.8 |
| Thr | 2.9 | 2.0 | 2.2 | 6.2 | 11.4 | 2.3 |
| Cit | 18.4 | 7.6 | 8.5 | 10.7 | 10.9 | 3.5 |
| Ala | 29.8 | 16.3 | 27.8 | 42.0 | 312.6 | 13.9 |
| Glu | 15.9 | 19.9 | 19.5 | 15.2 | 172.2 | 12.7 |
| 1MHis | 0.0 | 0.0 | 0.0 | 0.0 | 0.0 | 0.0 |
| 3MHis | 0.0 | 0.0 | 0.0 | 0.4 | 0.0 | 0.0 |
| Car | 2.0 | 0.9 | 2.9 | 1.5 | 2.0 | 4.7 |
| Ans | 0.2 | 0.3 | 0.1 | 1.2 | 0.3 | 0.0 |
| Hcit | 0.0 | 0.0 | 0.1 | 0.2 | 0.0 | 0.0 |
| Arg | 32.4 | 14.1 | 4.7 | 17.4 | 19.5 | 8.3 |
| Aad | 0.7 | 0.4 | 0.6 | 0.6 | 2.5 | 0.4 |
| GABA | 5.7 | 2.7 | 3.9 | 13.4 | 78.1 | 16.6 |
| bAib | 1.3 | 1.0 | 3.8 | 0.7 | 2.1 | 0.8 |
| Abu | 1.2 | 0.9 | 0.2 | 1.0 | 7.9 | 0.6 |
| Hyl | 29.3 | 24.7 | 25.4 | 23.5 | 29.9 | 26.7 |
| Pro | 91.2 | 72.1 | 35.5 | 74.2 | 124.9 | 12.4 |
| Orn | 39.8 | 10.9 | 9.1 | 43.8 | 15.9 | 6.8 |
| Cth | 0.7 | 0.0 | 0.3 | 0.0 | 0.0 | 0.6 |
| Cys | 0.0 | 0.4 | 0.0 | 0.0 | 0.0 | 0.3 |
| Lys | 23.4 | 5.3 | 11.0 | 34.7 | 20.0 | 11.2 |
| Val | 6.2 | 4.7 | 5.8 | 10.4 | 25.2 | 4.3 |
| Tyr | 52.9 | 17.2 | 12.7 | 37.7 | 29.4 | 6.2 |
| Hcy | 0.0 | 0.0 | 0.0 | 0.0 | 0.0 | 0.3 |
| Ile | 2.3 | 1.5 | 1.6 | 3.6 | 9.6 | 2.5 |
| Leu | 9.5 | 7.4 | 5.2 | 9.2 | 33.0 | 4.0 |
| Phe | 16.7 | 8.5 | 7.5 | 15.8 | 16.9 | 4.8 |
| Trp | 0.3 | 0.6 | 1.0 | 0.6 | 2.4 | 0.6 |
| **Sample No.** | **HC2** | **HC3** | **HC4** | **HC5** | **HC6** | **HC7** |
| PSer | 2.4 | 3.2 | 2.3 | 3.7 | 1.4 | 4.3 |
| PEtN | 0.7 | 6.1 | 11.8 | 0.5 | 2.1 | 0.6 |
| Tau | 17.0 | 97.7 | 32.6 | 90.4 | 57.9 | 24.3 |
| Asn | 0.0 | 0.2 | 0.0 | 0.1 | 0.1 | 0.1 |
| Ser | 3.0 | 40.8 | 4.7 | 16.2 | 12.2 | 7.6 |
| Hyp | 0.7 | 0.9 | 0.7 | 0.4 | 0.4 | 0.3 |
| Gly | 160.3 | 175.9 | 35.6 | 40.4 | 92.3 | 150.4 |
| Gln | 8.7 | 20.1 | 3.1 | 16.0 | 22.1 | 22.6 |

**Table 4.4** (continued)

| | | | | | | |
|---|---|---|---|---|---|---|
| Asp | 3.1 | 19.0 | 3.3 | 18.7 | 10.3 | 6.9 |
| EtN | 1.0 | 33.6 | 6.7 | 18.4 | 20.8 | 8.3 |
| His | 1.9 | 22.8 | 3.3 | 12.0 | 21.4 | 13.6 |
| Thr | 2.2 | 12.0 | 1.2 | 3.8 | 7.3 | 11.3 |
| Cit | 7.8 | 28.7 | 2.7 | 16.0 | 16.1 | 10.4 |
| Ala | 29.1 | 62.9 | 5.5 | 32.8 | 35.2 | 76.8 |
| Glu | 10.1 | 14.3 | 3.1 | 17.9 | 19.2 | 63.7 |
| 1MHis | 0.0 | 0.1 | 0.0 | 0.0 | 0.0 | 0.0 |
| 3MHis | 0.0 | 3.5 | 0.0 | 0.0 | 0.0 | 0.0 |
| Car | 0.8 | 0.4 | 0.8 | 1.9 | 0.1 | 1.4 |
| Ans | 0.9 | 0.5 | 0.0 | 0.3 | 0.1 | 0.2 |
| Hcit | 0.0 | 0.0 | 0.0 | 0.0 | 0.0 | 0.0 |
| Arg | 0.5 | 46.0 | 8.9 | 24.5 | 5.0 | 1.0 |
| Aad | 0.4 | 1.0 | 0.5 | 0.6 | 1.1 | 0.6 |
| GABA | 4.5 | 3.6 | 1.7 | 2.2 | 2.8 | 3.7 |
| bAib | 0.1 | 0.4 | 0.5 | 0.6 | 0.8 | 0.6 |
| Abu | 4.3 | 0.5 | 0.3 | 0.9 | 3.3 | 13.8 |
| Hyl | 24.2 | 26.0 | 24.7 | 23.7 | 20.5 | 20.6 |
| Pro | 26.0 | 175.8 | 16.6 | 88.7 | 27.6 | 53.8 |
| Orn | 19.8 | 76.9 | 4.1 | 17.0 | 56.3 | 23.6 |
| Cth | 0.2 | 0.5 | 0.3 | 0.3 | 0.6 | 0.0 |
| Cys | 0.4 | 0.0 | 0.6 | 0.3 | 0.0 | 0.0 |
| Lys | 11.0 | 90.3 | 4.5 | 31.3 | 30.0 | 41.2 |
| Val | 25.5 | 28.3 | 0.8 | 13.1 | 21.7 | 60.6 |
| Tyr | 10.7 | 57.3 | 5.5 | 26.1 | 44.7 | 48.7 |
| Hcy | 0.0 | 0.0 | 0.0 | 0.0 | 0.6 | 0.4 |
| Ile | 11.5 | 25.1 | 0.7 | 5.0 | 14.2 | 22.5 |
| Leu | 26.5 | 76.0 | 3.2 | 17.5 | 35.8 | 49.8 |
| Phe | 7.5 | 63.3 | 3.5 | 15.9 | 28.1 | 29.0 |
| Trp | 0.4 | 2.1 | 0.2 | 1.8 | 2.2 | 7.3 |
| **Sample No.** | **HC8** | **HC9** | **HC10** | **HC11** | **HC12** | **HC13** |
| PSer | 2.2 | 1.9 | 2.5 | 1.4 | 1.1 | 0.5 |
| PEtN | 29.7 | 61.0 | 8.0 | 0.6 | 0.2 | 5.9 |
| Tau | 256.0 | 33.3 | 18.1 | 15.9 | 11.8 | 67.8 |
| Asn | 0.0 | 0.1 | 0.1 | 0.1 | 0.0 | 0.1 |
| Ser | 36.3 | 9.1 | 5.7 | 2.8 | 2.7 | 19.2 |
| Hyp | 0.6 | 0.5 | 0.1 | 0.1 | 0.2 | 0.6 |
| Gly | 59.9 | 44.7 | 28.6 | 39.3 | 110.4 | 82.9 |
| Gln | 24.6 | 12.6 | 6.6 | 19.4 | 9.9 | 27.7 |
| Asp | 18.9 | 9.9 | 5.0 | 6.6 | 1.9 | 16.6 |
| EtN | 62.1 | 26.1 | 16.3 | 8.6 | 1.0 | 16.5 |
| His | 18.4 | 14.8 | 4.0 | 8.1 | 4.6 | 15.7 |

(continued)

**Table 4.4** (continued)

| | | | | | | |
|---|---|---|---|---|---|---|
| Thr | 22.6 | 3.6 | 0.7 | 7.1 | 1.9 | 4.9 |
| Cit | 21.5 | 9.3 | 3.0 | 3.9 | 2.7 | 22.0 |
| Ala | 74.7 | 20.2 | 5.1 | 58.6 | 11.1 | 60.7 |
| Glu | 33.9 | 15.8 | 6.6 | 41.4 | 7.6 | 30.2 |
| 1MHis | 0.0 | 0.0 | 0.0 | 0.0 | 0.0 | 0.0 |
| 3MHis | 0.0 | 0.0 | 0.0 | 0.0 | 0.0 | 0.0 |
| Car | 0.7 | 0.9 | 0.9 | 3.8 | 1.4 | 0.2 |
| Ans | 0.8 | 1.0 | 0.1 | 0.4 | 1.0 | 0.2 |
| Hcit | 0.1 | 0.0 | 0.0 | 1.2 | 0.0 | 0.0 |
| Arg | 35.2 | 14.8 | 10.2 | 0.6 | 1.6 | 31.3 |
| Aad | 0.7 | 0.7 | 0.2 | 0.5 | 0.3 | 0.8 |
| GABA | 2.5 | 8.5 | 1.0 | 3.0 | 0.8 | 3.3 |
| bAib | 0.8 | 1.0 | 0.8 | 0.4 | 0.1 | 2.2 |
| Abu | 0.5 | 1.9 | 0.2 | 17.3 | 1.3 | 0.8 |
| Hyl | 26.9 | 29.1 | 27.4 | 23.2 | 22.5 | 28.5 |
| Pro | 36.3 | 27.4 | 10.1 | 15.7 | 58.2 | 129.9 |
| Orn | 13.7 | 9.2 | 4.5 | 17.0 | 7.1 | 47.0 |
| Cth | 0.3 | 0.0 | 0.3 | 0.3 | 0.3 | 0.0 |
| Cys | 0.0 | 0.3 | 0.3 | 0.0 | 0.0 | 0.0 |
| Lys | 63.2 | 17.0 | 8.1 | 10.2 | 13.4 | 33.7 |
| Val | 48.2 | 8.2 | 1.4 | 34.9 | 22.6 | 19.9 |
| Tyr | 43.2 | 21.4 | 5.6 | 30.8 | 32.1 | 28.1 |
| Hcy | 0.3 | 0.0 | 0.0 | 0.0 | 0.0 | 0.0 |
| Ile | 41.4 | 3.3 | 1.5 | 14.6 | 16.2 | 8.2 |
| Leu | 115.1 | 6.7 | 3.1 | 32.4 | 49.0 | 24.4 |
| Phe | 38.3 | 16.4 | 4.2 | 19.0 | 30.7 | 17.4 |
| Trp | 3.8 | 0.7 | 0.6 | 2.3 | 1.6 | 1.3 |

17 gastric cancer patients (*GC*) and 13 healthy subjects (*HC*) were used and 38 molecules were screened out

## 4.4 Diagnosis of Gastric Cancer Using the Identified Biomarkers

Out of all the screened molecules, the eight identified biomarkers were upregulated in saliva of gastric cancer patients. To provide comparison of the regulation of the biomarkers, the scatter plot of the eight metabolites with median and standard deviation (SD) was provided by comparing gastric cancer patients with the healthy subjects (Fig. 4.2c). In order to evaluate the gastric cancer diagnostic power of the eight biomarkers in saliva, receiver operating characteristic (ROC) analyses were performed. As eight biomarkers were all upregulated in gastric cancer, the value of state variable was set to be 1 (Fig. 4.3a and Table 4.7). The results showed that as a single biomarker for gastric cancer diagnosis, 3-aminoisobutanoate had an area under curve (AUC) of 0.819 and an accuracy of 0.814. The other seven metabolites, 4-hydroxyproline, aspartate, histidine, glutamate, carnosine, α-aminoadipate, and 4-aminobutanoate, provided the AUC values of 0.794, 0.724, 0.726, 0.713, 0.731,

**Table 4.5** The identified biomarkers in saliva and the related pathways

| Metabolites | KEGG entry | *p*-value | FC | Related pathways |
|---|---|---|---|---|
| 4-Hydroxyproline | C01157 | 0.006 | 2.530 | Arginine and proline metabolism |
| Aspartate | C00049 | 0.038 | 1.872 | Alanine, aspartate, and glutamate metabolism<br>Glycine, serine, and threonine metabolism<br>Cysteine and methionine metabolism |
| Histidine | C00135 | 0.036 | 2.116 | Histidine metabolism |
| Glutamate | C00025 | 0.049 | 2.115 | Arginine and proline metabolism<br>Alanine, aspartate, and glutamate metabolism |
| Carnosine | C00386 | 0.032 | 1.428 | Histidine metabolism<br>Beta-alanine metabolism |
| α-Aminoadipate | C00956 | 0.037 | 1.814 | Lysine biosynthesis and degradation |
| 4-Aminobutanoate | C00334 | 0.031 | 2.401 | Beta-alanine metabolism<br>Arginine and proline metabolism<br>Alanine, aspartate, and glutamate metabolism |
| 3-Aminoisobutanoate | C03284 | 0.003 | 3.118 | Valine, leucine, and isoleucine degradation |

*FC* represents fold change. The *p*-value was calculated using Mann–Whitney *U* test. The pathways were from KEGG database

0.724, and 0.733, respectively. And the accuracy values of the seven metabolites were 0.776, 0.681, 0.708, 0.729, 0.738, 0.688, and 0.738, respectively.

Moreover, the 25 saliva independent samples from 15 gastric cancer patients and 8 healthy subjects were used to demonstrate the utility of the identified biomarkers for diagnosis in gastric cancer patients. The metabolite of glutamate had the highest accuracy of 0.838 in diagnosis of gastric cancer. The other seven metabolites, 4-hydroxyproline, aspartate, histidine, carnosine, α-aminoadipate, 4-aminobutanoate, and 3-aminoisobutanoate had accuracy values of 0.608, 0.592, 0.742, 0.575, 0.742, 0.646, and 0.804.

The results proposed that the eight identified biomarkers had the potential in gastric cancer diagnosis. And these metabolites would have significant application prospect in clinical diagnosis of gastric cancer.

## 4.5 Saliva Markers for Gastric Cancer Detection

In this study, 38 types of amino acids in saliva were screened out as metabolites, and the molecular concentration in saliva was obtained. Eight molecules were identified as biomarkers in gastric cancer diagnosis, which had higher concentrations in gastric cancer patients than that in healthy subjects. The eight metabolites were mapped into eight metabolic pathways, in which three pathways were associated with

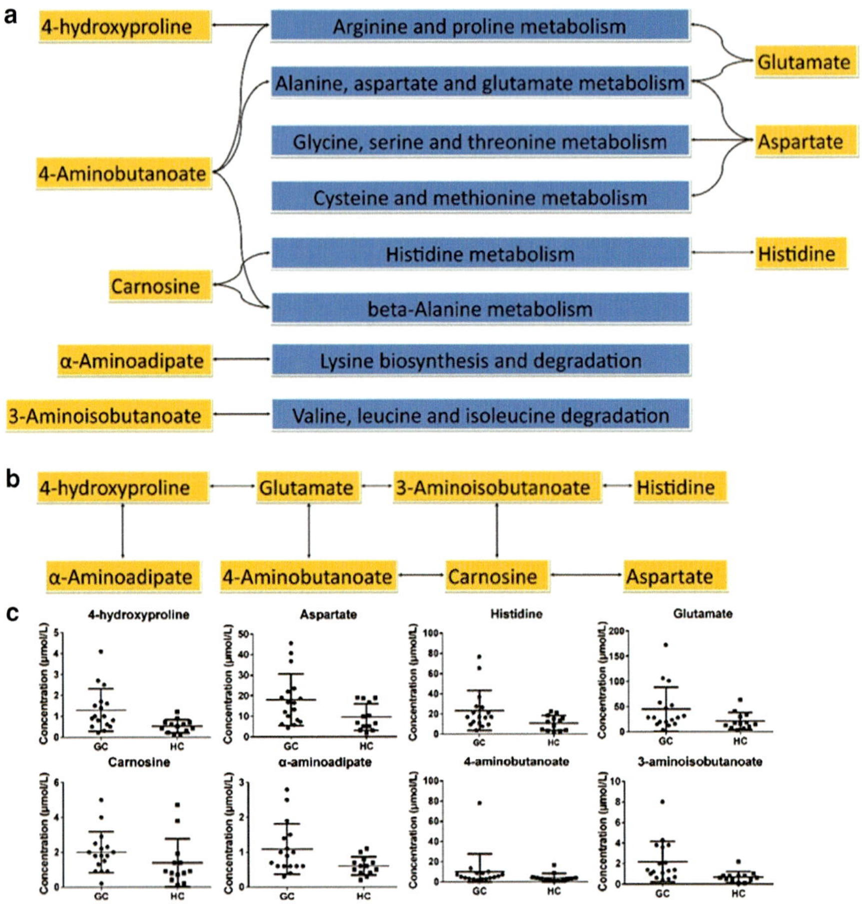


**Fig. 4.2** The molecules and associated pathways. (**a**) The line meant the metabolites were mapped in the pathways. (**b**) The Pearson correlation coefficients were calculated. The line represents high correlation between two metabolites. Pearson correlation coefficient is larger than 0.6; at the same time *p*-value is less than 0.05. (**c**) The *y*-axis is the concentration of the metabolites in saliva. GC and HC represent gastric cancer patients and healthy subjects, respectively

central carbon metabolism in cancer. The results showed that the eight metabolites, especially glutamate, 4-hydroxyproline, and 4-aminobutanoate, extracted from saliva were highly correlated with cancer metabolism. Furthermore, the diagnosis accuracy using the eight metabolites was calculated with the 23 independent saliva samples. The glutamate had the highest accuracy of 0.838 in diagnosis of gastric cancer. The results suggested that the eight identified biomarkers had the potential in gastric cancer diagnosis and had significant clinical value for the noninvasive diagnosis of gastric cancer.

Compared to the gastric cancer diagnosis by single metabolite, the gastric cancer diagnosis by combining more metabolites should be more precise. At last, we found

**Table 4.6** The Pearson correlation coefficient between the molecular metabolic levels

| | Hyp | Asp | His | Glu | Car | Aad | GABA | bAib |
|---|---|---|---|---|---|---|---|---|
| Hyp | 1 | 0.396* | 0.445* | **0.680**** | 0.502** | **0.669**** | 0.578** | 0.342 |
| Asp | 0.396* | 1 | 0.486** | 0.50** | **0.708**** | 0.05 | 0.434* | 0.314 |
| His | 0.445* | 0.486** | 1 | 0.478** | 0.537** | 0.370* | 0.496** | **0.634**** |
| Glu | 0.680** | 0.50** | 0.478** | 1 | **0.914**** | 0.117 | **0.788**** | **0.745**** |
| Car | 0.502** | 0.708** | 0.537** | 0.914** | 1 | −0.053 | **0.769**** | **0.699**** |
| Aad | 0.669** | 0.05 | 0.370* | 0.117 | −0.053 | 1 | 0.031 | 0.118 |
| GABA | 0.578** | 0.434* | 0.496** | 0.788** | 0.769** | 0.031 | 1 | 0.490** |
| bAib | 0.342 | 0.314 | 0.634** | 0.745** | 0.699** | 0.118 | 0.490** | 1 |

The *p*-values of Pearson correlation coefficients are shown. *$p<0.05$; **$p<0.01$. *Hyp* is 4-hydroxyproline, *Asp* is aspartate, *His* is histidine, *Glu* is glutamate, *Car* is carnosine, *Aad* is α-aminoadipate, *GABA* is 4-aminobutanoate, *bAib* is 3-animoisobutanoate

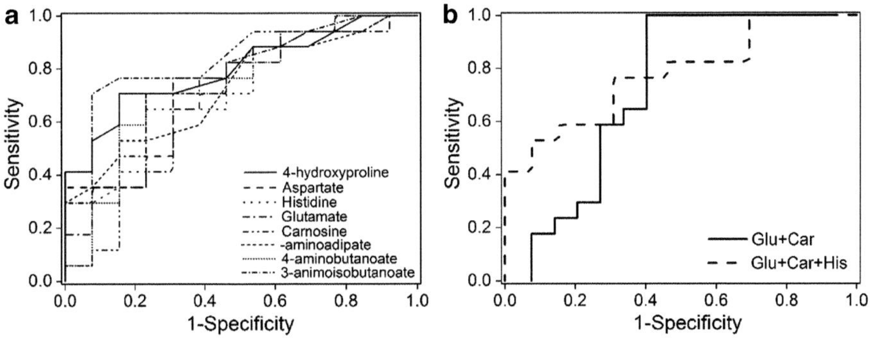


**Fig. 4.3** ROC analyses of the identified biomarkers in saliva of gastric cancer patients. (**a**) eight amino acids represents gastric cancer diagnosis with sensitivity and specificity. (**b**) Glu + Car represents the gastric cancer diagnosis by combining the two molecules of glutamate and carnosine. Glu + Car + His represents the gastric diagnosis by combing the three molecules of glutamate, carnosine, and histidine

**Table 4.7** ROC curve analysis of the identified biomarkers in saliva of gastric cancer patients

| Metabolite | AUCs | SN | SP | AC | SN-I | SP-I | AC-I |
|---|---|---|---|---|---|---|---|
| 4-Hydroxyproline | 0.794 | 0.706 | 0.846 | 0.776 | 0.467 | 0.750 | 0.608 |
| Aspartate | 0.724 | 0.824 | 0.538 | 0.681 | 0.933 | 0.250 | 0.592 |
| Histidine | 0.726 | 0.647 | 0.769 | 0.708 | 0.733 | 0.750 | 0.742 |
| Glutamate | 0.713 | 0.765 | 0.692 | 0.729 | 0.800 | 0.875 | **0.838** |
| Carnosine | 0.731 | 0.706 | 0.769 | 0.738 | 0.400 | 0.750 | 0.575 |
| α-Aminoadipate | 0.724 | 0.529 | 0.846 | 0.688 | 0.733 | 0.750 | 0.742 |
| 4-Aminobutanoate | 0.733 | 0.706 | 0.769 | 0.738 | 0.667 | 0.625 | 0.646 |
| 3-Aminoisobutanoate | 0.819 | 0.706 | 0.923 | 0.814 | 0.733 | 0.875 | 0.804 |

*AUC* represents the area under the curve. *SN*, *SP*, and *AC* represent sensitivity, specificity, and accuracy, respectively. SN-I, SP-I, and AC-I represent sensitivity, specificity, and accuracy calculated by the independent set, respectively. *Hyp* is 4-hydroxyproline. *Car* is carnosine. *bAib* is 3-aminoisobutanoate. Hyp + Car + bAib represents the gastric cancer diagnosis by combining the three molecules of 4-hydroxyproline, carnosine, and 3-aminoisobutanoate. The standard error and 95 % confidence interval are shown in Table 4.8

**Table 4.8** ROC curve analysis of the identified biomarkers in saliva of gastric cancer

| Metabolite | 95% confidence interval | Std. Error |
|---|---|---|
| 4-Hydroxyproline | 0.635~0.954 | 0.081 |
| Aspartate | 0.541~0.907 | 0.094 |
| Histidine | 0.543~0.909 | 0.093 |
| Glutamate | 0.521~0.904 | 0.098 |
| Carnosine | 0.534~0.927 | 0.100 |
| α-Aminoadipate | 0.543~0.905 | 0.093 |
| 4-Aminobutanoate | 0.545~0.921 | 0.096 |
| 3-Animoisobutanoate | 0.664~0.974 | 0.079 |
| Hyp+Car+bAib | 0.775~1.000 | 0.059 |

The standard error is calculated under the nonparametric assumption. Hyp+Car+bAib represents the gastric cancer diagnosis by combining the three molecules of 4-hydroxyproline, carnosine, and 3-aminoisobutanoate

**Table 4.9** ROC curve analysis of the combined biomarkers

| | Glu + Car | Glu + Car + His |
|---|---|---|
| AUCs | 0.765 | 0.778 |
| 95 % confidence interval | 0.575–0.954 | 0.614–0.943 |
| Std. error | 0.097 | 0.084 |
| SN-I | 0.933 | 0.933 |
| SP-I | 0.750 | 0.875 |
| AC-I | 0.842 | 0.904 |

*Glu* is glutamate. *Car* is carnosine. *His* is histidine. Glu + Car represents the gastric cancer diagnosis by combining the two molecules of glutamate and carnosine. Glu + Car + His represents the gastric cancer diagnosis by combing the three molecules of glutamate, carnosine, and histidine. The standard error is calculated under the nonparametric assumption. SN-I, SP-I, and AC-I represent sensitivity, specificity, and accuracy calculated by the independent set, respectively

that by combining the metabolites of glutamate and carnosine, the diagnosis acquired an accuracy of 0.842 and a sensitivity of 0.933 (Table 4.9). Especially, by combining the three metabolites of glutamate, carnosine, and histidine, the gastric cancer diagnosis could acquire an accuracy of 0.904 and a sensitivity of 0.933 (Table 4.9). The three metabolites were associated with metabolic pathways of alanine, aspartate, and glutamate metabolism, arginine and proline metabolism, and histidine metabolism. The metabolic pathways of alanine, aspartate, and glutamate metabolism and arginine and proline metabolism were proved to have high correlation with cancer in previous study [38].

Based on the concentration of the metabolites, the metabolic levels with the development of gastric cancer were further analyzed. The eight metabolites could be grouped into three types (Fig. 4.4). First, the concentrations of the metabolites of 3-aminoisobutanoate, α-aminoadipate, and 4-hydroxyproline became higher, which

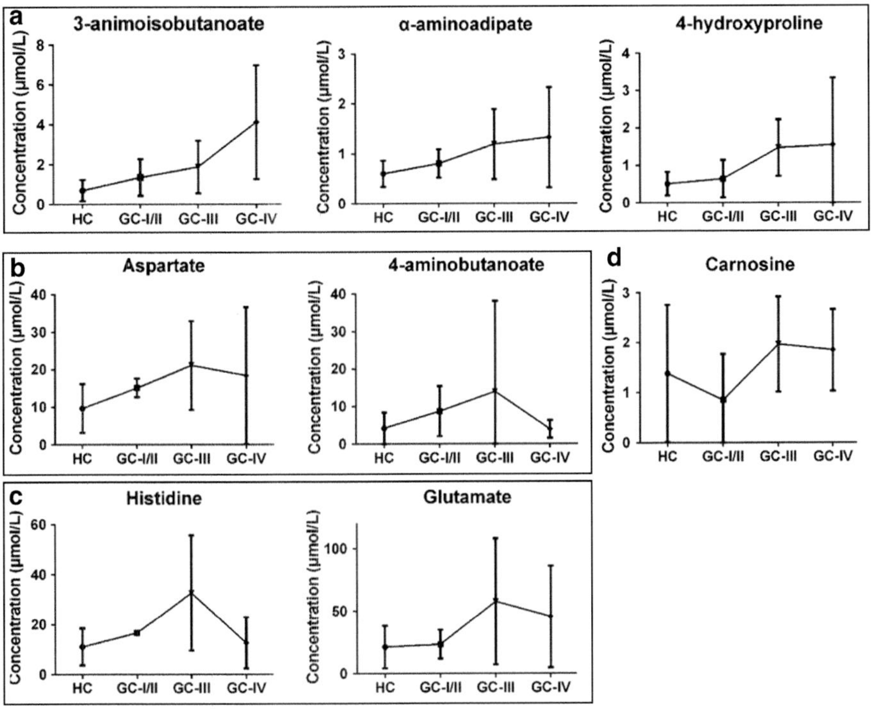


**Fig. 4.4** The metabolic changes with the development of gastric cancer. The HC represents the healthy saliva subjects. The *y*-axis is the concentration of the metabolites. (**a**) The concentration of 3-animoisobutanoate, α-aminoadipate and 4-hydroxyproline at different stage; (**b**) The concentration of aspartate, 4-aminobutanoate and carnosine at different stage; (**c**) The concentration of histidine and glutamate at different stage

meant that these molecules were upregulated along with the development of gastric cancer (Fig. 4.4a). Second, the metabolites of aspartate, 4-aminobutanoate, histidine, and glutamate were upregulated in GC-I/GC-II and GC-III and were downregulated in GC-IV compared to them in GC-I/GC-II and GC-III (Fig. 4.4b, c). Third, carnosine was downregulated in GC-I/GC-II and upregulated in GC-III and GC-IV. Furthermore, the ROC analyses of the metabolites in classifying the GC-III and GC-IV patients showed that the AUC values of 4-hydroxyproline, α-aminoadipate, and 3-aminoisobutanoate were 0.625, 0.528, and 0.708, respectively (Table 4.10). Using 4-hydroxyproline, α-aminoadipate, and 3-aminoisobutanoate, the accuracy in classifying the GC-III and GC-IV patients acquired 0.650, 0.800, and 0.625 tested by the independent samples. The results suggested that with the development of gastric cancer, different metabolites would have different regulated mechanisms, and it would help to classify gastric cancer patients into different stages.

In all, eight metabolites were identified as biomarkers in saliva for diagnosis of gastric cancer. By combining the metabolites of glutamate, carnosine, and histidine, the diagnosis accuracy of gastric cancer could acquire 0.904; at the same time, the

**Table 4.10** The ROC curve analysis in classify GC-III and GC-IV patients

| Metabolites | AUCs | 95% confidence interval | AC | Std. Error | AC-I |
|---|---|---|---|---|---|
| 4-Hydroxyproline | 0.625 | 0.215~1.000 | 0.650 | 0.209 | **0.650** |
| Aspartate | 0.611 | 0.216~1.000 | 0.760 | 0.202 | 0.325 |
| Histidine | 0.833 | 0.533~1.000 | 0.880 | 0.153 | 0.325 |
| Glutamate | 0.583 | 0.226~0.940 | 0.710 | 0.182 | 0.200 |
| Carnosine | 0.556 | 0.202~0.909 | 0.650 | 0.180 | 0.575 |
| α-Aminoadipate | 0.528 | 0.167~0.889 | 0.600 | 0.184 | **0.800** |
| 4-Aminobutanoate | 0.764 | 0.461~1.000 | 0.760 | 0.155 | 0.325 |
| 3-Animoisobutanoate | 0.708 | 0.372~1.000 | 0.760 | 0.172 | **0.625** |

AC-I is the accuracy calculated by the independent samples

sensitivity was 0.933, which was tested by independent samples. It gives us a new sight in diagnosis of gastric cancer that metabolites in saliva can help in clinical diagnosis of gastric cancer.

## References

1. Hartgrink HH, Jansen EPM, van Grieken NCT, van de Velde CJH. Gastric cancer. Lancet. 2009;374:477–90.
2. Torre LA, Bray F, Siegel RL, Ferlay J, Lortet-Tieulent J, Jemal A. Global cancer statistics, 2012. CA Cancer J Clin. 2015;65:87–108.
3. Siegel RL, Miller KD, Jemal A. Cancer statistics, 2015. CA Cancer J Clin. 2015;65:5–29.
4. Fock KM. Review article: the epidemiology and prevention of gastric cancer. Aliment Pharmacol Ther. 2014;40:250–60.
5. Malvezzi M, Bertuccio P, Levi F, La Vecchia C, Negri E. European cancer mortality predictions for the year 2013. Ann Oncol. 2013;24:792–800.
6. Yoon H, Kim N. Diagnosis and management of high risk group for gastric cancer. Gut Liver. 2015;9:5–17.
7. Karimi P, Islami F, Anandasabapathy S, Freedman ND, Kamangar F. Gastric cancer: descriptive epidemiology, risk factors, screening, and prevention. Cancer Epidemiol Biomarkers Prev. 2014;23:700–13.
8. Guggenheim DE, Shah MA. Gastric cancer epidemiology and risk factors. J Surg Oncol. 2013;107:230–6.
9. Wadhwa R, Song SM, Lee JS, Yao YX, Wei QY, Ajani JA. Gastric cancer-molecular and clinical dimensions. Nat Rev Clin Oncol. 2013;10:643–55.
10. Wu HH, Lin WC, Tsai KW. Advances in molecular biomarkers for gastric cancer: miRNAs as emerging novel cancer markers. Expert Rev Mol Med. 2014;16:e1.
11. Gonzalez CA, Agudo A. Carcinogenesis, prevention and early detection of gastric cancer: Where we are and where we should go. Int J Cancer. 2012;130:745–53.
12. Alfaro EE, Lauwers GY. Early gastric neoplasia: diagnosis and implications. Adv Anat Pathol. 2011;18:268–80.
13. Thrumurthy SG, Chaudry MA, Hochhauser D, Mughal M. The diagnosis and management of gastric cancer. BMJ (British Medical Journal). 2013;347:f6367.
14. Bentley-Hibbert S, Schwartz L. Use of imaging for GI cancers. J Clin Oncol. 2015;33:1729–36.
15. Dassen AE, Lips DJ, Hoekstra CJ, Pruijt JFM, Bosscha K. FDG-PET has no definite role in preoperative imaging in gastric cancer. Eur J Surg Oncol. 2009;35:449–55.

16. Hallinan JTPD, Venkatesh SK. Gastric carcinoma: imaging diagnosis, staging and assessment of treatment response. Cancer Imaging. 2013;13:212–27.
17. Shawihdi M, Thompson E, Kapoor N, Powell G, Sturgess RP, Stern N, et al. Variation in gastroscopy rate in English general practice and outcome for oesophagogastric cancer: retrospective analysis of Hospital Episode Statistics. Gut. 2014;63:250–61.
18. Asaka M, Mabe K. Strategies for eliminating death from gastric cancer in Japan. Proc Jpn Acad Ser B Phys Biol Sci. 2014;90:251–8.
19. Lee HY, Park EC, Jun JK, Choi KS, Hahm MI. Comparing upper gastrointestinal X-ray and endoscopy for gastric cancer diagnosis in Korea. World J Gastroenterol. 2010;16:245–50.
20. Axon A. Symptoms and diagnosis of gastric cancer at early curable stage. Best Pract Res Clin Gastroenterol. 2006;20:697–708.
21. Melton SD, Genta RM, Souza RF. Biomarkers and molecular diagnosis of gastrointestinal and pancreatic neoplasms. Nat Rev Gastroenterol Hepatol. 2010;7:620–8.
22. Atkinson AJ, Colburn WA, DeGruttola VG, DeMets DL, Downing GJ, Hoth DF, et al. Biomarkers and surrogate endpoints: preferred definitions and conceptual framework. Clin Pharmacol Ther. 2001;69:89–95.
23. Wang JB, Wang QW, Liu H, Hu B, Zhou W, Cheng YF. MicroRNA expression and its implication for the diagnosis and therapeutic strategies of gastric cancer. Cancer Lett. 2010;297:137–43.
24. Guilford PJ, Hopkins JBW, Grady WM, Markowitz SD, Willis J, Lynch H, et al. E-cadherin germline mutations define an inherited cancer syndrome dominated by diffuse gastric cancer. Hum Mutat. 1999;14:249–55.
25. Lin LL, Huang HC, Juan HF. Discovery of biomarkers for gastric cancer: a proteomics approach. J Proteomics. 2012;75:3081–97.
26. Barber M, Murrell A, Ito Y, Maia AT, Hyland S, Oliveira C, et al. Mechanisms and sequelae of E-cadherin silencing in hereditary diffuse gastric cancer. J Pathol. 2008;216:295–306.
27. Ueda T, Volinia S, Okumura H, Shimizu M, Taccioli C, Rossi S, et al. Relation between microRNA expression and progression and prognosis of gastric cancer: a microRNA expression analysis. Lancet Oncol. 2010;11:136–46.
28. Zhang JP, Song Y, Zhang CL, Zhi X, Fu HL, Ma Y, et al. Circulating MiR-16-5p and MiR-19b-3p as Two Novel Potential Biomarkers to Indicate Progression of Gastric Cancer. Theranostics. 2015;5:733–45.
29. Cohen M, Yossef R, Erez T, Kugel A, Welt M, Karpasas MM, et al. Serum Apolipoproteins C-I and C-III Are Reduced in Stomach Cancer Patients: Results from MALDI-Based Peptidome and Immuno-Based Clinical Assays. PLos One. 2011;e14540.
30. Zhang YX, Gao G, Liu HJ, Fu HL, Fan J, Wang K, et al. Identification of volatile biomarkers of gastric cancer cells and ultrasensitive electrochemical detection based on sensing interface of Au-Ag alloy coated MWCNTs. Theranostics. 2014;4:154–62.
31. Miyagi Y, Higashiyama M, Gochi A, Akaike M, Ishikawa T, Miura T, et al. Plasma free amino acid profiling of five types of cancer patients and its application for early detection. PLos One. 2011;6.
32. Jayavelu ND, Bar NS. Metabolomic studies of human gastric cancer: review. World J Gastroenterol. 2014;20:8092–101.
33. Hirayama A, Kami K, Sugimoto M, Sugawara M, Toki N, Onozuka H, et al. Quantitative metabolome profiling of colon and stomach cancer microenvironment by capillary electrophoresis time-of-flight mass spectrometry. Cancer Res. 2009;69:4918–25.
34. Abbassi-Ghadi N, Kumar S, Huang J, Goldin R, Takats Z, Hanna GB. Metabolomic profiling of oesophago-gastric cancer: a systematic review. Eur J Cancer. 2013;49:3625–37.
35. Chan AW, Gill RS, Schiller D, Sawyer MB. Potential role of metabolomics in diagnosis and surveillance of gastric cancer. World J Gastroenterol. 2014;20:12874–82.
36. Cristescu R, Lee J, Nebozhyn M, Kim KM, Ting JC, Wong SS, et al. Molecular analysis of gastric cancer identifies subtypes associated with distinct clinical outcomes. Nat Med. 2015;21:449–U217.
37. Li XZ, Yang TY, Lin JX. Spectral analysis of human saliva for detection of lung cancer using surface-enhanced Raman spectroscopy. J Biomed Opt. 2012;17:037003.
38. Kardos GR, Wastyk HC, Robertson GP. Disruption of Proline Synthesis in Melanoma Inhibits Protein Production Mediated by the GCN2 Pathway. Mol Cancer Res. 2015;13(10):1408–20.

# Chapter 5
# Metabolic Profiles and High-Accuracy Diagnosis Model Based on Plasma Metabolomics of Gastric Cancer Patients

**Cheng Shangli and Daxiang Cui**

## Abbreviations

| | |
|---|---|
| GC | Gastric cancer |
| GC-I | Stage I gastric cancer |
| GC-II | Stage II gastric cancer |
| GC-III | Stage III gastric cancer |
| HC | Healthy persons as control samples |
| OPLS-DA | Orthogonal projection to latent structure-discriminant analysis |
| PDV | Percentage difference value |
| TCA cycle | Tricarboxylic acid cycle |

## 5.1 Introduction

Gastric cancer (GC) is the fourth most common cancer and the second leading cause of cancer-related death worldwide, and the cases of men are twice than that of women [1]. The latest statistics reported 723,100 deaths and 951,600 new cases of GC all over the world in 2012 [2]. The mortality of GC had regional differences [2, 3]. The mortality rate of GC in the Eastern Asia was higher than other regions [4, 5]. The cause of GC was associated with the genetic factors, as well as environmental risk factors, including the infection of *Helicobacter pylori*, the intake of salt, the intake of alcohol, and the smoking of tobacco [6–9]. For the patients of late-stage

C. Shangli (✉) • D. Cui
Institute of Nano Biomedicine and Engineering, Shanghai Engineering Research Center for Intelligent Diagnosis and Treatment Instrument, National Center for Translational Medicine, Collaborative Innovational Center for System Biology, Shanghai Jiao Tong University, 800 Dongchuan Road, Shanghai 200240, P. R. China
e-mail: chengshangli@sjtu.edu.cn; dxcui@sjtu.edu.cn

© Springer Science+Business Media B.V. and Shanghai Jiao Tong University Press, Shanghai 2017
D. Cui (ed.), *Gastric Cancer Prewarning and Early Diagnosis System*, Translational Medicine Research, DOI 10.1007/978-94-024-0951-2_5

GC, the five-year survival rate was less than 24 % [10]. The five-year survival in Japan was 96.5 % better than European countries of 23 % and the USA of 29 % owing to the diagnosis of GC in the early stage [3, 11, 12]. How to improve the five-year survival of GC patients has become our concern.

The traditional methods for GC diagnosis were based on imaging methods, such as CT, PET/CT and multi-detector row spiral CT, endoscopy, and histological confirmation [13–16]. For detection of early GC, using endoscopy was more successful than using X-ray [17, 18]. From the statistics in Korea, the sensitivity and specificity in GC diagnosis using endoscopy were respectively 59.0 % and 96.3 %, which was higher than that using X-ray with the sensitivity of 42.1 % and 89.8 %, respectively [19]. As the symptoms of the patients with GC were always vague and it was hard for the diagnosis of GC in early stage [13, 20], most of the GC patients were found in the late stage, and diagnosis ratio of early GC patients in China was less than 20 %. How to find early GC is still a great challenge [20]. Thereby, how to find biomarkers associated with early GC is our concern.

The biomarkers that have the potential for early GC diagnosis which are often used as indicators of in vivo metabolisms are highly desirable [21, 22]. Up to date, the biomarkers such as SNP sites, plasma miRNAs, proteins, etc. have been found in GC research [23–25]. For example, (a) the germline mutation and hypermethylation of E-cadherin were found as a hereditary cause of GC [24, 26]; (b) the miRNA expressions of miR-125b, miR-199a, and miR-100 were shown to be upregulated in GC patients, and circulating MiR-16-5p and MiR-19b-3p were identified to be downregulated in GC patients, which can be used to indicate progress of gastric cancer [27, 28]; (c) proteins of apoC-I, apoC-III, CA19-9, and CRP were used for the prediction of GC and received the accuracy of ~74 % [29]. Furthermore, the biomarkers of small metabolic molecules were considered equally important as DNA mutation, miRNA, and protein in diagnosis of GC. For example, we screened out some volatile biomarkers from the breath metabolism of GC patients [30]. Up to date, many studies focused on small molecules, such as amino acids, nucleic acids, carbohydrates, lipids, and so on. The profiles of the small molecules could be used for the diagnosis and the discovery of the metabolic characteristics in GC [31–34]. By using the methods of metabolomics, the biomarkers from the tissue, serum, plasma, saliva, and urine were gradually found in GC patients [35]. Based on the biomarkers, development of novel diagnosis of GC has become one hot spot [31, 36].

In this study, 120 plasma samples were collected from 90 GC patients and 30 healthy persons. Metabolomics analyses of the small molecules in plasma were performed using $^{1}$H-NMR. Firstly, based on the metabolic analysis and the PDV analysis, proline was found to be a key biomarker in all stages of GC. Secondly, based on the regulation level of proline, two types of metabolic profiles were proposed, and the diagnosis model of GC was constructed with higher sensitivity of 82.2 % using independent testing set. The concrete flowchart was shown in Fig. 5.1. Our diagnosis model and metabolic profiles have the great potential in applications such as early gastric cancer diagnosis and subtype, as well as therapeutic effect monitoring in near future.

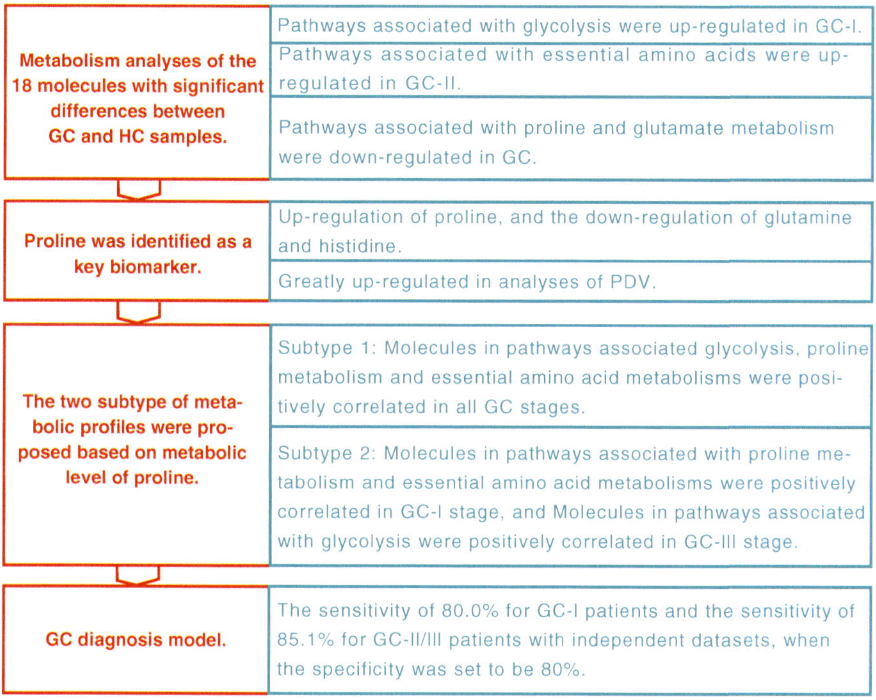


**Fig. 5.1** The flowchart showed the improvement of GC diagnosis based on the identification of the metabolic profiles

## 5.2 The Molecular Detection in Gastric Cancer Plasma

The tumor stage information followed the AJCC Cancer Staging Manual. At last, 90 plasma samples from GC patients and 30 plasma samples from HC persons were obtained. The average ages were 54 in GC-I patients, 55 in GC-II patients, 55 in GC-III patients, and 55 in HC persons. The range of the age and sex distribution was shown in Table 5.1. The age and sex distribution was not significantly different among GC-I, GC-II, and GC-III patients and HC samples, and it would lead the results without bias.

The molecules of the plasma samples were detected using $^{1}$H-NMR spectra. The analyses of OPLS-DA (Fig. 5.2) and ANOVA between GC-I, GC-II, and GC-III patients and HC showed significant difference under sevenfold cross-validation. At last, based on the $^{1}$H-NMR spectra, 30 metabolic molecules were identified in the GC and HC plasmas (Table 5.2). For each molecule, the integral of the peaks in $^{1}$H-NMR spectra was calculated. Then, the quantitative concentration and fold change of the metabolic molecules were obtained (Table 5.3). And 18 molecules were significantly different between GC and HC (Table 5.3).

**Table 5.1** The information of GC patients and healthy persons

| | Characteristic | GC-I | GC-II | GC-III | HC | *P* value |
|---|---|---|---|---|---|---|
| Total | Number | 30 | 30 | 30 | 30 | |
| | Percentage | 25 % | 25 % | 25 % | 25 % | |
| Sex | Male | 26 | 21 | 21 | 26 | $P = 0.34^a$ |
| | Female | 4 | 9 | 9 | 4 | |
| Age | Mean | 54 | 55 | 55 | 55 | $P = 0.954^b$ |
| | Range | 33–79 | 40–70 | 22–81 | 34–71 | |

[a]Two-side chi-square test
[b]Kruskal Wallis test

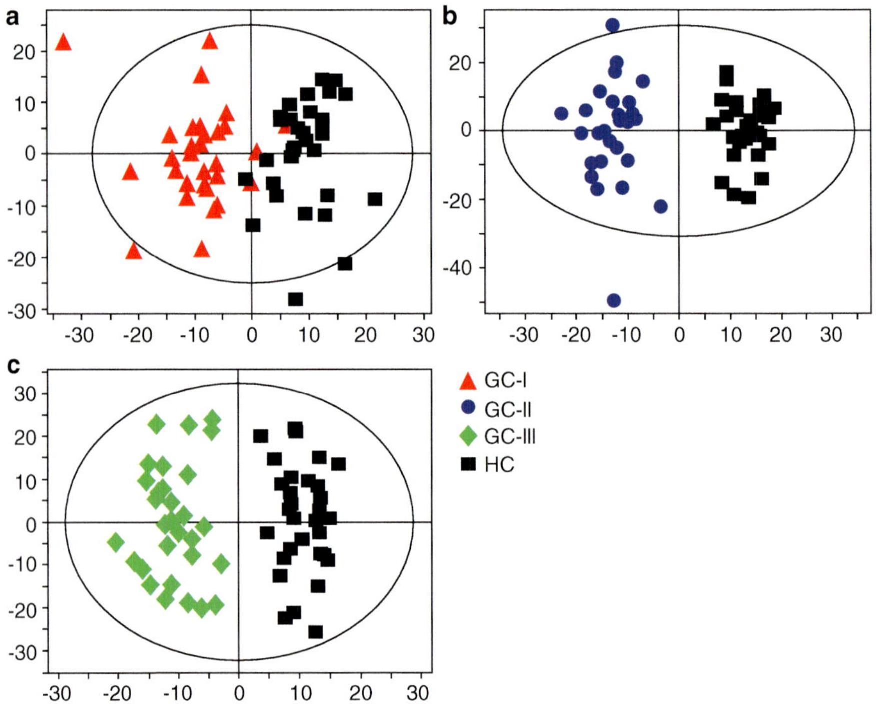


**Fig. 5.2** The OPLS-DA analyses among GC-I, GC-II, GC-III, and HC samples. The results of OPLS-DA were shown: (**a**) $R^2X = 0.231$; $Q^2 = 0.343$; $P = 9.947 \times 10^{-5}$. (**b**) $R^2X = 0.474$; $Q^2 = 0.741$; $P = 1.55 \times 10^{-12}$. (**c**) $R^2X = 0.404$; $Q^2 = 0.753$; $P = 1.82 \times 10^{-14}$

## 5.3 The Molecular Regulation Were Changed with the Development of GC and the Great Upregulation of Proline

In order to identify the regulation of the molecules and detect biomarkers in GC, 18 molecules, which were significantly different among HC, GC-I, GC-II, and GC-III stage, were further analyzed. The direct and indirect interactions among the

**Table 5.2** The integral of the molecular peaks in $^{1}$H-NMR spectra

| | HC | | GC-I | | GC-II | | GC-III | |
|---|---|---|---|---|---|---|---|---|
| | Mean value | STD | Mean value | STD | Mean value | STD | Mean value | STD |
| Fatty acid[0.86(b)] | 0.10369655 | 0.03164820 | 0.09854890 | 0.02427406 | 0.10852659 | 0.02696687 | 0.10215838 | 0.02228912 |
| Fatty acid[1.27(b)] | 0.19089815 | 0.07846849 | 0.18604952 | 0.07315102 | 0.21407002 | 0.09354821 | 0.19239304 | 0.06372045 |
| Fatty acid[1.55(m)] | 0.01627061 | 0.00626976 | 0.01662652 | 0.00622859 | 0.02053847 | 0.00883516 | 0.01744694 | 0.00546882 |
| Fatty acid[2.01(b)] | 0.03129745 | 0.00561454 | 0.02964196 | 0.00452220 | 0.03340166 | 0.00613040 | 0.03030597 | 0.00430750 |
| Fatty acid[5.28(b)] | 0.01610465 | 0.00658902 | 0.01500681 | 0.00483906 | 0.01788589 | 0.00681693 | 0.01681142 | 0.00509321 |
| Isoleucine | 0.00541263 | 0.00059279 | 0.00526479 | 0.00086631 | 0.00638601 | 0.00154317 | 0.00579857 | 0.00098752 |
| Leucine | 0.00846007 | 0.00088094 | 0.00828652 | 0.00136973 | 0.00920941 | 0.00151232 | 0.00783195 | 0.00149554 |
| Valine | 0.00678232 | 0.00069072 | 0.00682386 | 0.00111485 | 0.00699332 | 0.00121705 | 0.00604627 | 0.00121361 |
| 3-Hydroxyisobutyrate | 0.00097885 | 0.00012607 | 0.00096012 | 0.00016577 | 0.00097144 | 0.00016000 | 0.00087295 | 0.00011146 |
| Ethanol | 0.00537072 | 0.00065665 | 0.00592458 | 0.00147645 | 0.00658830 | 0.00237732 | 0.00567698 | 0.00084650 |
| Lactate | 0.08780144 | 0.01817264 | 0.07971235 | 0.02773955 | 0.10462535 | 0.05346516 | 0.07511280 | 0.02120974 |
| Alanine | 0.01010970 | 0.00199668 | 0.01154982 | 0.00212255 | 0.01078941 | 0.00235999 | 0.01065808 | 0.00289081 |
| Acetate | 0.00187773 | 0.00022844 | 0.00193074 | 0.00032587 | 0.00208229 | 0.00046462 | 0.00181926 | 0.00035959 |
| Proline | 0.00059833 | 0.00021348 | 0.00077957 | 0.00020172 | 0.00088961 | 0.00021773 | 0.00080783 | 0.00033745 |
| Glycoprotein | 0.03208833 | 0.00604621 | 0.03031499 | 0.00478270 | 0.03394763 | 0.00628753 | 0.03082978 | 0.00445821 |
| Glutamine | 0.00866511 | 0.00123441 | 0.00946838 | 0.00148595 | 0.00922446 | 0.00220174 | 0.00858508 | 0.00157764 |
| Acetone | 0.00343510 | 0.00149557 | 0.00364030 | 0.00151498 | 0.00387082 | 0.00177434 | 0.00305838 | 0.00160201 |
| Succinate | 0.00205228 | 0.00055522 | 0.00240035 | 0.00101397 | 0.00273028 | 0.00215243 | 0.00214348 | 0.00052390 |
| Creatine | 0.00462897 | 0.00070488 | 0.00447318 | 0.00044506 | 0.00456130 | 0.00051456 | 0.00406622 | 0.00057382 |
| Creatinine | 0.00216647 | 0.00023946 | 0.00193002 | 0.00034270 | 0.00191241 | 0.00024261 | 0.00179523 | 0.00029164 |
| Tyrosine | 0.00116049 | 0.00016801 | 0.00120236 | 0.00018929 | 0.00108884 | 0.00021668 | 0.00112205 | 0.00016892 |
| Choline | 0.00870735 | 0.00157143 | 0.01107342 | 0.00405340 | 0.00848084 | 0.00263082 | 0.00966197 | 0.00248982 |
| beta-D-glucose | 0.02124787 | 0.00500441 | 0.02905177 | 0.01099203 | 0.01831892 | 0.00625355 | 0.02048871 | 0.00511169 |

(continued)

**Table 5.2** (continued)

| | HC | | GC-I | | GC-II | | GC-III | |
|---|---|---|---|---|---|---|---|---|
| | Mean value | STD | Mean value | STD | Mean value | STD | Mean value | STD |
| Scyllo-inositol | 0.00297164 | 0.00065876 | 0.00329454 | 0.00063742 | 0.00261324 | 0.00062711 | 0.00293241 | 0.00057118 |
| Methanol | 0.00319568 | 0.00078358 | 0.00356888 | 0.00067861 | 0.00291102 | 0.00070714 | 0.00297209 | 0.00072459 |
| alpha-D-glucose | 0.01346397 | 0.00245074 | 0.01740261 | 0.00699116 | 0.01344439 | 0.00437389 | 0.01518221 | 0.00392522 |
| Glycine | 0.00290328 | 0.00046029 | 0.00308783 | 0.00102038 | 0.00248455 | 0.00055965 | 0.00296342 | 0.00092810 |
| Histidine | 0.00117016 | 0.00010756 | 0.00111540 | 0.00011432 | 0.00098207 | 0.00011975 | 0.00099240 | 0.00011022 |
| Phenylalanine | 0.00111609 | 0.00018762 | 0.00119179 | 0.00021110 | 0.00127289 | 0.00025277 | 0.00114724 | 0.00016288 |
| Formate | 0.00005984 | 0.00001846 | 0.00006722 | 0.00002245 | 0.00007619 | 0.00003510 | 0.00006640 | 0.00002906 |

The integral of the peaks for each molecule was calculated. The HC, GC-I, GC-II, and GC-III represented healthy persons, GC patients of stage I, GC patients of stage II, and GC patients of stage III

**Table 5.3** The fold change (FC) of the metabolic molecules among GC-I, GC-II, GC-III, and HC samples

| | | GC-I | GC-II | GC-III | Mean | |
|---|---|---|---|---|---|---|
| Character | Molecules | Fold change | | | FC | *P* value |
| Carbohydrate | alpha-D-glucose | 1.293 | 0.999 | 1.128 | 1.140 | **0.011** |
| | beta-D-glucose | 1.367 | 0.862 | 0.964 | 1.064 | **0.000** |
| Acid | Lactate | 0.908 | 1.192 | 0.855 | 0.985 | **0.009** |
| | Succinate | 1.170 | 1.330 | 1.044 | 1.181 | 0.404 |
| | Creatine | 0.966 | 0.985 | 0.878 | 0.943 | **0.001** |
| | Formate | 1.123 | 1.273 | 1.110 | 1.169 | 0.272 |
| | Acetate | 1.028 | 1.109 | 0.969 | 1.035 | 0.088 |
| | 3-Hydroxyisobutyrate | 0.981 | 0.992 | 0.892 | 0.955 | **0.016** |
| Anhydride | Creatinine | 0.891 | 0.883 | 0.829 | 0.868 | **0.000** |
| Alcohol | Ethanol | 1.103 | 1.227 | 1.057 | 1.129 | **0.031** |
| | Methanol | 1.117 | 0.911 | 0.930 | 0.986 | **0.002** |
| | Scyllo-inositol | 1.109 | 0.879 | 0.987 | 0.992 | **0.001** |
| Amino acid | Glycine | 1.064 | 0.856 | 1.021 | 0.980 | **0.006** |
| | Alanine | 1.142 | 1.067 | 1.054 | 1.088 | 0.053 |
| | Isoleucine | 0.973 | 1.180 | 1.071 | 1.075 | **0.000** |
| | Leucine | 0.979 | 1.089 | 0.926 | 0.998 | **0.009** |
| | Valine | 1.006 | 1.031 | 0.891 | 0.976 | **0.020** |
| | Histidine | 0.953 | 0.839 | 0.848 | 0.880 | **0.000** |
| | Phenylalanine | 1.068 | 1.140 | 1.028 | 1.079 | **0.025** |
| | Tyrosine | 1.036 | 0.938 | 0.967 | 0.980 | **0.047** |
| | Proline | 1.303 | 1.487 | 1.350 | 1.380 | **0.000** |
| | Glutamine | 1.093 | 1.065 | 0.991 | 1.050 | 0.107 |
| Other | Choline | 1.272 | 0.974 | 1.110 | 1.119 | **0.005** |
| | Acetone | 1.060 | 1.127 | 0.890 | 1.026 | 0.114 |
| | Glycoprotein | 0.945 | 1.058 | 0.961 | 0.988 | 0.142 |
| Fatty acid | Fatty acid[0.86(b)] | 0.950 | 1.047 | 0.985 | 0.994 | 0.597 |
| | Fatty acid[1.27(b)] | 0.975 | 1.121 | 1.008 | 1.035 | 0.809 |
| | Fatty acid[1.55(m)] | 1.022 | 1.262 | 1.072 | 1.119 | 0.172 |
| | Fatty acid[2.01(b)] | 0.947 | 1.067 | 0.968 | 0.994 | 0.107 |
| | Fatty acid[5.28(b)] | 0.932 | 1.111 | 1.044 | 1.029 | 0.366 |

The Kruskal-Wallis test was used to find the molecules with significant differences among GC-I, GC-II, GC-III, and HC samples

molecules had been extracted from the human metabolic pathways in KEGG database. Based on glycolysis (alpha-D-glucose, beta-D-glucose, lactate, acetate, and ethanol); pyruvate metabolism (formate); propanoate metabolism (acetone and methanol); inositol phosphate metabolism (scyllo-inositol); glycine, serine, and threonine metabolism (glycine and choline); alanine, aspartate, and glutamate metabolism (alanine); valine, leucine, and isoleucine metabolism (valine, leucine, isoleucine, and 3-hydroxyisobutyrate); histidine metabolism (histidine and glutamate); arginine and proline metabolism (proline, creatine, and creatinine);

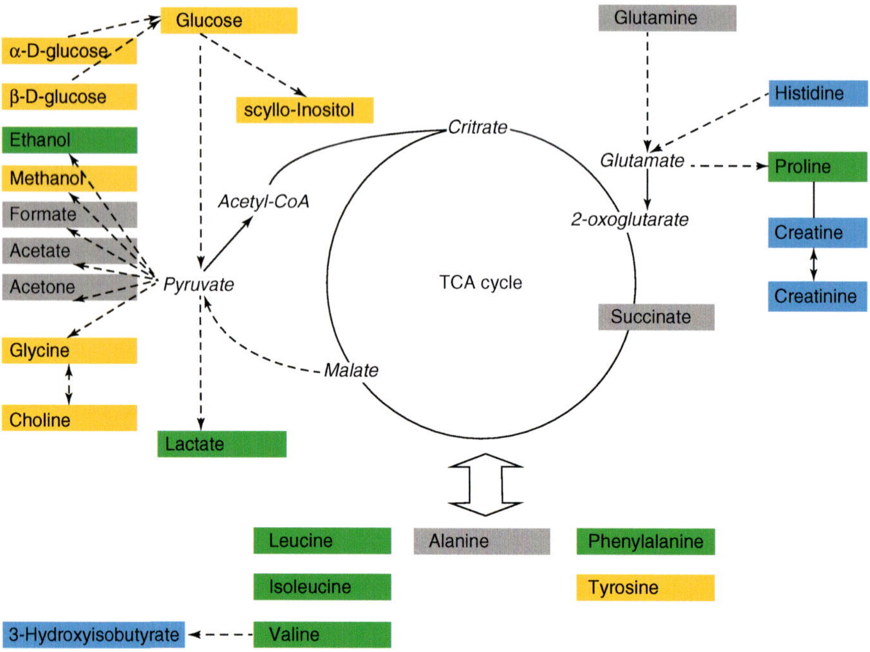


**Fig. 5.3** The metabolic map of the molecules. In the map, the different colors of the molecules represent the three modes of the molecular regulation. The molecule in the *yellow box* was upregulated in GC-I stage. The molecule in the *green box* was upregulated in GC-II stage. The molecule in the *blue box* had the trend of the downregulation in GC-I, GC-II, and GC-III stages. The molecules in the *gray box* had no significant differences among the different stages. The molecules with *italic* were not detected

phenylalanine metabolism (phenylalanine and tyrosine); and central carbon metabolism in cancer, the metabolic maps were constructed by indirect and direct interactions among the molecules (Fig. 5.3).

In the metabolic map, the molecules were mainly distributed in three parts: (1) the pathways associated with glycolysis (pyruvate as a hub molecule), (2) the metabolism associated with the essential amino acids, and (3) the pathways associated with glutamate and proline metabolism. Firstly, most of the molecules in pathways associated with glycolysis were upregulated in GC-I stage (Fig. 5.4a). And the molecules of ethanol and lactate, associated with glycolysis, also had higher concentration in GC-II stage. Secondly, the valine, leucine, and isoleucine metabolism was upregulated in GC-II stage (Fig. 5.4b). Thirdly, creatine and creatinine in arginine and proline metabolism and histidine metabolism, which were directly and indirectly interacted with glutamate, were downregulated (Fig. 5.4c). For each molecule, the average concentration was calculated in different GC stages (Fig. 5.4d).

However, proline was greatly upregulated in GC patients. Based on the central carbon metabolism in cancer, proline was found to be associated with arginine and proline metabolism, glutamine and glutamate, and histidine can be transformed to glutamate in histidine metabolism (Fig. 5.5a). The concentration of histidine in

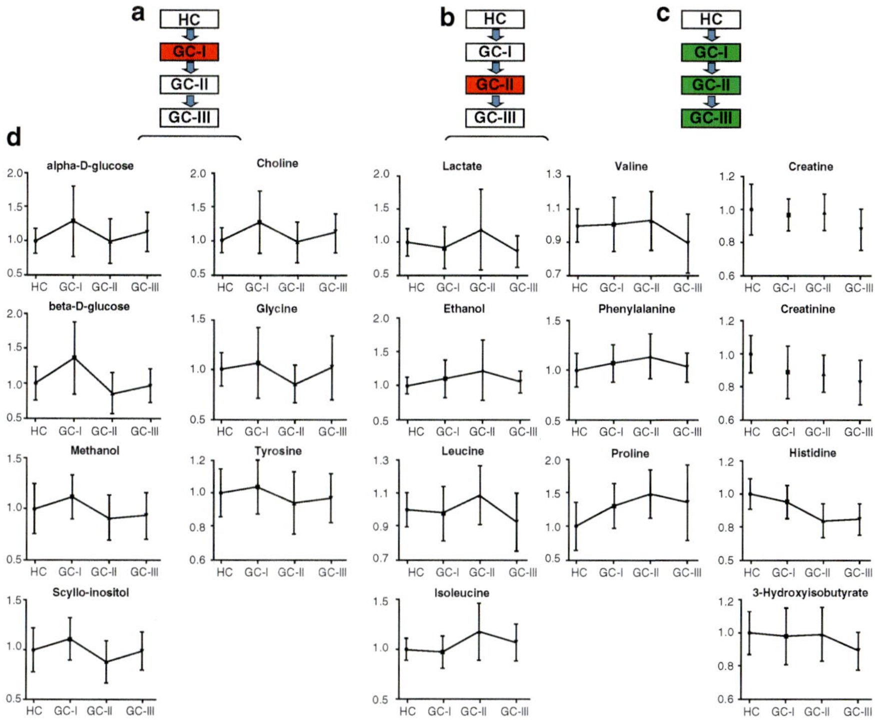


**Fig. 5.4** The molecular concentration in different GC stages. The three modes of metabolic regulation of the 18 molecules include upregulation in GC-I stage (**a**), upregulation in GC-II stage (**b**), and downregulation in GC-I, GC-II, and GC-III stages (**c**). The *red box* represents upregulation, and the *green box* represents downregulation. In (**d**), the *x*-axis represents the molecule in HC, GC-I, GC-II, and GC-III stages. The *y*-axis represents the concentration of the molecules

plasma was downregulated in GC patients. With the limited samples in this study, the glutamine was not significantly different between GC and HC samples. However, previous studies proved that glutamine was downregulated in plasma ($P < 0.001$) [31]. Glutamine is taken as an essential nutrient for cancer cell proliferation [43]. Through the pathway analyses in Fig. 5.5a, it was demonstrated that the pathway had a great consumption of glutamine and histidine. Thus, it leads to a downregulation of histidine and glutamine in plasma. The concentration of proline, as a downstream molecule in the glutamate pathway, in GC group was higher than that in HC group. In previous studies, proline was also found to be upregulated in both tissues and plasmas [33, 44, 45].

In summary, in the sight of metabolism, the pathways associated with glycolysis were dramatically upregulated in GC-I stage. The pathways associated with essential amino acids were dramatically upregulated in GC-II stage. The pathways associated with glutamate and proline metabolism were downregulated. However, since the greatly upregulation of proline in the downstream of the pathway associated glutamine and glutamate, it was suggested a key biomarker in GC.

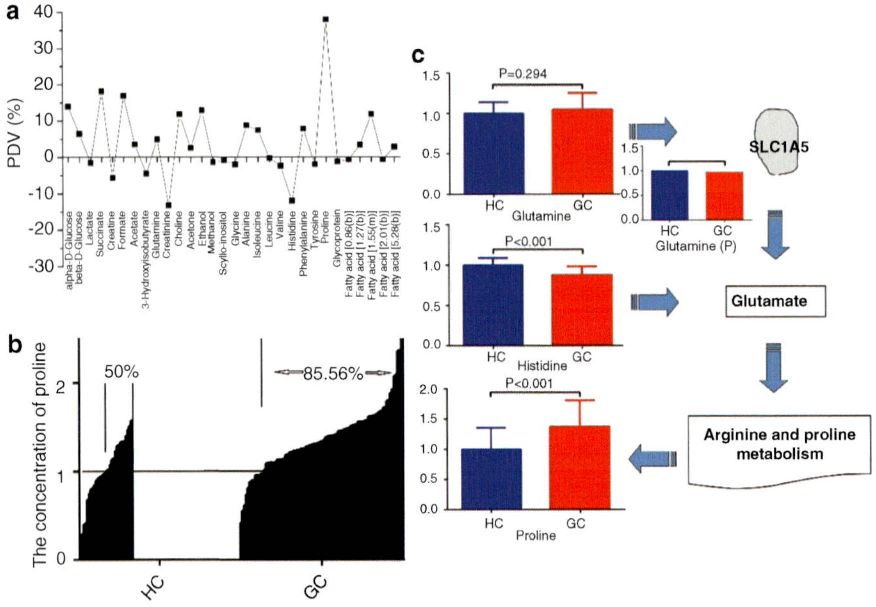


**Fig. 5.5** The PDVs of the 30 molecules and the concentration of proline in plasma. (**a**) The *x*-axis represents the 30 molecules, and the *y*-axis represents the PDVs. (**b**) The samples were sorted by the concentration of proline. All the HC and GC samples were used. (**c**) The glutamine (P) (Used the data from Miyagi et al. [31])

## 5.4 Two Subtypes of Metabolic Profiles by the Clustering Method Based on Proline

Through the analyses of the molecular regulations, it was found that the metabolic levels of the molecules in different GC stages were complicated and hard to extract the metabolic patterns. Thus, the PDVs of the molecules were calculated [37–39]. The results showed that proline had the highest PDV. It was also demonstrated that proline had great difference of regulation between HC and GC groups (Fig. 5.5b). Over 85 % of GC patients had a higher concentration of proline than the average (Fig. 5.5c).

Thus, along with the metabolism analyses, proline could be identified as a key biomarker in GC. However, the concentration of the proline was not greatly changed in ~15 % of GC patients. It was proposed that metabolic subtype of GC would exist besides the upregulation of proline. The clustering method based on proline could capture the significant dependencies between the subgroups and could extract the key metabolic molecules in the GC stage [40, 41].

In this case, based on the concentration of proline, all the samples were divided into two subgroups, named subgroup 1 and subgroup 2 (Fig. 5.6a). The concentration of proline of the samples in subgroup 1 was greater than the average concentration of proline, and the concentration of proline of the samples in subgroup 2 was lower

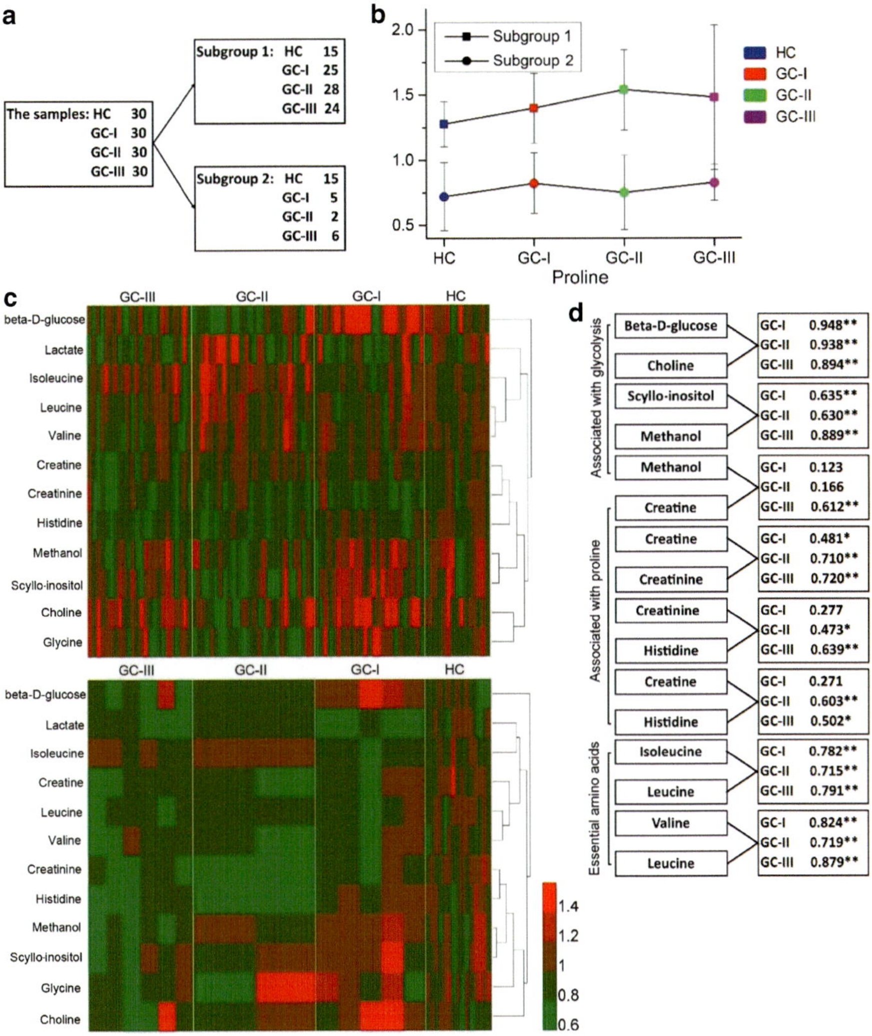


**Fig. 5.6** The metabolic profiles of the 12 molecules in subgroups 1 and 2. (**a**) The samples were divided into subgroup 1 and subgroup 2 by the concentration of proline. The average concentration of proline was calculated using the HC samples. In subgroup 1, the concentration of proline was higher than the average concentration, while the concentration of proline in subgroup 2 was lower than the average. (**b**) The concentration of proline in different stages of GC was shown. (**c**) The metabolic level of the 12 molecules was shown by heat map. (**d**) In subgroup 1, the Pearson correlation coefficient was calculated in different GC stages, while Pearson correlation coefficients of the molecules in subgroup 2 were in Fig. 5.8. The two molecules were set to be highly correlated when Pearson correlation coefficient was larger than 0.6. The *P* values of Pearson correlation coefficients were shown. *$P<0.05$; **$P<0.01$. The concentrations of the molecules were shown in Fig. 5.9

**Table 5.4** The metabolic differences among GC-I, GC-II, GC-III, and HC samples

| Subgroup 1 | *P* value | Subgroup 2 | *P* value |
|---|---|---|---|
| alpha-D-glucose | 0.051 | alpha-D-glucose | 0.245 |
| beta-D-glucose | **0.000** | beta-D-glucose | **0.025** |
| Lactate | 0.092 | Lactate | **0.008** |
| Succinate | 0.867 | Succinate | 0.289 |
| Creatine | **0.012** | Creatine | **0.013** |
| Formate | 0.813 | Formate | 0.401 |
| Acetate | 0.220 | Acetate | 0.418 |
| 3-Hydroxyisobutyrate | 0.106 | 3-Hydroxyisobutyrate | 0.050 |
| Glutamine | 0.364 | Glutamine | 0.176 |
| Creatinine | **0.007** | Creatinine | **0.002** |
| Choline | **0.018** | Choline | 0.147 |
| Acetone | 0.199 | Acetone | 0.427 |
| Ethanol | 0.133 | Ethanol | 0.140 |
| Methanol | **0.013** | Methanol | **0.029** |
| Scyllo-inositol | **0.003** | Scyllo-inositol | 0.121 |
| Glycine | **0.006** | Glycine | 0.463 |
| Alanine | 0.285 | Alanine | 0.188 |
| Isoleucine | **0.003** | Isoleucine | **0.027** |
| Leucine | **0.047** | Leucine | **0.098** |
| Valine | **0.032** | Valine | 0.206 |
| Histidine | **0.000** | Histidine | **0.007** |
| Phenylalanine | 0.139 | Phenylalanine | 0.451 |
| Tyrosine | 0.186 | Tyrosine | 0.152 |
| Proline | **0.035** | Proline | 0.689 |
| Glycoprotein | 0.394 | Glycoprotein | 0.177 |
| Fatty acid[0.86(b)] | 0.661 | Fatty acid[0.86(b)] | 0.669 |
| Fatty acid[1.27(b)] | 0.853 | Fatty acid[1.27(b)] | 0.822 |
| Fatty acid[1.55(m)] | 0.365 | Fatty acid[1.55(m)] | 0.407 |
| Fatty acid[2.01(b)] | 0.322 | Fatty acid[2.01(b)] | 0.107 |
| Fatty acid[5.28(b)] | 0.490 | Fatty acid[5.28(b)] | 0.407 |

The Kruskal-Wallis test was used to find the molecules with significant differences [42]

than the average (Fig. 5.6b). In subgroup 1, the sample numbers of HC, GC-I, GC-II, and GC-III groups were 15, 25, 28, and 24, respectively. In subgroup 2, the sample numbers of HC, GC-I, GC-II, and GC-III groups were 15, 5, 2, and 6, respectively. With Kruskal-Wallis test, 12 molecules were significantly different in HC, GC-I, GC-II, and GC-III stages besides proline (Table 5.4).

Through the ROC curve analyses of the 12 molecules, the AUC of the ten molecules had increased (Table 5.5). It was demonstrated that the molecules had higher ability to classify the GC and HC samples than that before the clustering. In order to find out the metabolic profiles in the two metabolic subtypes of subgroups 1 and 2, Pearson correlation coefficients between molecules were calculated.

In subgroup 1, the molecules of beta-D-glucose, choline, methanol, scyllo-inositol, creatine, creatinine, isoleucine, leucine, and valine were highly positively

**Table 5.5** The AUC of the ROC curves for the molecules

| | Before the clustering | After the clustering | |
|---|---|---|---|
| | | Subgroup 1 | Subgroup 2 |
| beta-D-glucose | 0.511 | **0.584** | **0.574** |
| Lactate | 0.609 | 0.574 | **0.862** |
| Creatine | 0.606 | 0.525 | **0.821** |
| Creatinine | 0.796 | 0.739 | **0.877** |
| Choline | 0.606 | 0.591 | 0.544 |
| Methanol | 0.525 | **0.563** | 0.513 |
| Scyllo-inositol | 0.515 | **0.526** | **0.523** |
| Glycine | 0.603 | **0.616** | 0.579 |
| Isoleucine | 0.588 | 0.575 | 0.523 |
| Leucine | 0.519 | 0.514 | **0.759** |
| Valine | 0.559 | **0.600** | **0.708** |
| Histidine | 0.810 | **0.852** | 0.718 |
| Proline | 0.763 | 0.678 | 0.621 |

The ROC curve was receiver operating characteristic curve. The AUC was area under curve

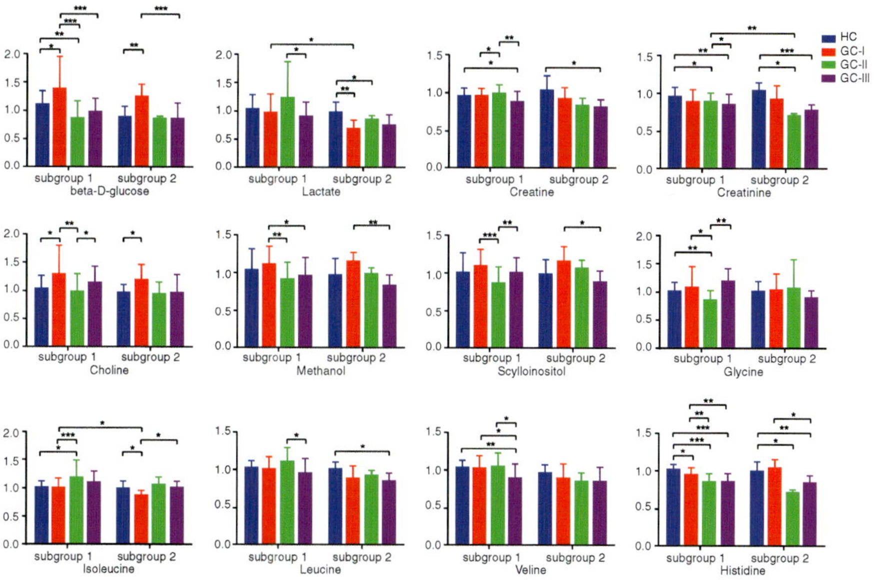


**Fig. 5.7** The concentration of the 12 molecules of different state in subgroup 1 and 2. The *P* values were shown $*P<0.05$; $**P<0.01$; $***P<0.001$

correlated in GC-I, GC-II, and GC-III stages (Fig. 5.6c, d). The molecules of creatine, methanol, creatinine, and histidine had positive correlations in GC-II or GC-III stage. In the sight of the metabolic level, the glycolysis was positively correlated and upregulated in GC-I stage, and they were downregulated in GC-II and GC-III stages compared to that in GC-I stage (Figs. 5.6c, 5.7, and 5.8a). For the essential amino acid metabolism, it had positively correlated upregulation in GC-II

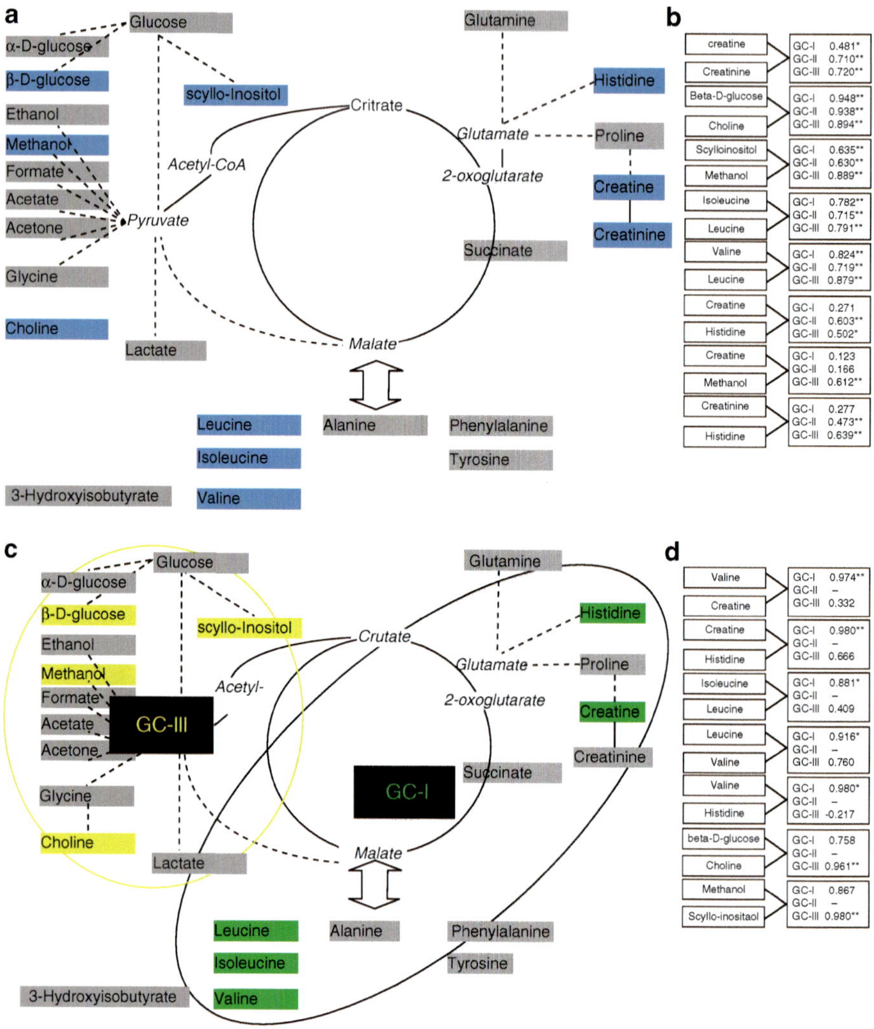


**Fig. 5.8** The metabolic profiles of the 12 molecules in subgroup 1 and 2. The Pearson correlation coefficient was calculated in different GC stages in subgroup 1 and 2. (**a**, **c**) The figures showed the metabolic profiles in subgroup 1 and subgroup 2. (**b**, **d**) The Pearson correlation coefficient was calculated. The two molecules were set to be highly correlated when Pearson correlation coefficient was larger than 0.6. $*P<0.05$; $**P<0.01$

stage. Histidine, creatine, and creatinine were also downregulated with positive correlation.

In subgroup 2, the regulation of creatine, histidine, valine, isoleucine, and leucine were positively correlated in GC-I stage. The molecules of beta-D-glucose, choline, methanol, and scyllo-inositol had positively correlated regulations in GC-III stage (Fig. 5.8c). The results showed that the metabolism associated with

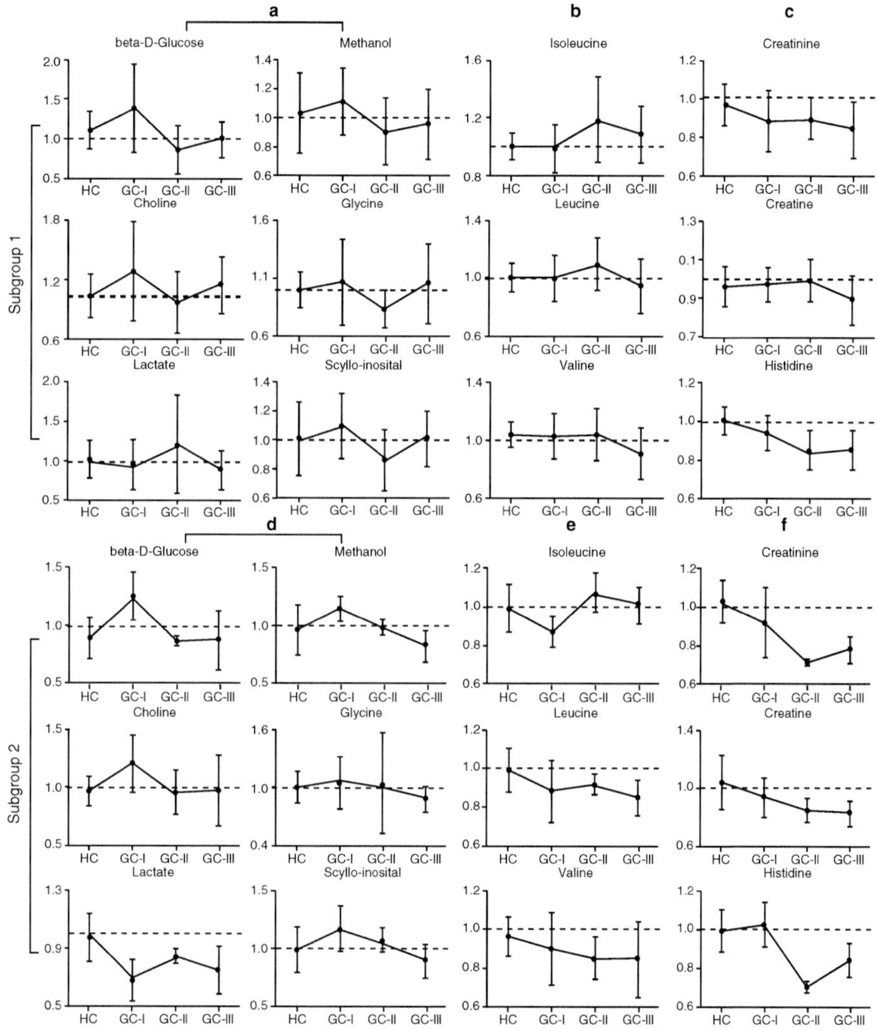


**Fig. 5.9** The molecular concentration in different GC stage in subgroup 1 and 2. (**a**, **d**) The molecules were in the metabolic pathways associated with glycolysis. (**b**, **e**) The molecules were associated with essential amino acids. (**c**, **f**) The molecules were associated with glutamine, glutamate, and proline metabolism

essential amino acids and proline was downregulated with positive correlation, while histidine was upregulated (Figs. 5.6c, 5.7, and 5.8d). The glycolysis, including the molecules of beta-D-glucose, choline, methanol, and scyllo-inositol, was upregulated in GC-III stage with positive correlation.

In brief, by the clustering of the samples, two subtypes of metabolic profiles were identified. When proline was upregulated, the molecular regulations were positively correlated in different metabolic pathways, including pathways associated with glycolysis, proline metabolism, and essential metabolism (Fig. 5.8a). Interestingly, when proline was downregulated, the molecular regulations were correlated in different GC stages. In GC-I stage, the pathways associated with proline and essential amino acid metabolism were positively correlated, while the pathways associated with glycolysis were positively correlated in GC-III stage (Fig. 5.8c).

## 5.5 The Improvement of GC Diagnosis by Clustering Based on the Regulation of Proline

In this study, proline was found to be the key molecule to divide the samples into two subgroups. Then, besides proline, 12 molecules were extracted as key molecules with Kruskal-Wallis test (Table 5.4). In this study, the diagnosis model for GC-I and GC-II/GC-III was created in subgroup 1 and subgroup 2, respectively, which was named after model-II. The GC-II/GC-III and HC samples were taken as the training data in diagnosis of GC-I, while GC-I and HC samples were taken as the training data in diagnosis of GC-I/GC-III. The results provided 12 molecules as candidate molecules for GC diagnosis. As a comparison, the diagnosis model-I was created without clustering of the samples (Fig. 5.10a).

In diagnosis of GC-I patients, the specificity of the diagnosis model-II was set to be 80.0 %. Then, 20 out of 25 GC-I patients were identified with six biomarkers in subgroup 1, while four out of five GC-I patients were diagnosed with five biomarkers in subgroup 2 (Figs. 5.10b, c and 5.11a, b, Table 5.6). In diagnosis of GC-II/GC-III patients, the specificity was also set to be 80.0 %. Then, 43 out of 52 GC-II/GC-III patients with eight biomarkers and seven out of eight GC-II/GC-III patients were diagnosed with six biomarkers in subgroups 1 and 2, respectively (Figs. 5.10d, e and 5.11c, d, Table 5.6). In all, the GC diagnosis model-II had a sensitivity of 80.0 % in the diagnosis of GC-I patients and a sensitivity of 85.1 % in the diagnosis of GC-II/III patients, when the specificity was set to be 80.0 %. In diagnosis model-I, the sensitivity in diagnosis of GC-I patients was 73.3 %, and the sensitivity in diagnosis of GC-II/GC-III patients was 65.0 % when the specificity was set to be 80.0 % (Fig. 5.10f, g and Table 5.6). The results showed the accuracy was improved in diagnosis of GC after the clustering based on the concentration of proline.

In this study, the metabolic characteristics of GC in plasma were analyzed. The metabolic level of proline was greatly changed in GC patients, and the higher concentration of proline was detected in plasma. Moreover, histidine and glutamine were downregulated in our previous results [31]. Glutamine and glutamate have been recognized as key biomarkers in different kinds of cancers, which were presented in different metabolic levels in cancer patients [46–48]. By the central carbon metabolism, glutamine and glutamate were involved in TCA cycle and proline metabolism (Fig. 5.5c). Moreover, histidine can be transformed to glutamate in histidine metabolism. In addition, proline was in the metabolic downstream in the metabolism associated with glutamine, glutamate, and histidine. Along with the higher concentration of proline in plasma, it was suggested that the biosynthesis of

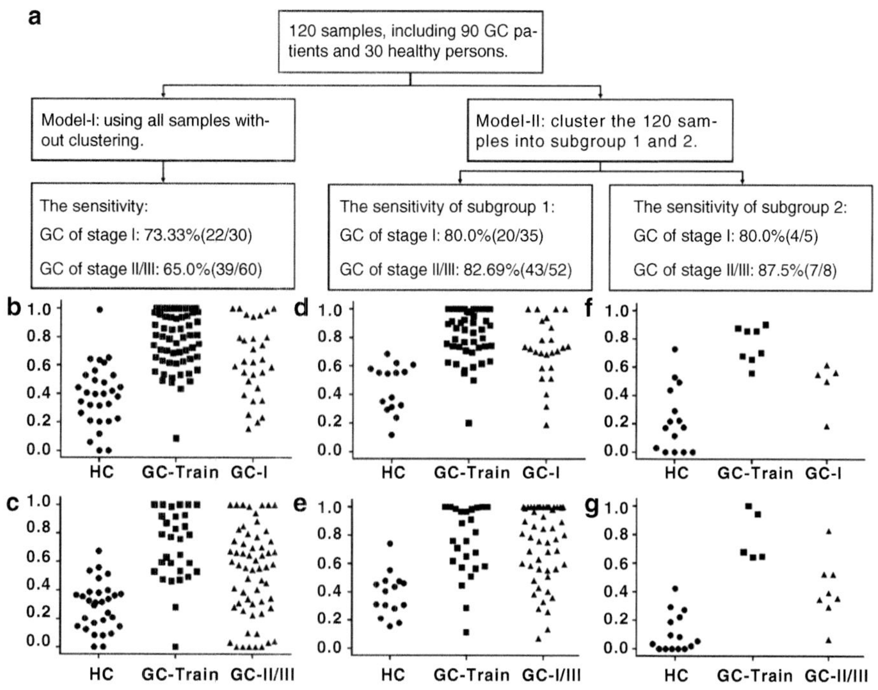


**Fig. 5.10** The cluster of the samples and the sensitivity of GC diagnosis. (**a**) For the model-I, the samples were not clustered. For the model-II, the samples were clustered and divided into two subgroups. In the model, the specificity was set to be 80.0 %, and the sensitivity was calculated using the independent data set. (**b–g**) The "y value" was the possibility in the diagnosis model. The larger the "y" is, the higher the possibility of being diagnosed of GC

proline would be upregulated in GC, which would have great consumption of the upstream amino acids in metabolic pathways, including glutamate, glutamine, and histidine. Thus, the upregulation of proline and downregulation of glutamine and histidine were found in blood plasma of GC. PYCR2 and ALDH18A1 were the major genes in the biosynthesis of proline from glutamate to proline. In a previous study on melanoma, the biosynthesis of proline could impair the development of melanoma. The knockdown of ALDH18A1 can lead to the decrease of delta-1-pyrroline-5-carboxylate synthase and affected the process of proline biosynthesis [49]. In summary, from our results of upregulation of proline, it was proposed that the regulation of proline was enhanced, and proline can be used as the key biomarker in GC.

However, although 85 % of GC patients had upregulation of proline, the metabolic level of proline in the other 15 % of GC patients was similar to the healthy persons. With the clustering of the samples based on the concentration of proline, the different metabolic profiles were found out in the two subgroups. In subgroup 1, in which proline was upregulated, the regulation of the molecules was positively correlated among different pathways. In subgroup 2, when the biosynthesis of

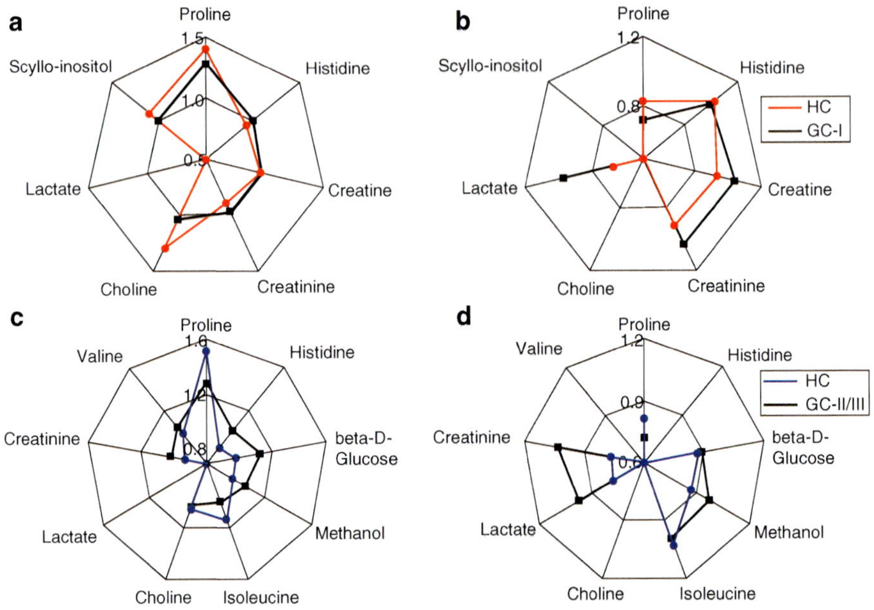


**Fig. 5.11** The diagnosis profile of GC-I and GC-II/III. In the figure, the radar chart displayed the concentration of the molecules used in the diagnosis model. (**a**, **b**) The diagnosis profile in subgroup 1. (**c**, **d**) The diagnosis profile in subgroup 2

**Table 5.6** The molecular profile and the parameters used in the GC diagnosis model

| | | | Model II | | | |
|---|---|---|---|---|---|---|
| | Model I | | Subgroup 1 | | Subgroup 2 | |
| Molecules | Stage I | Stage II/III | Stage I | Stage II/III | Stage I | Stage II/III |
| Constant | 2.430 | 2.361 | 2.147 | 3.814 | 2.984 | 0.384 |
| beta-D-glucose | 0.000 | 0.113 | 0.000 | −0.386 | 0.000 | 0.729 |
| Lactate | 0.153 | −0.306 | 0.000 | 0.000 | −1.197 | −0.707 |
| Creatine | 0.214 | 0.000 | 1.337 | 0.000 | 0.421 | 0.000 |
| Creatinine | −0.839 | −1.064 | −0.933 | −0.930 | −0.273 | −1.156 |
| Choline | 0.172 | 0.219 | 0.031 | 0.781 | 0.000 | 0.000 |
| Methanol | 0.000 | 0.732 | 0.000 | 0.378 | 0.000 | 0.859 |
| Scyllo-inositol | 0.413 | 0.000 | 0.254 | 0.000 | 0.000 | 0.000 |
| Glycine | 0.000 | 0.000 | 0.000 | 0.000 | 0.000 | 0.000 |
| Isoleucine | 0.000 | −0.795 | 0.000 | −0.965 | 0.000 | 0.077 |
| Leucine | 0.000 | 0.000 | 0.000 | 0.000 | 0.000 | 0.000 |
| Valine | 0.000 | 0.735 | 0.000 | 0.423 | 0.000 | 0.000 |
| Histidine | −2.162 | −1.713 | −2.356 | −2.751 | −1.791 | 0.000 |
| Cutoff | 0.529 | 0.393 | 0.584 | 0.476 | 0.463 | 0.247 |
| SN | 73. % | 65.0 % | 80.0 % | 82.6 % | 80.0 % | 87.5 % |
| *Total SN* | 67.7 % | | 82.2 % | | | |
| *Total AC* | 73.8 % | | 81.1 % | | | |

Through the logistic regression, a constant value and a weight value for each molecule were calculated. *SN* is the sensitivity. *AC* is the accuracy

proline was not enhanced, the molecular regulation was positively correlated among different GC stages (Fig. 5.8a, c). Thus, we proposed two subtype metabolic regulations in GC. The diagnosis model based on the clustering had higher accuracy in diagnosing GC and in classifying different stages of GC. Through the diagnosis model, it was suggested that GC patients had different metabolic characteristics, and the metabolic characters were changed from GC-I to GC-II/GC-III stage. Furthermore, two molecules in subgroup 1 (2/7 = 28.5 %) and three molecules in subgroup 2 (3/7 = 42.8 %) were downregulated in GC-I stage. In GC-II/GC-III stage, five molecules (5/9 = 55.7 %) in subgroup 1 and four molecules in subgroup 2 (4/9 = 44.7 %) were downregulated (Fig. 5.9). These results demonstrated a trend in metabolic regulation that more molecules were downregulated with the development of GC in blood plasma.

Specially, in all stages of GC, proline, histidine, creatine, and creatinine, which were associated with the metabolism of proline, were significant in the diagnosis of GC (Fig. 5.11). In the early stage of GC, the molecules, which were associated with glycolysis and the metabolism of glucose, were important. The molecules, associated with essential amino acid metabolism, were added into the metabolic profiles in the advanced stage of GC (Fig. 5.11).It was suggested that in the beginning of the GC, the metabolic changes were on the aspect of energy metabolism. With the development of GC, the metabolisms on the amino acids were up- or downregulated. The development of GC followed this pattern. This can be well used to classify the stage of GC.

In summary, we found that the molecules upregulated in early GC were mostly associated with glycolysis; the essential amino acids were mainly upregulated in GC stages II and III. The molecules associated with proline metabolism were downregulated in the course of progression of GC, while only proline was greatly upregulated in GC stages I, II, and III; proline was a key biomarker for GC diagnosis. In addition, proline integrated with 13 molecules associated with glycolysis; proline and the essential amino acid metabolisms were successfully used to construct the GC diagnosis model, which has the sensitivity of 82.2 % at the same time the specificity is set to be 80.0 %. Moreover, the model can be used to distinguish GC stage I from GC stages II and III and then acquire the sensitivity of 80 % for the classification of GC-I stage samples and 83.3 % for GC-II/GC-III samples, exhibiting clinical translational prospect. Further work will focus on clinical verification to have a deeper comprehension in GC metabolic characteristics.

## References

1. Hartgrink HH, Jansen EPM, van Grieken NCT, et al. Gastric cancer. Lancet. 2009;374:477–90.
2. Torre LA, Bray F, Siegel RL, et al. Global cancer statistics, 2012. CA Cancer J Clin. 2015;65:87–108.
3. Siegel RL, Miller KD, Jemal A. Cancer statistics, 2015. CA Cancer J Clin. 2015;65:5–29.
4. Fock KM. Review article: the epidemiology and prevention of gastric cancer. Aliment Pharmacol Ther. 2014;40:250–60.

5. Malvezzi M, Bertuccio P, Levi F, et al. European cancer mortality predictions for the year 2013. Ann Oncol. 2013;24:792–800.
6. Yoon H, Kim N. Diagnosis and management of high risk group for gastric cancer. Gut Liver. 2015;9:5–17.
7. Karimi P, Islami F, Anandasabapathy S, et al. Gastric cancer: descriptive epidemiology, risk factors, screening, and prevention. Cancer Epidemiol Biomarkers Prev. 2014;23:700–13.
8. Guggenheim DE, Shah MA. Gastric cancer epidemiology and risk factors. J Surg Oncol. 2013;107:230–6.
9. Wadhwa R, Song SM, Lee JS, et al. Gastric cancer-molecular and clinical dimensions. Nat Rev Clin Oncol. 2013;10:643–55.
10. Wu HH, Lin WC, Tsai KW. Advances in molecular biomarkers for gastric cancer: miRNAs as emerging novel cancer markers. Expert Rev Mol Med. 2014;16:e1.
11. Gonzalez CA, Agudo A. Carcinogenesis, prevention and early detection of gastric cancer: where we are and where we should go. Int J Cancer. 2012;130:745–53.
12. Alfaro EE, Lauwers GY. Early gastric neoplasia: diagnosis and implications. Adv Anat Pathol. 2011;18:268–80.
13. Thrumurthy SG, Chaudry MA, Hochhauser D, et al. The diagnosis and management of gastric cancer. BMJ. 2013;347:f6367.
14. Bentley-Hibbert S, Schwartz L. Use of imaging for GI cancers. JClin Oncol. 2015;33:1729–35.
15. Dassen AE, Lips DJ, Hoekstra CJ, et al. FDG-PET has no definite role in preoperative imaging in gastric cancer. Eur J Surg Oncol. 2009;35:449–55.
16. Hallinan JTPD, Venkatesh SK. Gastric carcinoma: imaging diagnosis, staging and assessment of treatment response. Cancer Imaging. 2013;13:212–27.
17. Shawihdi M, Thompson E, Kapoor N, et al. Variation in gastroscopy rate in English general practice and outcome for oesophagogastric cancer: retrospective analysis of Hospital Episode Statistics. Gut. 2014;63:250–61.
18. Asaka M, Mabe K. Strategies for eliminating death from gastric cancer in Japan. Proc Jpn Acad Ser B-Phys Biol Sci. 2014;90:251–8.
19. Lee HY, Park EC, Jun JK, et al. Comparing upper gastrointestinal X-ray and endoscopy for gastric cancer diagnosis in Korea. World J Gastroenterol. 2010;16:245–50.
20. Axon A. Symptoms and diagnosis of gastric cancer at early curable stage. Best Pract Res Clin Gastroenterol. 2006;20:697–708.
21. Melton SD, Genta RM, Souza RF. Biomarkers and molecular diagnosis of gastrointestinal and pancreatic neoplasms. Nat Rev Gastroenterol Hepatol. 2010;7:620–8.
22. Atkinson AJ, Colburn WA, DeGruttola VG, et al. Biomarkers and surrogate endpoints: preferred definitions and conceptual framework. Clin Pharmacol Ther. 2001;69:89–95.
23. Wang JB, Wang QW, Liu H, et al. MicroRNA expression and its implication for the diagnosis and therapeutic strategies of gastric cancer. Cancer Lett. 2010;297:137–43.
24. Guilford PJ, Hopkins JBW, Grady WM, et al. E-cadherin germline mutations define an inherited cancer syndrome dominated by diffuse gastric cancer. Hum Mutat. 1999;14:249–55.
25. Lin LL, Huang HC, Juan HF. Discovery of biomarkers for gastric cancer: a proteomics approach. J Proteomics. 2012;75:3081–97.
26. Barber M, Murrell A, Ito Y, et al. Mechanisms and sequelae of E-cadherin silencing in hereditary diffuse gastric cancer. JPathol. 2008;216:295–306.
27. Ueda T, Volinia S, Okumura H, et al. Relation between microRNA expression and progression and prognosis of gastric cancer: a microRNA expression analysis. Lancet Oncol. 2010;11:136–46.
28. Zhang JP, Song Y, Zhang CL, et al. Circulating MiR-16-5p and MiR-19b-3p as two novel potential biomarkers to indicate progression of gastric cancer. Theranostics. 2015;5:733–45.
29. Cohen M, Yossef R, Erez T, et al. Serum apolipoproteins C-I and C-III are reduced in stomach cancer patients: results from MALDI-based peptidome and immuno-based clinical assays. PLos One. 2011;6:e14540.

30. Zhang YX, Gao G, Liu HJ, et al. Identification of volatile biomarkers of gastric cancer cells and ultrasensitive electrochemical detection based on sensing interface of Au-Ag alloy coated MWCNTs. Theranostics. 2014;4:154–62.
31. Miyagi Y, Higashiyama M, Gochi A, et al. Plasma free amino acid profiling of five types of cancer patients and its application for early detection. PLos One. 2011;6:e24143.
32. Jayavelu ND, Bar NS. Metabolomic studies of human gastric cancer: review. World J Gastroenterol. 2014;20:8092–101.
33. Hirayama A, Kami K, Sugimoto M, et al. Quantitative metabolome profiling of colon and stomach cancer microenvironment by capillary electrophoresis time-of-flight mass spectrometry. Cancer Res. 2009;69:4918–25.
34. Abbassi-Ghadi N, Kumar S, Huang J, et al. Metabolomic profiling of oesophago-gastric cancer: a systematic review. Eur J Cancer. 2013;49:3625–37.
35. Chan AW, Gill RS, Schiller D, et al. Potential role of metabolomics in diagnosis and surveillance of gastric cancer. World J Gastroenterol. 2014;20:12874–82.
36. Cristescu R, Lee J, Nebozhyn M, et al. Molecular analysis of gastric cancer identifies subtypes associated with distinct clinical outcomes. Nat Med. 2015;21:449–U217.
37. Deng WK, Wang YB, Liu ZX, et al. Heml: a toolkit for illustrating heatmaps. PLos One. 2014;9, e111988.
38. Anbazhagan R, Tihan T, Bornman DM, et al. Classification of small cell lung cancer and pulmonary carcinoid by gene expression profiles. Cancer Res. 1999;59:5119–22.
39. Tan YX, Yin PY, Tang L, et al. Metabolomics study of stepwise hepatocarcinogenesis from the model rats to patients: potential biomarkers effective for small hepatocellular carcinoma diagnosis. Mol Cell Proteomics. 2012;11:M111.010694.
40. Cheng SL, Lian BF, Liang J, et al. Site selectivity for protein tyrosine nitration: insights from features of structure and topological network. Mol Biosyst. 2013;9:2860–8.
41. Lee TY, Lin ZQ, Hsieh SJ, et al. Exploiting maximal dependence decomposition to identify conserved motifs from a group of aligned signal sequences. Bioinformatics. 2011;27:1780–7.
42. Peng HC, Long FH, Ding C. Feature selection based on mutual information: Criteria of max-dependency, max-relevance, and min-redundancy. IEEE Trans Pattern Anal Mach Intell. 2005;27:1226–38.
43. Cassago A, Ferreira AP, Ferreira IM, et al. Mitochondrial localization and structure-based phosphate activation mechanism of Glutaminase C with implications for cancer metabolism. Proc Natl Acad Sci U S A. 2012;109:1092–7.
44. Song H, Peng JS, Yao DS, et al. Serum metabolic profiling of human gastric cancer based on gas chromatography/mass spectrometry. Braz J Med Biol Res. 2012;45:78–85.
45. Ikeda A, Nishiumi S, Shinohara M, et al. Serum metabolomics as a novel diagnostic approach for gastrointestinal cancer. Biomed Chromatogr. 2012;26:548–58.
46. Korangath P, Teo WW, Sadik H, et al. Targeting Glutamine Metabolism in Breast Cancer with Aminooxyacetate. Clin Cancer Res. 2015;21:3263–73.
47. Hassanein M, Qian J, Hoeksema MD, et al. Targeting SLC1a5-mediated glutamine dependence in non-small cell lung cancer. Int J Cancer. 2015;137:1587–97.
48. Budczies J, Pfitzner BM, Gyorffy B, et al. Glutamate enrichment as new diagnostic opportunity in breast cancer. Int J Cancer. 2015;136:1619–28.
49. Kardos GR, Wastyk HC, Robertson GP. Disruption of Proline Synthesis in Melanoma Inhibits Protein Production Mediated by the GCN2 Pathway. Mol Cancer Res. 2015.

# Chapter 6
# Identification of Volatile Organic Compound Biomarkers Associated with Gastric Cancer Cells and Their Ultrasensitive Electrochemical Detection

**Yixia Zhang and Daxiang Cui**

## Abbreviations

| | |
|---|---|
| CNTs | Carbon nanotubes |
| GC/MS | Gas chromatography/mass spectrometer |
| GCE | Glass carbon electrode |
| MWNTs | Multiwalled carbon nanotubes |
| SPME | Headspace solid-phase microextraction |
| VOCs | Volatile organic compounds |

## 6.1 Introduction

Gastric carcinoma is currently the fourth most common cancer in the world and the second most common cancer in China [1]. It is still the second leading cause of cancer-related death in the world [2, 3] and remains difficult to cure effectively, primarily because most patients present with advanced stage [4]. Therefore, how to recognize gastric cancer cells in early stage is a great challenge for early diagnosis and therapy of patients with gastric cancer. We have tried to establish a screening and prewarning system for early gastric cancer since 2005 [5] and tried to find early gastric cancer cells in vivo by multimode targeted imaging techniques [6, 7]. However, our efforts were hampered by a difficult problem, specifically, the lack of specific biomarkers for early gastric cancer. Therefore, to look for specific

Y. Zhang (✉) • D. Cui
Institute of Nano Biomedicine and Engineering, Shanghai Engineering Research Center for Intelligent Diagnosis and Treatment Instrument, National Center for Translational Medicine, Collaborative Innovational Center for System Biology, Shanghai Jiao Tong University, 800 Dongchuan Road, Shanghai 200240, P. R. China
e-mail: zhangyx@sjtu.edu.cn; dxcui@sjtu.edu.cn

© Springer Science+Business Media B.V. and Shanghai Jiao Tong University Press, Shanghai 2017

D. Cui (ed.), *Gastric Cancer Prewarning and Early Diagnosis System*, Translational Medicine Research, DOI 10.1007/978-94-024-0951-2_6

biomarkers associated with early gastric cancer has become our concern. One promising technology was the analysis of volatile metabolites released from cancer cells [8, 9]. Volatile organic compounds (VOCs) released from cancer cell metabolism were considered as very important indicators for biochemical processes occurring in cancer cells [10, 11] and may be capable of prognosticating and diagnosing early-stage cancer. Compared with genomics and proteomics, VOCs embody pathway feedback mechanisms which indicate the potential pathophysiological progress in cancer cells [12, 13]; VOC biomarkers may stand for current progress status of cancer cells; therefore, looking for VOC biomarkers associated with gastric cancer may be helpful to solve early gastric cancer prewarning and diagnosis.

Up to date, human cancers including lung cancer, breast cancer, melanoma, and colon carcinoma have confirmed to exit specific VOC biomarkers [14, 15]. Gas biosensors based on VOC detection have been being developed. For example, Peng et al. reported a method to diagnose early lung cancer by examining lung cancer VOCs' biomarkers using gold nanoparticles [16]. Buszewski et al. reported the correlations of VOCs among *H. pylori* infection, gastric cancer tissues, and expired breath [17, 18]. However, so far few specific VOC biomarkers associated with gastric cancer cells were reported.

Up to now, VOC isolation and identification are mainly done by using GC-MS, which can analyze multiple compounds simultaneously with a high sensitivity. However, GC-MS analysis requires complicated procedures for VOC sample collection. Pre-concentration of sample with SPME is also time-consuming and expensive. To develop a noninvasive, quick, and portable sensor to realize ultrasensitive detection of VOCs has become a hot spot. Electrochemical biosensor has received considerable attention because of its simple, fast response and relatively low cost compared with MS-based methods. Especially, more and more nanomaterials are synthesized increasingly, which provide new chance for enhancing the performance of electrochemical biosensor. Carbon nanotubes (CNTs) are a class of stiff, stable, and hollow nanomaterials with many unique properties such as mechanical, physical, and chemical properties and have been being broadly used for biosensor development [19, 20]. In particular, their fast responsibility, high selectivity, and reversibility have made CNTs as one kind excellent electrochemical sensing material. Many studies show that carbon-based metal nanocomposites have excellent electrical properties superior to pure carbon or metal species [21, 22]. For example, Espinosa et al. [23] reported that multiwalled carbon nanotubes (MWNTs) covered with gold or silver nanoclusters as gas-sensitive materials could improve the selectivity of sensors, which highly suggest that nanocomposites of CNTs/nanoparticles could be used to enhance the performance of electrochemical biosensors. Metal nanoparticles, especially bimetallic nanoparticles, own synergistic catalytic abilities [24, 25], exhibiting great potential in applications such as electrochemical biosensors. It is possible that gold-containing bimetallic nanoparticles may enhance the catalytic activity and selectivity of electrochemical reaction [26]. Few reports are associated with the electrochemical sensors for detection of volatile biomarkers of gastric cancer cells.

Herein, we reported some potential volatile organic compound biomarkers associated with gastric cancer cells that were identified by GC-MS analysis, the sample preparation method which was optimized for measurements of VOCs by HS-SPME-GC/MS. Among these VOC markers, we selected two kinds of VOC markers such as 3-octanone and butanone as target molecules, prepared MCNTs covered with Au-Ag composites as a sensing film, and established their ultrasensitive electrochemical detection system. Identified VOCs associated with gastric cancer cells and established electrochemical assay system may have great potential in applications such as early diagnosis and prognosis of gastric cancer in the near future.

## 6.2 Identification of VOC Biomarkers Associated with Gastric Cancer Cells

Pre-concentration sampling time may affect the patterns of chromatogram of VOCs. As shown in Fig. 6.1a, optimal pre-concentration sampling time was 45 min (HS-SPME), which was strongly preferred to perform GC-MS analysis in our study. As shown in Fig. 6.1b, cell culture flasks protected with Parafilm can substantially improve the profile of chromatograms of volatile metabolites. The optimal cell culture time was 22 h. With the optimized conditions, eight volatile compounds were identified as shown in Table 6.1.

Figure 6.2 showed the patterns of VOC chromatogram among MGC-803 cells, GES-1 cells, and control. Two of them, peak2 (Fig. 6.2a) and peak6 (Fig. 6.2b), only presented in the headspace of MGC803 cells but absent in that of GES-1 cells; further identification analysis showed that they were compounds 3-octanone and butanone, respectively. In addition, three compounds, formic acid propyl ester (peak1), 1.4-butanediol (peak3), and dodecane, 2, 6, 11-trimethyl (peak8), only existed in the headspace of GES-1 cells medium.

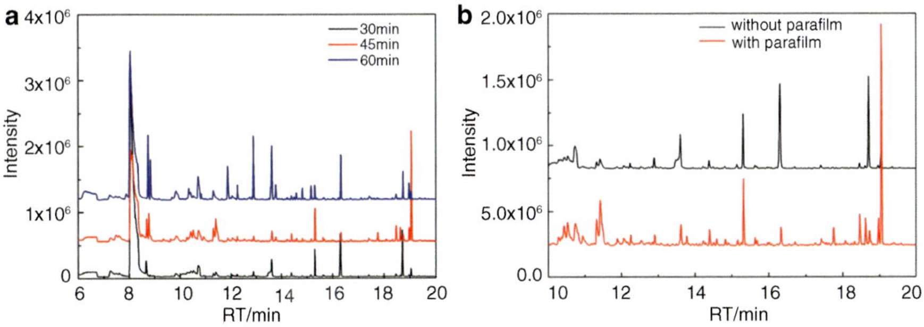


**Fig. 6.1** The patterns of chromatogram of VOCs from gastric cancer cell MGC803 influenced by (**a**) different headspace extraction time; (**b**) MGC-803 cells cultured with and without Parafilm

**Table 6.1** Summary of different VOCs between MGC-803 cell and GES-1 cell

| Peak | RT | Compounds | GES-1 | MGC-803 |
|---|---|---|---|---|
| 1 | 9.1 | Formic acid propyl ester | + | – |
| 2 | 10.4 | 3-Octanone | – | + |
| 3 | 10.8 | 1.4-butanediol | + | – |
| 4 | 11.6 | 4-Isopropoxybutanol | ++ | + |
| 5 | 12.75 | Nonanal | ++ | + |
| 6 | 18.08 | Butanone | – | + |
| 7 | 18.32 | 1-Butanol,4-butoxy | ++ | + |
| 8 | 18.84 | Dodecane, 2,6,11-trimethyl | + | – |

Notes: + present VOCs; – absent VOCs; different numbers of + stands for different concentrations

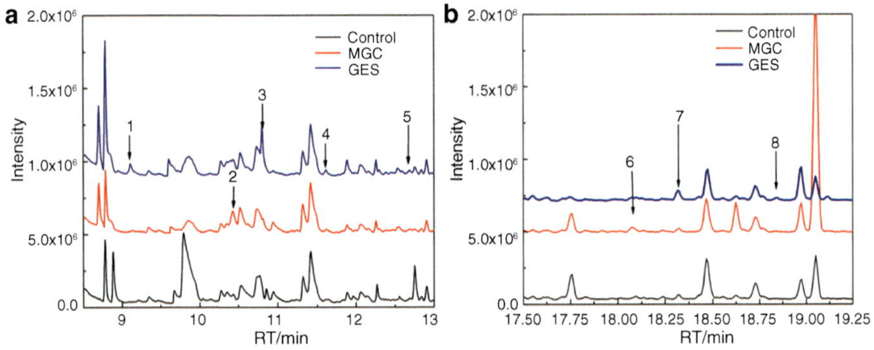


**Fig. 6.2** Comparison of patterns of VOC chromatogram in MGC-803 cell, GES-1 cell, and medium only: (**a**) the retention time of chromatogram at a range of 8–13 min, (**b**) the retention time of chromatogram at a range of 17–20 min

The amounts of volatile compounds such as 4-Isopropoxybutanol (peak4), nonanal (peak5), and 1-butanol, 4-butoxy (peak7) in the headspace of MGC-803 cells are higher than that in GES-1 cells. The ratio of relative peak area of the three compounds between GES-1 cells and MGC-803 cells were, respectively, 4-isopropoxybutanol $\leq$0.31, nonanal $\leq$0.36, and 1-butanol (4-butoxy) $\leq$0.40.

Besides the VOC biomarkers mentioned above, we also found some unknown concrete component biomarkers; data was not shown here.

## 6.3 Preparation of MWNTs/Au-Ag Modifying GCE

As shown in Fig. 6.3a, the surface of MWNTs was covered with Au-Ag nanoparticles; Fig. 6.3b furtherly confirmed that those nanoparticles on the surface of MWNTs were composed of carbon, gold, and silver elements, which highly suggest

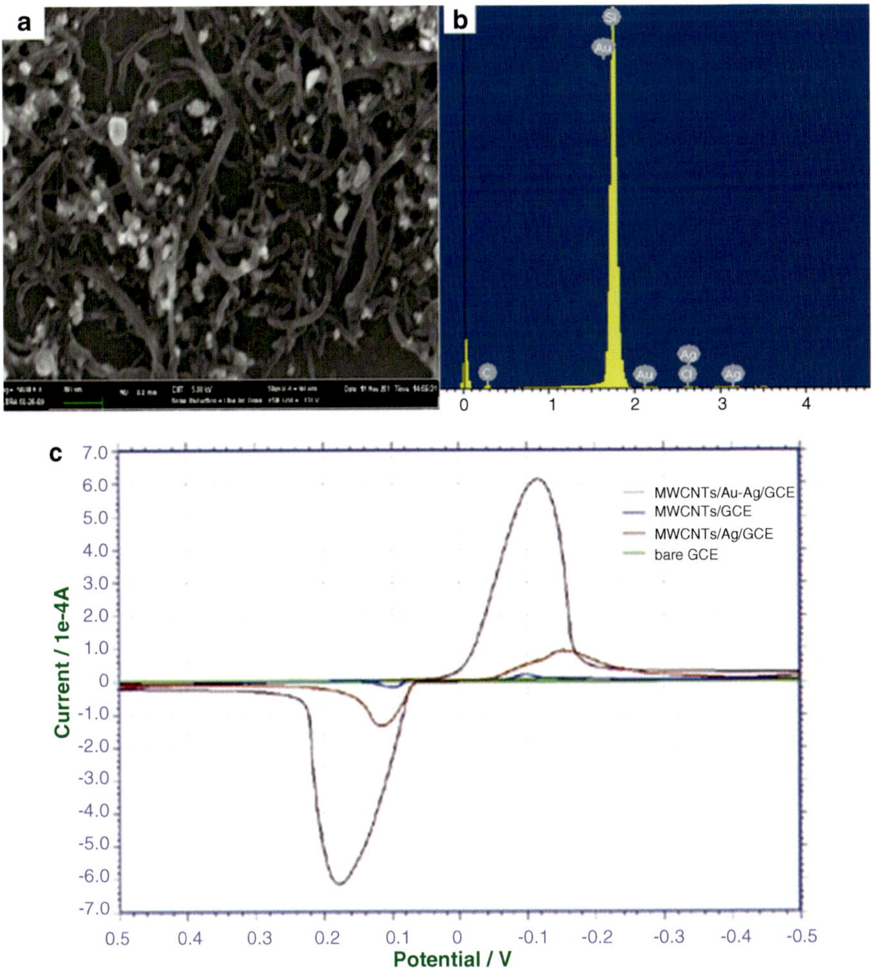


**Fig. 6.3** Characterization of MWNTs covered with Au-Ag nanoparticles. (**a**) The SEM image of MWNTs covered with Au-Ag nanoparticles, (**b**) the corresponding EDX spectra of MWNTs/Au-Ag nanoparticles, (**c**) cyclic voltammograms of MWNTs/Au-Ag/GCE, MWNTs/GCE, MWNTs/Ag/GCE, and bare GCE in 0.1 M KCl solution

that Au-Ag nanoparticles were successfully loaded onto the surface of MWNTs. Figure 6.3c showed the cyclic voltammograms of MWNTs/Au-Ag/GCE, MWNTs/GCE, Ag/MWNTs/GCE, and bare GCE in 0.1 M KCl solution; a pair of well-defined redox peaks was observed (−116 and 183 mv) on MWNTs/Au-Ag/GCE and displayed maximal detection signal, which highly suggested that MWNTs covered with Au-Ag nanoparticles as sensing film exhibited evident advantage over other electrodes.

## 6.4 Electrochemical Sensing System and Electrochemical Test

In order to detect the VOC biomarkers associated with gastric cancer cells, we set up an electrochemical sensing system based on MWNTs/Au-Ag/GCE; its schematic was shown in Fig. 6.4, which was mainly composed of counter electrode Pt, work electrode, and magnetic blender. Working electrode surface was shown in no. 5, composed of thin film covered with Au-Ag nanoparticles.

We selected two volatile biomarkers such as 3-octanone and butanone as detection target molecules and then finished their electrochemical test. The cathodic peak currents of modified GCE were directly proportional to the scan rates at a range of 20–100 mv (Fig. 6.5b, $I_p = -3.5035x + 1.4451$, $R = 0.9989$). The CV profiles of working electrode were recorded, when it was exposed to various concentrations of butanone (1 μL–65 μL). As the dose of butanone was varied, a gradual descend of the anodic peak current ($I_p$) was observed with a negative shift of potential (Fig. 6.5c). The oxidation peak current is linear with the concentrations of butanone from 0 to 0.055% (v/v) (Fig. 6.5d, $I_p = -9.8886x + 4.4592$, $R = -0.9986$). Meanwhile, as shown in Fig. 6.5e, the catalyzed oxidation of 3-octanone was carried out. Current response ($I_p$) exhibited a good linear relationship with the concentrations of 3-octanone in the range of 0–0.0025% (v/v) (Fig. 6.5f: $I_p = -1.8889x + 3.09$, $R = -0.9956$). Those results mentioned above highly indicated that butanone and 3-octanone were electrochemically active and adsorption-controlled process on the modified electrode. Meanwhile, the data suggested that the sensitivity of this electrochemical assay system for detecting butanone is 5.2 times higher than that of 3-octanone.

Regarding the other VOC interference or influence on the test results of 3-octanone and butanone, we used $CO_2$ to fill the work electrode surface; we did not

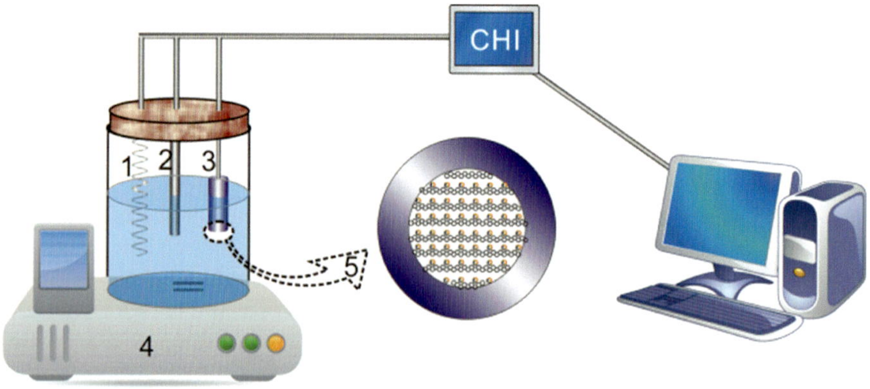


**Fig. 6.4** The schematic diagram of electrochemical sensing system. *1* Counter electrode Pt; *2* reference electrode SCE; *3* work electrode, modified glass carbon electrode; *4* magnetic blender; *5* the construction of the surface of working electrode. Copyright permission from Zhang et al. [31]

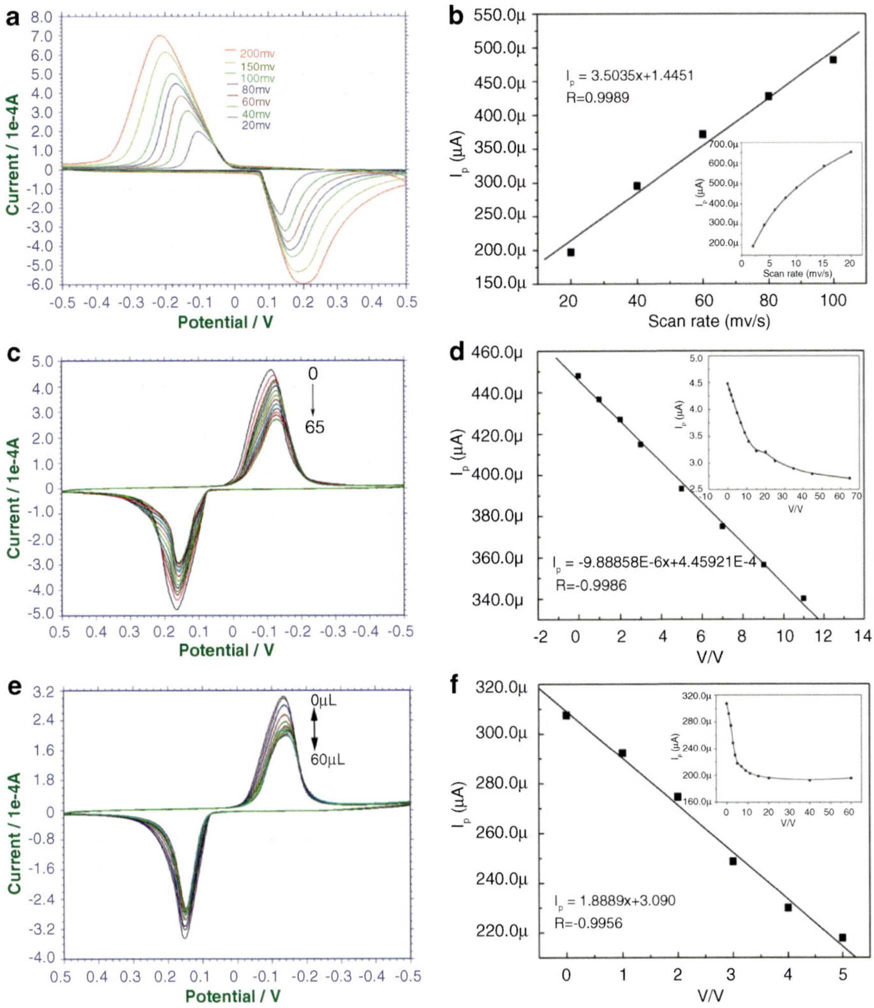

**Fig. 6.5** (**a**) Cyclic voltammograms at MWNTs/Au-Ag/GCE electrode in 0.1 M KCl with various scan rates. (**b**) Displays the linear plot of between the anodic peak current and the scan rate. (**c**) Cyclic voltammograms of MWNTs/Au-Ag/GCE was exposed to butanone. At a range of 0–65 μL in 20 mL 0.1 M KCl solution at the scan rate of 50 mv/s. Each data point was obtained from five replicate measurements. (**d**) Shows the linear calibration plot of the peak current against the concentration of butanone. The inset is the line plot picture of the peak current (*Ip*) against 0 μL–65 μL of butanone. (**e**) Reveals the typical cyclic voltammograms plots upon the addition of varying amounts of 3-octanone. All potentials are given versus SCE. (**f**) Display the calibration plots of the anodic peak current response versus concentration of 3-octanone. Copyright permission from Zhang et al. [31]

observe the obvious changes of peak currents of 3-octanone and butanone, which highly suggest that this electrochemical sensing system based on MWNTs/Au-Ag/GCE owns good selective specificity. Further work to improve the detection specificity is under way.

## 6.5 Significance and Potential Mechanism

We firstly identified VOC biomarkers to distinguish gastric cancer MGC803 cells from gastric mucous GES-1 cells by HS-SPME/GC-MS. We also optimized the method of cell sample preparation and pre-concentration conditions for GC-MS analysis. It is well known that most volatile biomarkers were intermediate products of cell metabolisms; the cell culture period has significant influence on the profile of cell VOC chromatograms. Because VOCs are volatile with very low concentrations, the cell culture flask protected with Parafilm can avoid omission of volatile biomarkers and environmental interferences. Therefore, we optimized the culture time and culture condition. Under optimal condition, with the same headspace-to-liquid volume ratio, the magnitude of the extracting effect of VOCs was bigger than that with lower headspace volume. According to the protocol reported in paper [28], 10 mL cell medium was introduced into 20 mL headspace vials for HS-SPME, which obtained good results. Furthermore, extracting time is also an important factor. Shorter sampling time (30 min) causes the lower extractable absorption of analytes, as a result of the reduced sensitivity of SPME. Longer sampling time (60 min) enables competitive absorption, which could result in lower efficiency of extracting compounds and some valuable biomarkers may be lost. Finally, 45 min HS-SPME sampling time was strongly preferred in this work to achieve the good sensitivity.

We also observed that the amount of alcohols or aldehydes in VOCs from MGC-803 cells was higher than VOCs from GES-1 cells. This phenomenon should be attributed to higher activity of aldehyde dehydrogenase in cancer cells [29, 30]. Alcohols or aldehydes were oxidized to corresponding carboxylic acids in an alcohol-dehydrogenase-independent pathway; carboxylic acids were further metabolized as precursors of synthesis cell membrane, which provide more substances to meet with the rapid growth requirement of cancer cells.

With the aim of investigating the feasibility of detecting the VOC biomarkers associated with gastric cancer cells, we developed an electrochemical assay based on nanocomposites of MWNTs/Au-Ag/GCE. By comparison the amperometric response of MWNTs/Au-Ag/GCE, MWNTs/Ag/GCE, MWNTs/GCE, and bare GCE and the peak current of MWNTs/Au-Ag/GCE ($E_p = -116$ mv, $I_p = 602.9$ μA) is nearly 26 times stronger than MWNTs/Ag/GCE ($E_p = -176$ mv, $I_p = 23.21$ μA). Moreover, the peak potential exhibited a positive shift of 60 mv (vs SCE). MWNTs, gold or silver nanoparticles could substantially increase the surface area of the electrode, so the response current of modified electrode is stronger than that of the bare electrode. The better electronic conductive characteristics of working electrode were attributed to the synergistic effect between MWNTs and Au-Ag nanoparticles. The peak current increased linearly with the scan rate by using CV, which indicated that butanone and 3-octanone were electrochemically active and adsorption-controlled process on the modified electrode. We also used $CO_2$ as interference gas to fill the work electrode surface; we did not observe the obvious changes of peak currents of 3-octanone and butanone, which highly suggest that this electrochemical sensing system based on MWNTs/Au-Ag/GCE owns good selective specificity. Further work to improve the detection specificity is under way.

In summary, we firstly identified VOC biomarkers to distinguish gastric cancer MGC803 cells from gastric mucous GES-1 cells by HS-SPME/GC-MS. It was established that an electrochemical sensing system based on MWNTs/Au-Ag/GCE, which can be used for ultrasensitive detection of volatile biomarkers associated with gastric cancer cells at room temperature. The specific volatile biomarkers of gastric cancer cells and the well-adapted electrochemical system have great potential in applications such as screening and prewarning of early gastric cancer in near future.

## References

1. Jemal A, Siegel R, Ward E, Hao Y, Xu J, Murray T, et al. Cancer statistics, 2008. CA Cancer J Clin. 2008;58:71–96.
2. Bondy M. Cancer epidemiology and prevention. JAMA. 2009;301:1074.
3. Xu AG, Li SG, Liu JH, Gan AH. Function of apoptosis and expression of the proteins Bcl-2, p53 and C-myc in the development of gastric cancer. World J Gastroenterol. 2001;7:403–6.
4. Wang K, Ruan J, Qian Q, Song H, Bao C, Zhang X, et al. BRCAA1 monoclonal antibody conjugated fluorescent magnetic nanoparticles for in vivo targeted magnetofluorescent imaging of gastric cancer. J Nanobiotechnology. 2011;9:23.
5. Cui DX, Zhang L, Yan XJ, Zhang LX, Xu JR, Guo YH, et al. A microarray-based gastric carcinoma prewarning system. World J Gastroenterol. 2005;11:1273–82.
6. Kong Y, Chen J, Gao F, Li W, Xu X, Pandoli O, et al. A multifunctional ribonuclease-A-conjugated CdTe quantum dot cluster nanosystem for synchronous cancer imaging and therapy. Small. 2010;6:2367–73.
7. Gao G, Wu HX, Gao WJ, Zhang YX, Huang P, Cui DX. Preparation of FeCO3-Fe3O4 nanoparticles and flower-like assemblies via a one-step hydrothermal method. Crystengcomm. 2011;13:6950–4. doi:10.1039/c1ce05994g.
8. Miekisch W, Schubert JK, Noeldge-Schomburg GFE. Diagnostic potential of breath analysis–focus on volatile organic compounds. Clin Chim Acta. 2004;347:25–39.
9. Dubowski KM. Breath analysis as a technique in clinical chemistry. Clin Chem. 1974;20:966–72.
10. Burke D, Halpern B, Malegan D, McCairns E, Danks D, Schlesinger P, et al. Profiles of urinary volatiles from metabolic disorders characterized by unusual odors. Clin Chem. 1983;29:1834.
11. Amann A, Poupart G, Telser S, Ledochowski M, Schmid A, Mechtcheriakov S. Applications of breath gas analysis in medicine. Int J Mass Spectrom. 2004;239:227–33.
12. Nicholson JK, Wilson ID. Understanding 'global' systems biology: metabonomics and the continuum of metabolism. Nat Rev Drug Discov. 2003;2:668–76.
13. Modak AS. Breath tests for evaluating enzyme activity in personalized medicine. Drug Metab Rev. 2007;39:95–9.
14. Phillips M, Cataneo RN, Saunders C, Hope P, Schmitt P, Wai J. Volatile biomarkers in the breath of women with breast cancer. J Breath Res. 2010;4:026003.
15. Peng G, Hakim M, Broza YY, Billan S, Abdah-Bortnyak R, Kuten A, et al. Detection of lung, breast, colorectal, and prostate cancers from exhaled breath using a single array of nanosensors. Br J Cancer. 2010;103(4):542–51.
16. Peng G, Tisch U, Adams O, Hakim M, Shehada N, Broza YY, et al. Diagnosing lung cancer in exhaled breath using gold nanoparticles. Nat Nanotechnol. 2009;4:669–73.
17. Buszewski B, Ulanowska A, Ligor T, Jackowski M, Klodzinska E, Szeliga J. Identification of volatile organic compounds secreted from cancer tissues and bacterial cultures. J Chromatogr B Analyt Technol Biomed Life Sci. 2008;868:88–94.

18. Ligor T, Szeliga J, Jackowski M, Buszewski B. Preliminary study of volatile organic compounds from breath and stomach tissue by means of solid phase microextraction and gas chromatography mass spectrometry. J Breath Res. 2007;1:016001.
19. Compton OC, Nguyen SBT. Graphene oxide, highly reduced graphene oxide, and graphene: versatile building blocks for carbon-based materials. Small. 2010;6:711–23.
20. Zhao X, Zhang B, Ai K, Zhang G, Cao L, Liu X, et al. Monitoring catalytic degradation of dye molecules on silver-coated ZnO nanowire arrays by surface-enhanced Raman spectroscopy. J Mater Chem. 2009;19:5547–53.
21. Ma Y, Ali SR, Dodoo AS, He H. Enhanced sensitivity for biosensors: multiple functions of DNA-wrapped single-walled carbon nanotubes in self-doped polyaniline nanocomposites. J Phys Chem B. 2006;110:16359–65.
22. Zhao Q, Nardelli MB, Lu W, Bernholc J. Carbon nanotube-metal cluster composites: a new road to chemical sensors? Nano Lett. 2005;5:847–51.
23. Espinosa E, Ionescu R, Bittencourt C, Felten A, Erni R, Van Tendeloo G, et al. Metal-decorated multi-wall carbon nanotubes for low temperature gas sensing. Thin Solid Films. 2007;515:8322–7.
24. Mallin MP, Murphy CJ. Solution-phase synthesis of sub-10 nm Au-Ag alloy nanoparticles. Nano Lett. 2002;2:1235–7.
25. Shibata T, Bunker BA, Zhang Z, Meisel D, Vardeman II CF, Gezelter JD. Size-dependent spontaneous alloying of Au-Ag nanoparticles. J Am Chem Soc. 2002;124:11989–96.
26. Valden M, Lai X, Goodman DW. Onset of catalytic activity of gold clusters on titania with the appearance of nonmetallic properties. Science. 1998;281:1647.
27. Schomburg I, Hofmann O, Baensch C, Chang A, Schomburg D. Enzyme data and metabolic information: BRENDA, a resource for research in biology, biochemistry, and medicine. Gene Funct Dis. 2000;3:109–18.
28. Grote C, Pawliszyn J. Solid-phase microextraction for the analysis of human breath. Anal Chem. 1997;69:587–96.
29. Shin HW, Umber BJ, Meinardi S, Leu SY, Zaldivar F, Blake DR, et al. Acetaldehyde and hexanaldehyde from cultured white cells. J Transl Med. 2009;7:1–11.
30. Patel M, Lu L, Zander DS, Sreerama L, Coco D, Moreb JS. ALDH1A1 and ALDH3A1 expression in lung cancers: correlation with histologic type and potential precursors. Lung Cancer. 2008;59:340–9
31. Zhang YX, Gao G, Liu HJ, Fu HL, Fan J, Wang K, Chen YS, Li BJ, Zhang CL, Zhi X, He L, Cui D. Theranostics 2014;4(2):154–62. doi:10.7150/thno.7560

# Chapter 7
# A Breath Analysis Based on SERS Sensor to Distinguish Patients with Early and Advanced Stages of Gastric Cancer from Healthy People

**Yunsheng Chen and Daxiang Cui**

## 7.1 Introduction

Gastric cancer (GC) is one kind of common tumor affecting about one million people per year, and the mortality ranks the second among all tumors [1]. The highest incidence is in Northeast Asia, especially in China with 42 % of worldwide cases [2]. Histologically, GC could be typically classified into early-stage gastric cancer (EGC) and advanced-stage gastric cancer (AGC). EGC is defined as that in which tumor invasion is confined to the mucosa or submucosa, irrespective of the presence of regional lymph node metastasis, and AGC is defined as the cancer with invasion to the muscularis propria or deeper gastric wall [3]. As with other cancers, GC prognosis is related to stage: 5-year survival rate is more than 62.8 % for EGC but below 24 % for AGC [4]. Besides early-stage diagnosis, an accurate preoperative staging also can reduce the mortality [5]. However, due to the lack of defined risk factors, GC is facing more challenges in its staging system, such as time-bulky, problematic diagnosis, and great pain for patients by gastroscopy or biopsy [2a, 6]. Unfortunately, delayed diagnosis and inadequacies of staging system lead to an increased mortality [7]. Therefore, it is very pressing to develop a rapid and noninvasive method that is capable of realizing not only to early diagnose but also to stage GC. Recently, various genetic and protein biomarkers in blood or urine have been developed, but their applicability is limited by suboptimal sensitivity or specificity, expensive cost, or lack of validation in large-scale clinical trials [8]. At this point, breath analysis provides an alternative method to solve current problem.

Y. Chen (✉) • D. Cui
Institute of Nano Biomedicine and Engineering, Shanghai Engineering Research Center for Intelligent Diagnosis and Treatment Instrument, National Center for Translational Medicine, Collaborative Innovational Center for System Biology, Shanghai Jiao Tong University, 800 Dongchuan Road, Shanghai 200240, Peoples Republic of China
e-mail: yunshengchen@sjtu.edu.cn; dxcui@sjtu.edu.cn

© Springer Science+Business Media B.V. and
Shanghai Jiao Tong University Press, Shanghai 2017

D. Cui (ed.), *Gastric Cancer Prewarning and Early Diagnosis System*,
Translational Medicine Research, DOI 10.1007/978-94-024-0951-2_7

The ancient Greek physicians knew that the specific odors in breath were associated with certain diseases [9]. Since 1971, breath has been for the first time confirmed as a complex mixture containing around 3000 volatile organic compounds (VOCs) [10]. The special VOCs are generated by cancer cells or tissues with changes of metabolism or oxidative stress; therefore, they can serve as cancer VOC biomarkers without affected by clinical symptoms [11]. As different VOC biomarker patterns can reflect different pathological states, breath analysis becomes a noninvasive diagnostic tool by recognizing the patterns of biomarker. Up to now, it has been focused on prewarning and evaluating various cancers, such as lung cancer [10a] and GC [12]. However, traditional breath analysis based on mass spectrometry (MS)-based approaches faces a great challenge during the clinical applications, such as expensive equipment, professional staffs, and long-time procedures [9]. Recently, the simpler and cheaper sensors have become the novel breath analysis approaches. The premise with the sensors is that selectively adsorbed VOC biomarkers can cause a sensitive response in the conductivity [13], mass [14], or color [15]. Carbon-based solid-phase microextraction (SPME) fibers are widely used as the adsorption matrix in identifying biomarkers, meaning that carbon materials can provide the same selectivity for the verified biomarkers [16]. In our previous works, an electrochemical sensor modified by carbon nanotubes was developed to detect VOCs of GC cells [17]. However, human breath is much more complex, so it is necessary to develop a more efficient sensor. Among various carbon materials, reduced graphene oxide (RGO, referred to as graphene) is an excellent matrix for selectively immobilizing organic molecules [18]. Meanwhile, surface-enhanced Raman scattering (SERS) is the best candidate to detect the trace biomarkers because it can detect almost any adsorbate down to a single molecule [19]. Importantly, biomarkers generate their specific Raman signatures, so they can be distinguished as different patterns by their Raman signatures, such as fingerprints. Therefore, SERS can sensitively recognize the patterns of biomarkers without sensor array construction as reported papers [11]. Fortunately, RGO can be competent as an attractive SERS sensor matrix for two reasons [20]. First, it has a clear Raman-enhancing effect due to its fluorescence quenching [21]. Second, it is apt at fabrication a film nanostructure, which is well accepted for SERS detection [22]. Therefore, RGO-based SERS sensor is our expectation.

To obtain an excellent enhancement efficiency, Au nanoparticles (AuNPs) should be densely and well dispersed on RGO based on electromagnetic mechanism (EM) [23]. However, the formed AuNPs were coated by various stabilizers, such as surfactants and polymers, to avoid the aggregation on RGO in reported papers [24]. Unfortunately, the Raman signals of the organic stabilizers will greatly interfere with the ones of biomarkers. Therefore, it is necessary to develop a clean SERS sensor by in situ synthesis of AuNPs on RGO for breath analysis, which has not been reported before but has been overcome in our works.

Herein, 14 VOC biomarkers in breath have been identified to distinguish the patients with EGC and AGC from healthy people by GC-MS coupled with SPME. Meanwhile, a breath analysis approach based on a clean SERS sensor has also been developed to recognize the different VOC biomarker patterns both in

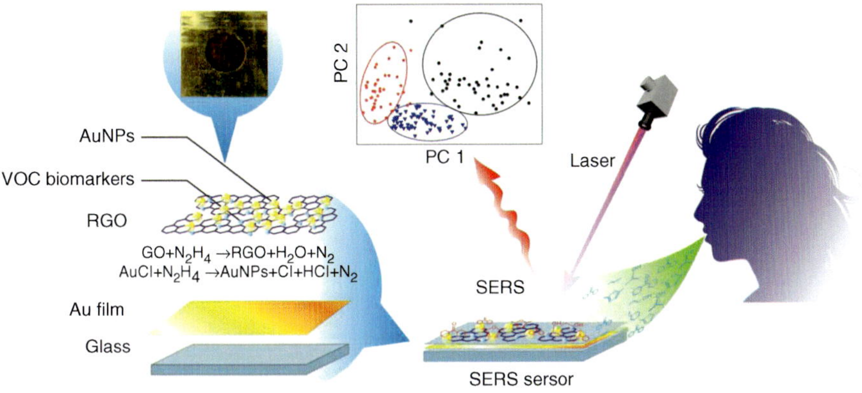


**Scheme 7.1** Schematic diagrams of SERS sensor and overview of the processes involved in breath. Copyright permission from Chen et al. [38]

simulated and real breath samples. Schematic diagrams of SERS sensor and breath analysis are shown in Scheme 7.1: RGO uniformly spreads onto Au film and selectively adsorbs the VOC biomarkers. AuNPs without organic stabilizing agents are facilely densely and well dispersed on RGO, endowing SERS sensor with an effective detection of adsorbed biomarkers. Hydrazine vapor adsorbed in graphene oxide (GO) plays as the reductant in two chemical reactions: the reduction of GO to RGO at first and then the in situ formation of AuNPs on RGO which provided reaction sites and worked as stabilizing agent. The clean SERS sensor can greatly insure the selectivity and sensitivity of the identified VOC biomarkers. The approach described in this paper has successfully analyzed breath samples with simple operation, rapid detection, low cost, and reliability. The mentioned VOC biomarkers and breath analysis approach in this work can not only diagnose GC but also distinguish EGC and AGC.

## 7.2 The Identification of Biomarkers

We identified the VOC biomarkers for GC in breath samples and determined their relative concentrations by GC-MS coupled with SPME at first. Figure 7.1a was the representative chromatograms illustrating different VOCs in the different breath samples. Although tens of peaks appeared in the chromatogram, several huge peaks attributed to the interference of baseline and chromatogram column. After analysis, 14 peaks with enough intensity displayed significant differences in different breath samples, so they were verified as VOC biomarkers. Forward stepwise discriminant analysis identified 14 VOC biomarkers corresponding to the peaks (Fig. 7.1b). Isoprene (peak 3) and menthol (peak 12) just appeared in at least 80 % of health breath samples but barely existed in EGC breath samples. However, the pivalic acid (peak 9) had an opposite result, that is, it just appeared in at least 80 % of GC breath

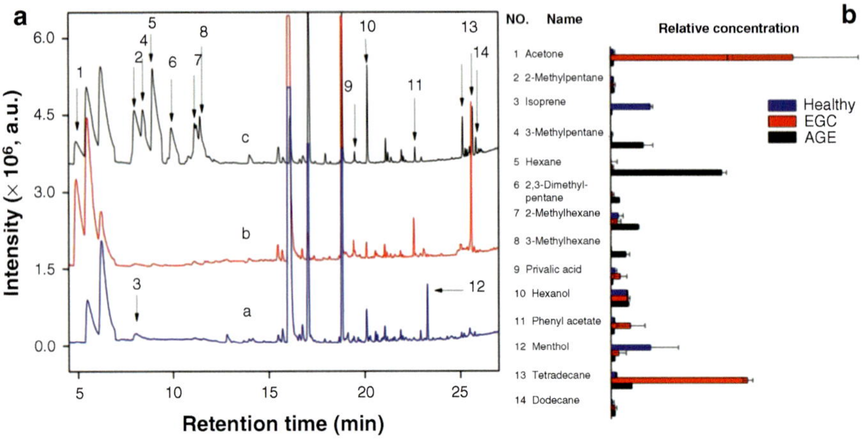

**Fig. 7.1** (**a**) The representative chromatogram illustrating different volatiles in the breath from the healthy people (a), the patients with EGC (b) and AGC (c). Peaks exceeding the range of scale were cut off. (**b**) The identified VOC biomarker patterns of the healthy people, the patients with EGC and AGC. The peak numbers correspond to **a**. The relative concentrations were calculated by dividing the peak areas of other compounds and the one of hexanol. Copyright permission from Chen et al. [38]

samples. On the concentrations of acetone (peak 1) and tetradecane (peak 13), at least 70 % of EGC samples were more than the ones in AGC samples. 2-Methylpentane (peak 2), 3-methylpentane (peak 4), hexane (peak 5), 2,3-dimethylpentane (peak 6), 2-methylhexane (peak 7), 2-methylhexane (peak 8), and dodecane (peak 14) were detected in at least 65 % of AGC samples. The 14 VOC biomarkers could be classified into saturated and unsaturated hydrocarbons and oxygen-containing compounds. Saturated hydrocarbons were the end products of lipid peroxidation, which was responsible for damage of tissues by cancer [25]. It was the reason why the most saturated hydrocarbons only existed in the breath of GC penitents. Unsaturated hydrocarbon, isoprene, was one of the most abundant hydrocarbons in breath and was formed by a normal enzymatic or nonenzymatic pathway. The pathway could be damaged in the GC tissue due to the reactive oxygen species (ROS). So it could explain why isoprene had a high concentration in healthy people but a very low concentration in EGC penitent and could not be found in AGC penitent. Hydrocarbons showed a low solubility in the blood and hence were excreted in the breath within minutes after their formation; thus they were considered as the most reliable cancer VOC biomarkers. Oxygen-containing compounds, including acetone, pivalic acid, hexanol, phenyl acetate, and menthol, were found in the breath. Acetone was produced by decarboxylation of acetoacetate which was derived from lipolysis [25]. For the penitents with EGC, due to changes in metabolic states, an increase in the rate of fatty acid oxidation resulted in the formation of acetone. During GC progression, the obviously decreased concentration of acetone attributed to the loss of the most fatty acid in AGC penitent. Therefore, the high concentration of acetone in EGC and a low concentration in AGC were also reliable. Hexanol and menthol were the products of hydrocarbon metabolism [25]. However, their metabolism was prone to be affected by confounding factors in the body. The biochemical pathways of pivalic acid and phenyl acetate were still unknown. The VOC biomarker patterns were important criterions to distinguish the patients with EGC and AGC from healthy people, varying widely among different people and genders. In this work, hexanol (peak 10) could be detected in more than 80 % breath samples with a similar concentration. Therefore, it can act as the internal standard to simplify the result of patterns of VOC biomarkers. The relative concentrations calculated by dividing the peak areas of other compounds and the one of hexanol were used to distinguish the patients with EGC and AGC from the healthy people. The results were also shown in Fig. 7.1b; the relative concentrations of VOC biomarkers differ dozens of times. In a word, the 14 VOC biomarkers can be selectivity adsorbed by carbon materials, and three VOC biomarker patterns could feasibly and reliably describe a big difference between the patients with EGC and AGC from the healthy people. As a result, we turned our attention to develop a breath analysis approach to detect these patterns using the RGO-based SERS sensor.

## 7.3 Characterization of the SERS Sensor

The chemical reactions in preparing SERS sensor were verified by X-ray photoelectron spectroscopy (XPS) and UV–Vis spectroscopy. The XPS spectra of the GO film, RGO film, and prepared SERS sensor corresponded to curve *a*, *b*, and *c* in Fig. 7.2a, respectively. An N1s peak at 399.6 eV that originated from the adsorbed molecule hydrazine was observed in curve *b*, indicating that RGO had adsorbed hydrazine from vapor with an excellent adsorbability [26]. The N1s peak disappeared in curve *c*, because hydrazine was consumed to in situ formed AuNPs and non-adsorbed $N_2$. At the same time, the C1s spectra of GO (Fig. 7.2c), RGO (Fig. 7.2d), and RGO/AuNPs (Fig. 7.2e) were fit into four main peaks: C–C, C–O, O–C=O, and C=O. The increase in C/O atomic ratio from 1.90 in GO to 4.31 in RGO and RGO/AuNPs indicates that some oxygen-containing functional groups were removed after the reduction. Furthermore, in Fig. 7.2d, no peak corresponding to the C–N bonds (285.54 eV) could be found, indicating that hydrazine was adsorbed with a non-covalent interaction. Therefore, the hydrazine could keep enough reducing capacity to react with $AuCl_4^-$. The same conclusion was further verified by UV–Vis spectroscopy (Fig. 7.2b). The absorption peak at 233 nm in curve *a* was redshifted toward 262 nm in curves *b* and *c* after reaction, indicating that GO had been reduced by hydrazine vapor. Compared with curve *b*, curve *c* with the same peak at 262 nm and a new peak at 561 nm indicated that AuNPs had been formed in RGO film. Moreover, the obvious redshift of the plasmon band of AuNPs from 540 nm in reported to 561 nm indicated that AuNPs adhered on RGO, which was in consistence with the electron microscope in the following [27].

Scanning electron microscope (SEM) and transmission electron microscope (TEM) images described the SERS sensor surface. GO could form a uniform film

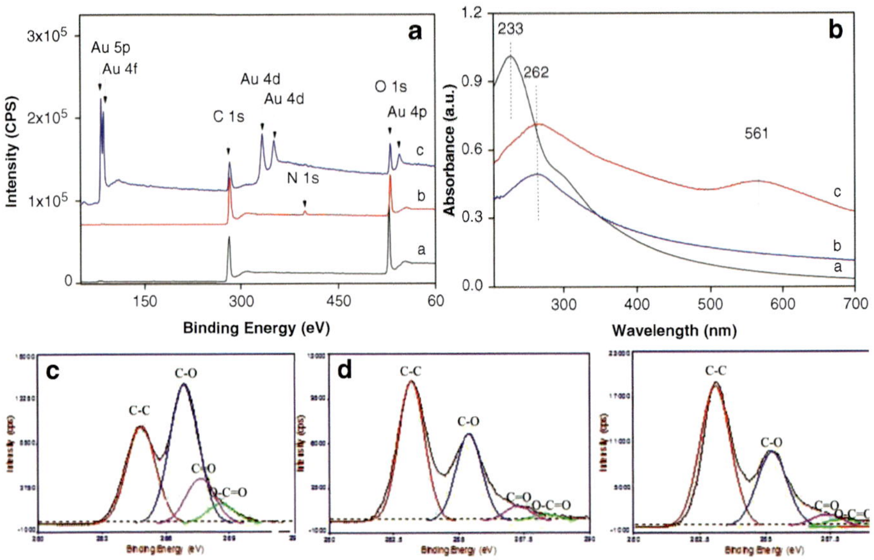


**Fig. 7.2** (**a**) XPS spectrum of GO film (curve *a*), RGO film (curve *b*), and prepared SERS sensor (curve *c*). (**c**) The C1s XPS spectra for GO film. (**d**) The C1s XPS spectra for RGO film. (**e**) The C1s XPS spectra for prepared SERS sensor. (**b**) UV–Vis spectrum of GO (curve *a*), RGO (curve *b*), and prepared SERS sensor (curve *c*). Copyright permission from Chen et al. [38]

from lamellar structures with typical wrinkles of GO both in the images of SEM (Fig. 7.3a) and TEM (Fig. 7.3c). After reactions, the prepared SERS sensor was also observed by SEM (Fig. 7.3b) and TEM (Fig. 7.3d). AuNPs with a diameter range of 40–50 nm were densely and well dispersed on RGO both in SEM and TEM images. This was because the structure could attribute to RGO evenly adsorbing hydrazine vapor, providing the nucleation and binding sites for every AuNPs. Besides, two points should be noteworthy. First, the wrinkles of RGO appeared around the AuNPs, which meant they can provide a large surface site for selectively adsorbing the biomarkers (insert of Fig. 7.3c). In general, the closer the distance between biomarkers to AuNPs, the stronger the strength of their SERS response. Second, RGO provided the reaction site and worked as stabilizing agent for AuNP formation. These well-dispersed AuNPs as an array field were in favor of improving the enhancement in SERS [28]. In addition, high-resolution TEM image showed lattice fringes of the AuNPs that matched the lattice structure of crystal Au (111) (insert of Fig. 7.3d). In a word, the prepared SERS sensor showed the high biomarker-adsorbed and SERS-enhanced nanostructures by morphological observations.

Generally, more RGO could adsorb and enrich more biomarkers from breath, and more AuNPs on RGO could create more "hot spot" with a higher enhancement ability [29]. Therefore, thermal gravimetric analysis (TGA) and Brunauer-Emmett-Teller (BET) were performed to evaluate the AuNP percentage and influence on adsorbability. In Fig. 7.4a, the TGA curves of RGO film (curve a) and prepared SERS sensor (curve b) exhibited weight loss between 25 and 100 °C, which was due

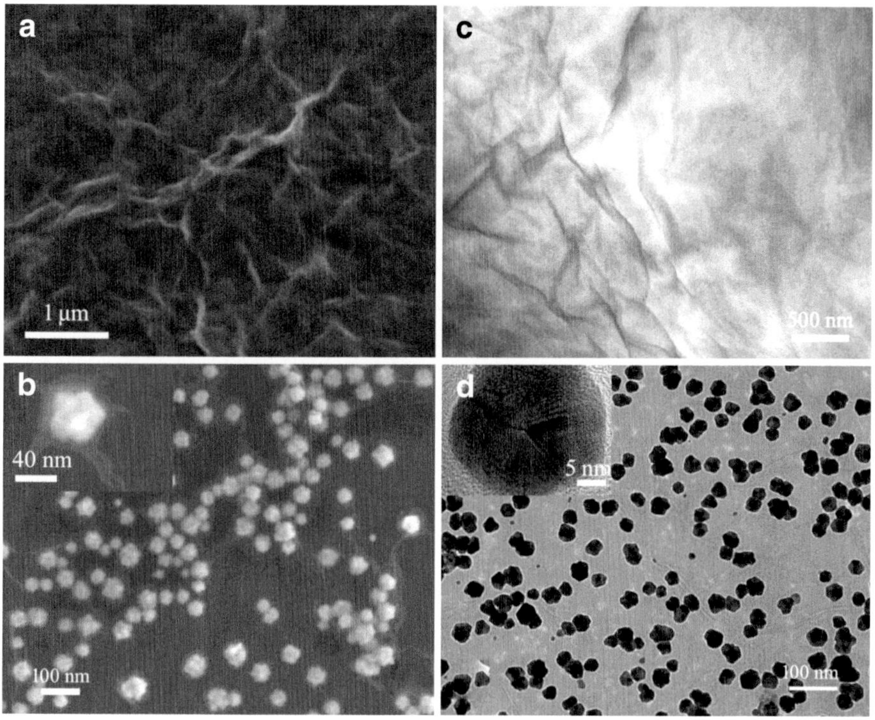


**Fig. 7.3** SEM images of GO (**a**) and prepared SERS sensor (**b**). Insert of B: the enlarged view of RGO and AuNPs in SERS sensor. TEM images of GO (**c**) and prepared SERS sensor (**d**). Insert of D: high-resolution TEM image of AuNPs in SERS sensor. Copyright permission from Chen et al. [38]

to the removal of water. In addition, there were two same regions of noticeable weight loss in 150–250 °C and 550–600 °C. The former was ascribed to the removal of oxygen functional groups from RGO surface, while the latter was related to the combustion of carbon constituting RGO [30]. It was noticeable that hydrazine was adsorbed and reacted with the top layer of GO film, and residual oxygen functional groups existed in the bottom of RGO film. Taking into consideration of the results of both XPS and TGA, the top layer of RGO film had been fully reduced. It was a very important phenomenon for two reasons: (a) the sensor had a better adsorbability for biomarker with the decrease of oxygen functional groups [31] and (b) the oxygen functional groups of GO had Raman signals to interfere the breath analysis, while they had been removed making blank sensor that has a clear SERS spectrogram after GO reduced (shown in Fig. 7.4b). Furthermore, TGA analysis also showed that more than 32 wt.% of AuNPs was on the sensor, which ensured a considerable enhancement for SERS. Moreover, BET analysis was used to evaluate the change of surface areas. $S_{BET}$ of RGO (60.5436 $m^2/g$) was slightly higher than that of RGO/AuNPs (48.6483 $m^2/g$). The lower surface area of RGO/AuNPs might be due to the high content of AuNPs. Therefore, BET analysis indicated that AuNPs on RGO did not affect the adsorption behavior of RGO, and the prepared SERS sensor

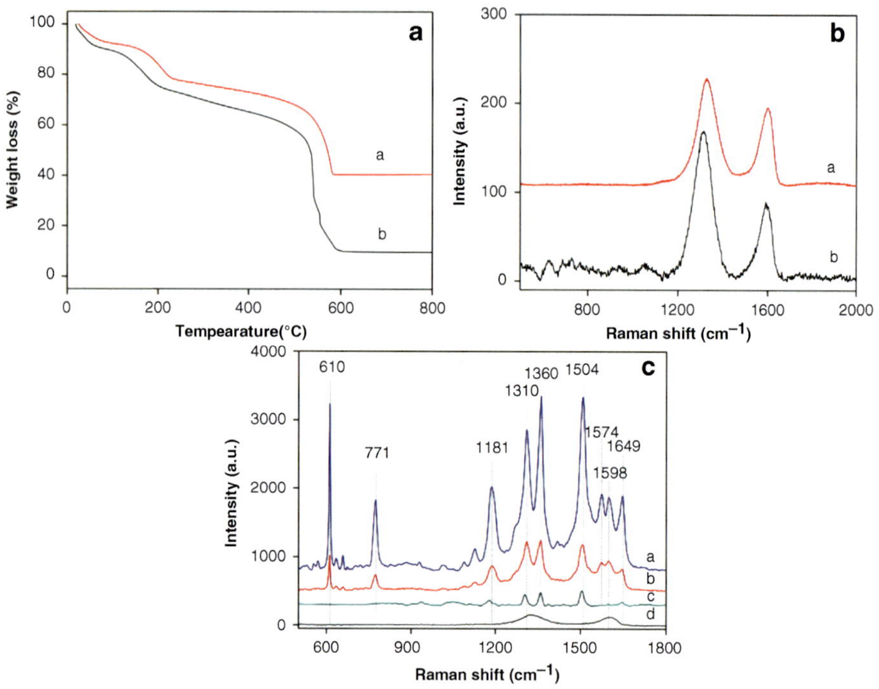


**Fig. 7.4** (**a**) TGA curves of RGO (*a*), RGO/AuNPs of SERS sensor (*b*). (**b**) The Raman spectrum of GO (*b*) on Au film substrate and SERS spectrum of blank prepared SERS sensor (*a*). (**c**) SERS spectra of ($10^{-4}$ M) R6G (*a*) and ($10^{-6}$ M) R6G (*b*) on SERS sensor, Raman spectra of bulk R6G on au film substrate (*c*), and the control on SERS sensor (*d*). Copyright permission from Chen et al. [38]

could act as the carbon-based SPME fiber to selectively adsorb the VOC biomarkers from breath. As expected, the conclusions were also verified by the next experiments.

The SERS activity was generally assessed based on the enhancement factor (EF), which was greatly concerned to SESR detection. In this work, R6G was chosen as an analyte because it has been well characterized by SERS and Raman spectroscopy. As shown in Fig. 7.4c, 1 μL R6G solution ($10^{-4}$ and $10^{-6}$ M) was deposited on SERS sensor. Several strong characteristic bands at 1649, 1598, 1574, 1504, 1360, 1310, 1181, 771, and 610 $cm^{-1}$ were observed. Finally, EF, defined as the ratio of inelastic scattering intensity per molecule between the presence and absence of the SERS substrate, could be written as the following relationship:

$$EF = \left(I_{SERS} / I_{bulk}\right)\left(N_{bulk} / N_{SERS}\right)$$

$I_{SERS}$ was the enhanced intensity of the adsorbed R6G molecules on the SERS sensor. $I_{Ref}$ was the spontaneous Raman intensity from the bulk R6G molecules under the laser spot on the blank Au film subtract. R6G in a top single molecule layer contributed to Raman signal, so R6G in other additional molecule layers should be neglected. Therefore, $N_{SERS}$ was the number of the single-layer molecules covering the SERS sensor under the laser spot. $N_{Ref}$ was the number of the bulk molecules excited by laser on the surface of Au film-covered glass subtract. According to the reported paper, $N_{SERS}$ was $1 \times 10^2$ molecules and $N_{Ref}$ was approximately $1.5 \times 10^7$ molecules [32]. Hence, EF, about $6.34 \times 10^6$, endowed SERS sensor with the excellent SERS enhancement for breath analysis in the following. Compared to AuNPs with the same diameter in reported paper, the EF was acceptable [33]. Moreover, the reproducibility was determined at three independently fabricated SERS sensor with an RSD of 5.73 %.

## 7.4 Raman Analysis Based on SERS Sensor

The SERS sensor was used in breath analysis by simulated breath and real breath samples. Simulated breath was supposed to having a composition similar to real breath samples, so it was used to build a model for real breath analysis [13]. In theory, analytes generated their specific Raman signature bands in Raman spectra, and the band intensities were linearly proportional to concentrations. As shown in Fig. 7.5a, the processed Raman spectra of simulated breath were obtained by subtracting the spectrum of control according to a reported method, and the detail was shown in *S2* in *Supplementary Material* [34]. Fourteen Raman bands of the simulated breath displayed obvious differences, and they were associated with the 14 VOC biomarkers (Table 7.1, and the Raman spectra of biomarker standards were shown in figure S2). Subsequently, the areas of these 14 Raman peaks carried out a statistical analysis by SPSS 18 (SPSS Inc, IL, USA). Figure 7.5b showed principal component 1 (PC1) and principal component 2 (PC2) for three groups, which accounted for >90 % variance. The results yielded well-defined clusters for simulated breath of healthy people (a), the patients with EGC (b) and AGC (c). Hence, the 14 bands were verified as the fingerprints of biomarker patterns to distinguish EGC and AGC from healthy people, and can be used in breath analysis based on SERS sensor.

The representative SERS spectra of simulated and real breath samples were shown in Fig. 7.5c. About 87 %, 73 %, and 82 % of samples, from healthy people, the patients with EGC and AGC, respectively, had useful bands in their SERS spectra. Due to the influences of two great bands corresponding to D and G bands of RGO, two parts of SERS spectra had been removed [35–37]. The areas of the fingerprint bands in these spectra also carried out the statistical analysis. Figure 7.5d showed PC1 and PC2 of both simulated and real breath samples both for three groups, which accounted for >85 % variance. Three clusters were well confined corresponding to the simulated breath of healthy people (c), the patients with EGC (b) and AGC (a). Meanwhile, three clusters were also well confined corresponding to real breath of healthy people (c′), the patients with EGC (b′) and AGC (a′). These clusters had the same distribution trend as the real breath clusters, which also verified that 14 bands as fingerprints were acceptable. However, the real breath clusters

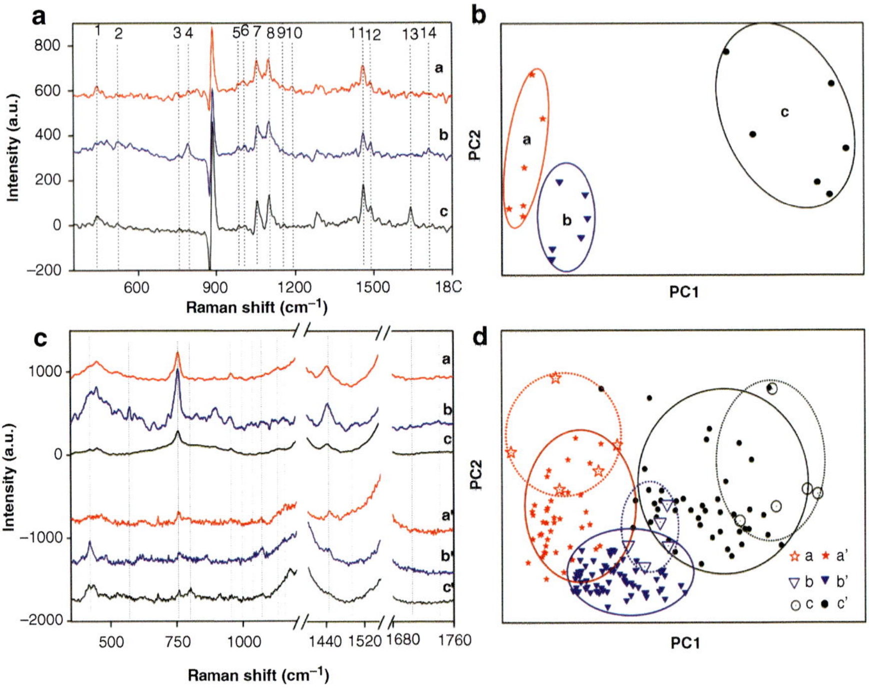


**Fig. 7.5** (**a**) The processed Raman spectra of VOC biomarker patterns. The biomarker patterns had the same components in breath samples of the healthy people (*c*), the patients with EGC (*b*) and AGC (*a*). The relations between 14 bands and 14 VOC biomarkers were described in Table 7.1. (**b**) PCA of the dataset of biomarker patterns of the healthy people (area *c*), the patients with EGC (area *b*) and AGC (area *a*). Each data point in PCA corresponded to the area of 14 bands in the processed Raman spectra. (**c**) SERS spectra of simulated and real breath samples of the healthy people (*c*, *c'*), the patients with EGC (*b*, *b'*) and AGC (*a*, *a'*). (**d**) PCA of the dataset of simulated and real breath samples of the healthy people (*c*, *c'*), the patients with EGC (*b*, *b'*) and AGC (*a*, *a'*). Each data point in PCA corresponded to the area of 14 bands in SERS spectra. Copyright permission from Chen et al. [38]

were more intensive than simulated breath ones, due to the lower biomarker concentrations. Interestingly, the patients of GC had tighter distribution than healthy people, which might be attributed to the fact that the cancer tissues could continuously generate identified VOC biomarkers. In brief, the breath analysis approach based on SERS sensor could be successful in diagnosing and staging GC.

## 7.5 Conclusion

In our work, 14 VOC biomarkers in human breath have been identified to distinguish the patients with EGC and AGC from the healthy people. Furthermore, a breath analysis approach based on a clean SERS sensor has also been developed to detect the biomarkers both in simulated and real breath samples. Utilizing hydrazine

**Table 7.1** The relation between the 14 band as fingerprints and corresponding VOC biomarkers

| Band no. | Band position ($cm^{-1}$) | VOC biomarker | Band no. | Band position ($cm^{-1}$) | VOC biomarker |
|---|---|---|---|---|---|
| 1 | 443 | 2-Methylpentane, 3-methylpentane, 2-methylhexane, 3-methylhexane, menthol | 8 | 1103 | Menthol, hexanol, hexane |
| 2 | 526 | Acetone, isoprene, pivalic acid, phenyl acetate | 9 | 1162 | 3-Methylpentane, 3-methylhexane, pivalic acid, phenyl acetate, 2,3-dimethylpentane |
| 3 | 755 | Pivalic acid, menthol | 10 | 1191 | 2,3-Dimethylpentane, phenyl acetate |
| 4 | 785 | Acetone, 2-methylpentane, 2,3-dimethylpentane | 13 | 1459 | 2-Methylpentane, 3-dimethylpentane, 3-methylpentane, 3-methylhexane, tetradecane hexanol, 2-methylhexane, menthol, dodecane, hexane |
| 5 | 986 | Acetone, isoprene, 2,3-dimethylpentane, 3-methylpentane, 3-methylhexane | 14 | 1487 | Pivalic acid, hexane |
| 6 | 1010 | Hexane, phenyl acetate | 15 | 1642 | Isoprene, pivalic acid |
| 7 | 1067 | 2-Methylpentane, 2-methylhexane, hexanol, tetradecane, dodecane | 16 | 1714 | Acetone |

vapor adsorbed in GO, SERS sensor is facilely prepared by in situ forming AuNPs on RGO film. The RGO film selectively adsorbed and concentrated identified VOC biomarkers as SPME fiber. AuNPs well dispersing on RGO endow SERS sensor with an effective detection and recognition of adsorbed VOC biomarkers. Fourteen Raman bands associated with biomarkers are selected as the fingerprints of biomarker patterns to distinguish the persons in different states by SERS sensor. The mentioned VOC biomarkers and breath analysis approach can not only diagnose GC but also distinguish the persons with EGC and AGC with the advantages of simple operation, low cost, and reliability. Therefore, the novel breath analysis method described in this work has a great potential for the clinical application in diagnosis and stage determination of GC in the near future. Moreover, we believe that the established SERS sensor and breath analysis approach can become a universal strategy for different diseases with different VOC biomarkers. Optimizing the SERS sensor and extending its application will be our next research focus, and the relevant work has been undergoing. Results will be reported in due course.

# References

1. Hartgrink HH, Jansen EP, van Grieken NC, van de Velde CJ. Gastric cancer. Lancet. 2009;374(9688):477–90.
2. (a) Washington K. 7th edition of the AJCC cancer staging manual: stomach. Ann Surg Oncol. 2010;17(12):3077–9; (b) Yamaoka Y, Kato M, Asaka M. Geographic differences in gastric cancer incidence can be explained by differences between Helicobacter pylori strains. Intern Med. 2008;47(12):1077–83.
3. (a) Saito Y, Suzuki H, Imaeda H, Matsuzaki J, Hirata K, Tsugawa H, Hibino S, Kanai Y, Saito H, Hibi T. The tumor suppressor microRNA-29c is downregulated and restored by celecoxib in human gastric cancer cells. Int J Cancer. Journal international du cancer. 2013;132(8):1751–60; (b) Ooki A, Yamashita K, Kikuchi S, Sakuramoto S, Katada N, Watanabe M. Phosphatase of regenerating liver-3 as a prognostic biomarker in histologically node-negative gastric cancer. Oncol Rep. 2009;21(6):1467–75.
4. Alberts SR, Cervantes A, van de Velde CJ. Gastric cancer: epidemiology, pathology and treatment. Ann Oncol Off J Eur Soc Med Oncol/ESMO. 2003;14(Suppl 2):ii31–6.
5. Liao SR, Dai Y, Huo L, Yan K, Zhang L, Zhang H, Gao W, Chen MH. Transabdominal ultrasonography in preoperative staging of gastric cancer. World J Gastroenterol. 2004;10(23):3399–404.
6. Hundahl SA, Phillips JL, Menck HR. The national cancer data base report on poor survival of U.S. Gastric carcinoma patients treated with gastrectomy: fifth edition American joint committee on cancer staging, proximal disease, and the "different disease" hypothesis. Cancer. 2000;88(4):921–32.
7. (a) Yasui W, Oue N, Aung PP, Matsumura S, Shutoh, M, Nakayama H, Molecular-pathological prognostic factors of gastric cancer: a review. Gastric Cancer Off J Int Gastric Cancer Assoc Jpn Gastric Cancer Assoc. 2005;8(2):86–94; (b) Zheng Z, Yu Y, Lu M, Sun W, Wang F, Li P, Zhang Y, Lin L, Huang P, Chen J, Zhang H, Xie Z, Dong Xda E. Double contrast-enhanced ultrasonography for the preoperative evaluation of gastric cancer: a comparison to endoscopic ultrasonography with respect to histopathology. Am J Surg. 2011;202(5):605–11.
8. (a) Dancey JE, Bedard PL, Onetto N, Hudson TJ. The genetic basis for cancer treatment decisions. Cell. 2012;148(3):409–20; (b) Ren H, Du N, Liu G, Hu HT, Tian W, Deng ZP, Shi JS. Analysis of variabilities of serum proteomic spectra in patients with gastric cancer before and after operation. World J Gastroentero. 2006;12(17):2789–92; (c) Wu C, Luo ZW, Chen XY, Wu CQ, Yao DK, Zhao P, Liu LJ, Shi B, Zhu L. Two-dimensional differential in-gel electrophoresis for identification of gastric cancer-specific protein markers. Oncol Rep. 2009;21(6):1429–37.
9. Kim KH, Jahan SA, Kabir E. A review of breath analysis for diagnosis of human health. Trac Trend Anal Chem. 2012;33:1–8.
10. (a) Hakim M, Broza YY, Barash O, Peled N, Phillips M, Amann A, Haick H. Volatile organic compounds of lung cancer and possible biochemical pathways. Chem Rev. 2012;112(11):5949–66; (b) Amal H, Ding L, Liu BB, Tisch U, Xu ZQ, Shi DY, Zhao Y, Cheng J, Sun RX, Liu H, Ye SL, Tang ZY, Haick H. The scent fingerprint of hepatocarcinoma: in-vitro metastasis prediction with volatile Organic Compounds (VOCs). Tumor Biol. 2012;33:61; (c) Peng G, Hakim M, Broza YY, Billan S, Abdah-Bortnyak R, Kuten A, Tisch U, Haick H. Detection of lung, breast, colorectal, and prostate cancers from exhaled breath using a single array of nanosensors. Brit J Cancer. 2010;103(4):542–51.
11. Konvalina G, Haick H. Sensors for breath testing: from nanomaterials to comprehensive disease detection. Acc Chem Res. 2014;47(1):66–76.
12. Xu ZQ, Broza YY, Ionsecu R, Tisch U, Ding L, Liu H, Song Q, Pan YY, Xiong FX, Gu KS, Sun GP, Chen ZD, Leja M, Haick H. A nanomaterial-based breath test for distinguishing gastric cancer from benign gastric conditions. Br J Cancer. 2013;108(4):941–50.
13. Peng G, Tisch U, Adams O, Hakim M, Shehada N, Broza YY, Billan S, Abdah-Bortnyak R, Kuten A, Haick H. Diagnosing lung cancer in exhaled breath using gold nanoparticles. Nat Nanotechnol. 2009;4(10):669–73.

14. Huang CY, Song M, Gu ZY, Wang HF, Yan XP. Probing the adsorption characteristic of metal-organic framework MIL-101 for volatile organic compounds by quartz crystal microbalance. Environ Sci Technol. 2011;45(10):4490–6.
15. Mazzone PJ, Hammel J, Dweik R, Na J, Czich C, Laskowski D, Mekhail T. Diagnosis of lung cancer by the analysis of exhaled breath with a colorimetric sensor array. Thorax. 2007;62(7):565–8.
16. (a) Ligor T, Ligor M, Amann A, Ager C, Bachler M, Dzien A, Buszewski B. The analysis of healthy volunteers' exhaled breath by the use of solid-phase microextraction and GC-MS. J Breath Res. 2008;2(4):046006; (b) Miekisch W, Fuchs P, Kamysek S, Neumann C, Schubert JK. Assessment of propofol concentrations in human breath and blood by means of HS-SPME-GC-MS. Clin Chim Acta. 2008;395(1–2):32–7.
17. Zhang YX, Gao G, Liu HJ, Fu HL, Fan J, Wang K, Chen YS, Li BJ, Zhang CL, Zhi X, He L, Cui DX. Identification of volatile biomarkers of gastric cancer cells and ultrasensitive electrochemical detection based on sensing interface of Au-Ag alloy coated MWCNTs. Theranostics. 2014;4(2):154–62.
18. Gao WH, Chen YS, Xi J, Lin SY, Chen YW, Lin YJ, Chen ZG. A novel electrochemiluminescence ethanol biosensor based on tris(2,2 '-bipyridine) ruthenium (II) and alcohol dehydrogenase immobilized in graphene/bovine serum albumin composite film. Biosens Bioelectron. 2013;41:776–82.
19. (a) Blackie EJ, Le Ru EC, Etchegoin PG. Single-molecule surface-enhanced Raman spectroscopy of nonresonant molecules. J Am Chem Soc. 2009;131(40):14466–72; (b) Nie SM, Emery SR. Probing single molecules and single nanoparticles by surface-enhanced Raman scattering. Science. 1997;275(5303):1102–6.
20. Xu WG, Mao NN, Zhang J. Graphene: a platform for surface-enhanced Raman spectroscopy. Small. 2013;9(8):1206–24.
21. (a) Williams G, Seger B, Kamat PV. TiO2-graphene nanocomposites. UV-assisted photocatalytic reduction of graphene oxide. ACS nano. 2008;2(7):1487–91; (b) Balapanuru J, Yang JX, Xiao S, Bao QL, Jahan M, Polavarapu L, Wei J, Xu QH, Loh KP. A graphene oxide-organic dye ionic complex with DNA-sensing and optical-limiting properties. Angew Chem Int Edit. 2010;49(37):6549–53.
22. Yang LB, Li P, Liu HL, Tang XH, Liu JH. A dynamic surface enhanced Raman spectroscopy method for ultra-sensitive detection: from the wet state to the dry state. Chem Soc Rev. 2015;44(10):2837–48.
23. (a) Lee J, Novoselov KS, Shin HS. Interaction between metal and graphene: dependence on the layer number of graphene. Acs Nano. 2011;5(1):608–12; (b) Zhou HQ, Qiu CY, Liu Z, Yang HC, Hu LJ, Liu J, Yang HF, Gu CZ. Sun LF. Thickness-dependent morphologies of gold on N-Layer graphenes. J Am Chem Soc. 2010;132(3):944–6.
24. Fan Z, Kanchanapally R, Ray PC. Hybrid graphene oxide based ultrasensitive SERS probe for label-free biosensing. J Phys Chem Lett. 2013;4(21):3813–8.
25. Haick H, Broza YY, Mochalski P, Ruzsanyi V, Amann A. Assessment, origin, and implementation of breath volatile cancer markers. Chem Soc Rev. 2014;43(5):1423–49.
26. Bult JB, Crisp R, Perkins CL, Blackburn JL. Role of dopants in long-range charge carrier transport for p-type and n-type graphene transparent conducting thin films. ACS Nano. 2013;7(8):7251–61.
27. (a) Haiss W, Thanh NTK, Aveyard J, Fernig DG. Determination of size and concentration of gold nanoparticles from UV–vis spectra. Anal Chem. 2007;79(11):4215–21; (b) Huang J, Zhang LM, Chen BA, Ji N, Chen FH, Zhang Y, Zhang ZJ. Nanocomposites of size-controlled gold nanoparticles and graphene oxide: Formation and applications in SERS and catalysis. Nanoscale. 2010;2(12):2733–8.
28. Tong LM, Zhu T, Liu ZF. Approaching the electromagnetic mechanism of surface-enhanced Raman scattering: from self-assembled arrays to individual gold nanoparticles. Chem Soc Rev. 2011;40:1296–304.
29. Vinodgopal K, Neppolian B, Lightcap IV, Grieser F, Ashokkumar M, Kamat PV. Sonolytic design of graphene-Au nanocomposites. Simultaneous and sequential reduction of graphene oxide and Au(III). J Phys Chem Lett. 2010;1(13):1987–93.

30. Guo YQ, Sun XY, Liu Y, Wang W, Qiu HX, Gao JP. One pot preparation of reduced graphene oxide (RGO) or Au (Ag) nanoparticle-RGO hybrids using chitosan as a reducing and stabilizing agent and their use in methanol electrooxidation. Carbon. 2012;50(7):2513–23.
31. Geng X, Niu L, Xing Z, Song R, Liu G, Sun M, Cheng G, Zhong H, Liu Z, Zhang Z, Sun L, Xu H, Lu L, Liu L. Aqueous-processable noncovalent chemically converted graphene-quantum dot composites for flexible and transparent optoelectronic films. Adv Mater. 2010;22(5): 638–42.
32. Yi Z, Chen SJ, Chen Y, Luo JS, Wu WD, Yi YG, Tang YJ. Preparation of dendritic Ag/Au bimetallic nanostructures and their application in surface-enhanced Raman scattering. Thin Solid Films. 2012;520(7):2701–7.
33. Yang MX, Chen T, Lau WS, Wang Y, Tang QH, Yang YH, Chen HY. Development of polymer-encapsulated metal nanoparticles as surface-enhanced Raman scattering probes (vol 5, pg 198, 2009). Small. 2009;5:198–202.
34. Beier BD, Berger AJ. Method for automated background subtraction from Raman spectra containing known contaminants. Analyst. 2009;134(6):1198–202.
35. Hummers WS, Offeman RE. Preparation of graphitic oxide. J Am Chem Soc. 1958;80(6):1339.
36. Becerril HA, Mao J, Liu Z, Stoltenberg RM, Bao Z, Chen Y. Evaluation of solution-processed reduced graphene oxide films as transparent conductors. ACS Nano. 2008;2(3):463–70.
37. Czolkos I, Hannestad JK, Jesorka A, Kumar R, Brown T, Albinsson B, Orwar O. Platform for controlled supramolecular nanoassembly. Nano Lett. 2009;9(6):2482–6.
38. Chen YS, Zhang YX, Pan F, Liu J, Wang K, Zhang CL, et al. Breath analysis based on surface-enhanced raman scattering sensors distinguishes early and advanced gastric cancer patients from healthy persons. ACS Nano. 2016;10:8169–79.

# Chapter 8
# The Application of Immunochromatographic Analysis in Early Detection of Gastric Cancer

**Kan Wang and Daxiang Cui**

## 8.1 Immunochromatographic Analysis Technology

As early as thousands of years ago, people were aware of infectious diseases and of the phenomenon that survivors of disease epidemics did not suffer the disease again. This phenomenon is called immunity. The immune response comprises two basic constituents – antigen and antibody. Antigens are classes of substances that stimulate the animal immune system to produce an immune response; humoral antibody or lymphoid effector cell–mediated immunity is induced by antigens. Antibodies are glycoproteins produced by B-lymphocytes when the animal body responds to an antigen [1]. Immunoassays are qualitative or quantitative methods used to analyze and detect trace substances. In an immunoasssay, antigen and antibody serve as the detecting agents, and we utilize their special binding reactions to analyze the molecules contained in the detected material [2]. The specificity of the binding between the antigen and antibody is a characteristic of the immune reaction. Conventional immunodetection analysis technology is mainly used to detect an antigen or antibody in in-vitro serological reactions, and the technology is also used in serological detection techniques. With the rapid development of modern immunological detection techniques, many new methods and technologies have emerged [3]. There are many advantages of immunoassay technology, such as greater specificity, higher sensitivity, and easier operation [4–6]; this technology is widely used in many medical detection areas. With the development of science and technology, immunological detection techniques will be automated, easy to operate [7, 8], and useful for

K. Wang (✉) • D. Cui
Institute of Nano Biomedicine and Engineering, Shanghai Engineering Research Center for Intelligent Diagnosis and Treatment Instrument, National Center for Translational Medicine, Collaborative Innovational Center for System Biology, Shanghai Jiao Tong University, 800 Dongchuan Road, Shanghai 200240, P. R. China
e-mail: wk_xa@sjtu.edu.cn; dxcui@sjtu.edu.cn

© Springer Science+Business Media B.V. and Shanghai Jiao Tong University Press, Shanghai 2017

D. Cui (ed.), *Gastric Cancer Prewarning and Early Diagnosis System*, Translational Medicine Research, DOI 10.1007/978-94-024-0951-2_8

multivariate analysis [9–11], providing high sensitivity [12–15], high specificity [16–18], and quantification [12, 16, 19, 20].

In clinical diagnostics, the point-of-care test (POCT) is a rapidly developing area and is regarded as one of the main driving forces in the in vitro diagnostics market [21]. POCTs are usually used as an auxiliary method and can be carried out at the patient's bedsideusing non-central laboratory testing methods. That is to say, POCTs can be directly operated in a hospital emergency department or in a doctor's office, at the bedside, and even in a patient's home, without the necessity for various immunology, pathology, cytology, and biochemistry tests, and other complex testing processes that must go through the hospital's central laboratory. Detection is instantly available; the detection method is easy to operate, highly efficient, and inexpensive, and because of all these advantages the POCT has great market demand. At present, commercial POCT methods are mainly immunochromatographic assay tests; this kind of detection method is called a lateral flow test [22]. Immunochromatographic, developed in the late 1980s, is a new rapid detection and analysis technology that achieves detection by chemiluminescence principles [23]. Immunochromatographic technology can be divided into two categories. One category is based on an enzymatic reaction, with the color strength indicating the quantitative result [24]. The other category uses latex particles, colloidal gold, colloidal selenium, and liposomes, and other coloring markers. Immunoreactions occurred on the nitrocellulose membrane. Monoclonal antibody and secondary antibody were coated separately to serve as the test line and control line, respectively. Quantitative assay is mainly performed using readers that measure the reflected light or fluorescence on the test line and control line [25]. Figure 8.1 shows the ratios of various coloring markers used in immunochromatographic strips [26].

The relationship between the principle of immunological experiments and thin-layer chromatography underlies the immunochromatographic test, which is easy to operate. The test results can be obtained in a specified time (such as 10 or 20 min) after the addition of the sample [27]. As shown in Fig. 8.2 [28], the strip used in the immunochromatographic test usually contains five parts, a sample pad, a marker pad, cellulose nitrate film (nitrocellulose membrane), an absorbing pad, and a backing plate. The sample pad is mainly used for filtering particulate matter in the sample, adjusting the pH value, and binding material interfacing the chromatographic reaction, which material may be cellulose, fiberglass, rayon, or other filtration media; the marker pad is mainly used for loading and releasing stable chromatographic markers in the reaction, which may be in forms such as glass fibers, polyester fibers, or rayon; the nitrocellulose membrane is mainly used for solidifying the antigen or antibody and providing places for the chromatographic reaction. Usually, monoclonal antibody and secondary antibody were coated separately to serve as the test and control line by using contact or non-contact dispensing system or noncontacting point meter in antigen or antibody membrane linear curing test line (test line) and quality control line (the control line); the absorbent pad, which is usually made of high-density cellulose, is mainly used for controlling the continuous flow direction of the chromatographic reaction sample; the backing plate (made of polystyrene or other plastic material) is mainly used to provide a rigid support for the immunochromatographic strip [29].

After the sample (e.g., serum, urine) is added to the immunochromatographic strip of the sample pad, the marker pad and sample will react to form a complex, and then,

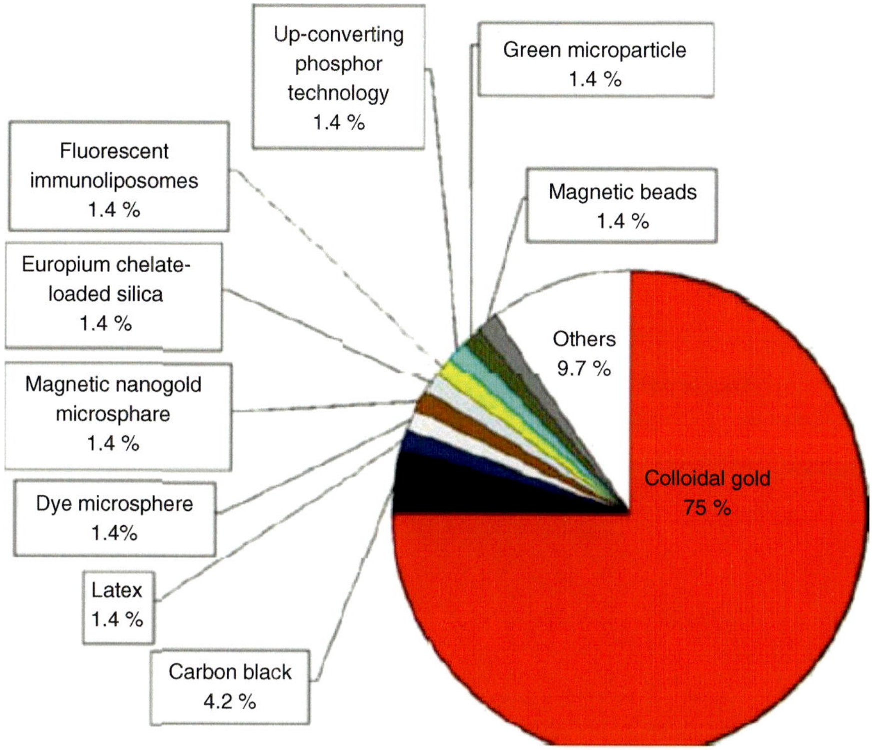


**Fig. 8.1** The colored particles used in the immunochromatographic test paper [26]

under capillary action, the composite flows through the nitrocellulose membrane to the absorbent pad. When the liquid sample migrated to the conjugate pad by capillary action, the probes bonded with the target analytes. The complexes were then released from the conjugate pad and migrated to the nitrocellulose membrane, then the complexes will be captured at test line and control line, respectively, while the target molecules and the marker complex show a color that can be detected by the naked eye or with appropriate equipment. Because of its simplicity, effectiveness, and low cost, the market demand for immunochromatographic tests in the POCT is strong. A series of analyses and marker methods based on this detection mode are still being developed and used for analysis in medical testing [12, 30–33], environmental testing [34, 35], agricultural testing [36, 37], and food-quality testing [38–40].

## 8.2 The Application of Immunochromatographic Analysis Technology

Immunochromatographic analysis technology maintains the advantage of the special binding between antigen and antibody in immune test technology. The immunochromatographic method eliminates the tedious pipetting and washing step of immune

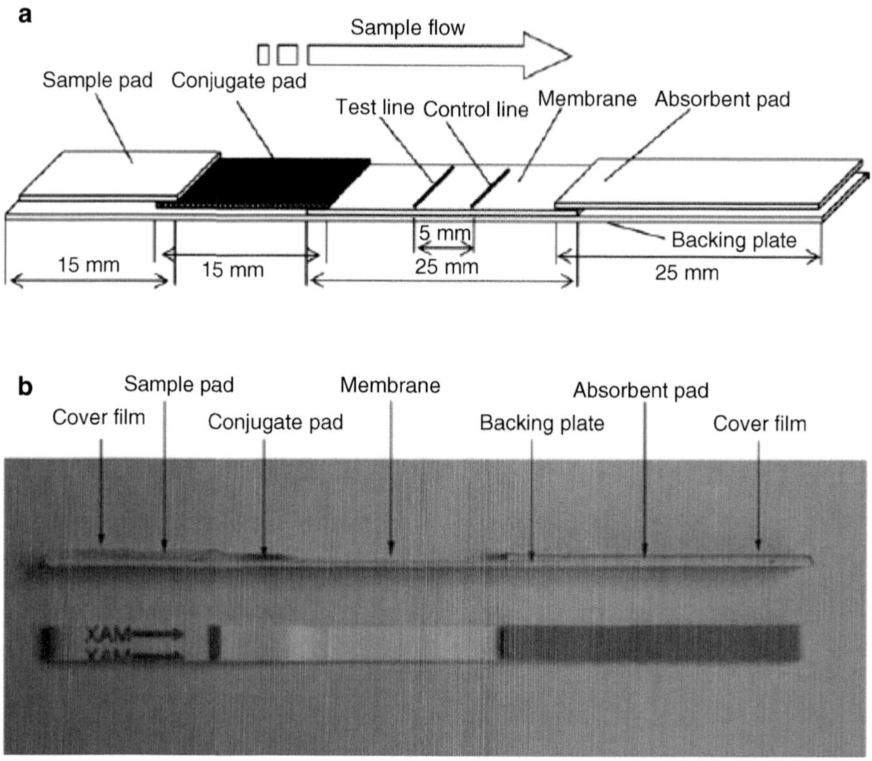


**Fig. 8.2** Immunochromatographic test strip structure [28]. (**a**) Schematic representation of the lateral flow strip. (**b**) Lateral (top) and over (bottom) view of a lateral flow strip. The outmost layers of the ends of the ends of the strip are plastic cover films

test technology. Thus, its operation is simple and fast (under 20 min to get the result), and staff do not need training, as the test can even be used without instrumentation or with just simple devices. This feature is very suitable for field testing, and the test is now widely used in medicine, agriculture, animal husbandry and veterinary medicine, environmental testing, and food supervision.

### *8.2.1 Application in the Medical Field*

Immunochromatographic analysis technology was first used to test human chorionic gonadotropin (HCG), and this approach is now very mature and widely used in pregnancy testing. Immunochromatographic analysis technology is very significant in clinical diagnosis, and with a typical POCT, testing is so fast that we can ensure maintenance of the patients' healthy state or the implementation of timely rescue, thus saving lives. The myoglobin (Mb) [41], creatine phosphokinase isoenzyme

(CK-MB) [42], cardiac troponin I (cTnI) [43], and cardiac troponin T (cTnT) [44] are most use as biochemical index for diagnose the acute myocardial infarction (AMI). Immunochromatographic analysis technology is also widely used in the testing of fecal blood [45]. Clavijo et al. [46] utilized immunochromatographic analysis technology to test Brucella-specific antibody in serum; through experimental comparison, the detection sensitivity was higher than that of the traditional enzyme-linked immunosorbent assay (ELISA) detection method.

#### 8.2.1.1 Detection of Pathogenic Microorganisms

Takeda et al. [53] utilized a colloidal gold immunochromatographic strip to directly test for *Escherichia coli* in a water sample, and compared results with those for ELISA (concordance rate of 87.2 %), and those for a culture method (concordance rate of 89.10 %). The time taken for chromatography was 5 min, and this was shorter than that for the other test methods. In addition, immunochromatographic test technology was also used to detect pathogens such as *Vibrio cholerae* and adenovirus, as well as hepatitis B surface antigen and *Helicobacter pylori* antigens. Immunochromatographic techniques are low cost and can easily and quickly detect various microbial indicators in patients for whom there is a suspicion of microbial disease.

#### 8.2.1.2 Detecting Human Hormone or Hormone-Related Proteins

Since the first use of an immunochromatographic system for testing HCG, researchers have developed many different systems to detect different hormone and protein targets. For example, Osikwicz et al. [54] utilized colloidal selenium to test HCG; Hedstrom et al. [55] utilized immunochromatographic to test urine trypsinogen-2 in patients with acute pancreatitis; and Muleret et al. [143] utilized an avidin-biotin immunochromatographic technique to detect cTnI. The immunochromatographic technique for detecting disease-related proteins, compared with the previous technique of chemiluminescence or other means of detection, has greatly improved the detection rate and this is very important for timely diagnosis and treatment.

#### 8.2.1.3 Parasitic Disease Diagnosis

Because of the epidemic and highly infectious nature (and other characteristics) of many parasitic diseases, there is a great need for simple, rapid, efficient, inexpensive diagnostic methods that can be used on a large scale in the field. Thus, immunochromatographic techniques have been increasingly and widely used in the diagnosis of parasitic diseases. Currently, immunochromatographic diagnostic methods have been established to test for plasmodium and for schistosomiasis, leishmaniasis, amebiasis, filariasis, trypanosomiasis, and cryptosporidiosis [47].

#### 8.2.1.4 Blood and Urine Drug Monitoring

Buecher et al. [126], using colloidal gold as a label, has detected prescription and illicit drugs (opiates, barbiturates, tetrahydrocannabinol (THC), amphetamine, cocaine, benzodiazepines, and phencyclidine (PCP)) in the urine. The detection sensitivity and specificity of that test method was very consistent with the results of gas chromatography/mass spectrometry (GC/MS) and an enzyme-multiplied immunoassay technique (EMIT).

The application of immunochromatographic analysis technology has been widely researched in China. For example, Zhao et al. utilized a *Chlamydia trachomatis* immunochromatographic strip to monitor the incidence of *C. trachomatis*, Ureaplasmaurealyticum (Uu), and other infections and to analyze the sensitivity and resistance to antimicrobial drugs of the organisms; Li et al. [48] have investigated the application of immunochromatographic analysis technology for the rapid detection of influenza viruses. In addition, the detection of hepatitis B surface antigen [49], Brucella [50], *H. pylori* antigen [51], and other pathogens is also a research hot-spot of immunochromatographic technology. Yang Yonghong [52] and others utilized mycobacterium-secreted protein antigen on a coated nitrocellulose membrane to detect tuberculosis antibody in serum and pleural effusions.

### *8.2.2 Drug Detection*

The worldwide prevalence of illicit drug use is increasingly becoming an important factor threatening social stability and people's health. Immunochromatographic detection has been widely used for drugs such as cocaine, marijuana, heroin, and others, as well as for the detection of coffee, and there are already many commercial drug-testing kits.

### *8.2.3 Live Animal Quarantine, Meat Distribution Process Monitoring, and Testing of Pesticide and Veterinary Drug Residues in Food*

Immunochromatographic methods are widely used in animal quarantine for the detection of antigens or antibodies in diseases such as swine fever, porcine reproductive and respiratory syndrome (PRRS), parvovirus infection, infectious bursal disease, Newcastle disease, rabies, and diseases caused by other infectious pathogens. With the development of industrialization, more and more pesticides, hormones, and antibiotics are being used for crop cultivation and animal husbandry. The residues of harmful substances in food can seriously affect people's health and life. Because of the widespread circulation of food, it is imperative to use a simple, fast, and low-cost method for the detection of these substances. Currently, immunochromatographic technology is being applied for many food-safety testing procedures, such as the detection of antibiotic residues, pesticide residue testing, and hormone testing.

### *8.2.4 Environmental Detection*

For various reasons, there may be harmful microbes, hormones, and other substances in the environment, and the timely detection of suspect agents (in water or air) is necessary. There have been numerous recent studies on environmental immunochromatographic detection. With the growing concern about the impact of environmental changes on human health, the previous time-consuming testing methods will be increasingly replaced by simple and economical immunochromatographic methods. So far, immunochromatographic detection technology has been applied in various areas of our lives. But the technology is not a panacea; there are many areas that need to be developed, and research on immunochromatographic technology in various areas of application is a concern for many scholars and businesses. Ongoing innovative research is necessary so that this technology can be used to better serve human life.

## 8.3 Immunochromatographic Based on Nanotechnology

The term "nanomaterials" refers to materials with a diameter between 1 and 100 nm, and at least one dimension in the nanoscale among three-dimensional spatial scales [56]. In the regional transition from the microscopic to the macroscopic world, when the size of a material is reduced to the nanometer level, it will show a series of unique properties, which are obviously different from those of body substances–revealed by optical, electrical, magnetic, thermal, and catalytic performance, and chemical activity [57]–such as quantum size effect [58], small molecule effect [59], surface effects [60], and the macroscopic quantum tunneling effect [61]. In recent years, with the continued deepening combination of nanotechnology and bioanalytical techniques, nanotechnology has become a hot issue and is at the forefront of technology in the field of international biomedical analysis. The optical properties of nanoparticles are the main features of immunoassays; in comparison to body materials, they display great enhancement of light absorption ; the light reflectance is low, so they are good photovoltaic, solar thermal conversion materials [62–64]. Researchers are interested in the luminescent properties of nanoparticles, from the ultraviolet to the visible range. Thanks to the rapid development of nanotechnology in the twentieth century, the breakthrough development of nanoparticle-labeled has opened the world of nanolabeled-imaging diagnostics [65, 66]. Research on nanoparticles, beginning with gold (Au) as the representative noble metal nanoparticle, and then cobalt (Co), representing magnetic nanomaterials, and later cadmium selenide (CdSe), representing the semiconductor quantum point, tagged functional nanoparticles have been a hot topic for analysis in nanoscale biomedical research.

### 8.3.1 Application of Metal Nanomaterials

In immunochromatographic studies, the most representative metal nanoparticles are gold nanoparticles. They not only have the surface effects and quantum size effects of nanoparticles in general, but they also show good biocompatibility. Nanogold refers to tiny gold particles, generally with a diameter between 1 and 100 nm. With a negative charge on the particle surface due to the electrostatic repulsion of gold nanoparticles, they remain stable in water, forming a stable colloid, and they are also known as colloidal gold [67]. Colloidal gold consists of negatively charged hydrophobic colloidal gold particles formed by the reduction of gold chloride acid (HAuCl4) to gold particles. Colloidal gold varies in color; the smallest colloidal gold particles (2–5 nm) are orange and yellow; medium size (10–20 nm) particles are red-wine-colored, and larger particles (30–80 nm) are purple. The molar extinction coefficient of colloidal gold is $2.4\times108$ $M^{-1}$ $cm^{-1}$, which is three times more than that of the best organic dye [68]. Because of the high charge on the surface of colloidal gold particles, the particles can be adsorbed onto and combined with polymer materials, such as proteins. Thus, proteins are adsorbed onto the surface of colloidal gold particles, and we obtain colloidal gold-labeled proteins. Since the colloidal gold particles have a high electron density, the label on the solid support of can render a visible red-to-purple color when the aggregates reach a certain density [24, 69]. In 1990, Beggs et al. [70] first reported the qualitative detection of HCG in maternal urine and serum by the use of a colloidal gold immunochromatographic method. Müller-Bardorff et al. [71] developed a cardiac troponin T (cTnT) strip, with a detection limit of up to 0.4 ng/mL. Today, nanogold immunochromatographic technology has matured, and many companies have produced such test systems, resulting in hundreds of test items, including alpha-fetoprotein (AFP) [72], carcinoembryonic antigen (CEA) [73] and other tumor marker series; cardiac troponin I (cTnI) [74]; creatine phosphokinase isoenzyme (CK-MB) [42] and other cardiac infarction (MI) series; AIDS hepatitis A virus (HIV) [75], second liver two half-and-half [76], hepatitis A virus (HAV) [77], *H. pylori* (HP) [78], and other infectious disease series; amphetamine (AMP), morphine (OPI) [79], cocaine (COC) [80], methamphetamine (MET) [81], and other drug series; HCG [82], luteinizing hormone (LH) [83], and other hormone series; and rubella virus (RV) [84], *Herpes simplex* virus (HSV) [85], and other prenatal and postnatal care series. Figure 8.3 shows a schematic of a nanogold immunochromatographic strip for the detection of the binding protein c-jun; this method can achieve rapid detection [86], in 10 min. The nanogold immunochromatographic test method is suitable for clinical laboratories, field-site diagnosis, and home self-testing, and is the most widely used POCT method in the twenty-first century [87–90].

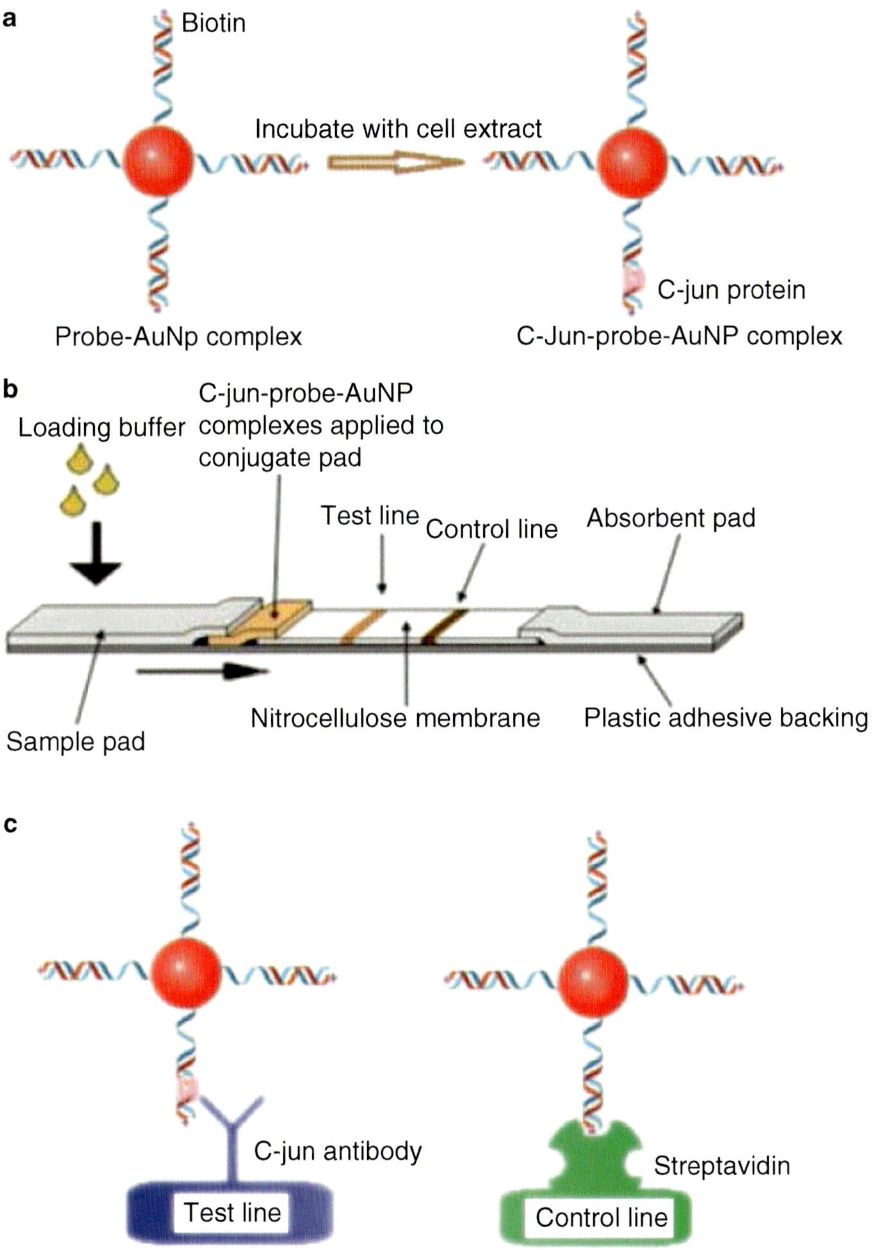


**Fig. 8.3** Schematic of the biosensor design. (**a**) Probe-AuNP complexes are incubated with cell lysate, and the c-jun protein is captured by the DNA-binding sequence on the AuNPs. (**b**) Schematic of the lateral flow biosensor. (**c**) C-jun-probe-AuNP complexes are captured by antibody on the test zone and the excess complexes are captured by streptavidin on the control zone [86]

As sensitivity is limited in the traditional gold nanoparticle-based immunochromatographic strip, many researchers began to look to new methods to increase the sensitivity of immunochromatographic detection, such as the use of enzymatic markers, silver-staining enhancement, chemiluminescence, or fluorescence [91]. In the immunogold-silver staining method, the catalytic reduction of silver ions in colloidal gold solution results in the accumulation of silver ions on the surfaces of the gold particles. Due to the electrostatic effect and large surface-to-volume ratio, gold solution could efficiently enrich silver ions being as catalysis, and thus enhance the redox reaction on the surface of gold probes. The radius of the obtained particles is increased by 10–50 times compared with immunogold alone, thereby improving the detection sensitivity. Immunogold silver-staining is widely used in DNA detection, and its sensitivity is up to 50 pmol/L [92].

### *8.3.2 Application of Magnetic Nanomaterials*

Magnetic microspheres [93] are complex combinations of a magnetic metal oxide (typically an iron oxide such as $Fe_3O_4$.) with nonmagnetic material, such as organic polymers, biological macromolecules, or an inorganic material contained in the interior and on the surface of the spheres, after functional modification, resulting in spherical particles (microspheres) having the surface features and magnetic response of nanoparticles. The term "nanomagnetic beads (magnetic nanospheres)" refers to magnetic microspheres (whose diameter is in the nanoscale range); these are also known as magnetic nanoparticles or magnetic nanobeads. The surface effect, quantum effect, small size effect, and macroscopic quantum tunneling effect, and other characteristics of the magnetic nanoparticles explain their unique optical, magnetic, electrical, thermal, mechanical, and chemical activity. The magnetic material used often has a multidomain structure in order to reduce the demagnetizing field energy of the system. When the size is in the single-domain range, the particles generate high coercivity, along with a low Curie temperature. When the magnetic temperature increases to a certain point (the Curie temperature), the particle can be converted to ferromagnetic or paramagnetic material. In ferromagnetic material, it is difficult for its own magnetic field to be changed, but in paramagnetic material , its own magnetic field can easily be changed with a change in the external magnetic field [94]. When the size of magnetic particles is sufficiently small (nanometer), arbitrariness of the magnetic poles can occur at room temperature. That is, the particles show strong magnetism in the presence of an applied magnetic field, and they no longer exhibit magnetism or residual magnetism when an external magnetic field is not present. This phenomenon is known as superparamagnetic behavior [95], and particles with such properties are called superparamagnetic nanoparticles.

Magnetic nanoparticles are used as biometric markers, and the detection of magnetic signals can provide highly sensitive quantitative results in immunochromatographic assays. As a result, the technology has broad prospects for development. The Dynal Biotech ASA has developed a number of commercial magnetic polymer microspheres, Dynal beads, and these have been successfully used in microbiology, immunology, cancer research, and other fields [95]. From the perspective of improving sensitivity and quantification, magnetic nanoparticles form an ideal marking material due to the small background interference. With the help of an apparatus that can detect magnetic signals, quantitative detection data can be obtained from magnetic-based strips. Puertas et al. [96] achieved good results with an apparatus that conjugated the magnetic particles with an antibody to detect HCG; in other respects, magnetic immunochromatographic tests have been used to detect the human papilloma virus [97], cTnI [98], cytokine interferon-gamma (IFN)-$\gamma$ [99], *E. coli* [100], and HIV [101].

### *8.3.3 Application of Quantum Dots*

Quantum dots (QDs) were first presented by Mark A. Reed in 1988. Quantum dots are nanocrystals made of a semiconductor material (size between 1 and 10 nm) and nanoparticles that are generally made with IIB ~ VIB or IIIB ~ VB group elements, showing a significant quantum confinement effect [102]. In the late 1990s, with the continuous improvement of QD preparation technology, the technology showed broad prospects for application in biomedical research [103–105].

Since the size of the QD is less than the Bohr radius, the quantum confinement effect endows special properties to the material, properties that are significantly different from those of most solid materials; the most important being the unique optical properties. The light-emitting principle of the QD is similar to that of light-emitting diodes; light-emitting diodes require electrical excitation, while QDs require only light excitation [106]. The molar extinction coefficient of the QD is $0.5–5 \times 106\ M^{-1}\ cm^{-1}$, 10- to 50-fold higher than that of organic dyes [107–109]. Table 8.1 shows the differences between QDs and conventional organic dyes [110]. Compared with conventional organic fluorescent dyes, QDs have a variety of unique

**Table 8.1** Comparison of organic fluorescent dyes and quantum dots [110]

| | Quantum dots | Organic fluorescent dyes |
|---|---|---|
| Excitation spectrum | Wide | Narrow |
| | UV light can excite all quantum dots | Close to the emission spectrum |
| Emission bandwidth | 20–40 nm | 50–100 nm |
| Fluorescence lifetime | 10–40 ns | <10 ns |
| Light stability | 50–10,000 times stronger than organic fluorescent dye | Determined by the type of fluorophore |
| Molar extinction coefficient | $0.5–5 \times 10^6\ M^{-1}cm^{-1}$ | $5–10 \times 10^8\ M^{-1}cm^{-1}$ |
| Detection sensitivity | +++++ | + |
| Multicolor coding ability | ++++ | +++ |
| Quantitative ability | +++++ | +++ |

advantages, such as high fluorescence luminance, strong anti-photobleaching capacity, adjustable emission size spectral components [111], and broad absorption spectra. By controlling the size and chemical composition of the QDs, the fluorescence emission spectrum can span from near ultraviolet–visible spectroscopy to the near infrared region, as shown in Fig. 8.4 [112–114].

Since two different groups, Alivisatos [103] and Nie [105], reported the use of QDs in biological studies in 1998, massive research work has sprung up around the application of QDs in biomedicine. The companies Nanosphere Inc. [115], Genicon Science Corp. [116], and Quantum Dots Corp. [117] are advanced in applying QDs to the field of biological immunoassays, and with these applications there is great expectation for the future employment of QDs. The most studied QDs in recent years are CdSe, CdS, and CdTe [118–120]. Analyzing the fluorescence intensity of QD-labeled probes that link directly to the target molecule enables the qualitative or quantitative detection of the target molecule. Lin et al. [121], using a QD-based lateral flow test strip (LFTS), achieved the rapid detection of prostate-specific antigen (PSA) in human serum. Through the preparation of water-soluble QDs, our laboratory has developed a QD-based LFTS and established a simple and sensitive single-target rapid detection method that has succeeded in the detection of syphilis leptospira antibody in clinical serum samples. The sensitivity and specificity of QDs, as a new fluorescence label for LFTS, have been proven in this preliminary work. Figure 8.5 shows, on the right, a gold nanoparticle-based LFTS; on the left is a QD-based LFTS. It is evident that the sensitivity of the QD-based LFTS is much higher than that of the gold nanoparticle-based LFTS [122].

## 8.4 Research Directions in Immunochromatographic Technology

Immunochromatographic technology has been widely employed in increasing numbers of fields for its simple and rapid properties. Since the 1990s, researchers have been constantly studying and improving the technology and have made remarkable achievements.

Recently, many efforts have been made to improve the performance of lateral flow immunoassays, in terms of better sensitivity, multiplexing capability, and quantification.

### *8.4.1 Research on Improving Sensitivity*

Detection sensitivity is improved mainly through changing the markers, adding color enhancement agents, and amplifying the immune response. Horton et al. [125], when testing mouse immunoglobulin G (IgG), used liquid silver to strengthen the dyeing result after colloidal gold was captured by the test line. The sensitivity was increased by 100 times. Muller et al. [143] introduced avidin and biotin to an immunochromatographic system to detect fetal cTnT. One molecule of avidin has

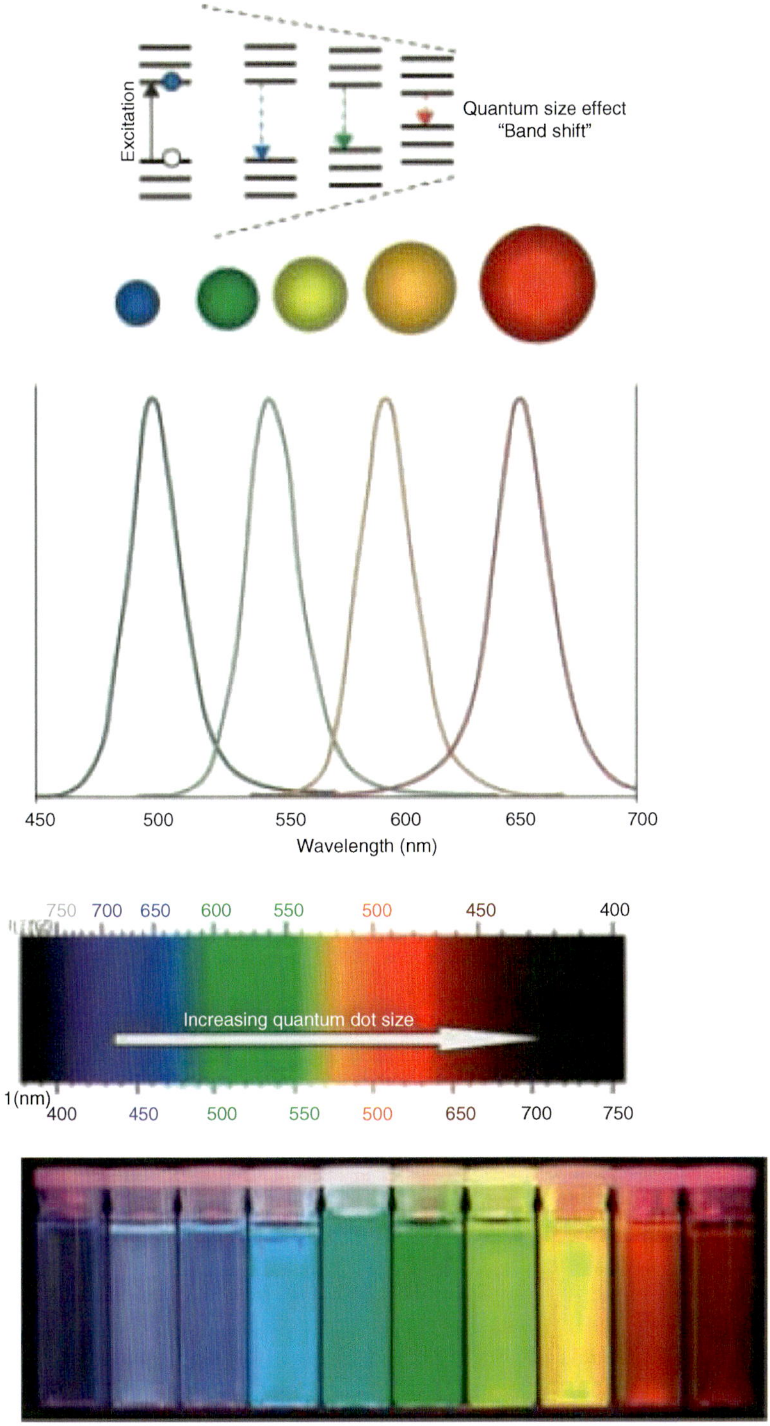


Fig. 8.4 Optical properties of quantum dot (QD) change with its diameter

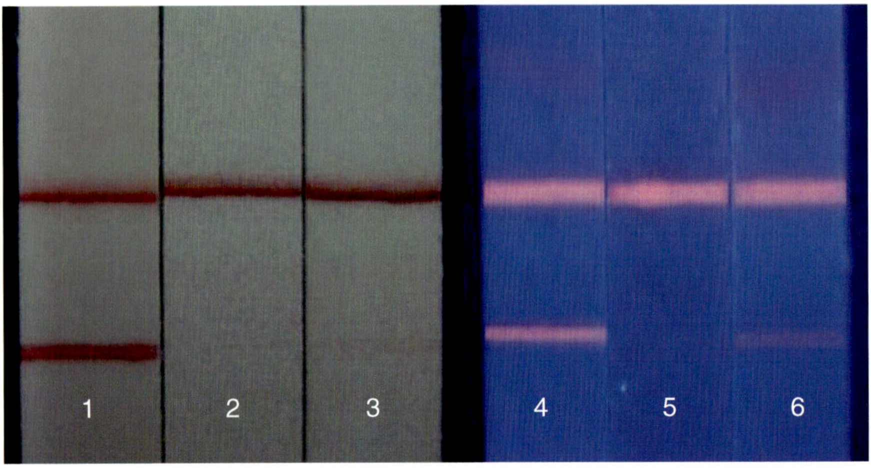


**Fig. 8.5** Test result of colloidal gold compared with QD-based lateral flow test strip (LFTS)

four binding sites that can bind with the biotin. So the signal intensity could be amplified greatly. Some researchers utilize liposomes as labels [123]; these offer higher sensitivity, owing to their strong plasticity and ability to set multiple antigens and antibodies in the membrane and to contain many indicators in the water phase space inside the membrane. Gred et al. [144] used europium microparticles as labels in an immunochromatographic system. Through the detection of eosinophil protein X and neutrophil lipocalin to reflect the concentration of eosinophil and neutrophil in blood. Magnetic nanoparticles, a new material developed in recent years, have a higher sensitivity in immunochromatographic assays, through testing magnetism, than the conventional luminescence method [124].

## *8.4.2 Development of Multivariate Analysis*

In some cases, we need to test a variety of indexes to make a comprehensive judgment. So the development of immunochromatographic assays also takes into account the need for multivariate analysis, i.e., the analysis of several substances in one strip; this not only improves detection efficiency but also reduces testing costs. Immunochromatographic assays have a high application value for testing indexes; this is well suited for multivariate analysis. For example, Hepatitis B virus (HBV) can be diagnosed by testing for preS1, preS2, hepatitis B surface antigen (HBsAg), Hepatitis B, envelope (env) antigen (HBeAg), and others; also, a variety of hormones, allergens, and cytokines can be analyzed according to different testing requirements. Emiliana Tjitra et al. [145], using an immunochromatographic assay, successfully developed a test strip that can simultaneously analyze *Plasmodium falciparum* and *Plasmodium vivax* with good sensitivity and specificity and clinical application value. Buechler et al. [126] utilized a membrane sprayed with several T

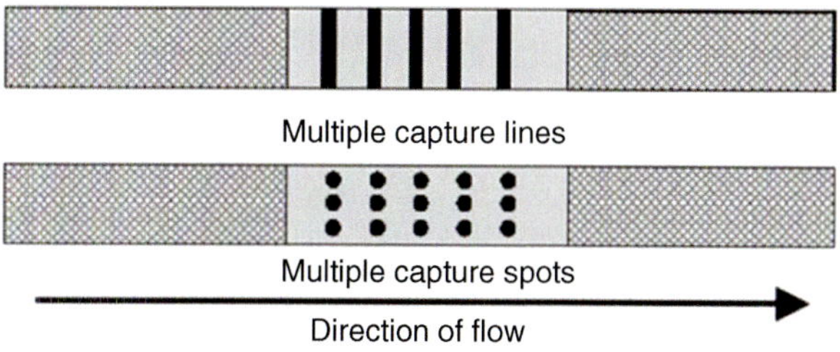


**Fig. 8.6** Principle of multiplex lateral flow immunoassay

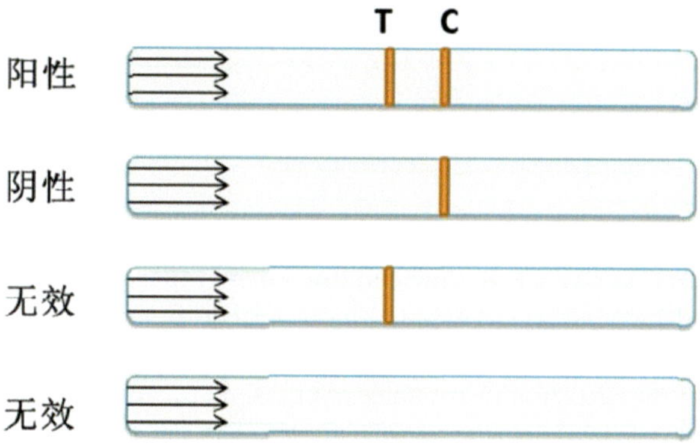


**Fig. 8.7** Example of qualitative detection

lines and succeeded in the simultaneous detection of seven illicit drugs in urine. Sterling et al. [146] employed different membranes, and enzymes in different places in the membrane, to separate and visualize the analyte; their system simultaneously tested hemoglobin, glucose, and cholesterol in blood (Fig. 8.6).

### 8.4.3 *Development from Qualitative to Quantitative Detection*

Immunochromatographic assays have the characteristics of easy operation and rapid, specific, and clear results; they are easy to assess and save, and there is no need for any instruments. Figure 8.7, from top to bottom, shows test results for in male and female subjects, and it also shows invalid findings. Obviously, qualitative detection is not sufficient on many occasions. Many test strips can be used to qualitatively analyze, in terms of HCG detection, whether or not a female is pregnant, but threatened abortion, ectopia, drug excretion, some obstetric and gynecological diseases, and postoperative treatment all need the quantitative detection of HCG, as well as the tracking of quantitative changes in this molecule. Recently, much research has focused on qualitative/quantitative immunochromatographic assays. Quantitative detection not only has significance for the timely diagnosis of acute disease but is also a very important means of establishing health standards in the relevant quality inspection departments. Tippkötter et al. [127] established a relation curve between the color values of the strips and the concentrations of the analytes in a rapid test of microcystin concentration (Fig. 8.8). In clinical medicine, quick and

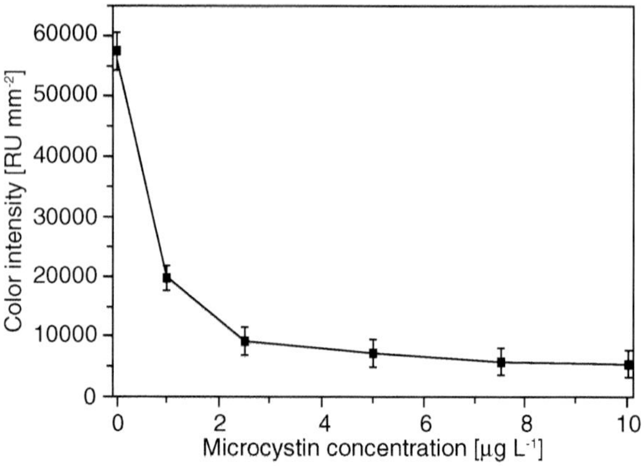


**Fig. 8.8** The relation curve between color values of the strips and the concentrations of microcystin

easy quantitative detection could help doctors to determine the pathogenetic conditions of patients in time and implement timely treatment, which is vitally important in the treatment of acute diseases such as myocardial infarction. Quantitative detection is not only based on the preparation of strips but also concerns the development of the detection instruments. At present, the main focus of research interest in quantitative testing is the use of a photoelectric detection method to transform the optical signal to an electrical signal and then a digital signal; thus obtaining a quantitative detection result. Alternatively, magnetic nanoparticle-based LFTS can be analyzed through magnetic flux-testing equipment to obtain quantitative results.

Drug abuse is an important factor in harming people's health. Recent quantitative methods for drug detection are accurate, but the instrumentation is expensive and needs strict experimental conditions. Researchers, both in China and abroad, are looking for a simple, fast, and accurate testing method. Gao Yueming et al. [147] developed an immune colloidal gold-based rapid quantitative testing method for drug detection. Because the content of some tested materials in the human body consistently changes within a certain range, the test showed only positive or negative results, which was unsatisfactory. Some researchers have succeeded in the quantitative detection of theophylline and cholesterol in blood, in tests based on enzyme colorimetry; quantitative results were obtained by comparing chromogenic heights on test strips with standard curves.

However, we note that simple and rapid quantitative detection is urgently needed in the fields of clinical medicine, environmental veterinary drug residues, and the detection of illicit drugs.

## 8.5 Research on Detection Systems

The field of immunochromatographic requires equipment that can be used for qualitative detection and that can improve the accuracy of tests and standard operation. Complete detection systems can help operators to achieve more accurate results and

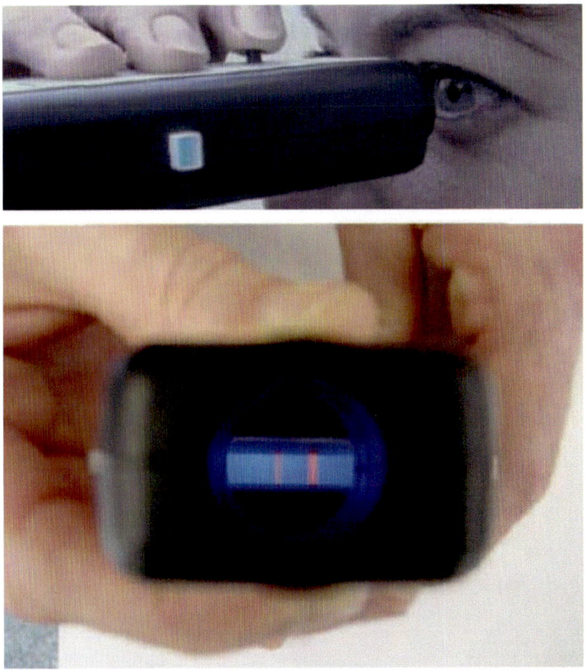

**Fig. 8.9** German ESE test strip qualitative detection equipment [128]

avoid environmental interference. Compared with the extent of the literature on methods of immunchromatographic detection, there are fewer articles about detection test devices. Qualitative test equipment developed by the German ESE (Jacques-Schiesser-Strasse 3, D-78333 Stockach, Germany) company is illustrated in Fig. 8.9. With the device, which provides and stimulates luminescence, the operator can observe the emission of light from a fluorescent strip on the immunoassay strip, but the device does not have the function of quantitative detection [128].

Both in China and abroad, there are fewer commercial quantitative devices than qualitative test devices. The qualitative test devices mainly use colloidal gold-labeled test strip detection. Many companies; for example, BBI (BBI Solutions 73 Ty Glas Avenue Cardiff, CF14 5DX, UK), and Boehringer Mannheim Corporation and Nycomed in Europe, have performed a great deal of research on the semiquantitative and quantitative detection of nanogold immunochromatographic tests. A test strip (based on nanogold technology) to be used in the immunochromatographic detection of troponin and myoglobin for diagnosing acute myocardial infarction was launched by Boehringer Mannheim. The simple measuring device used for this strip had passed through clinical testing. The precision and accuracy were consistent with the requirements of quantitative determination [129]. Chandler et al. [130] carried out quantitative detection with a nanogold immunochromatographic test strip; the concentration of ladder type was displayed on the test paper and then the color signal of the strip was converted to a digital signal by a portable instrument. A CD-ROM optical detector was designed by Suhyeon et al. [131] for the quantitative detection of

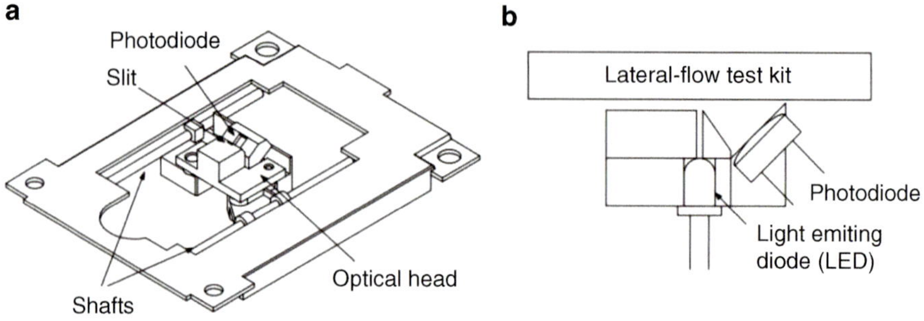


**Fig. 8.10** Schematic diagrams of (**a**) the mechanical parts of the strip reader and (**b**) cross-sectional view of the optical head [131]

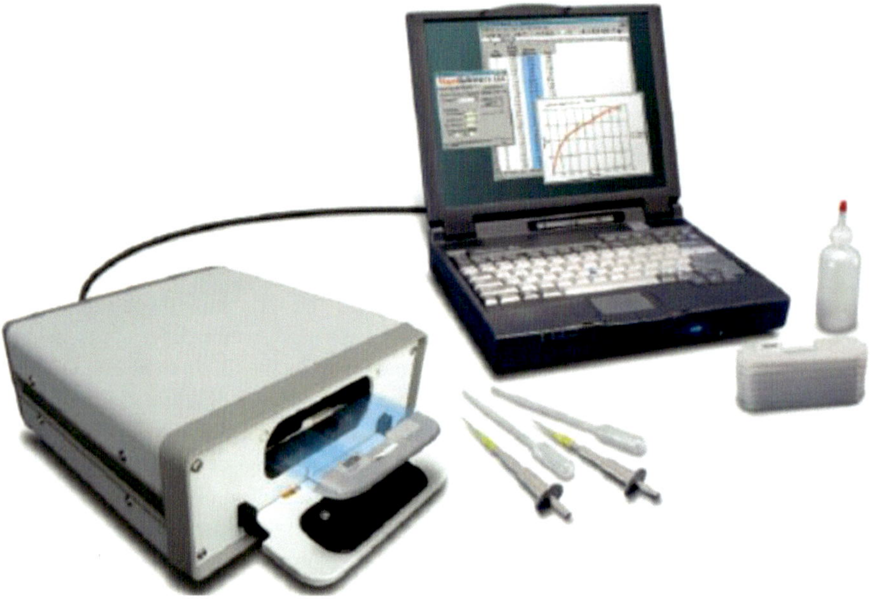

**Fig. 8.11** Immune magnetic strip detection system magnetic assay reader (MAR) [134]

hepatitis B virus (HBV); the mechanical components and optical probe of the detector are illustrated in Fig. 8.10. These researchers produced a cheap and fast detective system with the test sample being transmitted by the motor. Portable equipment used for the ultrasensitive detection of protein biomarkers was invented by Li et al. [132]; their device, based on QDs, achieved quantitative detection on the test strip, with a ceruloplasmin detection range of up to 1 ng/mL. Zou et al. [133] used a system of QD notation to achieve the quantitative detection of the insecticide trichloropyridinol.

MagnaBioSciences company developed a magnetic assay reader (MAR) magnetic strip for immunchromatographic detection. This represents a huge upgrade and challenge to the traditional nanogold-labeled or labeled latex particles such as

**Fig. 8.12** Embedded detection system [140]

those used on immunochromatographic strips [134]. The magnetic assay reader (MAR) (Fig. 8.11) for the quantitative detection of magnetic test strip is by measuring the magnetic flux generated at different zones on a membrane under the magnetic field. Compared with traditional optical test, magnetic test realizes quantitative measurement through the reading of magnetic signals that can be entirely captured by the devices, thereby increasing test sensitivity. The device has received a lot of attention since its introduction, and a large number of in-vitro applications based on the detection system have been published. The magnetic strip method is different from the traditional method (which is based on color depth discrimination and which is easily hampered by the presence of other colored material in the sample) and the sensitivity is high [135–139, 142].

The quantitative study of the immune system started late in China, and it mainly focused on the quantitative detection of nanogold immunochromatographic test strips. The Chinese Academy of Sciences and the Guangzhou Institutes of Biomedicine and Health applied immune layer technology to the detection of DNA-binding protein, and with a biological sensor using gold nanoparticles, detection was rapid, being completed within 10 min [86]. An embedded reading instrument (Fig. 8.12), developed by Nankai University, has high-resolution Complementary Metal Oxide Semiconductor (CMOS) image acquisition, and the detection accuracy is 1 ng/mL; the coefficient of variation (CV) is less than 3 % [140].

The Chinese Academy of Sciences Institute of Chemistry detected pesticide residues by using magnetic nanoparticles. There were corresponding curve between oxygen phosphorus methyl and optical density at the different concentrations, and the limit of detection was 1.7 mg/mL [141] (Fig. 8.13). The Shanghai Institute of Optical Precision Machinery performed quantitative detection of a gold tag strip using a reflection photometer. They deduced the quantitative detection of the cTnI equation. The correlation coefficient of the measured sample concentration and the optical detection value was 0.989. The CV was less than 5 %, as shown in Fig. 8.14, which shows the structure of the reflex photometer [74].

Fuzhou University developed testing equipment for HCG based on a Charge-coupled Device (CCD), colloidal gold-marked strip. Their correlation coefficients were 0.9975 and 0.9914 [142], respectively, for two different methods; i.e, determi-

**Fig. 8.13** Optical density profiles of the T-line and C-line recorded by using sample 1-ab (**a**) and sample 4-ab (**b**) after running a series of standard solutions with different paraoxon methyl concentrations [141]

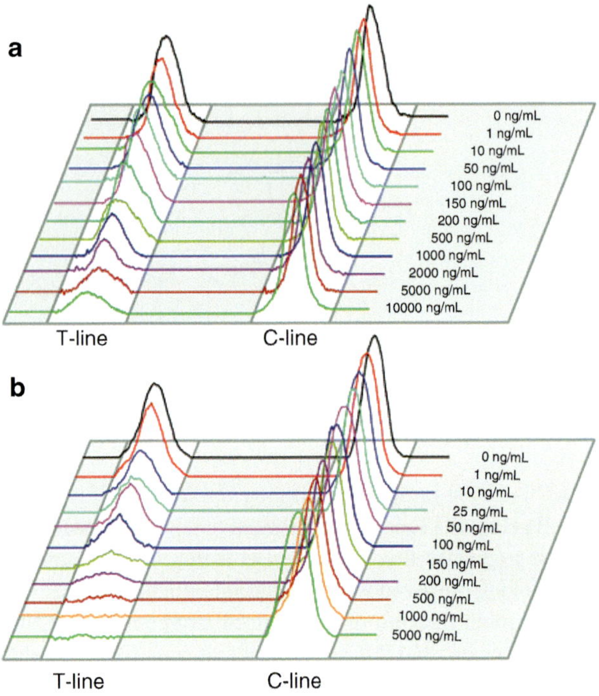


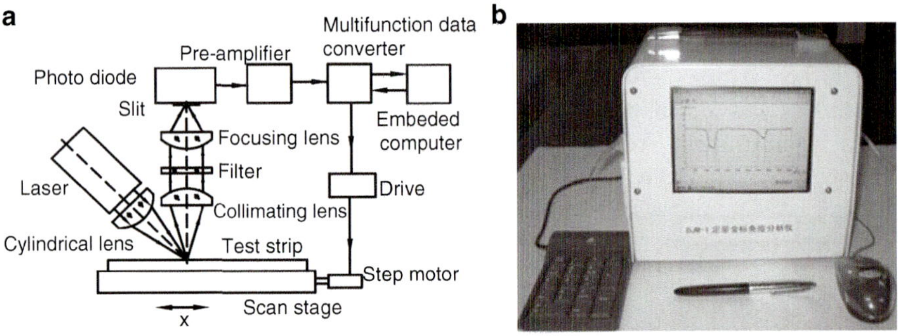


**Fig. 8.14** Structure of reflex photometer (diagram) [74]. (**a**) Structural illustration of the reflectance photometer. (**b**) Photograph of the reflectance photometer

nation of the concentration of the sample and detection of the optical density value. In China, there is little quantitative detection equipment to support the use of magnetic strips and QD strips. Shanghai Jiao Tong University developed a testing device based on CCD QD immunochromatographic test paper. The device has led to the detection of early gastric cancer markers, such as cytotoxin associated gene A (CagA) and urease, and it has been used for the quantitative detection of CEA,

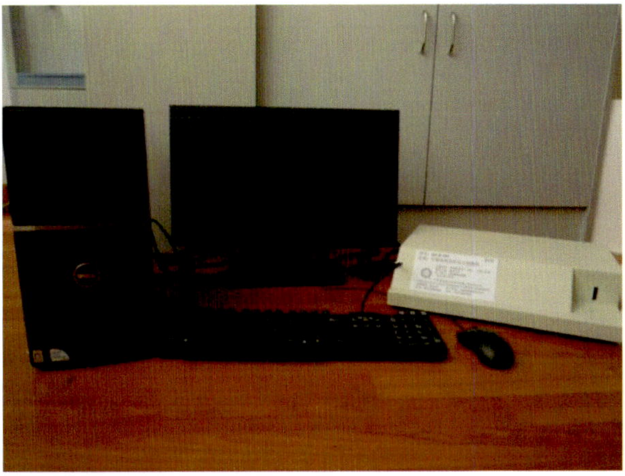

**Fig. 8.15** Immuno-chromatographic chip detector

**Fig. 8.16** Portable biological immunochromatographic detector

carbohydrate antigen 724 (CA724), and other markers; a medical device registration certificate has been obtained (Fig. 8.15).

At the same time, Shanghai Jiao Tong University developed a portable immune layer chip detector based on embedded technology (Fig. 8.16). The system, based on a confocal imaging principle, achieved more accurate data collection and analysis than QD immunochromatographic test strips, owing to the fluorescent signal obtained through the high-performance LED point light source, the low noise and high precision of the photoelectric diode, the mechanical movement module, and the positioning module. Single test time is about 1.5–2 min, with a fluorescent molecular/mu detection sensitivity of 0.4 $m^2$; spatial resolution is 1000 μm. The application of the embedded system greatly reduces the volume of the equipment and increases its portability. The system can provide rapid test results, so that effective treatment for major diseases can be initiated in a timely manner, thus saving lives. With portable devices, the deficiencies of large equipment can be avoided. Basic preventive screening can be performed for community residents; this equipment is more suitable for bedside testing requirements than other types of instruments, as has been shown in a community survey.

## References

1. Newman DJ, Thakkar H, Edwards RG, et al. Serum cystatin C measured by automated immunoassay: a more sensitive marker of changes in GFR than serum creatinine. Kidney Int. 1995;47(1):312–8.
2. Van Weeman B, Schuurs A. Immunoassay using antigen-enzyme conjugates. FEBS Lett. 1971;15:232–5.
3. Hales C, Randle P. Immunoassay of insulin with insulin-antibody precipitate. Biochem J. 1963;88(1):137–46.
4. Ahene AB, Morrow C, Rusnak D, et al. Ligand binding assays in the 21st century laboratory: automation. AAPS J. 2012;14(1):1–12.
5. Tung NH, Chikae M, Ukita Y, et al. Sensing technique of silver nanoparticles as labels for immunoassay using liquid electrode plasma atomic emission spectrometry. Anal Chem. 2012;84(3):1210–3.
6. Guitard J, Sendid B, Thorez S, et al. Evaluation of a recombinant antigen-based enzyme immunoassay for the diagnosis of noninvasive aspergillosis. J Clin Microbiol. 2012; 50(3):762–5.
7. Verch T, Bakhtiar R. Miniaturized immunoassays: moving beyond the microplate. Bioanalysis. 2012;4(2):177–88.
8. Zhang BY, Song HX, Chen T, et al. A microfluidic platform for multi-antigen immunofluorescence assays. Appl Mech Mater. 2012;108:200–5.
9. Strathmann FG, Borlee G, Born DE, et al. Multiplex immunoassays of peptide hormones extracted from formalin-fixed, paraffin-embedded tissue accurately subclassify pituitary adenomas. Clin Chem. 2012;58(2):366–74.
10. Flatley JE, Garner CM, Al-Turki M, et al. Determinants of urinary methylmalonic acid concentration in an elderly population in the United Kingdom. Am J Clin Nutr. 2012;95(3):686–93.
11. Arai M, Togo S, Kanda T, et al. Quantification of hepatitis B surface antigen can help predict spontaneous hepatitis B surface antigen seroclearance. Eur J Gastroenterol Hepatol. 2012;24(4):414–8.
12. Sloan JH, Ackermann BL, Carpenter JW, et al. A novel, high-sensitivity and drug-tolerant sandwich immunoassay for the quantitative measurement of circulating proteins. Bioanalysis. 2012;4(3):241–8.
13. Bayes-Genis A, de Antonio M, Galán A, et al. Combined use of high-sensitivity ST2 and NTproBNP to improve the prediction of death in heart failure. Eur J Heart Fail. 2012;14(1):32–8.
14. Mehta PK, Kalra M, Khuller GK, et al. Development of an ultrasensitive polymerase chain reaction–amplified immunoassay based on mycobacterial RD antigens: implications for the serodiagnosis of tuberculosis. Diagn Microbiol Infect Dis. 2012;72(2):166–74.
15. Maple P, Breuer J, Quinlivan M, et al. Comparison of a commercial Varicella Zoster glycoprotein IgG enzyme immunoassay with a reference time resolved fluorescence immunoassay (VZV TRFIA) for measuring VZV IgG in sera from pregnant women, sera sent for confirmatory testing and pre and post vOka vaccination sera from healthcare workers. J Clin Virol. 2012;23(3):201–7.
16. Smits GP, van Gageldonk PG, Schouls LM, et al. Development of a bead-based multiplex immunoassay for the simultaneously quantitative detection of IgG serum antibodies against Measles, Mumps, Rubella and Varicella Zoster. Clin Vaccine Immunol. 2012;19(3): 369–400.

17. Deutschbein T, Broecker-Preuss M, Flitsch J, et al. Salivary cortisol as a diagnostic tool for Cushing's syndrome and adrenal insufficiency: improved screening by an automatic immunoassay. Eur J Endocrinol. 2012;166(4):613–8.
18. Wang Z, Zong S, Li W, et al. SERS-fluorescence joint spectral encoding using organic-metal-QDs hybrid nanoparticles with a huge encoding capacity for high-throughput biodetection: putting theory into practice. J Am Chem Soc. 2012;134(6):2993–3000.
19. Connolly P, Hage CA, Bariola JR, et al. Blastomyces dermatitidis antigen detection by quantitative enzyme immunoassay. Clin Vaccine Immunol. 2012;19(1):53–6.
20. Lattanzio VMT, Nivarlet N, Lippolis V, et al. Multiplex dipstick immunoassay for semi-quantitative determination of Fusarium mycotoxins in cereals. Anal Chim Acta. 2012; 718(3):99–108.
21. Warsinke A. Point-of-care testing of proteins. Anal Bioanal Chem. 2009;393(5):1393–405.
22. Brooks DE, Devine DV, Harris PC, et al. RAMPTM: a rapid, quantitative whole blood immunochromatographic platform for point-of-care testing. Clin Chem. 1999;45(9):1676.
23. Shim WB, Yang ZY, Kim JS, et al. Development of immunochromatography strip-test using nanocolloidal gold-antibody probe for the rapid detection of aflatoxin B1 in grain and feed samples. J Microbiol Biotechnol. 2007;17(10):1629–37.
24. Zuk R, Ginsberg V, Houts T, et al. Enzyme immunochromatography—aquantitative immunoassay requiring no instrumentation. Clin Chem. 1985;31(7):1144–50.
25. Lingerfelt BM, Mattoussi H, Goldman ER, et al. Preparation of quantum dot-biotin conjugates and their use in immunochromatography assays. Anal Chem. 2003;75(16):4043–9.
26. Ngom B, Guo Y, Wang X, et al. Development and application of lateral flow test strip technology for detection of infectious agents and chemical contaminants: a review. Anal Bioanal Chem. 2010;397(3):1113–35.
27. Shim WB, Yang ZY, Kim JY, et al. Immunochromatography using colloidal gold-antibody probe for the detection of atrazine in water samples. J Agric Food Chem. 2006; 54(26):9728–34.
28. Zhang G, Guo J, Wang X. Immunochromatographic lateral flow strip tests. Methods Mol Biol. 2009;504(3):169–83.
29. Kolosova AY, De Saeger S, Sibanda L, et al. Development of a colloidal gold-based lateral-flow immunoassay for the rapid simultaneous detection of zearalenone and deoxynivalenol. Anal Bioanal Chem. 2007;389(7):2103–7.
30. Zong C, Wu J, Wang C, et al. Chemiluminescence imaging immunoassay of multiple tumor markers for cancer screening. Anal Chem. 2012;84(5):2410–5.
31. Guirgis BSS, Sá e Cunha C, Gomes I, et al. Gold nanoparticle-based fluorescence immunoassay for malaria antigen detection. Anal Bioanal Chem. 2012;402(3):1019–27.
32. Ren G, Yu Z, Ma S, et al. Determination of polybrominated diphenyl ethers and their methoxylated and hydroxylated metabolites in human serum from electronic waste dismantling workers. Anal Methods. 2011;3(2):408–13.
33. Wu D, Li R, Wang H, et al. Hollow mesoporous silica microspheres as sensitive labels for immunoassay of prostate-specific antigen. Analyst. 2012;137(3):608–13.
34. Schreier S, Doungchawee G, Triampo D, et al. Development of a magnetic bead fluorescence microscopy immunoassay to detect and quantify Leptospira in environmental water samples. Acta Trop. 2012;122(1):119–25.
35. Mandappa IM, Ranjini A, Haware DJ, et al. Immunoassay for the detection of lead ions in environmental water samples. Int J Environ Anal Chem. 2012;92(3):334–43.
36. Rossi CN, Takabayashi CR, Ono MA, et al. Immunoassay based on monoclonal antibody for aflatoxin detection in poultry feed. Food Chem. 2012;132(4):2211–6.
37. Martínez MA, Ballesteros S. Two Suicidal fatalities due to the ingestion of chlorfenvinphos formulations: simultaneous determination of the pesticide and the petroleum distillates in tissues by gas chromatography–flame-ionization detection and gas chromatography–mass spectrometry. J Anal Toxicol. 2012;36(1):44–51.
38. Reither K, Saathoff E, Jung J, et al. Evaluation of Diagnos TB AG, a flow-through immunoassay for rapid detection of pulmonary tuberculosis [technical note]. Int J Tuberc Lung Dis. 2010;14(2):238–40.

39. Qi YH, Shan WC, Liu YZ, et al. Production of the polyclonal antibody against sudan 3 and immunoassay of sudan dyes in food samples. J Agric Food Chem. 2012;60(9):2116–22.
40. Li X, Zhang G, Deng R, et al. Development of rapid immunoassays for the detection of ractopamine in swine urine. Food Addit Contam. 2010;27(8):1096–103.
41. Ecollan P, Collet JP, Boon G, et al. Pre-hospital detection of acute myocardial infarction with ultra-rapid human fatty acid-binding protein (H-FABP) immunoassay. Int J Cardiol. 2007;119(3):349–54.
42. Hampl J, Hall M, Mufti NA, et al. Upconverting phosphor reporters in immunochromatographic assays. Anal Biochem. 2001;288(2):176–87.
43. Xu QF, Xu H, Gu H, et al. Development of lateral flow immunoassay system based on superparamagnetic nanobeads as labels for rapid quantitative detection of cardiac troponin I. Mater Sci Eng C. 2009;29(3):702–7.
44. Kurihara T, Yanagida A, Yokoi H, et al. Evaluation of cardiac assays on a benchtop chemiluminescent enzyme immunoassay analyzer, PATHFAST. Anal Biochem. 2008;375(1):144–6.
45. Jun L, Wenquan Y, Nanhua W, et al. Analysis result of faecal occult blood test based on colloidal immunization method. Lab Med Clin. 2008;2:111–2. (in Chinese).
46. Clavijo E, Díaz R, Anguita A, et al. Comparison of a dipstick assay for detection of Brucella-specific immunoglobulin M antibodies with other tests for serodiagnosis of human brucellosis. Clin Vaccine Immunol. 2003;10(4):612–5.
47. Ruiming Z, Dandan T, Jie C, et al. Detection of mycoplasma and chlamydia in 292 cases of female urogenital tract infections and the antibiotic susceptibility analysis. Chin Prac Med. 2008;3(28):15–6. (in Chinese).
48. Yueyue L, Hangwei C, Ping W, et al. Application of gold immunochromatographic assay for a rapid detection of influenza viruses. J Fourth Mil Med Univ. 2008;29(18):1652–4. (in Chinese).
49. Saijun L, Yongjun Z. Optimization of production process of diagnostic reagent of colloidal gold for HBsAb. Prog Mod Biomed. 2008;8(002):289–92. (in Chinese).
50. Jingfeng T, Xiaoyan L, Xinglong W, et al. Development of colloidal gold-immunochromatographic assay for brucellosis. Chin J Biol. 2007;20(2):119–21. (in Chinese).
51. Chen TS, Chang FY, Lee SD. Serodiagnosis of Helicobacter pylori infection: comparison and correlation between enzyme-linked immunosorbent assay and rapid serological test results. J Clin Microbiol. 1997;35(1):184.
52. Yonghong Y. Application of gold immunchromatographic assay in rapid diagnosis of tuberculosis. Gansu Sci Technol. 2009;24(22):173–4. (in Chinese).
53. Takeda T, Yamagata K, Youhida Y, et al. Evaluation of immunochromatography-based rapid detection kit for fecal Escherichia coli O 157. Kansenshogaku Zasshi. 1998;72(8):834–9.
54. Osikwicz G, Beggs M, Brookhart P, et al. One-step chromatographicimmunoassay for qualitative determination of choriogonadotropin in urine. Clin Chem. 1990;36(9):1586.
55. Hedstrom J, Korvuo A, Kenkimaki P, et al. Urinary trypsinogen-2 test strip for acute pancreatitis. Lancet. 1996;347:729–31.
56. Wang J. Nanomaterial-based electrochemical biosensors. Analyst. 2005;130(4):421–6.
57. Wang J. Nanomaterial-based amplified transduction of biomolecular interactions. Small. 2005;1(11):1036–43.
58. Ekimov A, Efros AL, Onushchenko A. Quantum size effect in semiconductor microcrystals. Solid State Commun. 1985;56(11):921–4.
59. Jun Y, Huh YM, Choi J, et al. Nanoscale size effect of magnetic nanocrystals and their utilization for cancer diagnosis via magnetic resonance imaging. J Am Chem Soc. 2005;127(16): 5732–3.
60. Dingreville R, Qu J, Cherkaoui M. Surface free energy and its effect on the elastic behavior of nano-sized particles, wires and films. J Mech Phys Solids. 2005;53(8):1827–54.
61. Kawabata S, Kashiwaya S, Asano Y, et al. Effect of zero-energy bound states on macroscopic quantum tunneling in high-Tc superconductor junctions. Phys Rev B. 2005;72(5): 052506–9.

62. Yu I, Isobe T, Senna M. Optical properties and characteristics of ZnS nano-particles with homogeneous Mn distribution. J Phys Chem Solid. 1996;57(4):373–9.
63. Hao E, Bailey RC, Schatz GC, et al. Synthesis and optical properties of "branched" gold nanocrystals. Nano Lett. 2004;4(2):327–30.
64. Novak JP, Brousseau III LC, Vance FW, et al. Nonlinear optical properties of molecularly bridged gold nanoparticle arrays. J Am Chem Soc. 2000;122(48):12029–30.
65. Caruso F, Rodda E, Furlong DN, et al. DNA binding and hybridization on gold and derivatized surfaces. Sens Actuators B. 1997;41(1–3):189–97.
66. Zhao H, Lin L, Li J, et al. DNA biosensor with high sensitivity amplified by gold nanoparticles. J Nanopart Res. 2001;3(4):321–3.
67. Liqiang L, Chifang P, Zhengyu J, et al. Review on development and application of nanogold in rapid detection for food safety. Food Sci. 2007;28(5):348–52. (in Chinese).
68. Mirkin CA. Programming the assembly of two-and three-dimensional architectures with DNA and nanoscale inorganic building blocks. Inorg Chem. 2000;39(11):2258–72.
69. Glad C, Grubb AO. Immunocapillarymigration with enzyme-labeled antibodies: rapid quantification of C-reactive protein in human plasma. Anal Biochem. 1981;116(2):335–40.
70. Beggs M, Novotny M, Sampedro S. A selfperforming chromatographic immunoassay for the qualitative determination of human chorionic gonadotrophin (HCG) in urine and serum. Clin Chem. 1990;36(11):1084–5.
71. Müller-Bardorff M, Freitag H, Scheffold T, et al. Development and characterization of a rapid assay for bedside determinations of cardiac troponin T. Circulation. 1995;92(10):2869–75.
72. Zhang B, Tang D, Liu B, et al. Nanogold-functionalized magnetic beads with redox activity for sensitive electrochemical immunoassay of thyroid-stimulating hormone. Anal Chim Acta. 2011;711(20):17–23.
73. Perfézou M, Turner A, Merkoçi A. Cancer detection using nanoparticle-based sensors. Chem Soc Rev. 2012;41(7):2606–22.
74. Huang L, Zhang Y, Xie C, et al. Research of reflectance photometer based on optical absorption. Opt Int J Light Electron Optics. 2010;121(19):1725–8.
75. Wang S, Zhang C, Wang J, et al. Development of colloidal gold-based flow-through and lateral-flow immunoassays for the rapid detection of the insecticide carbaryl. Anal Chim Acta. 2005;546(2):161–6.
76. Lihua H. The application and evaluation of HBVM test card for rapid detection of serum hepatitis B marks. J Qiqihar Med Coll. 2001;22(12):1433–4. (in Chinese).
77. Meng Q, Shengli B, ZhingHeng D. The study of colloidal gold immunochromatography assay for the synchronous detection of hepatitis A and E virus-specific IgM antibody. Chin J Lab Diagn. 2003;7(2):90–2. (in Chinese).
78. Yuqin Q, Zhiying Z. Investigation and analysis on the positive rate of helicobacter pylori antibody in healthy people Journal of Chinese Modern Medicine. 2009;6(2):155. (in Chinese).
79. Yan J, Changlong L, Fengpign D. Preparation of colloid gold kit for one-step chromatography immunoassay to monitor the dopes of morphine and met-amphetamine simultaneously. Chin J Immunol. 2007;23(7):637–40. (in Chinese).
80. Kim GM, Wutzler A, Radusch HJ, et al. One-dimensional arrangement of goldnanoparticles by electrospinning. Chem Mater. 2005;17(20):4949–57.
81. Tyndall SJ, Walikonis RS. Report the receptor tyrosine kinase Met and its ligand hepatocyte growth factor are clustered at excitatory synapses and can enhance clustering of synaptic proteins. Cell Cycle. 2006;5(14):1560–8.
82. Kim HS, Pyun JC. Hyper sensitive strip test with chemi-luminescence signal band. Procedia Chem. 2009;1(1):1043–6.
83. Jianing L, Yibing L, Juanjuan J. Chemiluminescence immunoassay for luteinizing hormone. J Isot. 2010;23(1):28–33. (in Chinese).
84. Shengchu L. Early abortion and the detection of pathogens in preterm pregnant women. Guangxi Med J. 2007;29(4):496–7. (in Chinese).

85. Laderman EI, Whitworth E, Dumaual E, et al. Rapid, sensitive, and specific lateral-flow immunochromatographic point-of-care device for detection of herpes simplex virus type 2-specific immunoglobulin G antibodies in serum and whole blood. Clin Vaccine Immunol. 2008;15(1):159–63.
86. Fang Z, Ge C, Zhang W, et al. A lateral flow biosensor for rapid detection of DNA-binding protein c-jun. Biosens Bioelectron. 2011;21(1):192–6.
87. Glynou K, Ioannou PC, Christopoulos TK, et al. Oligonucleotide-functionalized gold nanoparticles as probes in a dry-reagent strip biosensor for DNA analysis by hybridization. Anal Chem. 2003;75(16):4155–60.
88. AziziM D. Comparison study of semiquantitative test strips for detecting human serum albumin (HSA) in urine specimens. J Diabetes Metab Disord (Formerly Iran J Diabetes Lipid Disord). 2011;10(1):1–5.
89. Dineva MA, Mahilum-Tapay L, Lee H. Sample preparation: a challenge in the development of point-of-care nucleic acid-based assays for resource-limited settings. Analyst. 2007;132(12):1193–9.
90. Omidfar K, Kia S, Larijani B. Development of a colloidal gold-based immunochromatographic test strip for screening of microalbuminuria. Hybridoma. 2011;30(2):117–24.
91. Holgate CS, Jackson P, Cowen PN, et al. Immunogold-silver staining: new method of immunostaining with enhanced sensitivity. J Histochem Cytochem. 1983;31(7):938–44.
92. Löning T, Henke RP, Reichart P, et al. In situ hybridization to detect Epstein-Barr virus DNA in oral tissues of HIV-infected patients. Virchows Arch. 1987;412(2):127–33.
93. Molday R, Yen S, Rembaum A. Application of magnetic microspheres in labelling and separation of cells. Nature. 1977;268(4):437–8.
94. Xinquan L, Meng J, Changyu L, et al. The current situation of studying SPIO and its application in the field of NSCs. J Clin Med Pract. 2003;7(3):232–5. (in Chinese).
95. Gangopadhyay S, Hadjipanayis G, Dale B, et al. Magnetic properties of ultrafine iron particles. Phys Rev B. 1992;45(17):9778–87.
96. Puertas S, Moros M, Fernández-Pacheco R, et al. Designing novel nano-immunoassays: antibody orientation versus sensitivity. J Phys D Appl Phys. 2010;43:474012–9.
97. Patel D, Huang SM, Baglia LA, et al. The E6 protein of human papillomavirus type 16 binds to and inhibits co-activation by CBP and p300. EMBO J. 1999;18(18):5061–72.
98. Hamm CW, Goldmann BU, Heeschen C, et al. Emergency room triage of patients with acute chest pain by means of rapid testing for cardiac troponin T or troponin I. N Engl J Med. 1997;337(23):1648–53.
99. Liu T, Khanna KM, Carriere BN, et al. Gamma interferon can prevent herpes simplex virus type 1 reactivation from latency in sensory neurons. J Virol. 2001;75(22):11178–84.
100. Seo K, Brackett R, Frank J. Rapid detection of Escherichia coli O157: H7 using immunomagnetic flow cytometry in ground beef, apple juice, and milk. Int J Food Microbiol. 1998;44(1–2):115–23.
101. Carbonari M, Cibati M, Cherchi M, et al. Detection and characterization of apoptotic peripheral blood lymphocytes in human immunodeficiency virus infection and cancer chemotherapy by a novel flow immunocytometric method. Blood. 1994;83(5):1268–77.
102. Drummen GP. Quantum dots—from synthesis to applications in biomedicine and life sciences. Int J Mol Sci. 2010;11(1):154–63.
103. Alivisatos AP. Semiconductor clusters, nanocrystals, and quantum dots. Science. 1996;271(5251):933–7.
104. Bruchez M, Moronne M, Gin P, et al. Semiconductor nanocrystals as fluorescent biological labels. Science. 1998;281(5385):2013–6.
105. Chan WCW, Nie S. Quantum dot bioconjugates for ultrasensitive nonisotopic detection. Science. 1998;281(5385):2016–8.
106. Watson A, Wu X, Bruchez M. Lighting up cells with quantum dots. Biotechniques. 2003;34(2):296–300.
107. Xing Y, Xia Z, Rao J. Semiconductor quantum dots for biosensing and in vivo imaging. IEEE Trans Nanobioscience. 2009;8(1):4–12.

108. Leatherdale C, Woo WK, Mikulec F, et al. On the absorption cross section of CdSe nanocrystal quantum dots. J Phys Chem B. 2002;106(31):7619–22.
109. Alivisatos AP. Perspectives on the physical chemistry of semiconductor nanocrystals. J Phys Chem. 1996;100(31):13226–39.
110. Rousserie G, Sukhanova A, Even-Desrumeaux K, et al. Semiconductor quantum dots for multiplexed bio-detection on solid-state microarrays. Crit Rev Oncol Hematol. 2010;74(1):1–15.
111. Smith AM, Gao X, Nie S. Quantum dot nanocrystals for in vivo molecular and cellular imaging. Photochem Photobiol. 2004;80(3):377–85.
112. Mansur HS. Quantum dots and nanocomposites. Wiley Interdiscip Rev Nanomed Nanobiotechnol. 2010;2(2):113–29.
113. Wehrenberg BL, Wang C, Guyot-Sionnest P. Interband and intraband optical studies of PbSe colloidal quantum dots. J Phys Chem B. 2002;106(41):10634–40.
114. Bailey RE, Nie S. Alloyed semiconductor quantum dots: tuning the optical properties without changing the particle size. J Am Chem Soc. 2003;125(23):7100–6.
115. Nanosphere Inc. (Commercial Ave Northbrook, Illinois, USA). www.nanosphere.us.
116. Genicon Science Corp (San Diego, Calif, US). http://www.geniconsciences.com/.
117. Quantum Dots Corp(San Marcos, Texas (Austin Metroplex), USA). www.qdots.com.
118. Ornberg RL, Harper TF, Liu H. Western blot analysis with quantum dot fluorescence technology: a sensitive and quantitative method for multiplexed proteomics. Nat Methods. 2005;2(1):79–81.
119. Yang H, Guo Q, He R, et al. A quick and parallel analytical method based on quantum dots labeling for ToRCH-related antibodies. Nanoscale Res Lett. 2009;4(12):1469–74.
120. Ho JAA, Wauchope R. A strip liposome immunoassay for aflatoxin B1. Anal Chem. 2002; 74(7):1493–6.
121. Lin YY, Wang J, Liu G, et al. A nanoparticle label/immunochromatographic electrochemical biosensor for rapid and sensitive detection of prostate-specific antigen. Biosens Bioelectron. 2008;23(11):1659–65.
122. Hao Y, Ding L, Rong H, et al. A novel quantum dots-based point of care test for syphilis. Nanoscale Res Lett. 2010;5(5):875–81.
123. Edwards KA, Baeumner AJ. Optimization of DNA-tagged dye-encapsulating liposomes for lateral-flow assays based on sandwich hybridization. Anal Bioanal Chem. 2006;386(5): 1335–43.
124. Sharma SK, Eblen BS, Bull RL, et al. Evaluation of lateral-flow Clostridium botulinum neurotoxin detection kits for food analysis. Appl Environ Microbiol. 2005;71(7):3935–41.
125. Horton J, Swinburne S, O'Sullivan M. A novel, rapid, single-step immunochromatographic procedure for the detection of mouse immunoglobulin. J Immunol Methods. 1991;140(1): 131–4.
126. Buechler KF, Moi S, Noar B, et al. Simultaneous detection of seven drugs of abuse by the TriageTM panel for drugs of abuse. Clin Chem. 1992;38(9):1678–84.
127. Tippkötter N, Stückmann H, Kroll S, et al. A semi-quantitative dipstick assay for microcystin. Anal Bioanal Chem. 2009;394(3):863–9.
128. Faulstich K, Haberstroh K, Gruler R, et al. Handheld and portable test systems for immunodiagnostics, nucleic acid detection and more. Soc Photo Opt Instrum Eng. 2008; 6945(1):69450H1–10.
129. Ming D. Study of nano-gold immunochromatographic quantitative assay based on photoelectric detection and processing. J Fuzhou Univ. 2005;4–5. (in Chinese).
130. Chandler J, Gurmin T, Robinson N. The place of gold in rapid tests. IVD Technol. 2000;6(2):37–49.
131. Kim S, Park JK. Development of a test strip reader for a lateral flow membrane-based immunochromatographic assay. Biotechnol Bioproc Eng. 2004;9(2):127–31.
132. Li Z, Wang Y, Wang J, et al. Rapid and sensitive detection of protein biomarker using a portable fluorescence biosensor based on quantum dots and a lateral flow test strip. Anal Chem. 2010;82(16):7008–14.

133. Zou Z, Du D, Wang J, et al. Quantum dot-based immunochromatographic fluorescent biosensor for biomonitoring trichloropyridinol, a biomarker of exposure to chlorpyrifos. Anal Chem. 2010;82(12):5125–33.
134. http://www.magnabiosciences.com/.
135. Wang Y, Xu H, Wei M, et al. Study of superparamagnetic nanoparticles as labels in the quantitative lateral flow immunoassay. Mater Sci Eng C. 2009;29(3):714–8.
136. Granade TC, Workman S, Wells SK, et al. Rapid detection and differentiation of antibodies to HIV-1 and HIV-2 using multivalent antigens and magnetic immunochromatography testing. Clin Vaccine Immunol. 2010;17(6):1034–9.
137. Peck RB, Schweizer J, Weigl BH, et al. A magnetic immunochromatographic strip test for detection of human papillomavirus 16 E6. Clin Chem. 2006;52(11):2170–2.
138. Workman S, Wells SK, Pau CP, et al. Rapid detection of HIV-1 p24 antigen using magnetic immuno-chromatography (MICT). J Virol Methods. 2009;160(1):14–21.
139. Handali S, Klarman M, Gaspard AN, et al. Development and evaluation of a magnetic immunochromatographic test to detect Taenia solium, which causes taeniasis and neurocysticercosis in humans. Clin Vaccine Immunol. 2010;17(4):631–7.
140. Jianchun M, Qing Y, Wenyuan Z, et al. Development and study of lateral flow test strip reader based on embedded system. IEEE Electron Meas Instrum. 2011;1:201–4.
141. Liu C, Jia Q, Yang C, et al. Lateral Flow Immunochromatographic assay for sensitive pesticide detection by using Fe3O4 nanoparticle aggregates as color reagents. Anal Chem. 2011;83(17):6748–84.
142. Yurong L, Nianyin Z, Min D. Study on the methodology of quantitative goldimmunochromatographic strip assay. IEEE Intell Syst Appl (ISA). 2010;2:1–4.
143. Muler-Barderff M, Freitag H, Scheffold T, et al. Development and characterization of a rapid assay for bedside determinations of Cardiac CardiacTroponin T. Circulation. 1995;92(10): 2869–75.
144. Rundstrom G, Ann J, Martensson O, et al. Lateral flow immunoassay using europium chelate microparticles and time-resolved fluorescence for eosinophils and neutrophic in whole blood. Clin Chem. 2007;53:342–8.
145. Tjitra E, Suprianto S, Dyer M, et al. Field evaluation of the ICT Malaria P.f/P.v immunochromatographic test for detection of Plasmodium falciparum and Plasmodium vivax in patients with a presumptive clinical diagnosis of malaria in eastern Indonesia. J Clin Microbiol. 1999;37(8):2412–7.
146. Sterling B, Kiang T, Subramanian K, et al. Simultaneous patient-side measurement of hemoglobin, glucose, and cholesterol in finger stick blood. Clin Chem. 1993;38(9):1658–64.
147. Yueming G, Min D, Zhenhua G, et al. Discussion of a method for quickly quantitative testing of drug. Shanghai J Biomed Eng. 2006;27(2):81–3. (in Chinese).

# Chapter 9
# Applications of Magnetic Nanoparticle-Based High-Throughput Single-Nucleotide Polymorphism Genotyping Platforms in Gastric Cancer Research

**Song Li, Hongna Liu, and Nongyue He**

## 9.1 Introduction

The completion of the Human Genome Project has set the stage for screening genetic mutations to identify disease genes on a genome-wide scale. Single-nucleotide polymorphisms (SNPs) are the most abundant forms of sequence variations between individuals and occur at about one per 500–1000 bp in the human genome. SNPs, because they are usually biallelic, are more amenable to automated detection; they are also regarded as ideal genetic markers in linkage disequilibrium analysis for identifying genetic factors associated with common diseases or adverse drug responses due to their accessible class of polymorphisms present and genetic stability. Clinical investigators have begun analyzing SNP alleles in population-based studies to identify loci that are statistically associated with a particular disease or phenotype. In these clinical studies, the number of samples available for analysis is important for data reliability. According to the reported research [1–3], a large amount of samples are required to detect a susceptibility allele with a high degree of confidence in a disease-associated study. To fulfill this need, the development of a new SNP detection platform that is quick, simple, high throughput, automated, and cost-effective is warranted.

S. Li (✉)
Hunan Key Laboratory of Green Packaging and Application of Biological Nanotechnology, Hunan University of Technology, Zhuzhou 412007, Peoples Republic of China

State Key Laboratory of Bioelectronics, Department of Biological Science and Medical Engineering, Southeast University, Nanjing 210096, Peoples Republic of China
e-mail: sosong1980@gmail.com

H. Liu • N. He
State Key Laboratory of Bioelectronics, Department of Biological Science and Medical Engineering, Southeast University, Nanjing 210096, Peoples Republic of China
e-mail: nyhe1958@163.com

© Springer Science+Business Media B.V. and Shanghai Jiao Tong University Press, Shanghai 2017

D. Cui (ed.), *Gastric Cancer Prewarning and Early Diagnosis System*, Translational Medicine Research, DOI 10.1007/978-94-024-0951-2_9

A number of techniques and commercial platforms have been developed for high-throughput SNP genotyping. The most commonly original SNP genotyping methods involve DNA sequencing [4], single-strand conformation polymorphism (SSCP) [5], restriction fragment length polymorphisms (RFLPs) [6], and the methods based on mass spectrometry [7, 8]. RFLP is a low-throughput, time-consuming, and labor-intensive method, relatively. Moreover, even if DNA sequencing methods have been fully developed during the past few years [9, 10] and the cost of human genome sequencing keeps decreasing, direct DNA sequencing technologies are not capable of detecting simultaneously SNP locus from a large amount of samples. All these techniques require multiple steps and are less well suited to high-throughput applications and automation. Fluorescence resonance energy transfer (FRET) is one of the most promising tools for multiplex SNP genotyping and has provided novel and unique detection systems [11], such as TaqMan assay [12], molecular beacons [13, 14], invader assay [15], and allele-specific polymerase chain reaction (PCR) with universal energy-transfer-labeled primer [16]. These techniques are powerful for screening SNPs and are therefore used in many fields, but the cost of analysis is relatively high because FRET-based methods usually require large amounts of fluorescently labeled probes.

As one of the most promising tools for nucleic acid detection, microarrays produce a "readout" that is especially suitable for high-throughput detection. In recent years, several high-throughput SNP detection methods are available using either the Affymetrix or the Illumina microarray platform; they can genotype thousands of SNPs distributed all over the genome. At the same time, in order to do some genome-wide association studies, these methods have emerged as a powerful new approach to identify common disease alleles. These approaches using the Affymetrix or the Illumina microarray platform permit comparative genome studies and afford different properties of genomic coverage [17, 18]; however, when identifying multiple individuals for limited SNP loci, they do not allow the simultaneous identification of these samples for genetic studies in a narrow range of prices, which requires development of some efficient and cost-effective methods for that purpose. Gene-specific hybridization combined with microarray platform is also a promising tool for high-throughput applications. The feasibility of the technique is often addressed by selecting randomly a number of SNPs associated with disease genes to demonstrate a wide application range. Ji et al. [19] have previously developed a microarray-based SNP detection method, based on dual-color fluorescence hybridization. The experiment successfully demonstrated that PCR products subjected to dual-color hybridization on a microarray could be used to analyze molecular markers from a large scale of samples. However, to obtain the genotyping results with high signal-to-noise ratio on the array, the amplicon always needs to be purified to remove nucleotides, enzymes, primers, etc. and made these methods laborious and time-consuming.

With the rapid advancing of nanotechnology in life science, nano-materials have become a focus of interest in various fields of biological and biomedical application. As one of the most important nanostructured materials, magnetic nanoparticles (MNPs) have been widely applied in biological detection due to their higher dispersion capability in aqueous solution and their higher separation efficiency in magnetic field. With high separation efficiency in magnetic fields, MNPs have great potential in automated and high-throughput nucleic acid analysis – especially when combined with commercial automation [20]. To fully exhibit the powerful capabilities of microarrays and MNPs, Li et al. have developed high-throughput genotyping methods based on MNPs and microarray detection [21–24]. The applications of these methods for gastric cancer-related SNP researches successfully demonstrated that SNPs from a large number of samples could be efficiently and automatically genotyped using MNPs as DNA carriers. This chapter will focus on MNP-based high-throughput SNP genotyping approaches.

## 9.2 Magnetic Nanoparticle-Based High-Throughput Genotyping Method with Dual-Color Fluorescence Hybridization

### *9.2.1 Method*

Biotin-labeled PCR products were captured on SA-MNPs. Then the DNA-MNP complexes are denatured and magnetically separated. Further the single-stranded DNA (ssDNA) on the MNPs were interrogated by hybridization with a pair of dual-color probes to determine SNP, and then the genotype of each sample can be simultaneously identified by scanning the microarray printed with the denatured fluorescent probes. A schematic outline of this method is shown in Fig. 9.1. As an example, the methylenetetrahydrofolate reductase (MTHFR) gene C677T polymorphism was selected as target. In principle, dual-color probe hybridization should produce different spot colors depending on the SNP genotypes, as clearly shown in

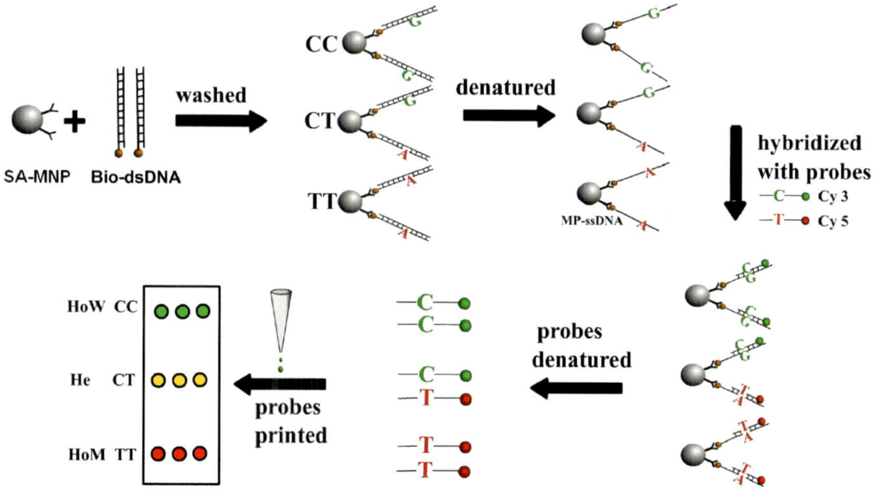


**Fig. 9.1** Scheme of SNP genotyping based on MNPs and dual-color fluorescence hybridization

Fig. 9.1. The homozygous wild type (HoW, CC), homozygous mutant type (HoM, TT), and heterozygote type (He, CT) should yield strongly green, red, and yellow spots, respectively.

### 9.2.2 *Preparation and Functionalization of Magnetic Nanoparticles*

Different kinds of magnetic nanoparticles, such as $\gamma$-$Fe_2O_3$ MNPs, $SiO_2$/(PMMA/$Fe_3O_4$) MNPs, and $Fe_3O_4$ MNPs, can be used as DNA carrier in high-throughput genotyping. These MNPs were prepared according to the previously published papers [21, 23, 30]. MNPs were incubated with 2 % 3-aminopropyl triethoxysilane (APTES) in 95 % ethanol at room temperature for 20 min. The amido-modified MNPs ($NH_2$-MNPs) were washed with 95 % ethanol and deionized water thrice, respectively. And then, the immobilization of streptavidin (SA) onto MNPs was performed by using glutaraldehyde for linkage between -$NH_2$ of APTES on MNP surface and streptavidin. Finally, streptavidin-covered MNPs (SA-MNPs) were magnetically washed with phosphate-buffered saline (PBS, 0.1 M, pH 7.4) buffer to remove excess unconjugated streptavidin and suspended in PBS buffer with the concentration of 4 mg/mL.

### 9.2.3 *Single-Stranded Amplicon Preparation*

Genomic DNA sample was extracted from peripheral blood samples by the use of commercial DNA extraction kit. PCR fragments containing SNP sites were amplified in a 96-well PCR plate with 30 μL of mixture containing buffer, template DNA, deoxynucleotide triphosphate (dNTP), Taq DNA polymerase, forward primer, and biotinylated reverse primer in a thermal cycler as described in the published paper [23]. After amplification, 80 μg of SA-MNPs were directly placed in each well of the 96-well PCR plate and mixed with the PCR mixture for 15 min so that the biotinylated PCR products can be captured on the MNPs by the linkage between the biotin and the streptavidin. The dsDNA-MNP complexes were magnetically separated and washed twice with PBS buffer. After washing, the dsDNA-MNP complexes were denatured in 25 μL of 0.1 M NaOH solution for 1–2 min and then magnetically washed twice with PBS buffer. The ssDNA-MNP complexes were stored in the plate at 4 °C until hybridization.

### 9.2.4 *SNP Genotyping*

A pair of 13 bp oligonucleotides labeled with Cy3 and Cy5 fluorophores were designed as hybridization probes, the nucleotide sequences of the probes are 5′-Cy3-CGGGAG*C*CGATTT-30 (677CC probe) and 5′-Cy5- CGGGAG*T*CGATTT-30 (677TT probe), and the underlined base represents the recognition site. In order to

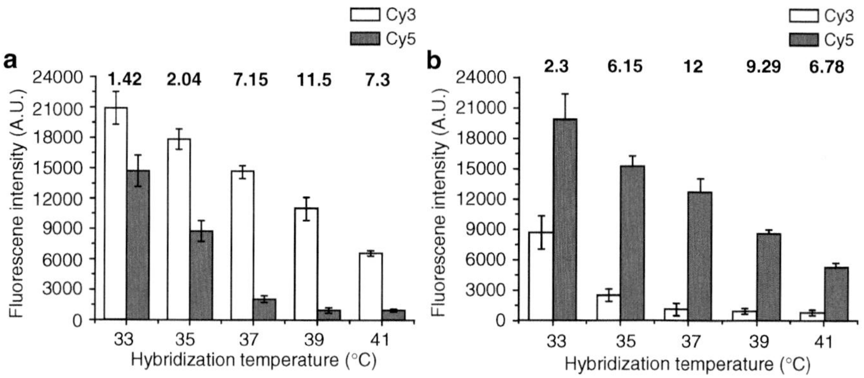


**Fig. 9.2** Optimization of hybridization temperature. (**a**) Fluorescent intensity of 677CC wild-type sample hybridized with dual-color fluorescence-labeled probes at different temperatures. (**b**) Fluorescent intensity of 677TT mutant-type sample hybridized with dual-color fluorescence-labeled probes at different temperatures

get a better discrimination, the recognition site is located in the middle of detection probe. The probes and the ssDNA-MNP complexes were suspended in 30 μL uni-hybridization solution (3:1 dilution v/v; Telechem). The hybridization was performed in the 96-well PCR plate by using a thermal cycler at optimized temperature for 1 h. Figure 9.2a, b shows the optimal hybridization temperature obtained for C677T locus. Fluorescent intensities of the denatured probes after hybridization using wild and mutant target oligonucleotides were detected, respectively. The highest signal ratio (match/mismatch) of wild and mutant-simulated targets could be obtained when hybridization temperature was performed at 39 °C and 37 °C, respectively. Based on these results, the hybridization of SNP genotyping was performed at the optimal temperature of 38 °C. After hybridization, the duplex DNA-MNP complexes were respectively rinsed and washed at room temperature successively with 2×SSC-0.1 % SDS, 0.1×SSC-0.1 % SDS, and deionized water (8 min for each wash) and resuspended in 20 μL 3×SSC buffer. Before arraying, the duplex DNA-MNP complexes were denatured at 95 °C for 5 min, kept on ice for 2 min, and then magnetically separated at the bottom of the well. The denatured probe solutions were printed directly from the 96-well PCR plate onto a cleaned glass slide without modification. After printing, the microarrays were snap-dried for 2 s on a hot plate (100 °C) and then scanned with an Axon 4100A microarray scanner, fitted with filters for Cy3 and Cy5. A representative result is shown in Fig. 9.3a.

### *9.2.5 Data Analysis*

The images acquired by the scanner are analyzed with software Genepix Pro 6.0. For each fluorescent image, the average pixel intensity within each circle is determined, and a local background using mean pixel intensity is computed for each

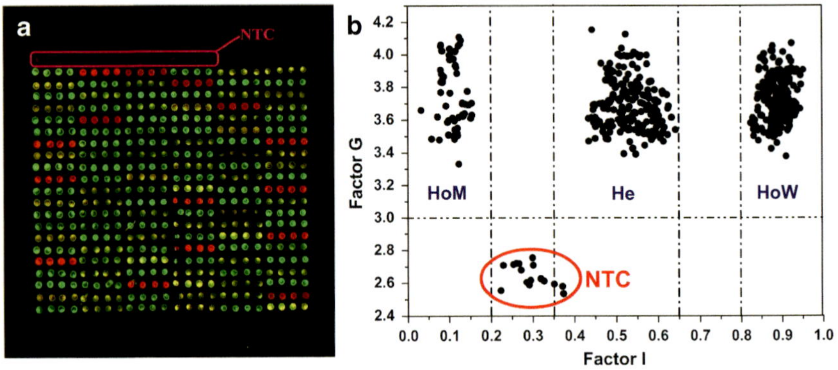


**Fig. 9.3** (**a**) The genotyping results of 126 different samples and four negative controls (NTC) assayed for the C677T locus of the MTHFR gene. Each sample is spotted four replicates in a column. (**b**) Cluster analysis of tag array hybridization results in 126 samples using factors I and G

spot. The net signal is determined by subtraction of this local background from the mean average intensity for each spot. Herein, we defined two factors to identify the genotype of each sample, the signal intensity factor $I$ ($I$=Log (signal (Cy3+Cy5))) and genotype factor $G$ ($G$=signal ratio (Cy3/Cy3+Cy5)). When the signal intensity factor I are over 3.0, which indicates that the summation of Cy3 and Cy5 signal intensities were over 1000 which is sufficiently above background signals, the boundaries for clusters are therefore defined as mutant (HoM), heterozygote (He), and wild type (HoW) when $G<0.2$, $0.35<G<0.65$, and $G>0.8$, respectively. If samples fall outside of these boundaries, they would be classified as ambiguous. An example result is shown in Fig. 9.3b.

## *9.2.6 Application in Gastric Cancer Research*

Cyclooxygenase-2 (COX-2) is an inducible isoform activated by cytokines and growth factors and is the rate-limiting enzyme in prostaglandin synthesis, catalyzing the conversion of arachidonic acid to prostaglandin E2 [26]. COX-2 is associated with proliferation and apoptosis markers and serves as an independent prognostic factor in gastric cancer. Enhanced expression of COX-2 has been observed in several forms of cancer [27], including gastric cancer and precancerous tissues [28]. A polymorphism in the promoter region of COX-2, COX-2 –765G>C polymorphism, may be a susceptibility marker for gastric cancer in patients. In Zhang's study [29], cyclooxygenase-2 (COX-2) gene promoter region –765G>C polymorphism of 118 gastric cancer cases and 120 normal control samples in northern Jiangsu of China were genotyped using a SNP genotyping method based on magnetic nanoparticles (MNPs) and dual-color fluorescence hybridization. The individuals with −765 GC+CC genotype group had a 1.808-fold (95 % confidence interval [95% CI]=1.050~3.113, $P$=0.033) increased risk of developing gastric

**Table 9.1** The comparison on the frequencies of –765G>C in COX-2 gene between patients and controls

| Genotype | GC patient | | Controls | | OR | 95 % CI | *P* |
|---|---|---|---|---|---|---|---|
| | *N* | % | *N* | % | | | |
| −765GG | 70 | 59.3 | 87 | 72.5 | 1.000(Ref) | – | – |
| −765GC | 39 | 33.1 | 28 | 23.3 | 1.731 | 0.971–3.087 | 0.063 |
| −765CC | 9 | 7.6 | 5 | 4.2 | 2.237 | 0.717–6.978 | 0.165 |
| −765GC+CC | 48 | 40.7 | 33 | 27.5 | 1.808 | 1.050–3.113 | 0.033* |

*Note*. GC: gastric cancer
$^*P<0.05$

**Table 9.2** The comparison on the allele frequency of –765G>C in COX-2 gene between patients and controls

| Group | Allele frequency | | | *P* | OR (95 % CI) |
|---|---|---|---|---|---|
| | G allele (*N*) | C allele (*N*) | Frequency (GAF/CAF) | | |
| GC patients | 179 | 57 | 0.758/0.242 | 0.029* | 1.701(0.840–3.446) |
| Controls | 202 | 38 | 0.842/0.158 | | |

*Note*. *GC* gastric cancer, *GAF* G allele frequency, *CAF* C allele frequency
$^*P<0.05$

carcinoma in comparison with those possessing –765GG genotype (Table 9.1). The frequency of variant allele C in patients was higher than that in control group ($P<0.05$, odds ratio [OR] = 1.701, 95 % CI = 0.840 ~ 3.446) (Table 9.2). Individuals carrying –765GC or –765CC genotype may have higher risk for the development of gastric cancer, and the –765C allele is associated with increased susceptibility to gastric cancer in northern Jiangsu area of China. This research also demonstrated that SNPs from a large number of samples could be rapidly and efficiently investigated by using MNP-based dual-color hybridization method.

### *9.2.7 Improved MNP-Based High-Throughput SNP Genotyping Method Using Tagged Probes*

Based on using MNPs as target DNA carrier and dual-color hybridization, the method described above can perform rapidly and effectively SNP genotyping from a large-scale sample. However, a different pair of fluorescent probes was still necessary when analyzing different SNP locus, so the cost will also be very high when dealing with thousands of DNA samples for multiplex SNP loci. To reduce the analysis cost, Li et al. developed an improved high-throughput genotyping method based upon tagged probes for scoring SNP markers. Instead of gene-specific probes, tagged probes were adopted in their research, with a part as allele-specific probe hybridizing with MNP-PCR and the other part as universal tags hybridizing with universal fluorescent detectors, respectively. By this way, each SNP locus only need one pair of unlabeled allele-specific tag probes, and

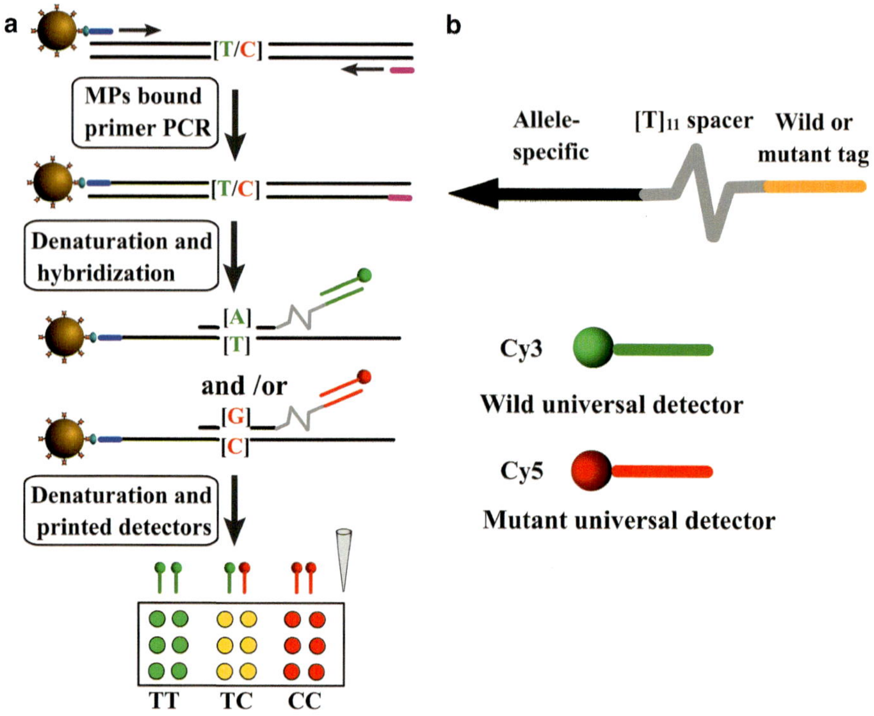


**Fig. 9.4** (**a**) Schematic of magnetic nanoparticles-based PCR amplification and detection. (**b**) Allele-specific labeling and wild or mutant type tag

one pair of given universal detectors can be applied to differ all marker analysis (Fig. 9.4). The cost of this approach is very competitive; if use this method to analyze ten different markers from over 1000 samples, the cost for reagents for per marker and per sample will be less than $0.08 dollars, which is only 10–20% of traditional methods. In this way, a slide could be used to score anything from different marker in large amounts of samples to thousands of markers, which is much more adaptable and versatile than the strategy based on Affymetrix or Illumina platform.

## 9.3 Magnetic Nanoparticle-Based High-Throughput Genotyping Method with Single-Base Extension

### *9.3.1 Method*

For gene-specific hybridization-based genotyping methods, selectivity and quantification are dependent upon the dissociation properties of the target DNA hybridized to the allele-specific probes. Thus to achieve SNP discrimination, hybridization temperatures must be optimized, and stringent wash steps must be included. Single-base extension (SBE) involves extension of a primer located adjacent to the position of a SNP. Using DNA polymerase in the presence of fluorescently labeled dideoxynucleotide

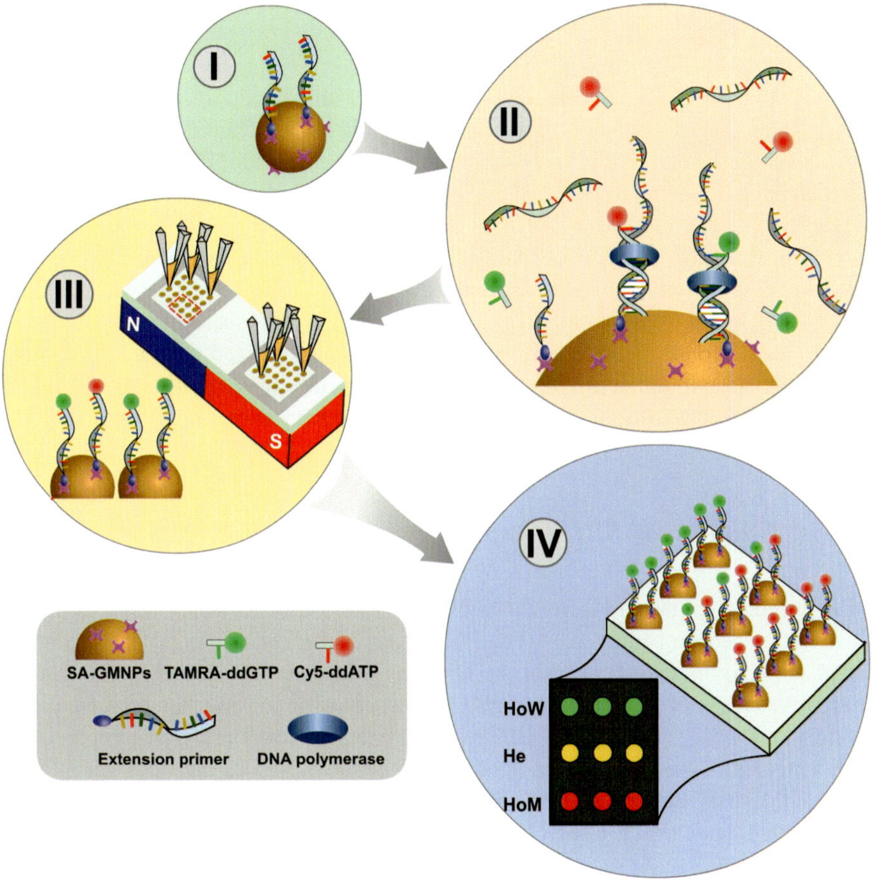


**Fig. 9.5** Schematic diagram showing the SNP genotyping method based on a GMNP array and dual-color SBE. (*I*) Biotinylated primers are captured by the SA-GMNPs. (*II*) ssPCR products are hybridized to primers, and dual-color, single-base extension is performed. (*III*) Fluorophore-GMNP complexes for each sample are immobilized on the glass slide. (*IV*) The genotype of each sample is determined by measuring the fluorescent intensity of the fluorophore-GMNP complexes on the array

triphosphates (ddNTPs) ensures that the polymerase will incorporate only a single nucleotide for the SNP site of interest. The SBE technique provides highly accurate and effective genotyping and has been widely used in various genotyping procedures. Li et al. developed a high-throughput and automated SNP genotyping method that uses a magnetic nanoparticle (MNP) array and dual-color, single-base extension (Fig. 9.1). The solid-phase extension primers were prepared by immobilizing biotinylated oligonucleotides onto gold magnetic nanoparticles coated with streptavidin (SA-GMNPs). In the presence of DNA polymerase, allele-specific extension using dual-color ddNTPs occurred along the stretch of hybridized target DNA, incorporating a fluorophore into the primer bound to the surface of the magnetic beads. The magnetic beads extended with fluorophores were spotted on a glass slide to fabricate a “bead array” for SNP discrimination. A schematic outline of this method is shown in Fig. 9.5

### 9.3.2 Preparation of Functional MNPs

Gold-coated magnetic nanoparticles (GMNPs) with low fluorescent background were synthesized according to published papers [21, 25]. The GMNPs were functionalized with carboxylic acid groups by reacting GMNPs with 200 mM 11-mercaptoundecanoic acid (MUA) in ethanol at room temperature. Next the GMNPs were activated with EDC for 30 min at 4 °C using slow-tilt rotation. After removing the supernatant using magnetic separation, the activated GMNPs were washed once with 500 μL cold, deionized water and a second time with 500 μL 25 mM MES (pH 6) as quickly as possible to avoid hydrolysis of the activated carboxylic acid groups. Streptavidin was covalently immobilized on the GMNPs by linking the streptavidin amine groups to the carboxylic acid groups on the GMNPs. The SA-GMNPs were found to be stable for at least 6 months when stored at 4 °C.

### 9.3.3 Dual-Color SBE Reaction

PCR products containing different SNP loci were amplified according to a previously published procedure [21]. The PCR products for dual-color SBE were incubated with exonuclease I and shrimp alkaline phosphatase at 37 °C for 90 min to inactivate excess PCR primers and dNTPs, respectively, followed by enzyme deactivation at 94 °C for 15 min. GMNP-bound extension primers were prepared by covalently linking 24 pmol of each of the biotinylated oligonucleotides with 80 μg of SA-GMNPs. Solid-phase SBE reaction was performed in a 20 μL reaction mixture containing relevant TAMRA-ddGTP and Cy5-ddATP, PCR buffer, unlabeled ddNTPs, thermo sequenase, and MNP-bound extension primer. The solid-phase SBE reactions for C677T were carried out in thermal cycler with published conditions [21]. After SBE reaction was completed, the SBE-GMNP complexes were separated using a neodymium-boron (Nd-B) magnet and washed for 5 min using sterilized $ddH_2O$. The remaining fluorophore-GMNP particles were suspended in 10 μL of sterilized $ddH_2O$ and stored at 4 °C until arraying.

### 9.3.4 Bead Array Fabrication and Signal Detection

Standard glass slides (75 mm × 25 mm × 1 mm) were used as the substrate for the GMNP array. Nd-B magnet was placed on the opposite side of the slide, and each fluorophore-GMNP reaction was printed in quadruplicate directly onto the slide to fabricate a "bead array." After printing, the "bead array" was snap-dried for 2 s on a hot plate set at 100 °C and directly scanned with a 4100A Microarray Analysis System (Axon). To test this method for high-throughput genotyping capacity, MTHFR gene C677T locus detection using 320 diverse DNA samples. Figure 9.6a shows the microarray results of the high-throughput analysis. Good discrimination between the two alleles was obtained for C677T locus. The complete array, spotted

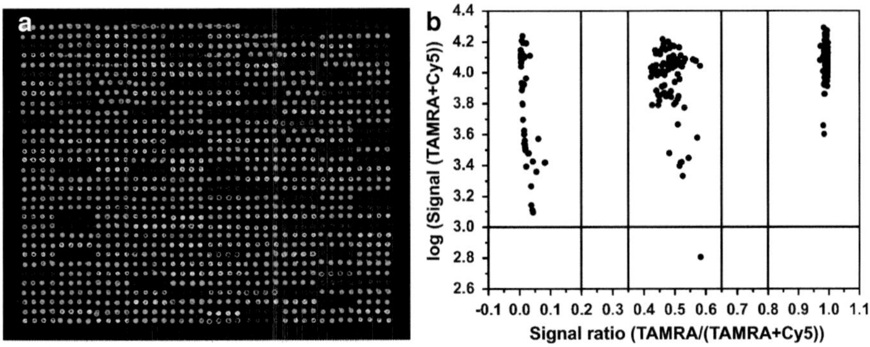


**Fig. 9.6** Results of SNP detection of 320 samples based on SA-GMNPs assayed for MTHFR C677T. (**a**) The fluorescence images of microarray assayed for MTHFR C677T locus. (**b**) Allelic-fraction scatter plot

in replicates of four, produced a fluorescence pattern that was highly reproducible between replicates. The three allelic states for the locus (C/C, green; C/T, yellow; T/T, red) were easily discriminated. Separating samples into allelic fractions allowed unequivocal genotype assignment for the three loci. Figure 9.6b illustrates the clear ratio differences between the three clusters. From left to right on the graph, G factor ranging from 0.03 to 0.075 were classified as homozygous mutant samples; G factor ranging from 0.96 to 0.98 can be classified as homozygous mutant samples. G-factor values for homozygous wild and mutant samples should be as close as possible to 1.0 and 0.0, respectively. All homozygous samples produced a match/mismatch ratio between 12.33 and 49. The middle cluster had G-factor values that ranged from 0.43 to 0.58, all of which were classified as heterozygotes.

### *9.3.5 Comparison of Dual-Color SBE with Dual-Color Hybridization*

In order to evaluate if SBE increases the quality of SNP genotyping results obtained by dual-color magnetic bead arrays, we compared the dual-color SBE and hybridization methods by genotyping two different SNP loci, rs1801133 and rs48412845. Homozygous wild-type and mutant DNA samples that were verified by sequencing were used to amplify the targets. Figure 9.7a shows the genotyping results obtained from wild-type and mutant samples analyzed for the rs1801133 locus using both methods. The homozygous wild-type samples produced strong green fluorescence using both methods, while mutant samples produced a strong red fluorescence. The genotype of the two samples can be easily discriminated using both methods, but the dual-color SBE method produced an S/N ratio of 44.15, whereas the hybridization method produced a wild/mutant ratio of 12.28. A bar graph (Fig. 9.8b) depicts the signal intensity of the genotyping results obtained from wild-type and mutant samples analyzed for rs48412845 locus with both methods. Using dual-color SBE (Fig. 9.8a, left), the wild-type and mutant samples can be easily discriminated by

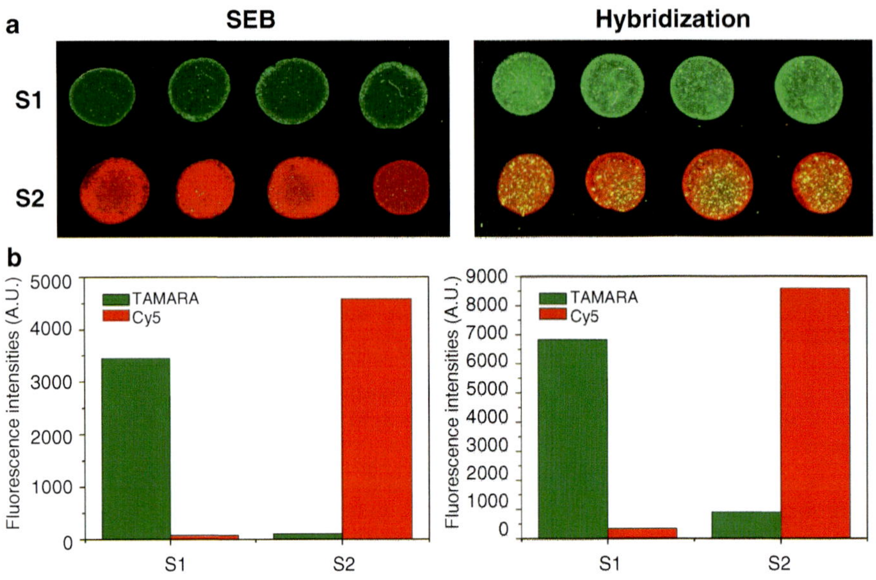


**Fig. 9.7** Results of SNP detection for MTHFR C677T using dual-color SBE and dual-color hybridization, respectively. (**a**) Fluorescence images of microarrays assayed for MTHFR C677T locus. (**b**) Relative fluorescence intensities for the two samples

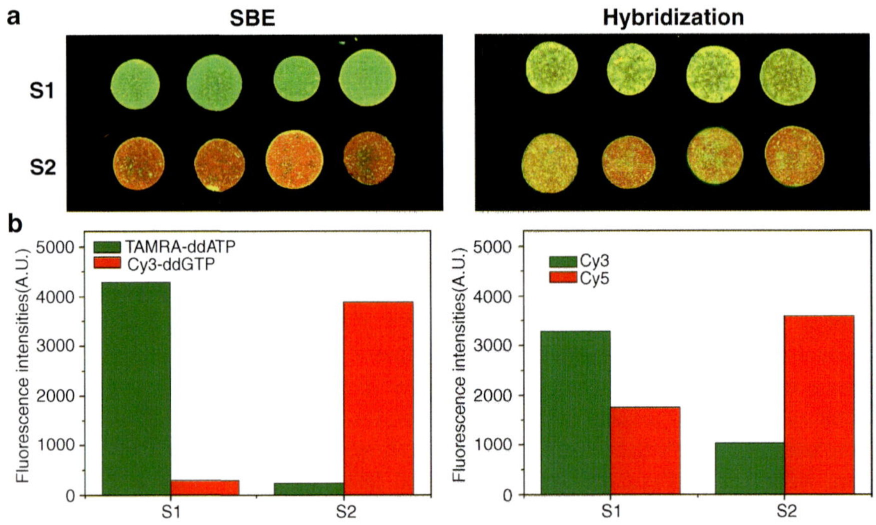


**Fig. 9.8** Results of SNP detection for rs48412845 locus using dual-color SBE and dual-color hybridization, respectively. (**a**) The fluorescence images of microarray assayed for rs48412845 locus. (**b**) Relative fluorescence intensities for the two samples

**Table 9.3** Frequency distribution (percent) of cancer cases and control by genotype and chi-square value (*P*-value) for testing the significance and independency of observed genotype frequency between gastric cancer cases and controls

| | Gastric cancer status | | |
|---|---|---|---|
| Genotype status | Control *n* (%) | Cases *n* (%) | Chi-square value (*P*-value) |
| *NEIL2 SNP* (*rs804270*): | | | |
| CC (normal allele) | 45 (38.14) | 4 (3.81) | 44.8 (<.0001) |
| GG (mutant allele) | 38 (32.2) | 72 (68.57) | |
| C/G (heterozygous) | 35 (29.66) | 29 (27.62) | |
| Total | 118 (52.91) | 105 (47.09) | |

the strong green and red fluorescence, respectively. However, using dual-color hybridization method (Fig. 9.8a, right), the wild-type and mutant samples cannot be easily discriminated. Using hybridization method, the wild/mutant ratios of the two homozygous samples are only 1.87 and 0.288, respectively. Taken together, these results suggest that the SBE method provides better SNP discrimination than hybridization. For gene-specific hybridization method to get better discrimination, the SNP loci need to be located in the middle of detection probe. However, the rs48412845 SNP locus is only 2 bp away from another SNP locus; another SNP locus in the detection probes causes inefficient hybridization and results in poor discrimination. Compared with techniques based on hybridization, SBE-based methods do not have limitation on probe designing. SBE-based method is compatible with multiplexed analysis and provides highly accurate genotyping results because primer extension with DNA polymerase can accurately distinguish single-nucleotide differences, even for difficult SNP loci.

### *9.3.6 Application in Gastric Cancer Research*

Elingarami et al. [30] conducted a hospital-based case–control study to evaluate the potential association between the polymorphism in a base excision repair (BER) gene (NEIL2) and susceptibility to gastric cancer in Chinese population. NEIL2 (rs804270) SNP was genotyped by single-base extension (SBE) assay based on magnetic nanoparticles (MNPs). The $X^2$ test was used to compare genotype frequency between patients and controls. The unconditional logistic regression analysis was used to estimate odds ratios (OR) and 95 % confidence intervals (CI) for genotype susceptibility to gastric cancer. The NEIL2 (rs804270) SNP was successfully genotyped by single-base extension (SBE) assay with the help of a detection platform based on GMNPs as DNA carriers. NEIL2 SNP rs804270 genotype frequency distribution among gastric cancer cases and controls is represented in Table 9.3. Genotypes in the NEIL2 SNP demonstrated conformance to Hardy–Weinberg equilibrium. Table 9.4 shows results from the unconditional logistic regression. As seen from Table 9.4, there was an increased risk for gastric cancer in

**Table 9.4** Unconditional logistic regression fitted for NEIL2 SNP genotypes separately for association of the genotypes with gastric cancer risk

| Genotype | Odds ratio | 95 % Confidence interval for odds ratio | *P*-value |
|---|---|---|---|
| NEIL2 SNP (rs804270): | | | |
| CC (wild type) | 1.0 (reference) | | 0.4539 |
| C/G (heterozygous) | 0.107 | (0.034, 0.334) | 0.0001 |
| GG (mutant allele) | 2.287 | (1.218, 4.293) | 0.0101 |

subjects with GG genotype in the NEIL2 (rs804270) SNP (OR =2.3, 95 % CI = 1.22–4.3, $p = 0.01$). Variant alleles in the NEIL2 (rs804270) SNP may independently influence susceptibility to gastric cancer in a northern Jiangsu Chinese population. GMNPs were successfully used as DNA carriers in SNP genotyping by single-base extension (SBE) assay.

## 9.4 Conclusion

To achieve high-throughput genotyping, the challenge lies in pairing the right assay chemistry with the right carriers to maximize efficiency with respect to accuracy, sensitivity, speed, and cost. Herein, we successfully demonstrated high-throughput SNP detection using MNPs as DNA carrier. With the advantage of easy separation, these assays can be automated, and, by means of the robotic platform for DNA extraction, SNP types present a real opportunity for the laboratory to expedite experiment processing as well as reduce reagents. The genotype results will be analyzed in a very simple and rapid way using conventional expression microarray software analysis. The combination of highly multiplexed assays with a highly parallel readout represents a relatively new but already successful paradigm in high-throughput analysis. One of the great promises of these approaches is that the ability to carry out genomic analyses flexibly, easily, inexpensively, accurately, and rapidly with high sensitivity. We believe that these approaches will be very useful in screening large amounts of markers in thousands of individual samples, which will contribute to large-scale genotyping studies, and help to create a new generation of routine diagnostic and prognostic tools.

## References

1. Risch N, Merikangas K. The future of genetic studies of complex human disease. Science. 1996;273:1516.
2. Jain KK. Biochips for gene spotting. Science. 2001;294:621.
3. Gresham D, Ruderfer DM, Pratt SC, Schacherer J, Dunham MJ, Botstein D, Kruglyak L. Genome-wide detection of polymorphisms at nucleotide resolution with a single DNA microarray. Science. 2006;311:932.

4. Alderborn A, Kristofferson A, Hammerling U. Determination of single-nucleotide polymorphisms by real-time pyrophosphate DNA sequencing. Genome Res. 2000;10:1249.
5. Gonen D, Veenstra-Vander Weele J, Yang Z, Leventhal BL, Cook Jr EH. High throughput fluorescent CE-SSCP SNP genotyping. Mol Psychiatr. 1999;4:339.
6. Lander ES, Botstein D. Mapping mendelian factors underlying quantitative traits using RFLP linkage maps. Genetics. 1989;121:185.
7. Braun A, Little DP, Köster H. Detecting CFTR gene mutations by using primer oligo base extension and mass spectrometry. Clin Chem. 1997;43:1151.
8. Tang K, Fu DJ, Julien D, Braun A, Cantor CR, Köster H. Chip-based genotyping by mass spectrometry. Proc Natl Acad Sci. 1996;96:10016.
9. Shendure J, Ji H. Next-generation DNA sequencing. Nat Biotechnol. 2008;26:1135.
10. Kircher M, Kelso J. High-throughput DNA sequencing–concepts and limitations. Bioessays. 2010;32:524.
11. Takatsu K, Yokomaku T, Kurata S, Kanagawa T. A new approach to SNP genotyping with fluorescently labeled mononucleotides. Nucleic Acids Res. 2004;32:e60.
12. De la Vega FM, Lazaruk KD, Rhodes MD, Wenz MH. Assessment of two flexible and compatible SNP genotyping platforms: TaqMan SNP genotyping assays and the SNPlex genotyping system. Mutat Res. 2005;573:111.
13. Barreiro LB, Henriques R, Mhlanga MM. High-throughput SNP genotyping: combining tag SNPs and molecular beacons. Methods Mol Biol. 2009;578:255.
14. Marras SA, Kramer FR, Tyagi S. Genotyping SNPs with molecular beacons. Methods Mol Biol. 2003;212:111.
15. Hall JG, Eis PS, Law SM, Reynaldo LP, Prudent JR, Marshall DJ, Allawi HT, Mast AL, Dahlberg JE, Kwiatkowski RW, de Arruda M, Neri BP, Lyamichev VI. Sensitive detection of DNA polymorphisms by the serial invasive signal amplification reaction. Proc Natl Acad Sci. 2000;97:8272.
16. Myakishev MV, Khripin Y, Hu S, Hamer DH. High-throughput SNP genotyping by allele-specific PCR with universal energy-transfer-labeled primers. Genome Res. 2001;11:163.
17. Lamy P, Andersen CL, Dyrskjot L, Torring N, Wiuf C. A Hidden Markov Model to estimate population mixture and allelic copy-numbers in cancers using Affymetrix SNP arrays. BMC Bioinformatics. 2007;8:434.
18. Steemers FJ, Chang W, Lee G, Barker DL, Shen R, Gunderson KL. Whole-genome genotyping with the single-base extension assay. Nat Methods. 2006;3:31.
19. Ji MJ, Hou P, Li S, He NY, Lu ZH. Microarray-based method for genotyping of functional single nucleotide polymorphisms using dual-color fluorescence hybridization. Mutat Res. 2004;548:97.
20. Matsunaga T, Maruyama K, Takeyama H. High-throughput SNP detection using nano-scale engineered biomagnetite. Biosens Bioelectron. 2007;22:2315.
21. Li S, Liu HN, Jia YY, Mou XB, Deng Y, Lin L, Liu B, He NY. An automatic high-throughput SNP genotyping approach based on universal tagged arrays and magnetic nanoparticles. J Biomed Nanotechnol. 2013;9:68.
22. Li S, Liu HN, Jia YY, Deng Y, Zhang LM, Lu ZX, He NY. A novel SNPs detection method based on gold magnetic nanoparticles array and single base extension. Theranostics. 2012;2:967.
23. Li S, Liu HN, Wang ZF, Hou P, Guo YF, He QG, He NY. Magnetic-particles-based high-throughput genotyping method with dual-color Fluorescence hybridization. Anal Biochem. 2006;359:277.
24. Li S, Liu HN, Wang ZF, Ji MJ, Guo YF, He NY, Dai YB. A novel in situ magnetic particle PCR based high-throughput genotyping method using universal tags. Prog Biochem Biophys. 2007;34:1107.
25. Coskunpinar E, Eraltan IY, Turna A. Cyclooxygenase-2 gene and lung carcinoma risk. Med Oncol. 2011;28:1436.
26. Brown JR, DuBois RN. Cyclooxygenase as a target in lung cancer. Clin Cancer Res. 2004;10:4266.

27. Zhang JT, Wang MW, Zhu ZL, Huo XH, Chu JK, Cui DS, Qiao L, Yu J. Increased expression of cyclooxygenase-2 in first-degree relatives of gastric cancer patients. World J Gastroenterol. 2005;11:4918.
28. Elingarami S, Deng Y, Fan J, Zhang YY, He NY. NEIL-2 single nucleotide polymorphism genotyping using single base extension on core-shell $Fe_3O_4$@$SiO_2$@Au magnetic nanoparticles and association of the genotypes with gastric cancer risk in Northern Jiangsu (China). Sci Adv Mater. 2014;6:899.
29. Zhang YY, Jia YY, Li S, Deng Y, Liu HN, He NY. Genotyping of −765G>C in COX-2 gene based on MNPs and dual-color fluorescence hybridization and its association with risk of gastric cancer in Northern Jiangsu of China. Sci Adv Mater. 2014;6(1146).
30. Mou XB, Li T, Wang JH, Ali Z, Zhang YY, Chen Z, et al. Genetic variation of BCL2 (rs2279115), NEIL2 (rs804270), LTA (rs909253), PSCA (rs2294008) and PLCE1 (rs3765524, rs10509670) Genes and their correlation to gastric cancer risk based on universal tagged arrays and $Fe_3O_4$ magnetic nanoparticles. J Biomed Nanotechnol. 2015;11:2057–66.

# Chapter 10
# Single-Cell Manipulation Technology for Cancer Research

**Shujing Lin, Di Chen, and Yao Xie**

## 10.1 Introduction

Since single cell sizes in a range from several micrometers to hundreds of micrometers were discovered, researchers began to observe and study them. From the mid-1900s, scientists and engineers started to have strong interest in manipulation of biological particle techniques, not just in observation of stationary microscopic particles [1]. In order to meet the demands of observation and analysis of cellular and intracellular components, much research has been done to find an effective way to manipulate cells at single-cell level by researchers [2–6].

With the microfluidic technology emerging in the 1980s, scientists have devoted a lot of work to improve the performance of microfluidic. In the early days, microfluidic approaches are used in various research fields such as molecular biology, genetic analysis, and proteomics, but not cell biology [7]. Considering advantages of microfluidics, such as precise control of the cellular environment and ease of analyzing cell at the single-cell level, it is very appropriate for cell biology. Hence, microfluidic technology was introduced into the field of cell biology by researchers step by step. After decades of development, microfluidic technology has been successfully used in the area of manipulation of fluid droplets, particles, single cell, and so on [8].

Nowadays, microfluidic technique plays a critical role in the cell manipulation field. Cell manipulation technology, which is applied to sort cells, separate cells, trap cells, isolate cells, move cells, recover cells, and so on [9–14], combined with microfluidic technique, makes the ways of manipulating cells at single-cell level

S. Lin • D. Chen (✉) • Y. Xie
Institute of Nano Biomedicine and Engineering, Shanghai Engineering Research Center for Intelligent Diagnosis and Treatment Instrument, National Center for Translational Medicine, Collaborative Innovational Center for System Biology, Shanghai Jiao Tong University, 800 Dongchuan Road, Shanghai 200240, Peoples Republic of China
e-mail: dchen@sjtu.edu.cn

© Springer Science+Business Media B.V. and Shanghai Jiao Tong University Press, Shanghai 2017

D. Cui (ed.), *Gastric Cancer Prewarning and Early Diagnosis System*, Translational Medicine Research, DOI 10.1007/978-94-024-0951-2_10

more precise and convenient than ever before. According to different cell manipulation force generation principles, cell manipulation technology can be divided into optical manipulation technique, magnetic manipulation technique, electrical manipulation technique, acoustic manipulation technique, and mechanical manipulation technique. This review will introduce the principles of each cell manipulation technique, overview the current situation of them, and then show the advantages and drawbacks of each cell manipulation technique at the end.

## 10.2 Single-Cell Manipulation Technology

### *10.2.1 Optical Manipulation Technique*

A concept of an optical trapping technique was first reported by Ashkin at Bell Labs in 1970; they have demonstrated that micro-sized particles had been accelerated and trapped in stable optical potential wells using only the force of radiation pressure from a continuous laser [15]. Shortly afterward, optical trapping and manipulation of viruses and bacteria by laser radiation pressure were demonstrated with single-beam gradient traps by Ashkin and his colleagues [16]. Since optical trapping technique, which also is known as "optical tweezers", was first reported, it makes many improvements in cell manipulation area.

Optical manipulation technique has found applications in both physics and biology. However, how do we use a laser beam to manipulate particles or bio-particles? An optical manipulation is formed by tightly focusing a laser beam with an objective lens of high numerical aperture (NA). A dielectric particle near or in the focus will experience a force due to a portion of the scattering of incident photon momentum that will transfer to the particle. This optical force has traditionally been decomposed into two components: scattering force and gradient force [17]. The scattering force can be imagined photons as fire horses pushing the particle moving toward the light propagation direction and the effective scattering force can be calculated. The scattering force of a sphere is

$$F_{\text{scatt}} = \frac{I_0 \sigma n_m}{c} \tag{10.1}$$

$$\sigma = \frac{128\pi^5 a^6}{3\lambda^4}\left(\frac{m^2 - 1}{m^2 + 2}\right)^2 \tag{10.2}$$

where $I_0$ is the intensity of the incident light; $\sigma$ is the scattering cross section of the sphere; $n_m$ and $n_p$ are the index of refraction of the medium and particle, respectively; $c$ is the speed of light in vacuum; $a$ is the radius of sphere; $m$ is the ratio of the index of refraction of the particle to the index of the medium ($n_p/n_m$); and $\lambda$ is the wavelength. The gradient force arises from the interaction of the induced dipole

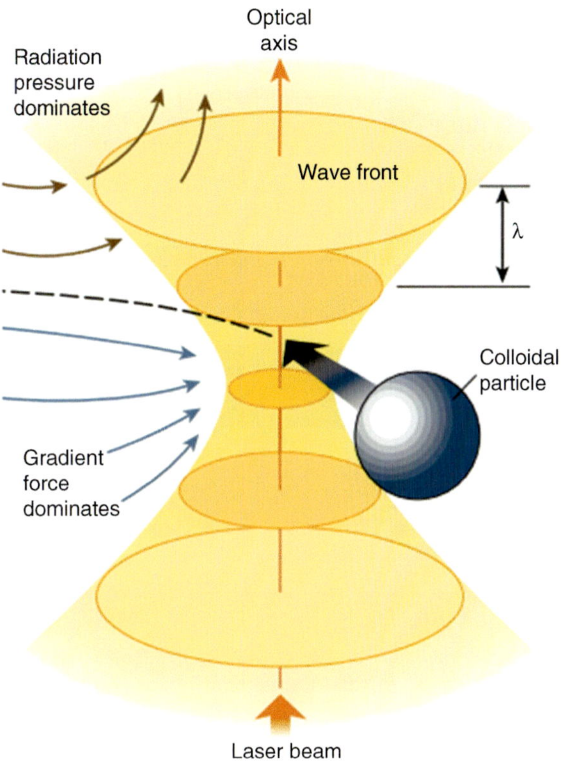


**Fig. 10.1** A diagram of optical force for particle trapping. A dielectric particle will experience a gradient force that pull it toward the focus and a radiation pressure that push it away from the focus, while these two forces are balance, the particle is trapped [18]

with an inhomogeneous field. The gradient force is proportional to the intensity gradient ($\nabla I_0$), as shown in Eq. (10.3):

$$F_{\text{grad}} = \frac{2\pi}{c} a^3 \left( \frac{m^2 - 1}{m^2 + 2} \right) \nabla I_0 \tag{10.3}$$

Although the theory of optical manipulation is still being developed, the basic principles are still the same. Stable trapping requires the gradient force to dominate and is achieved when the beam diverges rapidly enough away from the focal point. Intensity gradients in the converging beam pull small particle toward the focus, whereas the radiation pressure of the beam tends to blow them down the optical axis. Under the gradient force dominates condition, a particle can be trapped in three dimensions, near the focal point, as shown in Fig. 10.1 [18].

Based on this principle, researchers developed a kind of tool, which is called optical tweezers, to manipulate particles and bio-particles [19–22]. These particles and bio-particles include cells, bacteria, virus, microbeads, and so on. However, the usefulness of this technique has been limited, particularly in manipulating bio-particles, due to the manipulation laser beam that may cause some potential damage to the bio-samples. In order to characterize photodamage of optical manipulation

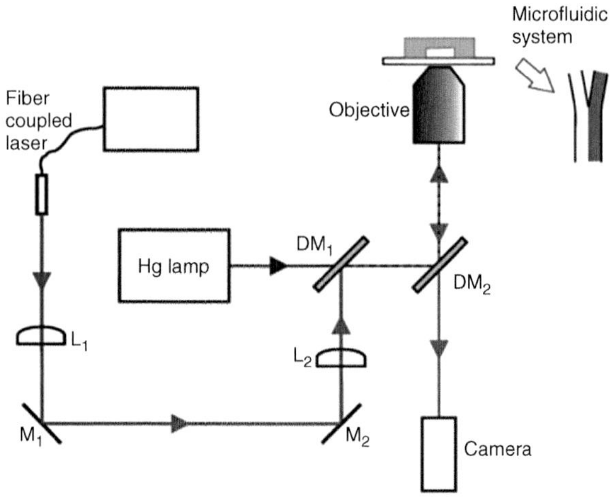


**Fig. 10.2** A schematic of experimental setup of the optical tweezers. In this design, the laser beam through an optical system for manipulation target cells, which are selected by a digital microscope system. This experimental optical tweezers can move the target cells from one side to another side in the microfluidics [25]

technique, Neuman characterized photodamage throughout the near-infrared region for optical trapping by using *Escherichia coli* as specimens [23]. They found that the action spectrum for photodamage minima is at 830 and 970 nm and maxima at 870 and 930 nm. A few years later, in order to quantify potentially harmful or misleading heating effects of this technique in biophysical experiments, Peterman demonstrated that the heating effect in water has non-negligible effects in typical biophysical experimental circumstances and it should be taken into account when laser powers of more than 100 mW are used [24].

Although optical manipulation technique has its limitation in the field of bio-particle manipulation, this technique is still a powerful tool to manipulate bio-particles due to its own advantages, such as high throughput and high accuracy. Eriksson and Enger created an environmental gradient between two media in a microfluidic system and used optical tweezers to move a single trapped cell repeatedly between the different environments for analyzing rapid and reversible cytological alterations in single cells; the schematic of the setup is shown in Fig. 10.2 [25]. The trapped object can be moved within the microfluidic system using the motorized microscope stage to change the position of the channel system relative to the fixed trap.

In particle sorting field, MacDonald demonstrated an optical sorter for microscopic particles that exploits the interaction of particles – biological or otherwise – with an extended, interlinked, dynamically reconfigurable, three-dimensional optical lattice. This concept of optical fractionation can both sort by size and refractive index. The sorting efficiency of this method was approximately 100 % [26]. After that, researchers have designed various structures of microfluidics combined with optical manipulation technique for sorting, trapping, and separating cells or particles (see Fig. 10.3). Kovac and Voldman designed a microfluidic chip containing a microwell array that can be passively loaded with mammalian cells via sedimentation shown in Fig. 10.3a. Subsequently, target cells were selected by microscopy and, then, used the scattering force from a focused infrared laser to levitate target cells from their wells into a flow field for collection. This work achieved

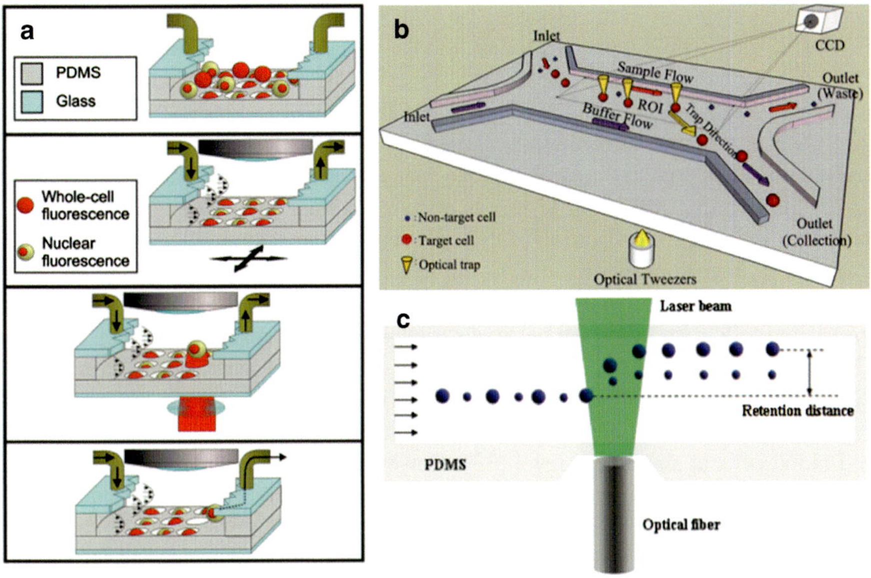


**Fig. 10.3** Various structures of optical manipulation devices. (**a**) Each single cell is trapped in a microwell, selecting the target cell by employing fluorescence method, and then push the target cell out of the microwell by using a optical manipulation technique. (**b**) Cell solution and buffer are injected into the microfluidic by syringe pump; while the sample flow pass the detection zone, target cells are detected by CCD detector; the target cells will be changed direction by optical tweezers. (**c**) Cells are separated according to the size of cells by optical force [27–29]

post-sort purities up to 89 % and up to 155-fold enrichment of target cells [27]. Wang and his workmates designed a microfluidic chip based on dynamic fluid and dynamic light pattern and a recognition capability of multiple cell features shown in Fig. 10.3b. The experiments of sorting yeast cells and human embryonic stem cells are demonstrated by this approach [28]. Kim designed a generic single-cell manipulation tool that integrates optical tweezers and microfluidic chip technologies for handling small cell population sorting with high accuracy shown in Fig. 10.3c [29].

Over the last few years, although optical manipulation technique combined with microfluidics has made a big advancement and been used for clinical applications, there are some issues and drawbacks, such as photodamage, that still need to be improved and solved.

### *10.2.2 Electrical Manipulation Technique*

Electrical manipulation technique, such as single-cell gel electrophoresis assay (SCGE) and dielectrophoresis (DEP) cell manipulation technique, also has been reported. SCGE was first presented by Ostling and Johanson in 1984; they used this method to study radiation-induced DNA damages in individual mammalian cells [30]. And the term "dielectrophoresis (DEP)" was coined by Pohl in 1951, who performed important early experiments with small plastic particles suspended in insulating dielectric liquids and found that the particles would move in response to the application of a nonuniform AC or DC electric field [31]. At that time, however, the DEP manipulation technique had not been applied to manipulate bio-particles and cells. Currently, nevertheless, DEP manipulation technique has been highlighted due to its potential in the selective spatial manipulation of particles and cells.

SCGE was used as a tool to assess genetic damage in exposed populations. It enables the detection of various DNA damage in individual cells with ease and speed. It is, therefore, well suited to the analysis of a large group in a population [32]. Miloshev has applied the SCGE assay on yeast cells treated with *Saccharomyces cerevisiae* cells with hydrogen peroxide, methyl methanesulfonate (MMS), and two DNA-damaging agents [2]. The human red blood cells were separated at the single-cell level by using capillary zone electrophoresis in 2000. This finding, which is reported by Tsuda, is the first report to present the fine separation of population of red blood cells at single-cell level [33]. However, SCGE is rarely found in single-cell manipulation field, because it is very difficult to manipulate individual cell at single-cell level and hard to recover single target cell.

At the same time, researchers developed DEP manipulation technique for single-cell manipulation. DEP, the force produced by acting a nonuniform electric field upon a neutral object, is shown to be a simple and useful technique for the study of cellular organisms [34]. The magnitude of DEP force is dependent on the size, shape, electrical property of the particle, and the electric field gradient. While the direction of the DEP force is the same as the electric gradient, this phenomenon is called as positive DEP (p-DEP). Conversely, it is called as a negative DEP (n-DEP). The DEP force can be expressed as Eq. (10.5)

$$\left\langle \overline{F}_{\mathrm{DEP}}(t) \right\rangle = 2\pi\varepsilon_m R^3 \left[ f_{\mathrm{CM}}(\omega) \right] \nabla E_{\mathrm{RMS}}^2 \tag{10.5}$$

where $\varepsilon_m$ is the medium permittivity, $R$ is the particle radius, and $E_{\mathrm{RMS}}$ is the root mean square magnitude of the electric field. $f_{\mathrm{CM}}$ is the Clausius–Mossotti (CM) factor. It is given by

$$f_{\mathrm{CM}} = \frac{\overline{\varepsilon}_p - \overline{\varepsilon}_m}{\overline{\varepsilon}_p + 2\overline{\varepsilon}_m} \tag{10.6}$$

where $\overline{\varepsilon}_p$ and $\overline{\varepsilon}_m$ are the complex permittivity of the particle and the medium, respectively. And $\overline{\varepsilon}$ is expressed as $\varepsilon + \sigma / j\omega$ . $\varepsilon$ is permittivity and $\sigma$ is conductivity of the particle and the medium and $j = \sqrt{-1}$ . The CM factor is comprised between $-0.5$ and 1. When the particle is less polarizable than the surrounding medium $\left( f_{\mathrm{CM}} < 0 \right)$, it will experience an n-DEP force pushing it toward regions of electric field minima. On the contrary, when the particle is more polarizable than the medium $\left( f_{\mathrm{CM}} > 0 \right)$, it will be attracted by a p-DEP force toward regions

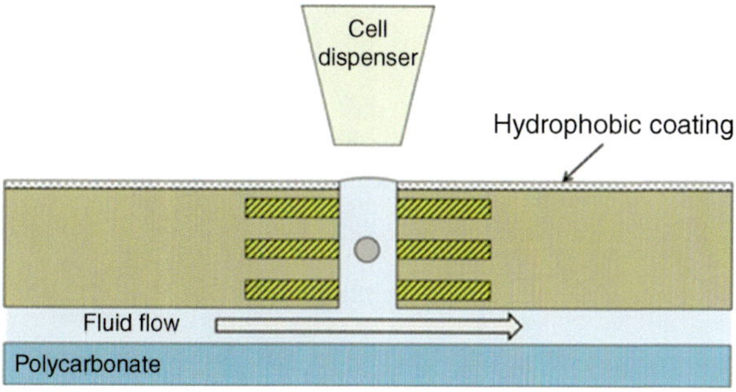


**Fig. 10.4** Cross section of a microwell consisting of a 125-μm hole drilled in a multilayer flexible PCB substrate. A fluid carrier is used to create a channel under the well which is filled by capillarity, while the hydrophobic coating on the top side allows the fluid to be kept within the hole and a meniscus consequently to form. A particle can be inserted manually or with the support of a cell dispenser from the top and will be trapped at the level of the second metal layer [35]

of field maxima [35]. Based on the theory of DEP, scientists and engineers designed and developed many devices for manipulation of cells. DEP cell manipulation technique is also a research highlight in cell manipulation field, due to the advantages of being harmless to the cell, pretreatment free, high precision, easy to manipulate an individual cell, and so on. Nowadays, DEP cell manipulation technique can achieve target cell separation [36], trapping/capturing [37], and release [38]. In order to obtain target particle or cell faster and more effective, a novel concept of active microwells, which can vertically trap and control single particles by means of n-DEP, was presented. The authors applied sinusoidal signals on the electrodes at frequencies ranging from 100 kHz to 1.5 MHz and amplitudes between 2 and 7 V. Particles are successfully trapped and levitated at the level of the central electrode in the middle of microwells with a diameter of 125 μm, as shown in Fig. 10.4 [35].

Different electrode shapes produce different electric field gradient. Therefore, various electrode shapes were designed for cell and particle manipulation, such as ring electrode, quadrupole, and so on. Thomas designed a ring electrode for single-cell immobilization, as shown in Fig. 10.5a, b. The analysis of the 80-μm diameter ring electrode showed that a 15.6-μm diameter latex particle could be held with a force of 23 pN when applied voltage of 5 V peak–peak on the electrode [39]. Another design uses quadrupole to trap cells by means of alternating current electrothermal effect (ACET), as shown in Fig. 10.5c, d [40]. This quadrupole method can trap single cells by adding voltage on different electrode groups.

Park presented a design, fabrication, and characterization of a microfluidic biochip with integrated actuation electrodes for manipulation of single cell and single micro-bead by DEP and sensing electrodes for the detection of the trapping particles by using the impedance detection method, as shown in Fig. 10.6 [41].

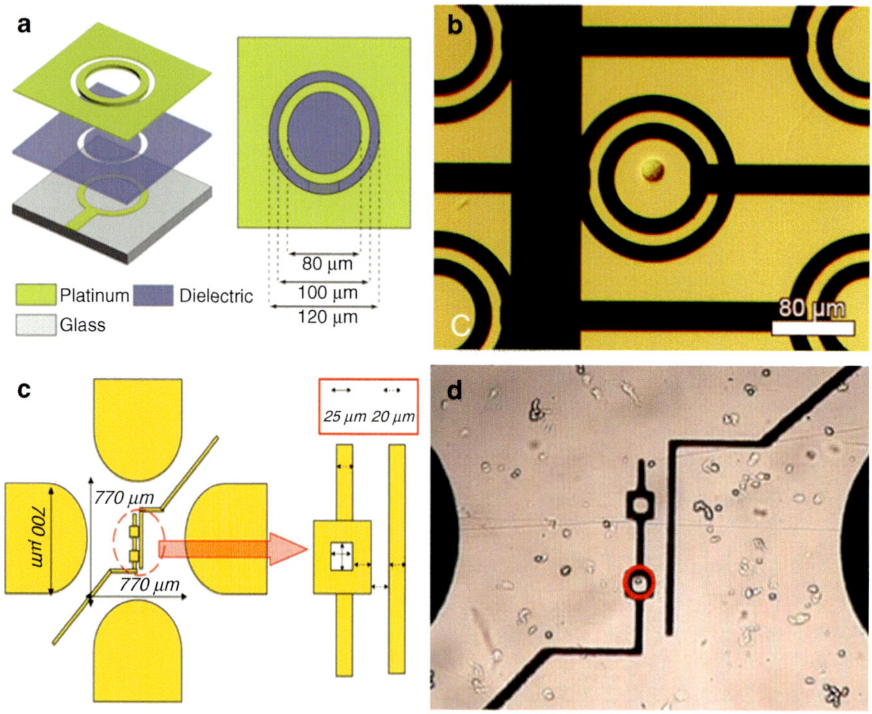


**Fig. 10.5** Different electrode shapes for cell manipulation. By applying voltage on the different electrodes to generate nonuniform electric field for cells trapping. (**a**) The ring traps were fabricated from two titanium/platinum layers with a benzocyclobutene (BCB) dielectric. An alternative design was also produced with dimensions 40/60/80 µm. (**b**) A single HeLa cell immobilised in a nDEP ring trap against a continuous flow (of DMEM culture medium) at 1.0 ml/min. Eletrical excitaion is 2.5 V peak at 20 MHz. (**c**) Schematic of the designed micro electrodes: the microwell electrodes are located in the center of the quadrupole electrodes. (**d**) Cell is trapped at the voltage of 9 Vpp. The *red circle* in the figures indicates the target cell [39, 40]

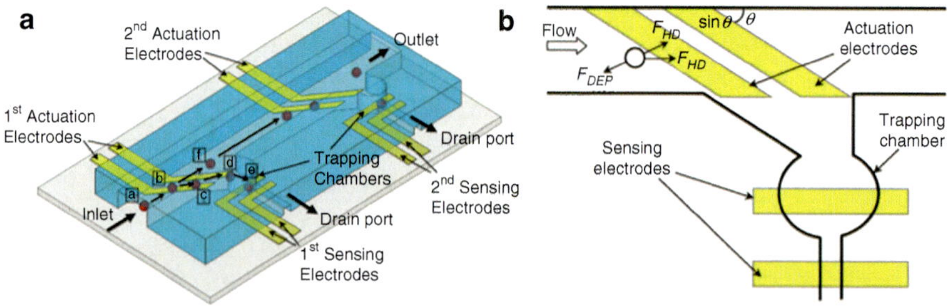


**Fig. 10.6** (**a**) Schematic diagram of proposed microfluidic chip. (**b**) Forces act on a particle, causing the particle moving into the trapping chamber for analyzing [41]

DEP cell manipulation technique combines with microfluidic, which contains a microwell array inside, for high-efficient single-cell trapping and analysis was reported in 2011 [42]. At the same time, a novel concept of micro-cavity array with multilayer electrodes for trapping and programmable releasing single cells was proposed [38, 43–45]. The model of multilayer electrodes DEP chip is shown in

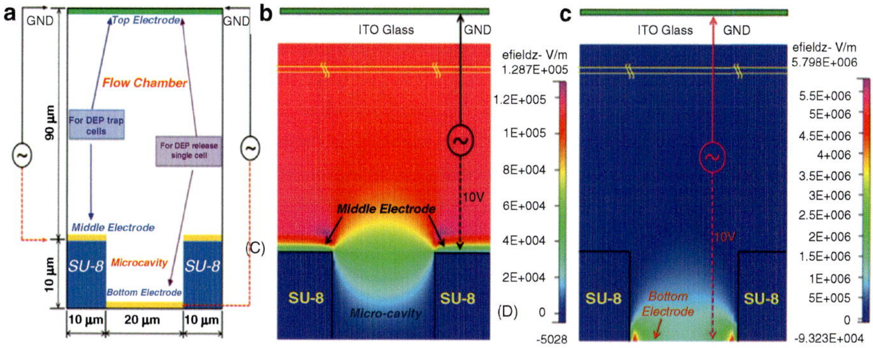


**Fig. 10.7** Model and simulation result of multilayer electrodes EDP chip. (**a**) In this design, by applying a AC voltage on the top and middle electrodes for trapping cells into microwell and by applying an AC voltage on top and bottom electrodes to release cell out of microwell. (**b–c**) show the gradient of electric field generate by applying voltage on different electrodes group [38, 43–45]

**Fig. 10.8** An experimental and fundamental structure of optoelectronic tweezers. Liquid-containing particles are sandwiched between a transparent electrode and a photoconductive electrode. Light patterns from a digital projector are imaged onto the device. The image together with an AC electrical bias generates "virtual electrodes" that in turn create dielectrophoretic trap in the illuminated areas [14]

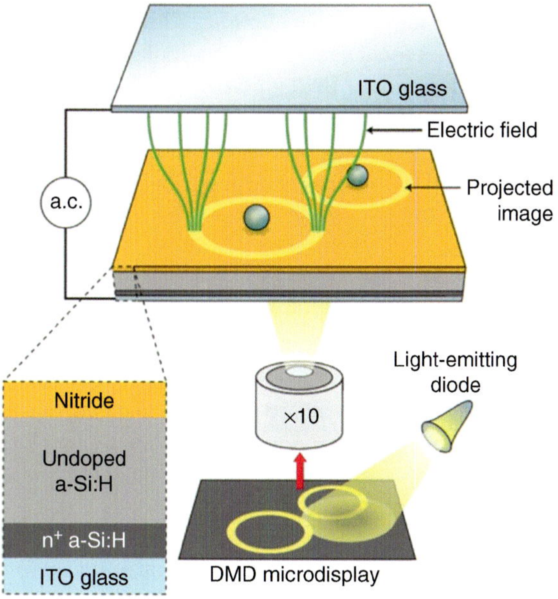


Fig. 10.7a. This design can trap cells by applying the AC signal to the middle and top electrodes; the electric field gradient was simulated by finite element analysis software, as shown in the Fig. 10.7b. Similarly, to release the trapped cells by applying an AC signal on the bottom and top electrodes, and the electric field gradient is shown in Fig. 10.7c.

Fabricating a large electrode array in a microfluidic chip may result in a complex and large electrode driver circuit. This will need to integrate both MEMS and CMOS process and lead to very high design costs. Hence, researchers proposed optoelectronic tweezers (OETs) based on the DEP principle. OET devices are conceptually similar to solar cells. Instead of generating photocurrents, the photo-generated carriers increase the conductance locally near the illuminated area, thus forming virtual electrodes, the experimental and fundamental structure of OETs shown in Fig. 10.8 [14].

On the OET basis, a massively parallel manipulation of single cells and microparticle method was presented. It has been demonstrated that parallel manipulation of 15,000 particles can be trapped on a 1.33 × 1.0 $mm^2$ area [46]. A few years later, the authors designed a device of phototransistor-based OETs for dynamic cell manipulation in cell culture media. This device demonstrated precise control of separation between two cells [47]. In order to obtain a single cell for sample preparation and analysis, Huang designed an OET platform with microfluidics [48]. Recently, a compact OET system combined with cell manipulation and analysis is presented. The trap strength and profile for two emission wavelengths have been measured, and the maximum trapping force of 13.1 and 7.6 pN was achieved by projected micro-LED devices emitting at $\lambda_{max}$ 520 nm and 450 nm, respectively [49]. In all, this OET system is very suited for single-cell manipulation and fluorescence imaging of immune cell.

#### 10.2.2.1 Magnetic Manipulation Technique

Magnetic manipulation technique, which is known as magnetic tweezers, emerged later than electrical and optical manipulation techniques. The magnetic force is generated by a magnetic field gradient. While this force is acting on a magnetic particle, it depends on the volume of the particle ($V$), the difference in magnetic susceptibilities, $\Delta\chi\left(\Delta\chi = \chi_p - \chi_m\right)$, between the particle ($\chi_p$), and surrounding buffer medium ($\chi_m$), as well as the magnetic flux density ($B$) and magnetic flux density gradient ($\nabla B$) [50, 76]:

$$F_{\mathrm{m}} = \frac{V \cdot \Delta\chi}{\mu_0} B \cdot \nabla B \qquad (10.7)$$

To our knowledge, there are two ways to create magnetic field. One way is applying permanent magnets to create magnetic field, and the other way is utilizing electromagnets. A system that utilizes electromagnets is more flexible and easier to control than the permanent one.

Based on this theory, the magnetic manipulation technique for particles was proposed. This technique with great advantages, for instance, magnetic manipulation technique, is not affected by particle surface charges, PH, ionic concentrations, or temperature. In recent years, it was widely applied in biology, such as to manipulate DNA or RNA molecules [51], micro- or nanoparticles [52], and single cell [53] and can be used for analysis of cell mechanics as well [54, 55]. In order to achieve precisely control, an electromagnet was assembled to generate a controllable magnetic gradient for DNA manipulation by Haber [56]. After that, based on the general circuit theory and magnetic bubble technology, Lim demonstrated a class of integrated circuits for executing sequential and parallel timed operations on an ensemble of single particles and cells [57]. In single-cell manipulation field, Liu reported a simple and straightforward approach to fabricate

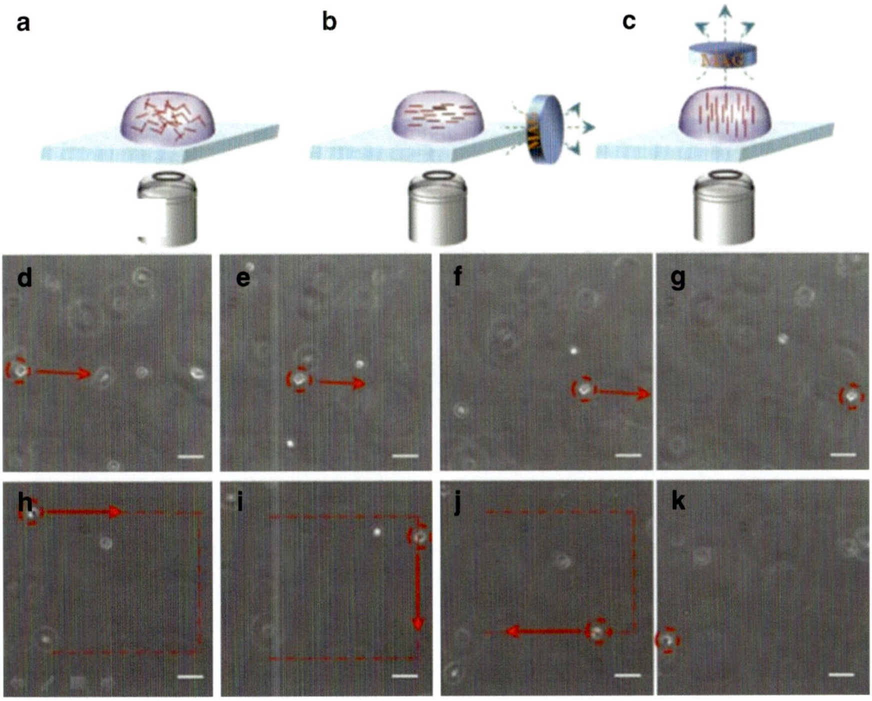


**Fig. 10.9** (**a–c**) Schematic of magnetic particle-doped nanofiber segments in water drops on a glass slide. (**d–k**) Showing cell movement path, controlled by an external magnetic field. The scale bar is 20 μm [58]

magnetic nanofiber segments for cell manipulation. They used NIH 3 T3 cells, which were cultured in a medium containing magnetic fibers, as assay samples; the result of the assays showed that cells can be conveniently manipulated with a magnet, as shown in Fig. 10.9 [58].

Ebrahimian injected magnetic particles of 1 μm diameter into barley cell vacuoles using a microinject system under microscopic control. By injecting the magnetic particles into cells, they achieved that manipulating cell at single-cell level by using magnetic tweezers [59]. In order to fabricate a microstructure that can be powered and controlled wirelessly in fluidic environments, a kind of biocompatible ferromagnetic microtransporters driven by external magnetic fields was described by Sakar. Those microtransporters were fabricated with SU-8 photoresist and magnetic nanoparticles [60]. Figure 10.9 (A) shows the shapes of the microtransporters. Figure 10.10b–d shows the process of manipulation of a target cell. (B) The orientation of the transporter is adjusted according to the position of the target cell. (C) With the application of an out-of-plane time-varying magnetic field, the transporter starts translating toward the target. The pulsing frequency is 100 Hz. (D) The target is engaged and transported out of the field of view. At the end, the authors demonstrated that the average velocity of the transporter can be up to 350 μm/s.

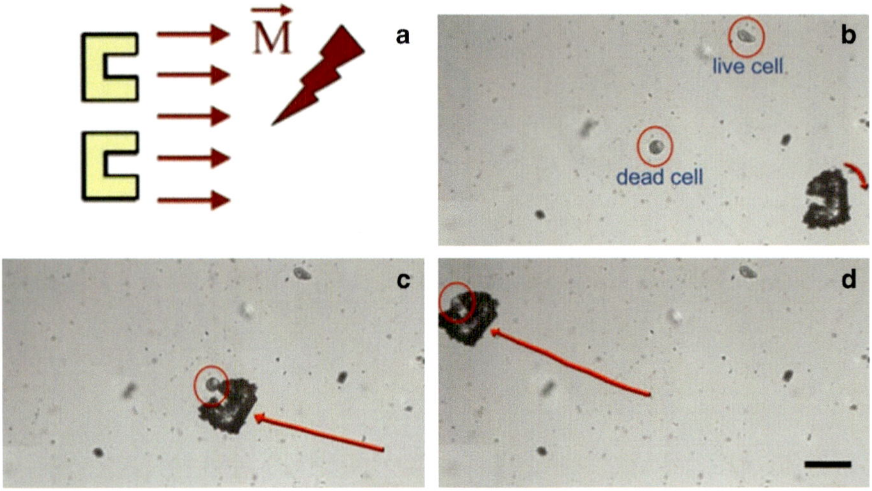


**Fig. 10.10** (**a**) The shapes of microtransporter. (**b–d**) Process of manipulation a target cells. (**b**) By controlling direction of the magnetic field to move the micotransporter toward the dead cell. (**c**) Moving the microtransporter to capture the dead cell by magnetic flied. (**d**) Moving away the dead cell by microtransporter. The scale bar is 100 μm [60]

Another application of magnetic manipulation technique is magnetic-activated cell sorting (MACS). MACS is a method for separation of various cell populations depending on their particular surface antigens. The procedure of MACS is that the cell solution, which the target cells are expressing their antigen to attach to the magnetic nanoparticle-coated antibodies, is transferred on a column placed in a strong magnetic field; the labeled cells are separated by the magnetic force from the magnetic nanoparticles. MACS method is extensively used in biotechnology for a wide range of application from in vitro diagnostics to cell-based therapies. MACS allow high-throughput separation of magnetically labeled target species [61]. Nowadays, circulating tumor cells (CTCs) have been studied as a mean of overcoming cancer. In order to select and isolate the rarity and heterogeneity of CTCs, a two-stage microfluidic chip was designed by Hyun [62]. The first stage (Fig. 10.11a, b) involves a microfluidic MACS chip to elute white blood cells (WBCs). The second stage (Fig. 10.11c, d) involves a geometrically activated surface interaction (GASI) chip for the selective isolation of CTCs.

Due to the physics of separation based on a single parameter – magnetization – the MACS is generally effective only for single-species target cell selection. In order to separate multi-species cells, a multi-target magnetic-activated cell sorter (MT-MACs), which separating cells depend on the magnetization of different magnetic nanoparticles, was presented by Adams, and the MT-MACs separation architecture is shown in Fig. 10.12 [63].

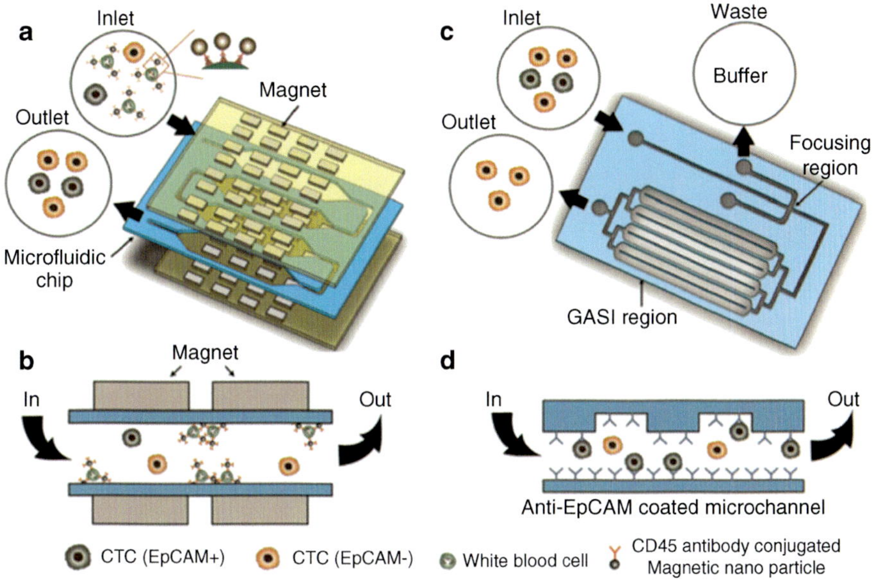


**Fig. 10.11** (**a**) Schematic diagram of a two-stage enrichment chip for isolation of CTCs. (**b**) White blood cells are trapped by magnetic field. (**c**) Microfluidic structure for CTCs separation. (**d**) CTCs(EpCAM+) are trapped by the antibody [62]

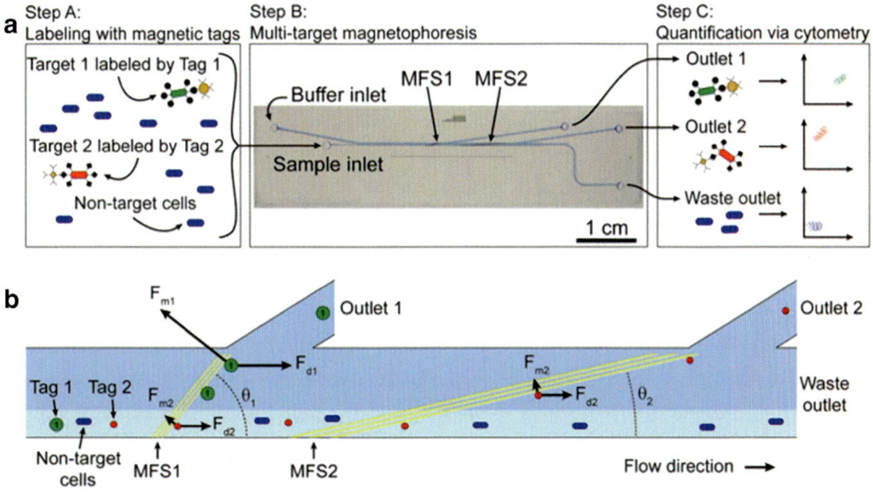


**Fig. 10.12** The MT-MACs separation architecture. (**a**) Step A, different target cells are labeled by different types of magnetic tag; step B, sample with different tags will be separate in the magnetic field; step C, quantification cell via a cytometry. (**b**) Forces act on target cells [63]

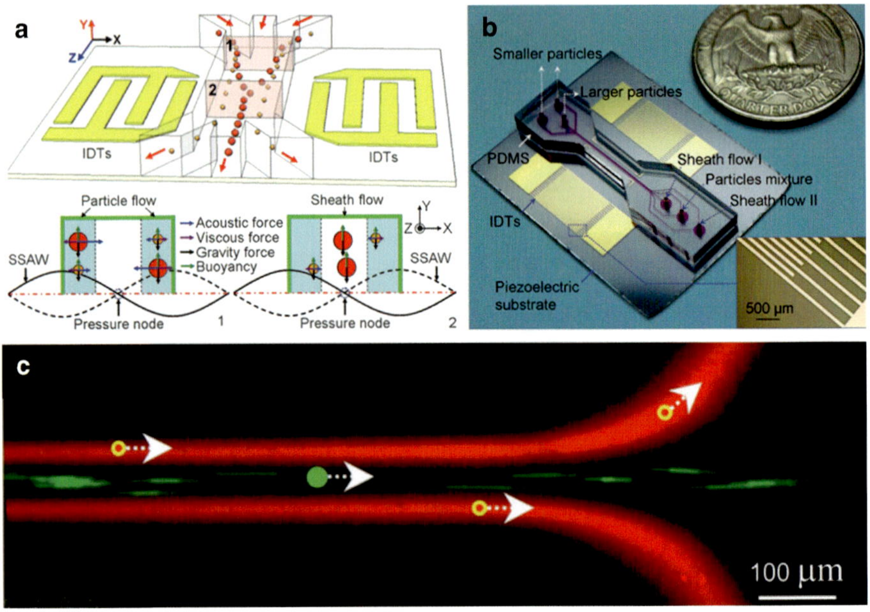


**Fig. 10.13** (**a**–**b**) Schematic of separation mechanism, (**c**) optical image of the design, (**d**) image of particle separation [67]

#### 10.2.2.2 Acoustic Manipulation Technique

In recent years, surface acoustic wave (SAW) techniques in microfluidics were a concern by researchers. Recent research demonstrates that SAW technique can be an effective method to manipulate microscale particles [64]. The best-known SAW is composed of a longitudinal and a vertically polarized shear component [65]. SAWs are generally produced by applying an appropriate voltage on the electrodes of piezoelectric material. The electrodes will generate propagating mechanical stress. A typical SAW device uses at least one set of metallic interdigital transducers (IDTs) fabricated on the surface of a piezoelectric substrate. The IDT then introduces the electric field, generating a SAW displacement amplitude on the order of 1 nm [66].

In the last few years, SAW technology combined with microfluidic technique has turned into a tool, which is known as acoustic tweezers, to manipulate particles. Liquid inside of the microfluidics is pressed by mechanical stress. Hence, a liquid pressure gradient in microfluidics can be controlled by controlling the IDTs. Based on this principle, a continuous particle separation in a microfluidic channel via standing surface acoustic waves was introduced by Shi, as shown in Fig. 10.13 [67]. The particle separation method is capable of separating virtually all kinds of particles with high separation efficiency and low power consumption, the separation mechanism and result of separation.

After a short while, a design of on-chip manipulation of single particles, cells, and organisms using SAW was presented by Ding [68]. Figure. 10.14 shows the (A)

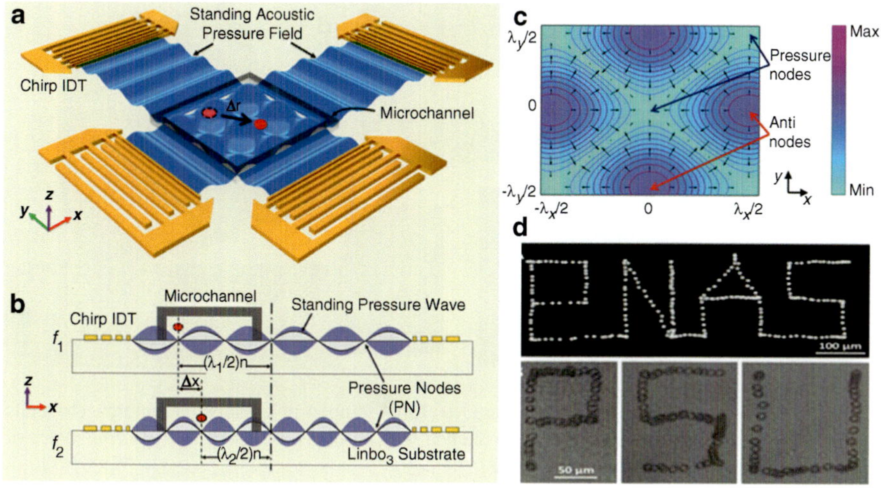


**Fig. 10.14** (**a–b**) Device structure and working mechanism of the acoustic tweezers. (**c**) The pressure nodes of solution inside this device, the particles will move to and stay at the minimum pressure nodes. (**d**) Result of polystyrene bead and cell manipulation, top one using polystyrene beads to write "PNAS" letters and the bottom one employing cells to write "PSU" letters [68]

schematic illustration of a microfluidic device with orthogonal pairs of chirped IDTs for generating standing SAW. (B) shows standing SAW field generated by driving chirped IDTs at frequency $f_1$ and $f_2$. When particles are trapped at the nth (it is an ordinal number) pressure node, they can be translated at a distance of $(\Delta\lambda/2)n$ by switching from $f_1$ to $f_2$. This relationship indicates that the particle displacement can be tuned by varying the pressure node where the particle is trapped. (C) shows simulated pressure field between adjacent pressure antinodes. (D) shows the author used a 10-μm fluorescent polystyrene bead to write the word "PNAS" and dynamic control of a bovine red blood cell to trace the letters "PSU."

## 10.3 Mechanical Manipulation Technique

Using mechanical methods to manipulate particles and bio-particles, such as capturing, sorting, and isolation, have emerged for several decades. These methods use mechanical forces, such as gravity, hydrodynamic, and suction to manipulate particles or bio-particles. Different microwell dimension arrays for large-scale single-cell trapping were designed by Rettig [69]. With this method, the parameters that maximize single-cell occupancy for two cell types, including microwell diameter, microwell depth, and settle time, were determined. The authors injected cell suspension into the microfluidics, which contains microwell array inside and then the cells will settle into the microwells due to gravity. They found that microwell with an aspect ratio of ~1 yield optimal single-cell occupancy. Other structures for cell

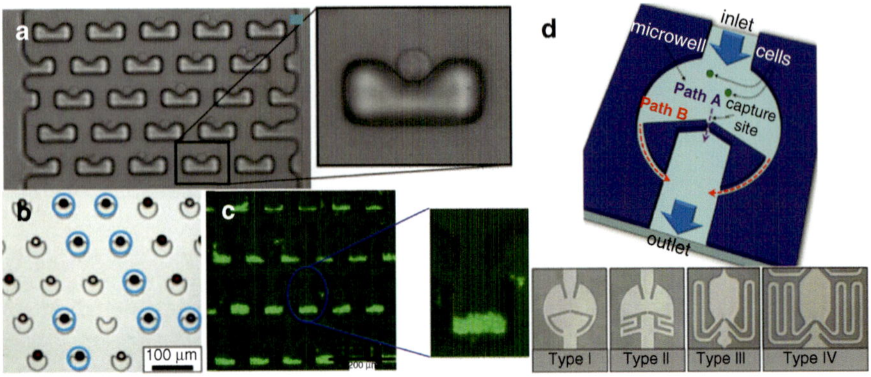


**Fig. 10.15** Different shapes for cell capture. (**a**) Cells trapped in U-shaped structures. (**b**) Cells trapped in V-shape. (**c**) Cells trapped in U-shaped with Y-shaped fluidic guide. (**d**) Four types of different structure for cells trapping [70–73]

manipulation based on hydrodynamics also were presented. A device that consists of physical U-shaped hydrodynamic trapping structure array for single-cell trapping and culture had been designed; the shape of this design is shown in Fig. 10.15a [70]. Researchers improved the trapping structure for high-efficient single-cell capture. In order to trap particles more efficiently, a geometrical V-cup barrier array was presented. The capture efficiency of this platform is nearly 100 %, and the trapping shape is shown in Fig. 10.15b [71]. Another structure for cell capture was proposed by Chen [72] . In this design, the first layer consists of spacers to create a small gap between the upper layer and the glass. The second layer is a sharp corner U-shaped compartment with sharp corners at the fore-end. And Y-shaped fluidic guide structures are designed on the top of each U-shaped capture structures, as shown in Fig. 10.15c. A highly efficient single-cell capture scheme using hydrodynamic guiding structures also was presented [73]. The authors designed four types of cell capture module and tested them for optimal structure. The capturing efficient of this single-cell capture chip is more than 80 % shown by the experiments. The structures for single-cell trapping were shown in Fig. 10.15d.

The structures were continually improved by researchers; however, those designs are only suited for one or several kinds of cells due to its manipulation principle dependent on both the geometric size of particles and capture structures. Thus, the method of suction for manipulation cell was presented. A micromanipulation method for single prokaryotic cells extracting was improved, as shown in Fig. 10.16 [74]. Figure 10.16a shows the workstation for manipulation of single cells, (1) joystick, (2) inverse microscope, (3) micromanipulator, (4) thermometer/hygrometer, and (5) Bactotip. (b–d) shows the schematic drawing of the isolation of a single bacterial cell.

Another design based on the method of suction was presented by Anis and his colleagues [75]. They developed a pico-liter pump and integrated it into a robotic manipulation system. This pico-liter pump can automatically select and transfer individual living cells of interest to analyze locations. The authors demonstrated that the pump aspirates and dispenses volumes of fluid between 500 pL and 250 nL at flow rates up to 250 nL/s. And then they successfully accomplished single-cell

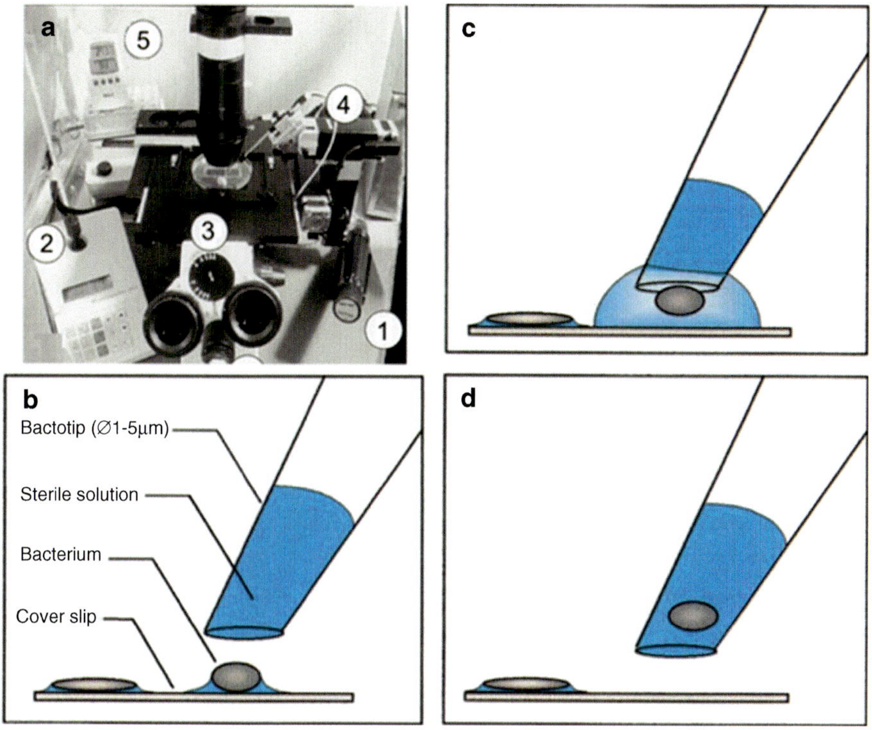


**Fig. 10.16** (**a**) A workstation and schematic drawing for isolation a single bacterial cell. (**b–d**) The procedure of isolation a target cell [74]

manipulation assay by using Barrett's esophagus cell. The six-axis robotic workstation schematic and assay image is shown in Fig. 10.17.

Furthermore, Flanders' team presented a methodology for fabrication of cellular force sensors composed of high aspect ratio cantilevered poly (3, 4-ethylenedioxythiophene) (PEDOT) fibers [77]. These fiber sensors can be used to characterize the dynamics of apical pseudopod-substrate adhesive contacts of *D. discoideum* cells. They have also shown that these cellular force probes may be positioned independently around the single cell and can interface with target cells without forming secondary contacts. Due to the characters of these probes, they may have the capacity of manipulating a single cell.

## 10.4 Conclusions and Outlook

In this review, we summarized the methods of single-cell manipulation technologies. Table 10.1 shows the advantages and drawbacks of each cell manipulation technique. Nowadays, however, various cell manipulation methods often require a set of parameters for further standardization and commercialization. Hence, scientists and engineers design various structures based on different principle of cell manipulation technology

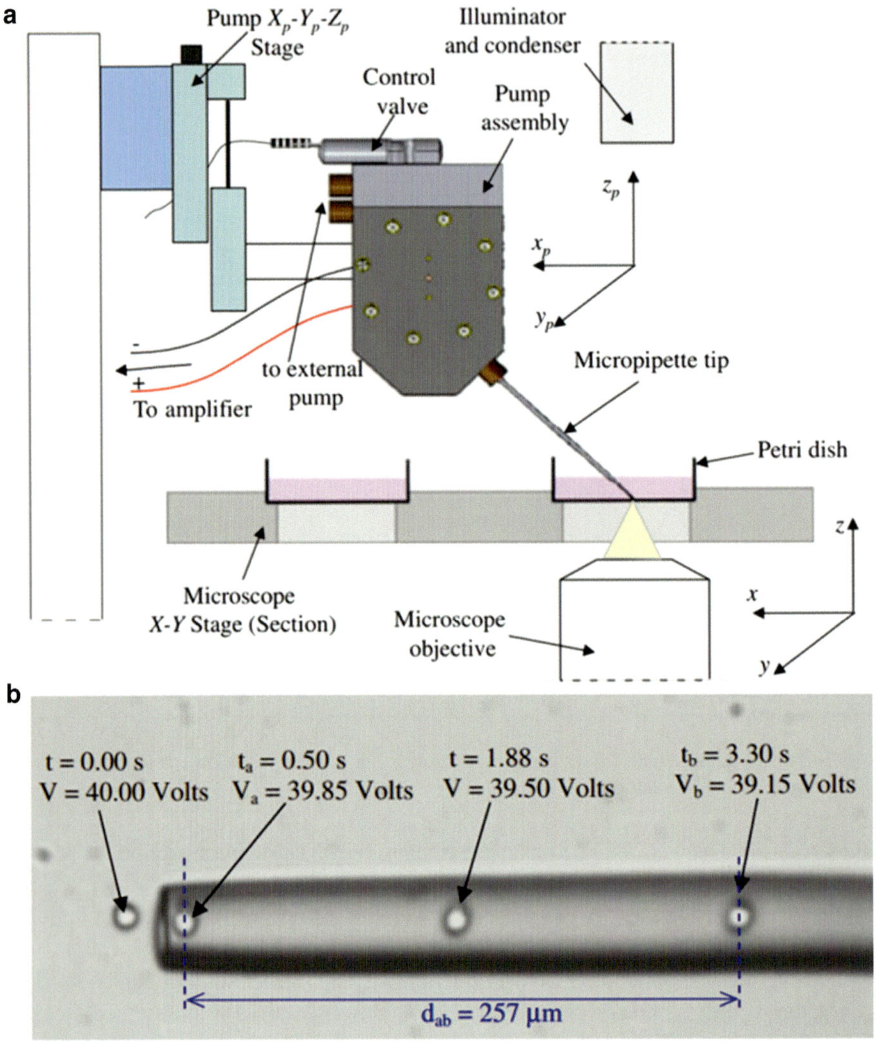


**Fig. 10.17** (**a**) Six-axis robotic workstation schematic and assay image. (**b**) According to this piezoactuator operate mode, by reducing the voltage on piezoactuator, it leads the pump to work and causes aspiration of a cell into the pipette[75]

to fulfill the purpose of high efficient, high throughput, low injury, and high precision of manipulation cells in single-cell level. Overall, cell manipulation technologies combined with microfluidics bring lots of benefit to biological research. To reach the requirement of exact preclinical samples for current clinical trials, single-cell manipulation technologies need to be further studied. And the capacity of cell manipulation devices, such as obtaining high-purity and high-recovery rate of target cells, high throughput high automaticity, being harmless to cells, and easy to use, is mainly concerned by designers and users in future.

**Table 10.1** Comparison of each cell manipulation technique

| Manipulation technique | | Advantages | Drawbacks |
|---|---|---|---|
| Optical | | High precision<br>High stability<br>High throughput | May harm cells<br>Device complexity<br>Device expensive |
| Electrical | SCGE | Harmless to cells<br>Device simple | Hard to manipulate cells in single-cell level<br>Low throughput |
| | DEP | Harmless to cells<br>High throughput<br>High precision<br>Device simple | Hard to control<br>Drive circuit is complex |
| Magnetic | | High precision<br>Device is simple | Magnetic material pre-label<br>Hard to control |
| Acoustic | | Harmless to cells<br>Low power consumption<br>Device is simple | Hard to control<br>Electrode drive is complex |
| Mechanical | Gravity | Device is simple | Only use for trapping |
| | Hydrodynamic | Device simple | Hard to manipulate cells in single-cell level |
| | Suction | Simple to operate | Device is complex<br>Hard to align |

# References

1. Yun H, Kim K, Lee WG. Cell manipulation in microfluidics [J]. Biofabrication. 2013;5(2):022001.
2. Miloshev G, Mihaylov I, Anachkova B. Application of the single cell gel electrophoresis on yeast cells [J]. Mutat Res. 2002;513(1-2):69–74.
3. Zabzdyr JL, Lillard SJ. New approaches to single-cell analysis by capillary electrophoresis [J]. Trends Anal Chem. 2001;20(9):467–76.
4. Chen J, Chen D, Yuan T, et al. A microfluidic chip for direct and rapid trapping of white blood cells from whole blood [J]. Biomicrofluidics. 2013;7(3):34106.
5. Lee SW, Kang JY, Lee IH, et al. Single-cell assay on CD-like lab chip using centrifugal massive single-cell trap [J]. Sensors Actuators A: Physical. 2008;143(1):64–9.
6. Wakamoto Y, Inoue I, Moriguchi H, et al. Analysis of single-cell differences by use of an on-chip microculture system and optical trapping [J]. Fresenius J Anal Chem. 2014; 371(2):276–81.
7. Whitesides GM. The origins and the future of microfluidics [J]. Nature. 2006;442(7101): 368–73.
8. Pushkarsky I, Tseng P, Murray C, et al. Research highlights: microfluidics and magnets [J]. Lab Chip. 2014;14(16):2882.
9. Di Martino D, Giuffre G, Staiti N, et al. Single sperm cell isolation by laser microdissection [J]. Forensic Sci Int. 2004;146(Suppl):S151–3.
10. Fuchs AB, Romani A, Freida D, et al. Electronic sorting and recovery of single live cells from microlitre sized samples [J]. Lab Chip. 2006;6(1):121–6.
11. Huang WE, Ward AD, Whiteley AS. Raman tweezers sorting of single microbial cells [J]. Environ Microbiol Rep. 2009;1(1):44–9.
12. Keithley RB, Metzinger MP, Rosado AM, et al. Manipulating ionic strength to improve single cell electrophoretic separations [J]. Talanta. 2013;111:206–14.

13. Wong X, Rosales C. Robust dielectrophoretic single-cell trap design using BEM [J]. Eng Anal Bound Elem. 2008;32(5):388–94.
14. Wu MC. Optoelectronic tweezers [J]. Nat Photonics. 2011;5(6):322–4.
15. Ashkin A. Acceleration and trapping of particles by radiation pressure [J]. Phys Rev Lett. 1970;24(4):156–9.
16. Ashkin A, Dziedzic JM. Optical trapping and manipulation of viruses and bacteria [J]. Science. 1987;235(4795):1517–20.
17. Neuman KC, Block SM. Optical trapping [J]. Review Sci Instrum. 2004;75(9):2787–809.
18. Grier DG. A revolution in optical manipulation [J]. Nature. 2003;424(14):810–6.
19. Goksör M, Enger J, Hanstorp D. Optical manipulation in combination with multiphoton microscopy for single-cell studies [J]. Appl Optics. 2004;43(25):4831–8.
20. Greulich KO, Pilarczyk G, Hoffmann A, et al. Micromanipulation by laser microbeam and optical tweezers: from plant cells to single molecules [J]. J Microsc. 2000;198(3):182–7.
21. Ando J, Bautista G, Smith N, et al. Optical trapping and surgery of living yeast cells using a single laser [J]. Rev Sci Instrum. 2008;79(10):103705.
22. Barak P, Rai A, Rai P, et al. Quantitative optical trapping on single organelles in cell extract [J]. Nat Methods. 2013;10(1):68–70.
23. Neuman KC, Chadd EH, Liou GF, et al. Characterization of photodamage to escherichia coli in optical traps [J]. Biophys J. 1999;77(5):2856–63.
24. Peterman EJG, Gittes F, Schmidt CF. Laser induced heating in optical traps [J]. Biophys J. 2003;84(2):1308–16.
25. Eriksson E, Enger J, Nordlander B, et al. A microfluidic system in combination with optical tweezers for analyzing rapid and reversible cytological alterations in single cells upon environmental changes [J]. Lab Chip. 2007;7(1):71–6.
26. MacDonald MP, Spalding GC, Dholakia K. Microfluidic sorting in an optical lattice [J]. Nature. 2003;426(6965):421–4.
27. Kovac JR, Voldman J. Intuitive, image-based cell sorting using optofluidic cell sorting [J]. Anal Chem. 2007;79(24):9321–30.
28. Wang X, Chen S, Kong M, et al. Enhanced cell sorting and manipulation with combined optical tweezer and microfluidic chip technologies [J]. Lab Chip. 2011;11(21):3656–62.
29. Kim SB, Yoon SY, Sung HJ, et al. Cross-type optical particle separation in a microchannel [J]. Anal Chem. 2008;80(7):2628–30.
30. Ostling O, Johanson KJ. Microelectrophoretic study of radiation-induced DNA damages in individual mammalian cells [J]. Biochem Biophys Res Commun. 1984;123(1):291–8.
31. Pohl HA. The motion and precipitation of suspensoids in divergent electric fields [J]. J Appl Phys. 1951;22(7):869.
32. Kassie F, Parzefall W, Knasmuller S. Single cell gel electrophoresis assay: a new technique for human biomonitoring studies [J]. Mutat Res-Rev Mutat. 2000;463(1):13–31.
33. Tsuda T, Yamauchi N, Kitagawa S. Separation of red blood cells at the single cell level by capillary zone electrophoresis [J]. Anal Sci. 2000;16(8):847–50.
34. Pohl HA, Crane JS. Dielectrophoresis of cells [J]. Biophys J. 1971;11(9):711–27.
35. Bocchi M, Lombardini M, Faenza A, et al. Dielectrophoretic trapping in microwells for manipulation of single cells and small aggregates of particles [J]. Biosens Bioelectron. 2009;24(5):1177–83.
36. Gascoyne PRC, Vykoukal J. Particle separation by dielectrophoresis [J]. Electrophoresis. 2002;23(13):1973–83.
37. Wang CC, Lan KC, Chen MK, et al. Adjustable trapping position for single cells using voltage phase-controlled method [J]. Biosens Bioelectron. 2013;49:297–304.
38. Chuang CH, Huang YW, Wu YT. Dielectrophoretic chip with multilayer electrodes and microcavity array for trapping and programmably releasing single cells [J]. Biomed Microdevices. 2012;14(2):271–8.
39. Thomas RS, Morgan H, Green NG. Negative DEP traps for single cell immobilisation [J]. Lab Chip. 2009;9(11):1534–40.
40. Jang LS, Huang PH, Lan KC. Single-cell trapping utilizing negative dielectrophoretic quadrupole and microwell electrodes [J]. Biosens Bioelectron. 2009;24(12):3637–44.

41. Park H, Kim D, Yun KS. Single-cell manipulation on microfluidic chip by dielectrophoretic actuation and impedance detection [J]. Sens Actuators B. 2010;150(1):167–73.
42. Kim SH, Yamamoto T, Fourmy D, et al. Electroactive microwell arrays for highly efficient single-cell trapping and analysis [J]. Small. 2011;7(22):3239–47.
43. Chuang CH, Huang YW, Wu YT. System-level biochip for impedance sensing and programmable manipulation of bladder cancer cells [J]. Sensors. 2011;11(12):11021–35.
44. Chuang CH, Hsu YM, Wei CH. The effects of microstructures on a dielectrophoretic chip for trapping particles [J]. Electrophoresis. 2009;30(17):3044–52.
45. Chuang CH, Huang YW, Wu YT, et al. Programmable dielectrophoretic chip for cell manipulations [J]. Jpn J Appl Phys. 2011;50(6):06GL11.
46. Chiou PY, Ohta AT, Wu MC. Massively parallel manipulation of single cells and microparticles using optical images [J]. Nature. 2005;436(7049):370–2.
47. Hsu HY, Ohta AT, Chiou PY, et al. Phototransistor-based optoelectronic tweezers for dynamic cell manipulation in cell culture media [J]. Lab Chip. 2010;10(2):165–72.
48. Huang KW, Wu YC, Lee JA, et al. Microfluidic integrated optoelectronic tweezers for single-cell preparation and analysis [J]. Lab Chip. 2013;13(18):3721–7.
49. Jeorrett AH, Neale SL, Massoubre D, et al. Optoelectronic tweezers system for single cell manipulation and fluorescence imaging of live immune cells [J]. Opt Express. 2014;22(2): 1372–80.
50. Pamme N. Magnetism and microfluidics [J]. Lab Chip. 2006;6(1):24–38.
51. Lipfert J, Hao X, Dekker NH. Quantitative modeling and optimization of magnetic tweezers [J]. Biophys J. 2009;96(12):5040–9.
52. Lansdorp BM, Tabrizi SJ, Dittmore A, et al. A high-speed magnetic tweezer beyond 10,000 frames per second [J]. Rev Sci Instrum. 2013;84(4):044301.
53. Chung YC, Chen PW, Fu CM, et al. Particles sorting in micro-channel system utilizing magnetic tweezers and optical tweezers [J]. J Magn Magn Mater. 2013;333:87–92.
54. Alenghat FJ, Fabry B, Tsai KY, et al. Analysis of cell mechanics in single vinculin-deficient cells using a magnetic tweezer [J]. Biochem Biophys Res Commun. 2000;277(1):93–9.
55. Bausch AR, Möller W, Sackmann E. Measurement of local viscoelasticity and forces in living cells by magnetic tweezers [J]. Biophys J. 1999;76(1):573–9.
56. Haber C, Wirtz D. Magnetic tweezers for DNA micromanipulation [J]. Rev Sci Instrum. 2000;71(12):4561–70.
57. Lim B, Reddy V, Hu X, et al. Magnetophoretic circuits for digital control of single particles and cells [J]. Nat Commun. 2014;5:3846.
58. Liu J, Shi J, Jiang L, et al. Segmented magnetic nanofibers for single cell manipulation [J]. Appl Surf Sci. 2012;258(19):7530–5.
59. Ebrahimian H, Giesguth M, Dietz KJ, et al. Magnetic tweezers for manipulation of magnetic particles in single cells [J]. Appl Phys Lett. 2014;104(6):063701.
60. Sakar MS, Steager EB, Kim DH, et al. Single cell manipulation using ferromagnetic composite microtransporters [J]. Appl Phys Lett. 2010;96(4):043705.
61. Miltenyi S, Muller W, Weichel W, et al. High gradient magnetic cell separation with MACS [J]. Cytometry. 1990;11(2):231–8.
62. Hyun KA, Lee TY, Lee SH, et al. Two-stage microfluidic chip for selective isolation of circulating tumor cells (CTCs) [J]. Biosens Bioelectron. 2014;67:86–92.
63. Adams JD, Kim U, Soh HT. Multitarget magnetic activated cell sorter [J]. Proc Natl Acad Sci U S A. 2008;105(47):18165–70.
64. Lin SC, Mao X, Huang TJ. Surface acoustic wave (SAW) acoustophoresis: now and beyond [J]. Lab Chip. 2012;12(16):2766–70.
65. Gedge M, Hill M. Acoustofluidics 17: theory and applications of surface acoustic wave devices for particle manipulation [J]. Lab Chip. 2012;12(17):2998–3007.
66. Ding X, Li P, Lin SC, et al. Surface acoustic wave microfluidics [J]. Lab Chip. 2013;13(18):3626–49.
67. Shi J, Huang H, Stratton Z, et al. Continuous particle separation in a microfluidic channel via standing surface acoustic waves (SSAW) [J]. Lab Chip. 2009;9(23):3354–9.

68. Ding X, Lin SCS, Kiraly B, et al. On-chip manipulation of single microparticles, cells, and organisms using surface acoustic waves [J]. Proc Natl Acad Sci U S A. 2012; 109(28):11105–9.
69. Rettig JR, Folch A. Large-scale single-cell trapping and imaging using microwell arrays [J]. Anal Chem. 2005;77(17):5628–34.
70. Di Carlo D, Wu LY, Lee LP. Dynamic single cell culture array [J]. Lab Chip. 2006;6(11):1445–9.
71. Burger R, Reith P, Kijanka G, et al. Array-based capture, distribution, counting and multiplexed assaying of beads on a centrifugal microfluidic platform [J]. Lab Chip. 2012;12(7): 1289–95.
72. Chen J, Chen D, Yuan T, et al. Microfluidic chips for cells capture using 3-D hydrodynamic structure array [J]. MicrosystTechnol. 2014;20(3):485–91.
73. Chung JH, Kim YJ, Yoon E. Highly-efficient single-cell capture in microfluidic array chips using differential hydrodynamic guiding structures [J]. Appl Phys Lett. 2011;98(12):123701.
74. Fröhlich J, König H. New techniques for isolation of single prokaryotic cells [J]. FEMS Microbiol Rev. 2000;24(5):567–72.
75. Anis Y, Houkal J, Holl M, et al. Diaphragm pico-liter pump for single-cell manipulation [J]. Biomed Microdevices. 2011;13(4):651–9.
76. Inglis DW, Riehn R, Sturm JC, et al. Microfluidic high gradient magnetic cell separation [J]. J Appl Phys. 2006;99(8):08K101.
77. Paneru G, Thapa PS, Flanders BN, et al. Long reach cantilevers for sub-cellular force measurements [J]. Nanotechnology. 2012;23(45):455105.

# Chapter 11
# Multifunctional Nanoprobes for Theranostics of Gastric Cancer

**Daxiang Cui**

## 11.1 Introduction

Gastric cancer (GC) is the fourth commonest cancer and the second leading cause of cancer-related mortality worldwide [1]. Gastric cancer is the second most common cancer and the third leading cause of cancer-related death in China [2]. Although gastroscopy has been broadly used for screening of early gastric cancer patients, gastric cancer remains very difficult to cure effectively, primarily because most patients present advanced stages of the diseases. Therefore, how to find early gastric cancer has become a great challenge.

In 1999, we first proposed the project of developing gastric cancer prewarning and early diagnosis system [3]. Since 1999, we try to realize gastric cancer prewarning and early diagnosis by screening early gastric cancer biomarkers, developing new detection method of gastric cancer biomarkers, and establishing gastric cancer prewarning database including molecular imaging database, biomarker database, and information treatment platform [4, 5]. Especially under the support from Chinese nanokey fundamental project (no.2010CB933900), we have achieved great advances.

Molecular imaging has been being made great advances and has been actively explored potential applications such as tumor early diagnosis, in vivo genotyping, targeted imaging and simultaneously therapy, and surgery navigation [6]. These advances highly depend on advances of multifunctional nanoprobes and also highly depend on advances of nanomaterials and nanotechnology.

D. Cui
Institute of Nano Biomedicine and Engineering, Shanghai Engineering Research Center for Intelligent Diagnosis and Treatment Instrument, National Center for Translational Medicine, Collaborative Innovational Center for System Biology, Shanghai Jiao Tong University, 800 Dongchuan Road, Shanghai 200240, Peoples Republic of China
e-mail: dxcui@sjtu.edu.cn

© Springer Science+Business Media B.V. and Shanghai Jiao Tong University Press, Shanghai 2017

D. Cui (ed.), *Gastric Cancer Prewarning and Early Diagnosis System*, Translational Medicine Research, DOI 10.1007/978-94-024-0951-2_11

Nanotechnology makes an important contribution toward cancer prevention, diagnosis, imaging, and treatment [7]. It not only provides unprecedented capability for carrying multiple diagnostic and therapeutic payloads in the same package but also facilitates targeting delivery into specific sites across complex biological barriers [8]. The multifunctional integrated system combines different properties such as tumor targeting, imaging, and selective therapy in an all-in-one system, which will provide more useful multimodal approaches in the battle against cancer.

Up to date, gastric cancer therapeutic methods mainly include surgery, radiation, and chemotherapies, which are generally very effective for early and in situ gastric cancers, but advanced and metastatic cases do not respond to chemo- or radiation therapies [9]. Resistance to chemotherapy-induced apoptosis is a major cause for the failure of conventional therapies. The current prognosis of gastric cancer is very poor with 5-year survivals of less than 24% [10]. Therefore, how to recognize, track, or kill early gastric cancer cells is a great challenge for patients with early gastric cancer.

Current studies show that gastric cancer is not particularly sensitive to traditional therapies, especially to chemotherapy agents, which seems to be closely related to numerous intrinsic or acquired properties of gastric cancer stem cells (CSCs) [11]. Despite the hypothesis of CSCs being challenged, growing evidence has been provided to support the existence of CSCs. The notion that CSCs give rise to GC and may be responsible for invasion, metastasis, and resistance to conventional treatment has profound implications for anticancer-targeted therapies [12]. Therefore, GCSC-based targeted therapy is destined to be one of the most effective anticancer strategies.

Immunological studies highly suggest that the reasons for metastasis and recurrence of GC can be summarized as follows [13]: (I) the intrinsic antigenicity weakness of tumor cells, immunological surveillance of the host, which cannot identify and eliminate the malignant cells that are distributed out of the resection field and the peripheral lymphoid organs and (II) immunological surveillance defect or dysfunction of the host. After the traditional therapy was done, the residual tumor cells escaped from immunosurveillance revived again. (III) The substantial toxicities of most traditional curative intents, which is often compromised by the extreme feeble immunity of the patient. Therefore, it is very necessary to develop new therapeutic strategies that could enhance the host immunosurveillance and/or improve immunogenicity of the tumor cell.

Since US President Obama proposed precision medicine project in January of 2015, how to realize precision medicine has become a hotspot. Genomics, big data, and molecular imaging have become key tools to realize precision medicine. Herein, we review the main advances of multifunctional nanoprobes for targeted imaging and therapy of gastric cancer in our team over past several years, explore the clinical translational prospects, and discuss the concepts, issues, approaches, and challenges, with the aim of improving the clinical application of multifunctional nanoprobes for targeted imaging and therapy of gastric cancer in the near future.

## 11.2 Multifunctional Fluorescent Magnetic Nanoprobes for Targeted Imaging and Therapy of Gastric Cancer

In 2007, we controllably prepared silica-coated CdTe quantum dots and superparamagnetic nanoparticle composites (FMNPs) with strong fluorescent signals and excellent magnetic properties [14]. We also observed that as-prepared nanoparticles own good biocompatibility and stability [15].

### *11.2.1 BRCAA1 Antibody-Conjugated FMNPs for Targeted Imaging*

In 1999, we screened out and cloned breast cancer-associated antigen 1 gene (BRCAA1 gene) from breast cancer cell line MCF-7cells [AF208045, also called AT-rich interactive domain-containing protein 4B (ARID4B)] and identified its antigen epitope peptide SSKKQKRSHK [16, 17]. We also prepared BRCAA1 polyclonal antibody and observed that the BRCAA1 protein exhibited overexpression in almost 65 % clinical specimens of gastric cancer tissues [18]. We also observed that BRCAA1 antigen is overexpressed in gastric cancer cell lines such as MKN-1, MKN-74, SGC-7901, KATO-III, and MGC803 cells [19]. Therefore, we predict that BRCAA1 protein may be one potential targeting molecule for in vivo gastric cancer cells.

Based on our previous work, we fully used the advantages of FMNPs and BRCAA1 antigen, prepared monoclonal antibody against BRCAA1 protein, and prepared BRCAA1 monoclonal antibody-conjugated fluorescent magnetic nanoprobes (BRCAA1-FMNPs), employed nude mice model loaded with gastric cancer of 5 mm in diameter and IVIS imaging system and magnetic resonance imaging system, and investigated the feasibility of as-prepared nanoprobes for in vivo targeted dual-mode imaging of gastric cancer.

As shown in Table 11.1, we successfully obtained two BRCAA1 monoclonal antibody cell lines with high specificity. As shown in Table 11.2, BRCAA1 monoclonal antibody-conjugated FMNP nanoprobes were successfully prepared with coupling ratio of more than 80 %.

**Table 11.1** Titers of BRCAA1 monoclonal antibodies in ascites fluid induced by hybridoma clone cells by ELISA

| | Antibody titer[a] | | | |
|---|---|---|---|---|
| Clone | BRCAA1 (C)-OVA[b] | BRCAA1 (C)-BSA[b] | BSA[b] | OVA[b] |
| S-200-5 | 1,024,000 | 1,024,000 | <1000 | <1000 |
| S-335-5 | 128,000 | 512,000 | <1000 | <1000 |

[a]The reciprocal of ascites fluid dilution, the first dilution of ascites fluid was 1:1,000
[b]The antigens were coated on ELISA plate

**Table 11.2** Coupling rate measurement of FMNP-BRCAA1 antibody

| | Total concentration of the anti-BRCAA1 antibody (ng/μL) | The concentration of BRCAA1 antibody in residual reaction mixture (ng/μL) | Coupling rate (%) |
|---|---|---|---|
| 1 | 1000.0 | 197.3 | 80.27 |
| 2 | 1000.0 | 191.2 | 80.88 |
| 3 | 1000.0 | 203.0 | 79.70 |

By monitoring real-time fluorescence intensity in the whole body, the tumor-targeting character of the anti-BRCAA1-FMNP probe was easily determined in the nude mice loaded with gastric cancer MGC803 cells. As shown in Fig. 11.1a, the whole animals produced fluorescent signals within 30 min postinjection of nanoprobes; the subcutaneous tumor tissues could be clearly delineated from the surrounding background tissue between 1 h and 12 h postinjection, with maximum contrast occurring at 6 h postinjection. Strong fluorescence signal was still detected in the tumor site at 6 h postinjection, which indicated that the anti-BRCAA1-FMNP nanoprobes were preferentially accumulated in the tumor tissues. Indeed based on the results in Fig. 11.1b, the higher tumor-to-background ratio (TBR) value highly suggested that as-prepared nanoprobes preferentially accumulated in tumor tissues compared to normal control tissues. This was confirmed in fluorescence images, which showed that the fluorescence signal of as-prepared nanoprobes in the tumor site was strongest among all mice organs as shown in Fig. 11.1c. In addition, after 12 h postinjection of anti-BRCAA1-FMNP nanoprobes, fluorescence intensity in tumor was still observed clearly, while the uptake of prepared nanoprobes in normal organs was not obvious. These data highly suggest that prepared nanoprobes can target highly efficient tumor tissues inside nude mice loaded with gastric cancer. We also observed that those nanoprobes in the whole mouse body almost completely disappeared at 12 h postinjection, we also detected the partial nanoprobes exited out from the cholecyst system, the time-dependent cholecyst clearance of nanoprobes highly suggest that as-prepared nanoprobes cannot stay inside nude mice for longer time, and thus, as-prepared nanoprobes own good biosafety.

In vivo MR imaging was performed on nude mice loaded with subcutaneous gastric cancer at 12 h postinjection. Representative images of T2 maps were shown in Fig. 11.2; after injecting the nanoprobes, a significant change in signal intensity was observed in some regions of tumors, indicating that there existed accumulation of the nanoprobes in tumor site as shown in Fig. b, as the arrow showed. As a control, after the mice model with gastric cancer were injected with FMNPs for 12 h, the mice were performed with MR imaging, which did not show intensive signal in tumor area (Fig. 11.2a).

In vitro evaluation of excised major tissues, including the liver, lung, spleen, kidney, and heart, as well as the tumor, indicated that the anti-BRCAA1-FMNP probes were mainly uptaken by the tumor tissues, which exhibited strong fluorescence signals, as shown in Fig. 11.1d, whereas other tissues including the liver, lung, spleen, and heart uptook anti-BRCAA1-FMNP nanoprobes very less, which furtherly indicates that as-prepared anti-BRCAA1-FMNP nanoprobes can target

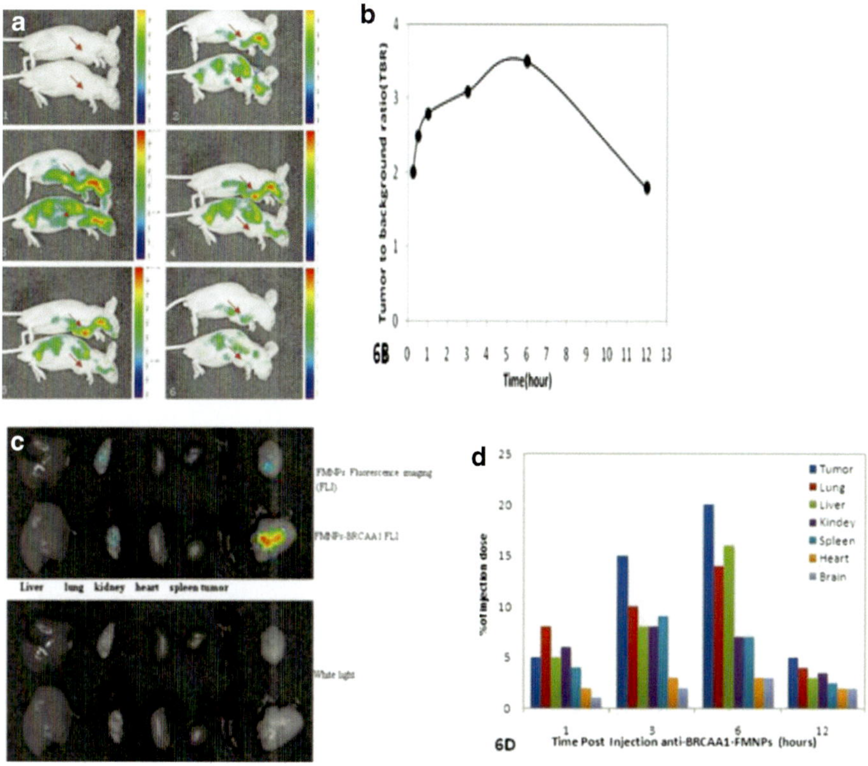


**Fig. 11.1** In vivo fluorescence images of tumor accumulation and tissue distribution for anti-BRCAA1-FMNP nanoprobes in MGC803 human gastric tumor-bearing athymic nude mice. (**a**) In vivo fluorescence images of athymic nude mice-bearing MGC803 human gastric tumor were obtained after injection of anti-BRCAA1-FMNP nanoprobes at different time point. The tumor location is specified with an arrow. A-1: 0 h, A-2:0.5 h, A-3:1 h, A-4:3 h, A-5:6 h, A-6:12 h. (**b**) Tissue to background [TBR (muscle) ratio] value. The TBR value was determined as follows: TBR = (tumor signal-background signal)/(background signal). (**c**) Ex vivo fluorescence images of dissected organs and tumor of mice-bearing MGC803 human gastric tumor sacrificed at 12 h after injection of anti-BRCAA1-FMNP nanoprobes. The fluorescence images of dissected organs and tumor were obtained using a fluorescence imaging system with a 630 nm emission filter. (**d**) Biodistribution of anti-BRCAA1-FMNPs in mice after intravenous injection. Several time points after injection, iron amounts in tissue samples were evaluated by ICP mass spectrometry (*n* = 3) (Permission is from Springer Press)

gastric cancer tissues. We also used HE staining to check all organs; no obvious damages were observed in important organs.

Therefore, we successfully prepared a novel anti-BRCAA1-FMNP nanoprobes, which can be used for in vivo two modal imaging such as fluorescent imaging and magnetic resonance imaging, and own an obviously specific targeting ability toward gastric cancer tissues with 5 mm in diameter during 0.5 and 12 h of postinjection and own good biocompatibility. The as-prepared multifunctional nanoprobes also can be used for hyperthermia therapy of gastric cancer under in vitro alternating

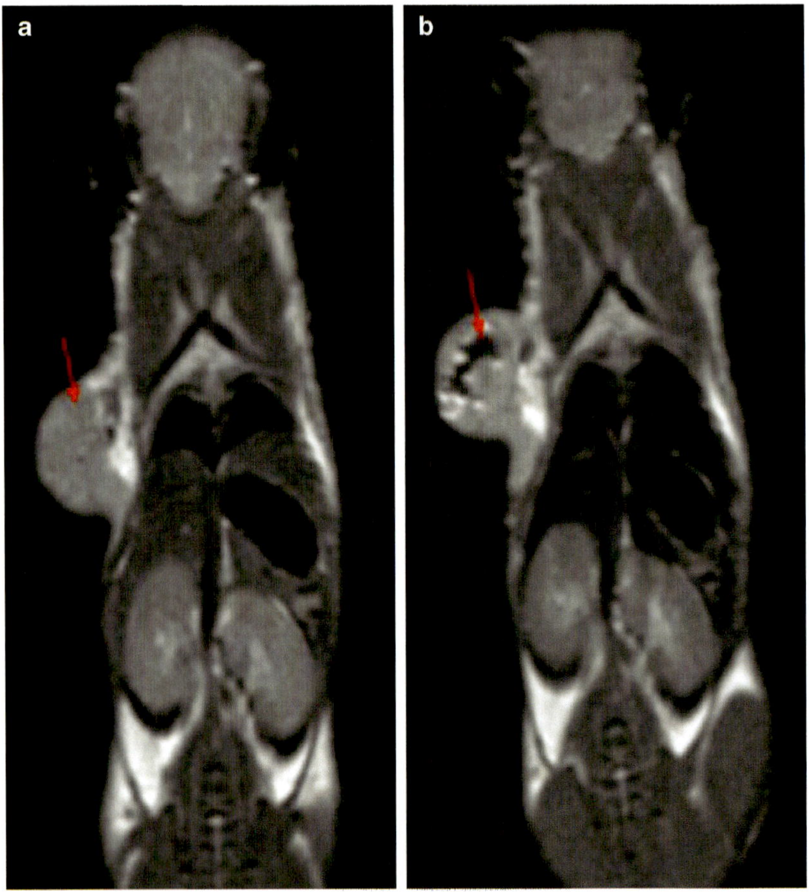


**Fig. 11.2** In vivo MR imaging of anti-BRCAA1-FMNP nanoprobes. (**a**) MR imaging of the mice model with gastric cancer injected with FMNPs for 12 h; (**b**) MR imaging of the mice model with gastric cancer injected with anti-BRCAA1-FMNPs for 12 h (Permission is from Springer Press)

magnetic field irradiation and have great potential in applications such as simultaneous imaging and targeting therapy of clinical gastric cancer in the near future [20].

### *11.2.2 HAI-178 Antibody-Conjugated FMNPs for Targeted Imaging and Therapy*

We also prepared another important nanoprobe, that is, HAI-178 antibody-conjugated fluorescent magnetic nanoprobes, which was successfully used for targeted imaging and hyperthermia therapy of gastric cancer. Dr. Jian Ni's group prepared specific monoclonal antibody against α-subunit of ATP synthase, named as HAI-178 antibody, and provided HAI-178 antibody to my group. Our primary studies showed that α-subunit of ATP synthase also exhibited overexpression in gastric cancer cells and 94.7 % clinical gastric cancer tissues; no or very low expression in normal gastric mucous tissues should be a potential biomarker with diagnosis value.

We prepared HAI-178 antibody-conjugated fluorescent magnetic nanoparticles (HAI-178-FMNPs), co-incubated with gastric cancer MGC803 cells and gastric mucous GES-1 cells. Gastric cancer-bearing nude mice models were established, were injected with prepared HAI-178-FMNPs via tail vein, and were imaged by magnetic resonance imaging (MRI) and small animal fluorescent imaging system. Results showed that α-subunit of ATP synthase exhibited high expression in 94.7 % gastric cancer tissues. Prepared HAI-178-FMNPs could target actively MGC803 cells, realized fluorescent imaging and magnetic resonance imaging of in vivo gastric cancer, and actively inhibited growth of gastric cancer cells. HAI-178 antibody-conjugated fluorescent magnetic nanoparticles own great potential in applications such as targeted imaging and simultaneous therapy of in vivo early gastric cancer cells in the near future [21].

### *11.2.3 FMNP-Labeled MSCs for Targeted Imaging and Hyperthermia Therapy*

We also prepared FMNP-labeled human MSCs and realized targeted imaging and hyperthermia therapy of gastric cancer. We firstly proposed the concept of stem cell nanotechnology, which is an emerging interdisciplinary field, which refers to the application of nanotechnology in stem cell research and development. Although great advances in the field of stem cells have been and being made, several obstacles must be overcome before their therapeutic application can be realized. These include the development of advanced techniques to understand and control functions of microenvironmental signals and novel methods to track and guide transplanted stem cells. The application of nanomaterials and nanotechnology in stem cells research and development exhibits attracting technological prospects, which provide a new chance to solve current problems that stem cells research and development meet [22, 23].

Mesenchymal stem cells (MSCs) are multipotent stem cells that can differentiate into a variety of cell types, including osteoblasts (bone cells), chondrocytes (cartilage cells), and adipocytes (fat cells). MSCs possess immunosuppressive or immunomodulatory properties and have the characteristics of home to the sites of active tumorigenesis. Thus, MSCs can be considered as a candidate cell type for cell-based tissue engineering, cancer therapeutics, and regenerative medicine applications. However, the distribution and final fate of MSCs inside human body are still not clarified well, which urgently need the novel labeling and in vivo tracking technology.

As shown in Fig. 11.3, MSCs were characterized. The silica-coated fluorescent superparamagnetic nanoparticles (FMNPs) were prepared, were evaluated for their effects on mesenchymal stem cells (MSCs), were used to label MSCs, and then were observed for the distribution and final sites of the labeled MSCs in vivo in nude models loaded with gastric cancer. Some chemoattractant cytokines were analyzed by ELISA and Western blotting. As shown in Fig. 11.4, MSCs were labeled with FMNPs

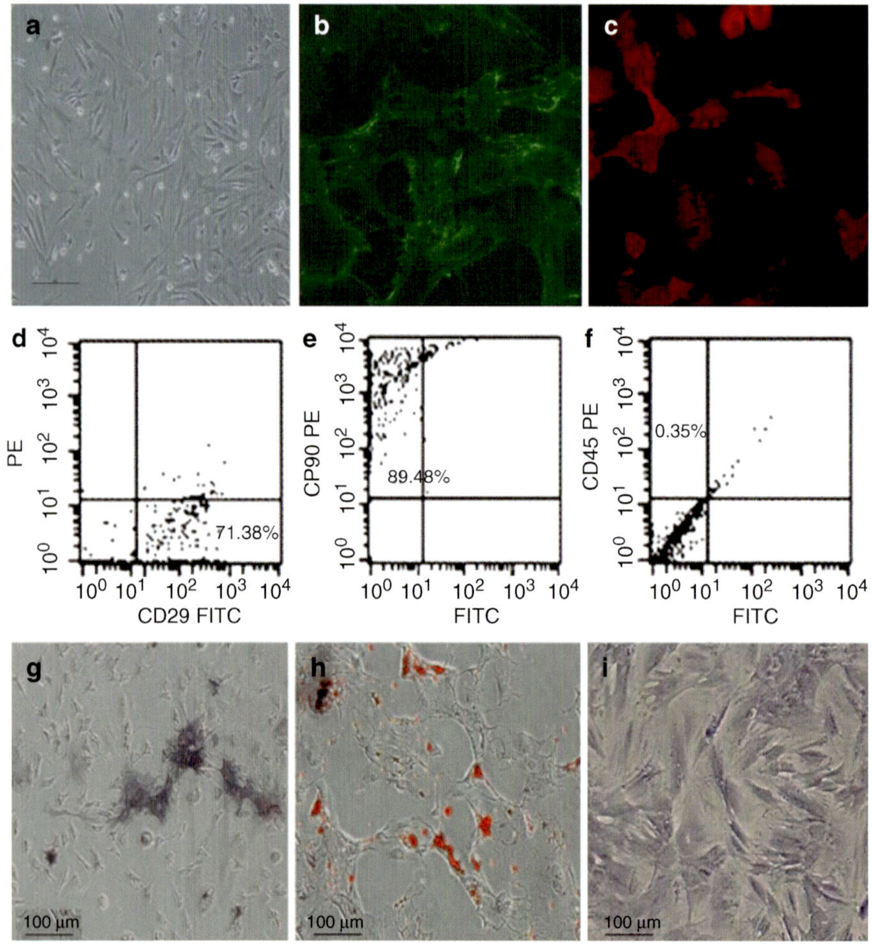


**Fig. 11.3** *Morphology and immunophenotypic characterization of MSCs.* (**a**) *The fibroblastic morphology of passage 3 MSCs* (magnification = ×100); (**b**) MSCs stained with FITC-conjugated CD29 antibody (×200); (**c**) MSCs stained with PE-conjugated CD90 antibody (×200);(**d**), (**e**), and (**f**) MSCs analyzed by FACS for the positive expression of CD29 (**d**) and CD90 (**e**) and negative expression of CD45 (**f**); (**g**) differentiated osteoblasts tested with alkaline phosphatase staining (×100); (**h**) differentiated adipocytes characterized by oil red O staining (×100); (**i**) differentiated chondrocytes verified by toluidine blue staining (×100) (Permission is from Springer Press)

efficiently and kept stable fluorescent signal and magnetic properties within 14 days, FMNP-labeled MSCs could target and image in vivo gastric cancer cells after being intravenously injected for 14 days, FMNP-labeled MSCs could significantly inhibit the growth of in vivo gastric cancer because of hyperthermia effects, and CCL19/CCR7 and CXCL12/CXCR4 axis loops may play key roles in the targeting of MSCs to in vivo gastric cancer. This FMNP-based labeling and tracking technology have great potential in applications such as labeling and tracking implanted cells, evaluating cell therapeutic effects, and recognizing and mapping early gastric cancer cells [24].

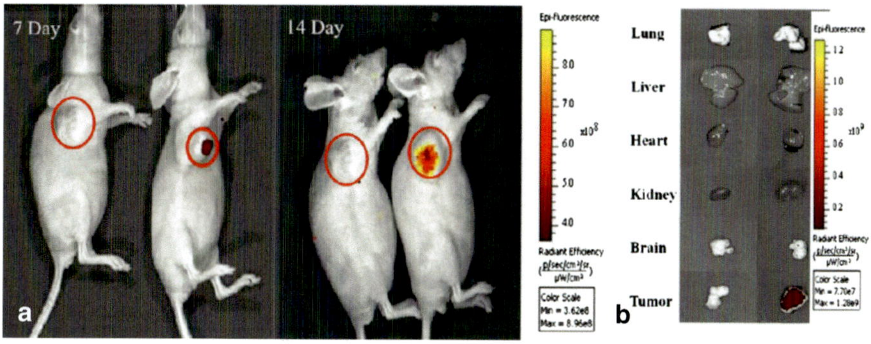


**Fig. 11.4** Fluorescent imaging of FMNP-labeled MSCs targeting gastric cancer cells in vivo. (**a**) The in vivo fluorescent images show that tumor sites of the mice in the test group had fluorescent signals after postinjection of FMNP-labeled MSCs at 7 and 14 days (*right*), and tumor sites of the mice in the control group had no fluorescent signal after postinjection of FMNPs at 7 and 14 days (*left*). (**b**) The fluorescent imaging of major organs show that no signal was detected in the tumor and organs of the control group (*left*), and obviously fluorescent signals were detected in the tumor tissues of the test group (*right*) (Permission is from Springer Press)

## 11.3 Multifunctional QD Probes for Fluorescent Imaging, Genotyping, and Therapy of Gastric Cancer

### *11.3.1 RGD-Conjugated RNase A-Associated dRQDs for Targeted Imaging and Therapy*

Colloidal semiconductor nanocrystals (NCs), also known as quantum dots (QDs), have gained immense attention due to their unique optical properties, including a broad absorption with narrow photoluminescence spectra, high quantum yields, low photobleaching, and size-dependent emission wavelength tenability [25]. These fascinating properties make QDs a promising candidate for biological imaging and labeling probes, with significant advantages over conventional fluorescent dyes. However, the toxicity associated with cadmium-containing QDs turned out to be a chief hurdle for in vitro and in vivo biomedical diagnostics [26]. For this reason, a variety of attempts have been made to reduce the toxicity of QDs and improve their biocompatibility, which has become a rapidly growing area of nanoscience research. Although the surface modification strategy proves to be prosperous, to select an appropriate surface coating material is not easy. More recently, due to the excellent biocompatibility and bioabsorbility, some proteins have been immobilized on the surface of QDs to produce low-toxicity protein–QD hybrid nanomaterials. Additionally, protein molecules have also been adopted to serve as the biomolecular templating agent to regulate the formation of inorganic nanostructures [27]. We reported the first application of bovine pancreatic ribonuclease A (RNase A) as a biomolecular templating agent for preparing CdTe QD clusters in aqueous phase via a chemical reduction approach. We have chosen RNase A due to its unique features.

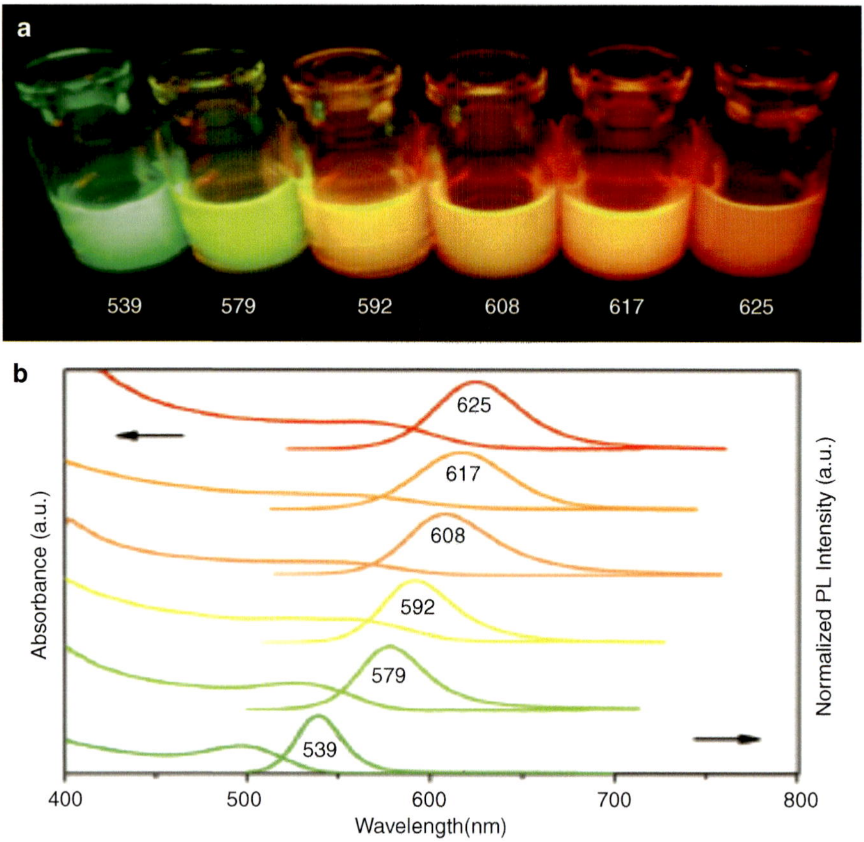


**Fig. 11.5** (**a**) Photograph of freshly prepared CdTe QD suspension with different emitting colors in the presence of RNase A as a biomolecular templating agent under a handheld UV lamp. (**b**) Corresponding UV–vis absorption and normalized photoluminescence (PL) spectra (Permission is from Wiley Press)

RNase A is a low-molecular-weight protein (124 residues, 13.7 kDa, pI 9.4) with a globular configuration (2.2 nm, 2.8 nm, 3.2 nm). The outstanding thermal stability of the protein is well known, thereby being able to endure the high temperature in our synthesis process. Recently, other groups have reported that RNase A was able to inhibit the development of cancerous cells depending on its enzymatic activity of degrading cellular RNA and has been used as a chemotherapeutic agent in clinical trials. It was found that RNase A inevitably lost its native conformation to form denatured RNase A (dRNase A)-capping CdTe QD clusters (dRQDs). Moreover, dRNase A not only effectively reduced the cytotoxicity of CdTe QDs but also unexpectedly induced a temporal evolution of the luminescence during a 50-day storage period under ambient conditions, as shown in Fig. 11.5. Figure 11.6 shows prepared RNase A-QDs composed of QD clusters; the potential synthesis mechanism is shown in Fig. 11.7. Inspired by these fascinating properties, dRQDs was associated

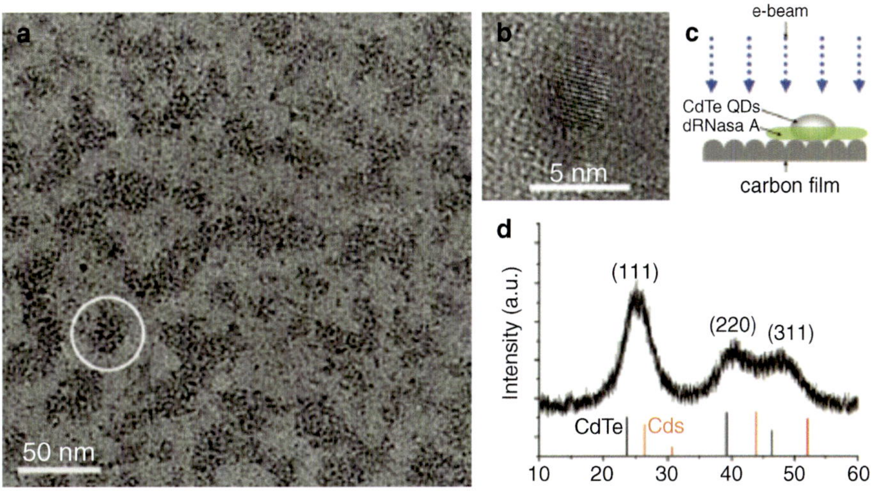


**Fig. 11.6** (**a**) TEM image of as-prepared dRQD539. A typical cluster is labeled with a *white circle*. (**b**) High-resolution TEM image (HR-TEM) of an individual CdTe QD. (**c**) Schematic diagram to explain the difference in image contrast in the HR-TEM image during imaging in the vertical section. (**d**) Powder X-ray diffraction (XRD) pattern of dRQD539. Standard diffraction lines of cubic CdTe and cubic CdS are shown for comparison (Permission is from Wiley Press)

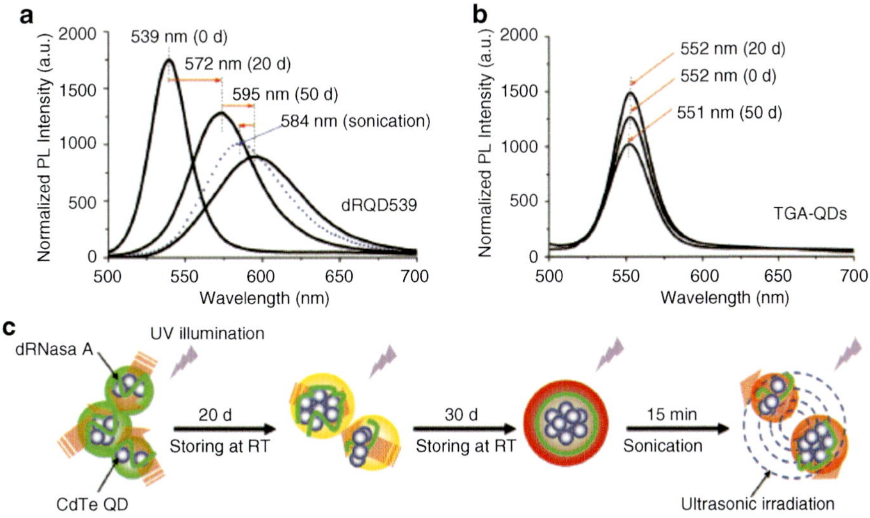


**Fig. 11.7** (**a**) Normalized photoluminescence (PL) spectra of dRQD539 (excitation wavelength λex = 400 nm). The three *solid lines* are the corresponding PL spectra of the same sample stored for 0 day (0 d), 20 days (20 d), and 50 days (50 d), respectively. The *blue dotted line* is the PL spectrum of the 50-day-stored dRQD539 after sonicating for ~15 min in an ultrasonic cell crusher. (**b**) Normalized PL spectra of TGA-capping CdTe QDs stored for 0, 20, and 50 d, respectively. (**c**) Schematic illustration of the potential mechanism for the red shift of emission peaks of dRQD539 during the storage period under ambient conditions (Permission is from Wiley Press)

with native RNase A by the physical absorption, which was then coupled with cyclic arginine–glycine–aspartic acid (RGD) peptide c (RGDfK) that served as the targeting biomolecule. A novel multifunctional nanosystem RGD-conjugated RNase A-associated dRQDs (RGD-R-dRQDs) was built up for synchronous targeted cellular imaging and therapeutic applications [28].

### *11.3.2 Her2 Antibody-Conjugated RQDs for Targeted Imaging and Therapy of In Situ Gastric Cancer*

One kind of multifunctional HER2 monoclonal antibody-conjugated RNase A-associated CdTe quantum dot cluster (HER2-RQD) nanoprobes was prepared, its cytotoxicity was evaluated. Subcutaneous gastric cancer nude mouse models and in situ gastric cancer SCID mouse models were established and were intravenously injected with HER2-RQD nanoprobes; the biodistribution and therapeutic effects of HER2-RQDs in vivo were evaluated. Results showed that HER2-RQD nanoprobes could selectively kill gastric cancer MGC803 cells, as shown in Fig. 11.8, could target imaging subcutaneous gastric cancer cells at 3 h postinjection and in situ gastric cancer cells at 6 h postinjection, and could inhibit the growth of gastric cancer tissues and extended survival time of gastric cancer-bearing mouse models, which is closely associated with destroying functional RNAs in cytoplasm by RNase A released from HER2-RQD nanoprobes, preventing protein synthesis and inducing cell apoptosis, as shown in Fig. 11.9. High-performance HER2-RQD nanoprobes exhibit great potential in applications such as in situ gastric cancer-targeted imaging and selective therapy in the near future [29].

### *11.3.3 BRCAA1 and Her2 Monoclonal Antibody-Conjugated PQDs for Targeted Imaging and Therapy*

In order to observe multi-target molecule imaging of gastric cancer, we chose the CdSe/ZnS (core/shell) quantum dots (QDs) as prototypical materials, synthesized one kind of new amphiphilic polymer including dentate-like alkyl chains and multiple carboxyl groups, and then used prepared amphiphilic polymer to modify QDs; resultant amphiphilic polymer engineered QDs (PQDs) were conjugated with BRCAA1 and Her2 monoclonal antibody, and prepared BRCAA1 antibody- and Her2 antibody-conjugated QDs were used for in vitro MGC803 cell labeling and in vivo targeted imaging of gastric cancer cells. Results showed that the PQDs exhibited good water solubility, strong photoluminescence (PL) intensity, and good biocompatibility. BRCAA1 antibody- and Her2 antibody-conjugated QD nanoprobes successfully realized targeted imaging of in vivo gastric cancer MGC803 cells. BRCAA1 antibody and

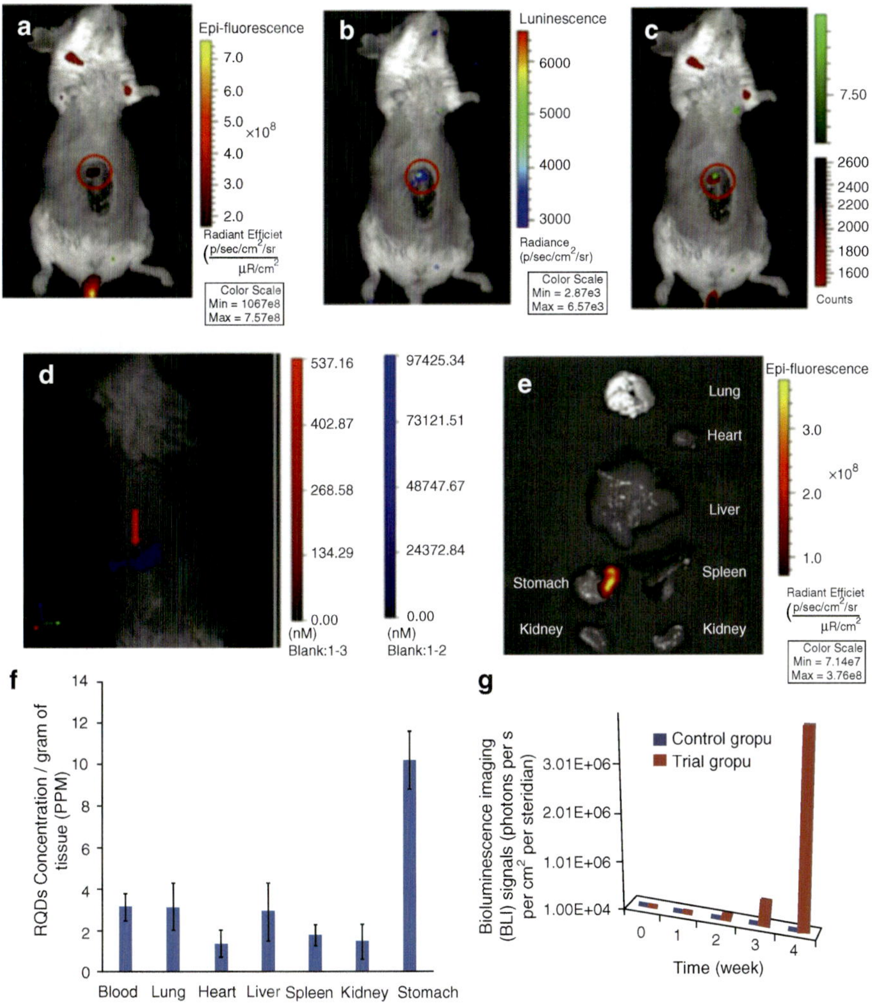


**Fig. 11.8** Optical images of HER2-RQD nanoprobe in in situ gastric cancer mice. (**a**) Fluorescence image of HER2-RQD nanoprobe targeted in situ gastric cancer tissue. (**b**) Bioluminescence image of in situ gastric cancer tissue. (**c**) Overlay image of fluorescence and bioluminescence imaging. (**d**) 3D image of HER2-RQD nanoprobes targeted gastric cancer tissue. (**e**) Ex vivo fluorescence image of the dissected organs of the mice sacrificed at 24 h postinjection of HER2-RQD nanoprobes. (**f**) Bio-distribution of HER2-RQDs in mice after intravenously injection. The amounts of RQDs in tissue samples were evaluated by ICP mass spectrometry (n ¼ 3). (**g**) Bioluminescence imaging signals of gastric tumor tissues in in situ gastric cancer mice after treated with HER2-RQDs for 4 weeks (Permission is from Elsevier Press)

Her2 antibody-conjugated PQDs own great potential in applications such as single cell labeling and in vivo tracking and targeted imaging and therapeutic effects' evaluation of in vivo early gastric cancer cells in the near future [30].

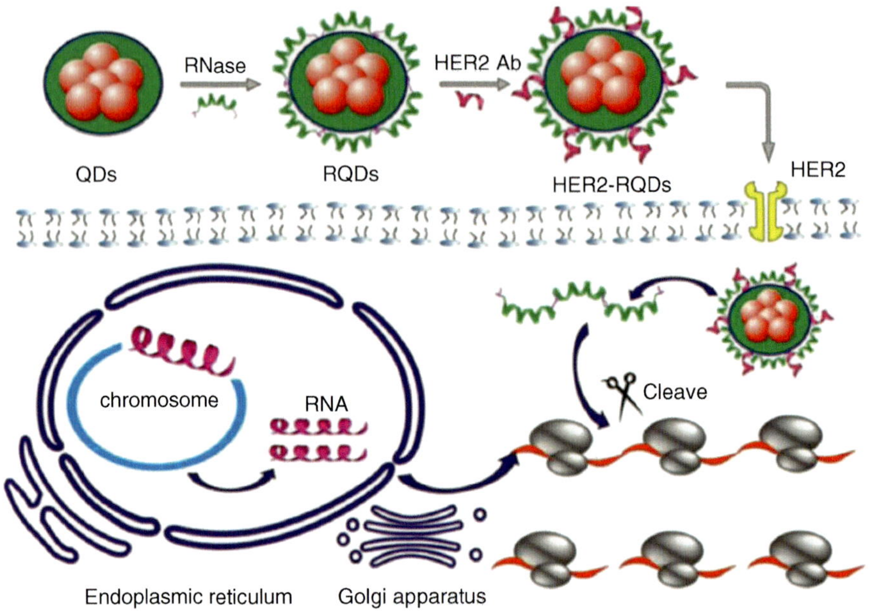


**Fig. 11.9** *Therapeutic mechanism of as-prepared nanoprobes:* HER2-RQD nanoprobes could specifically enter into gastric cancer cells, RNase A was released from the HER2-RQD nanoprobes under the acid environment, which degraded total RNAs inside gastric cancer cells, inhibiting RNA translation and protein synthesis, and finally induced tumor cells' apoptosis (Permission is from Elsevier Press)

## 11.4 Multifunctional Upper Conversion Nanoprobes for Targeted Imaging and Therapy of Gastric Cancer

Upconversion nanoparticles (UCNPs), especially lanthanide-doped nanocrystals, have stolen the limelight due to their seductive optical and chemical features such as low toxicity, large Stokes shifts, and high resistance to photobleaching and photochemical degradation [31, 32]. Moreover, near-infrared (NIR) excitation contributes to the auto-fluorescence minimization, a larger penetrating depth, and less harmfulness to cells compared with traditional ultraviolet (UV) excitation. On the other hand, the composites of UCNPs with other materials exert superior performance in a sense, broadening the range of biological applications from multimodality imaging and targeted therapy to biodetections and bioassays. My group firstly reported the OA/ionic liquid two-phase system combining the merits of thermal decomposition method, the ionic liquid (IL)-based strategy, and the two-phase approach which is introduced to synthesize high-quality lanthanide-doped NaGdF 4 upconversion nanocrystals with different crystal phases in OA phase and IL phase through a one-step controllable reaction. Oil-dispersible cubic-phase NaGdF 4:Yb, Er (Ho, Tm) nanocrystals with ultra-small size (~5 nm) and monodispersity are obtained in the OA phase of the two-phase system via an IL-based reaction. More importantly, water-soluble hexagonal-phase NaGdF 4:Yb, Er

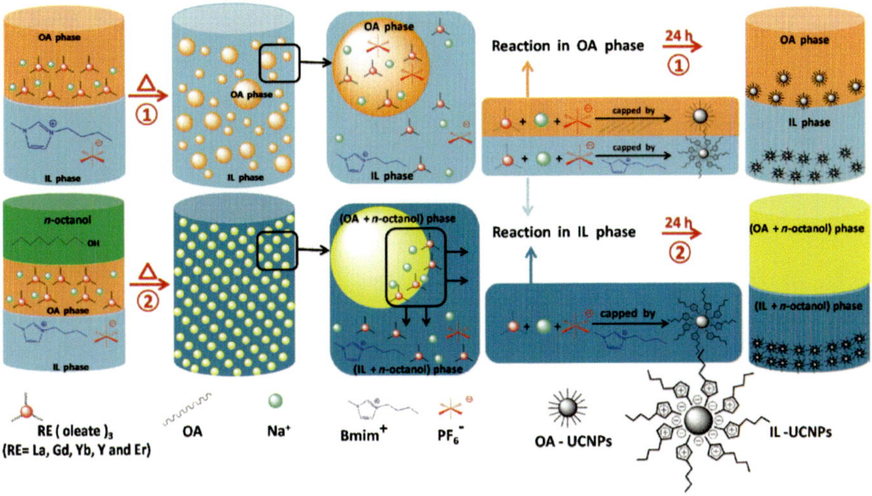


**Fig. 11.10** Schematic diagram showing the formation of RE fluoride nanocrystals in the OA/IL and n-octanol-induced two-phase systems (Permission is from Wiley Press)

nanocrystals are obtained in the same system simply by adopting an extremely facile method to complete the dual-phase transition (crystal-phase transition and OA phase to IL phase transition) simultaneously. The synthesized lanthanide-doped NaGdF4 upconversion nanocrystals are effective for dual-mode UCL imaging and CT imaging in vivo [33]. Based on this basis, we also improved the synthesis condition, developed phase- and size-controllable synthesis of hexagonal upconversion rare-earth fluoride nanocrystals through an oleic acid/ionic liquid two-phase system [34] (Fig. 11.10), and prepared nanoparticles that were successfully used as magnetic nanorattle materials for magnetism-directed targeting imaging and chemical therapy [35] (Fig. 11.11).

### 11.4.1 *Lanthanide-Doped NaGdF4 Upconversion Nanocrystals for Dual-Mode UCL Imaging and CT Imaging and Targeted Chemical Therapy (Figs. 11.11 and 11.12)*

Traditional chemotherapy is used in the clinical management of gastric cancer patients, often with severe side effects due to the cytotoxicity of the chemical drugs and/or the genetic heterogeneity of the tumors, that is, their phenotypic drug sensitivity or resistivity. Nanoengineered drug-delivery systems could improve the clinical performance of conventional chemotherapeutic agents and reduce overall toxicity by enhancing the specificity of the drug's delivery through tumor targeting and making elevated local dosages possible. A major goal in nanomedicine is to establish the multifunctional platforms to realize simultaneous diagnosis, targeted delivery and efficient therapy. In order to realize the dual-mode imaging, CT imaging and targeted chemical therapy of gastric cancer, as shown in (Fig. 11.11), the

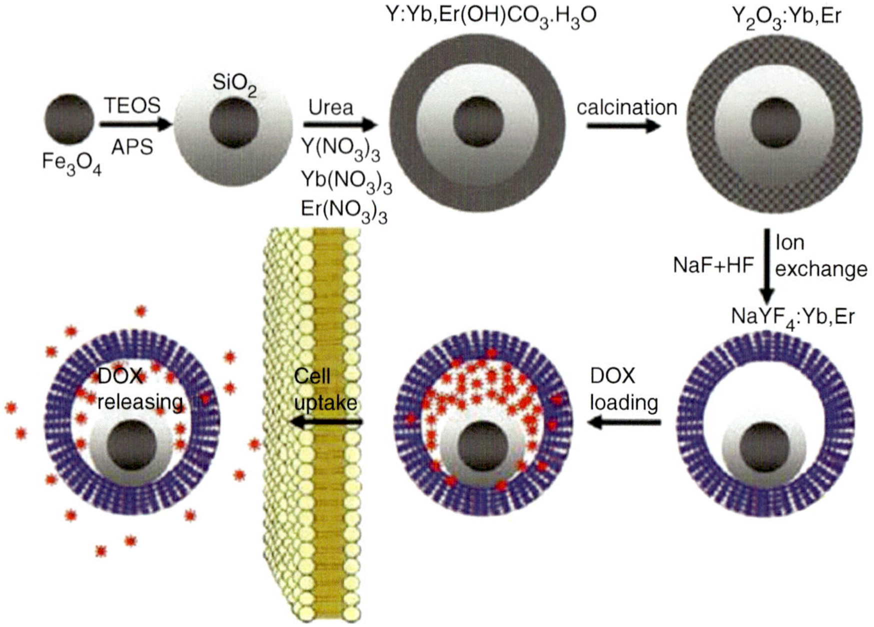


**Fig. 11.11** Synthetic procedure for the drug-loaded Fe3O4@SiO2@α-NaYF/Yb, Er nanorattles (DOX-MUC-F-NR)

nanorattles consisting of hydrophilic, rare-earth-doped $NaYF_4$ shells each containing a loose magnetic nanoparticle were synthesized through an ion-exchange process. The inner magnetic $Fe_3O_4$ nanoparticles are coated with a $SiO_2$ layer to avoid iron leaching in acidic biological environments. This multifunctional mesoporous nanostructure with both upconversion luminescent and magnetic properties has excellent water dispersibility and a high drug-loading capacity, were successfully used for MR imaging and fluorescent imaging. As shown in Fig. 11.12, the material emits visible luminescence upon NIR excitation and can be directed by an external magnetic field to a specific target, delivered chemical drugs to local tumor sites, reducing the side-effects, in vivo experiments showed that tumor tissues became shrinkage treated with the antitumor drug doxorubicin (DOX), confirming that prepared nanorattles own great potential in dual-mode imaging and targeted drug delivery and therapy [36].

### *11.4.2 Folic Acid-Conjugated Silica-Modified Upconversion Nanoprobes for Targeted UCL and CT Imaging*

Folic acid-conjugated silica-modified $LaF_3$:Yb,Tm upconversion nanoparticles (UCNPs@$SiO_2$-FA) with high La content in single particle were strategically designed and prepared for simultaneous targeting dual-mode imaging of upconversion luminescence (UCL) and X-ray computed tomography (CT) [37]. $LaF_3$ UCNPs

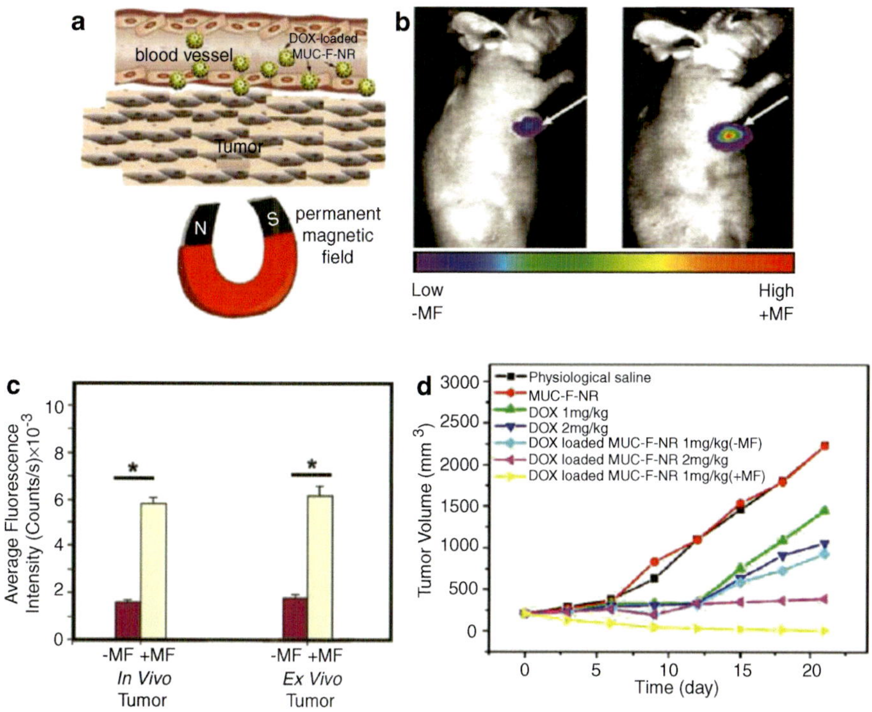


**Fig. 11.12** (**a**) Schematic illustration of targeting of DOX-loaded multifunctional drug carrier to tumor cells assisted by an externally applied magnetic field (MF). (**b**) Tumor location as defined by MUC-F-NR intensity increases with 1 h magnetic field treatment. Mice-bearing H22 xenograft tumor were injected with DOX-loaded MUC-F-NR (1 mg/kg) and subjected (+MF) or not subjected (MF) to the magnetic field for 1 h. At 24 h postinjection, mice were imaged in vivo. (**c**) The luminescence signal was measured from the whole tumor in vivo and ex vivo. (Excitation was provided by the CW infrared laser at 980 nm and upconversion luminescence signals were collected at 650 (10 nm). Fluence rates for 980 nm excitation light were 80 mW/cm$^2$.) (**d**) Tumor volume changes of saline-treated mice compared to mice treated with MUC-F-NR, DOX, and DOX-loaded MUC-F-NR over 21 days in the absence and presence of magnetic field. Data show mean ±SD (*n* =5,*p e 0.05) (Permission is from ACS Press)

were synthesized by a novel oleic acid (OA)/ionic liquid (IL) two-phase system. Afterward, folic acid molecule was covalently anchored on the surface of UCNPs with silane coupling agent. The UCNPs@$SiO_2$-FA exhibits good stability, water dispersibility and solubility, noncytotoxicity, good biocompatibility, highly selective targeting, excellent X-ray attenuation, and UCL emission under excitation at 980 nm. In vivo UCL and CT images of mice show the UCNPs@$SiO_2$-FA can be used in targeting dual-mode imaging. These results suggest that the as-prepared nanoprobe is a good candidate with excellent imaging and targeting ability for targeting dual-mode imaging of UCL and CT.

Multimodal contrast agent-based folic acid-conjugated silica-modified $LaF_3$:Yb,Tm upconversion nanoparticles (UCNPs@$SiO_2$-FA) with high La content in single particle were strategically designed and developed. We have dem-

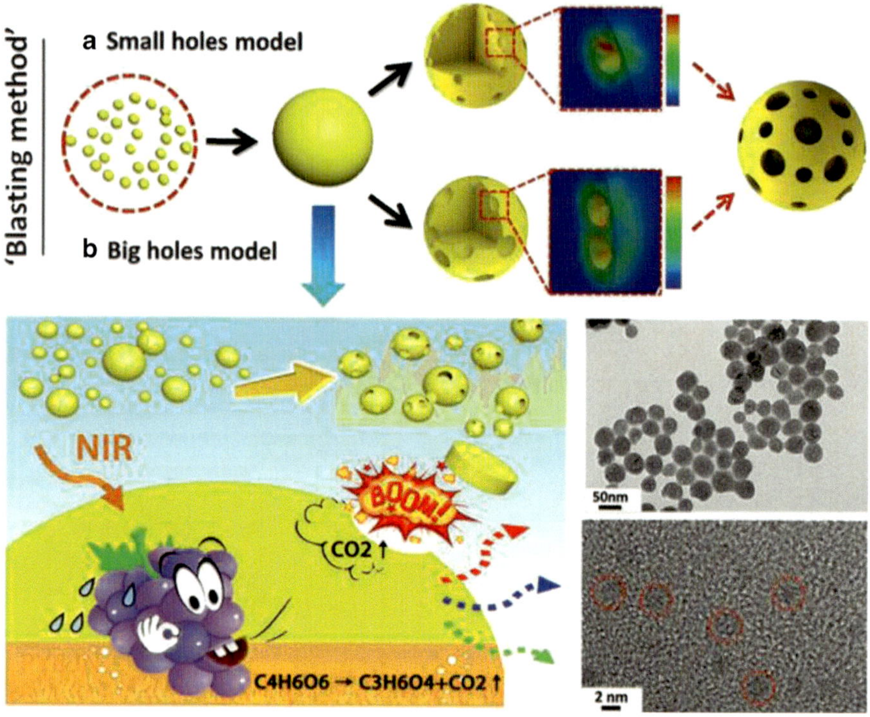


**Fig. 11.13** Illustration of the "blasting" method for fabricating porous UNs (Permission is from RSC Press)

onstrated that UCNPs@$SiO_2$-FA is a good candidate with excellent imaging and targeting ability for UCL imaging and CT imaging in vivo. The prepared UCNPs@$SiO_2$-FA exhibits good stability, water dispersibility and solubility, noncytotoxicity, good biocompatibility, highly selective targeting, excellent X-ray attenuation, and UCL emission under excitation at 980 nm. Our results indicated that the synthesized UCNPs@$SiO_2$-FA is effective for simultaneous targeting dual-mode imaging of UCL and CT. It could bring novel opportunities to the next generation of theranostic nanoprobes for simultaneous diagnosis and treatment in vivo.

We also developed a novel synthesis method of UCPs, hydrothermal system in which an anion induces the phase transition process to give simultaneous control over the size, morphology, phase, and emission properties. We first confirm that the crystal cell-oriented rotation driven by an anion in a hydrothermal system promoted the phase transition and the energy zones figure of the phase transition from cubic to hexagonal structure has been figured out. We have successfully applied the structural mechanics finite element calculations to validate the reaction process. We have also demonstrated that porous UNs can be rationally tuned in size (down to 15 nanometers), phase (cubic or hexagonal), and emission properties at precisely defined conditions and were effective for in vitro and in vivo CT imaging [38] (Figs. 11.13 and 11.14).

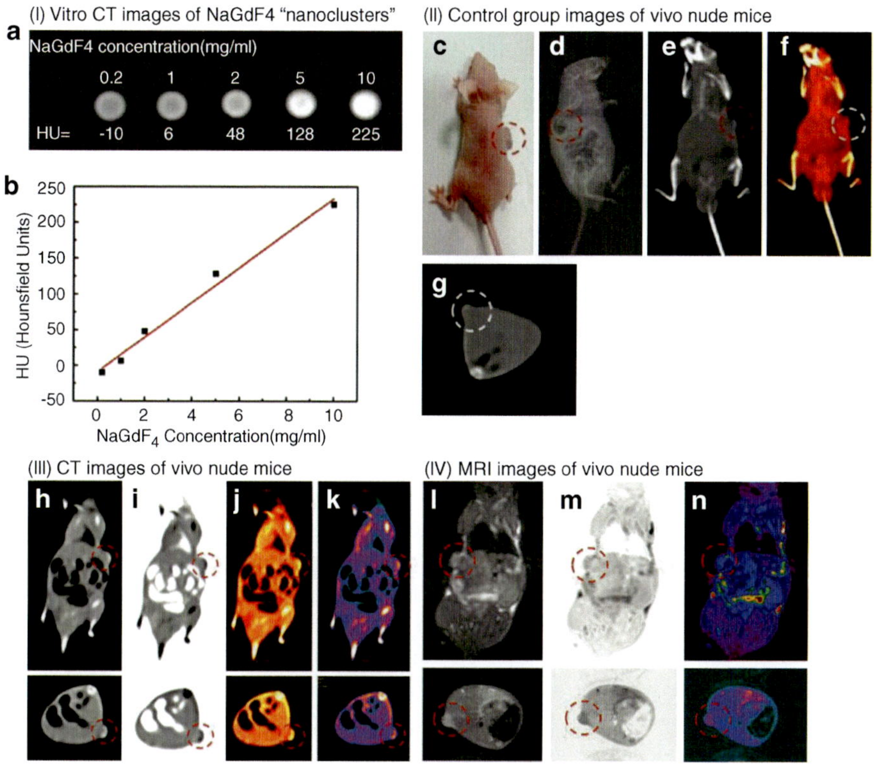


**Fig. 11.14** (**a**) In vitro CT images (*a*) of lanthanide-doped NaGdF4 upconversion "nanoclusters" (sub-5 nm) suspended in aqueous solution. CT attenuation (HU) plot (*b*) of NaGdF4 changes along with the concentration of each sample from 0.2 to 10 mg ml-1 to further investigate the CT contrast effect. (**b**) Images of control group before injection: (*c*) photograph of the nude mouse model loaded with gastric cancer MGC-803 cells; (*d*) X-ray image and (*e*)–(*g*) CT images of nude mouse as the control group. (**c**) CT images (*h*)–(*k*) and (**d**) MRI images (*l*)–(*n*) with NaGdF4 upconversion "nanoclusters" through intravenous injection under the function of passive targeting (Permission is from RSC Press)

## 11.5 Multifunctional Gold Nanoprobes for Targeted Imaging and Therapy

GNPs are within the same size domain as many biomaterials, including enzymes, antibodies, and protein receptors. Combined with the unique properties of materials in the nanoscale range, gold nanoparticles provide scope for making measurements more efficiently than with existing molecular materials. Since light absorption from biologic tissue components is minimized at NIR wavelengths, most NPs for in vivo imaging have been designed in order to strongly absorb in the NIR region so as to be used as effective contrast agents [39].

Optical coherence tomography (OCT) is an imaging modality that provides cross-sectional subsurface imaging of biological tissues at micrometer-scale resolution.

Briefly, the OCT system can generate a signal based on refractive index mismatches and scattering events. The extra scattering is achieved by using AuNPs, which possess both absorption and scattering properties in the NIR region and provide optical contrast for improved diagnostic imaging at higher light intensities. This property has been used in order to increase the contrast of tumors in mice. Au nanoshells accumulate preferentially in the tumor and so increase the contrast more than in other tissues [40]. In situ OCT based on spherical AuNPs in ophthalmology and dental imaging has also been reported [41]. Moreover, an extension of OCT – photothermal OCT – can fill the spatial niche for in vivo imaging of NP delivery [42].

X-ray, computed tomography (CT), and micro-CT techniques can offer excellent improvements in terms of medical diagnosis. The basic process of these techniques is to detect the X-rays that pass through a sample by exposing a charge-coupled device detector. Due to their high atomic number and electrodensity (e.g., 79 and 19.32 g/cm$^3$, respectively, vs the typical X-ray contrast agent, iodine, with values of 53 and 4.9 g/cm$^3$, respectively), AuNPs have higher attenuation coefficients and are proposed to be used as better contrast agents for X-ray imaging, CT, and micro-CT [43]. AuNPs have demonstrated greater contrast than the iodine-based agents that are currently used in the clinic, as well as reduced toxicity and prolonged circulation times. Moreover, imaging Au at 80–100 keV reduces the signal interference derived from bone absorption and therefore would allow for dose reduction. In addition, fluorescent bioimaging is also achieved with luminescent AuNCs, which can be used for multimodal fluorescence/CT imaging without the need for having two different elements. Via glutathione capping and folic acid conjugation, the as-prepared nanoprobe could target a xenotransplanted tumor model in nude mice, as evaluated by fluorescence/CT dual-mode imaging. Although it is not very common, X-ray fluorescent imaging of AuNPs represents a bioimaging technique with great potential. This method could be of great interest once there are imaging systems in place that have sufficient sensitivity to NP concentrations and, at the same time, are capable of distribution measurements at appropriate tissue depths for in vivo and in vitro studies. Consequently, Ricketts et al. reported that a high detecting sensitivity of AuNPs can be achieved using X-ray fluorescence, enabling a greater depth imaging in comparison with optical modalities.

Photoacoustic imaging (PAI) is a hybrid imaging technique; however, it has only become an efficient method of biomedical imaging in recent decades. The development for PAI accounts for the improvement in the machining process and innovative technology relating to the contrast agent. Briefly, after excitation by a laser, a photoabsorber (e.g., an internal photoabsorber, such as hemoglobin, or an external photoabsorber, such as photoacoustic contrast agents) is heated, leading to thermoelastic expansion. Such expansion can cause acoustic pressure elevations and therefore produce ultrasound waves (usually a rise of 1 mK can cause an 800 Pa pressure rise), which are then detected by a sound transducer before finally being reconstructed as photoacoustic images. NPs, especially AuNPs with NIR absorbance due to their intrinsic LSPR, are very promising for use as photoacoustic signal amplifiers. Two important commercialized, small animal PAI systems are MSOT (iThera Medical GmbH©, Munich, Germany) and Nexus 128 (Endra Life Sciences©, MI, USA). The iThera system is based on multispectral optoacoustic tomography, and Nexus 128 is based on hemispherical PAI. Despite the adoption of different strategies for 3D imaging, they are both excellent examples of focusing on a crucial problem in bioimaging and thus obtaining images with high resolution and even in-depth functional information. Unlike OCT, PAI is speckle-free. For example, polyethylene glycol (PEG)-coated Au nanoshells have been used as contrast agents in order to image their circulating distribution in the vasculature in the rat brain using PAI. The images of the distribution of nanoshells circulating in the vasculature of a rat brain achieved by deep-penetrating NIR light presented a gradual enhancement of the optical absorption in the brain vessels. Au nanorods and Au nanoprisms have also been extensively used in bioimaging due to their ability to have their maximum plasmon resonance tuned further into the NIR region. Although PAI is promising for applications in biomedical imaging, drug delivery, and PTT, there are some challenges for PAI based on the AuNP platform. First, although its application in cancer imaging has been abundant over the last 5 years, most of the results are obtained from subcutaneous cancer models and are confined within angiography. However, the advantage of PAI is its deep penetration at a high resolution and the potential for in situ cancer diagnostics. The reason for this is the limitation of the laser energy. According to the American National Standards Institute (ANSI), the energy for skin exposure is advised to be lower than 20 mJ/cm$^2$, but deep penetration is difficult to achieve in this range, so the detection limit needs to be improved by enhancing the NIR absorption of AuNPs. Au nanorods, nanocages, nanoprisms, or nanostars are particularly useful as PAI contrast agents. Second, the scan speed needs to be improved for PAI. Although this mainly needs the improvement of the mechanical drive, AuNPs at the range of 100 nm, which can prolong the circulation time, may solve this problem via another method. Diagnostics usually require the combination of anatomic imaging as well as functional imaging at the same time. The former can provide the precise location of organs, and the latter is for imaging the anatomical structures of the area of interest. This goal is achieved by either multimodal bioimaging or bioimaging techniques that can achieve both.

Theranostics refers to the combination of diagnostics and therapeutics. Due to their LSPR, the strong NIR light absorption of AuNPs makes them a great potential application in PTT, which possesses efficacy and noninvasiveness. There are numerous PTTs based on AuNPs (e.g., antibody- or small-molecule-conjugated AuNPs in tumor therapy). Combined with the aforementioned PAI, AuNPs have great potential to be used as theranostic platforms.

### *11.5.1 Folic Acid-Conjugated Gold Nanorods for Targeted Imaging and Therapy*

Multifunctional nanoprobes are designed to own various functions such as tumor targeting, imaging, and selective therapy, which offer great promise for the future of cancer prevention, diagnosis, imaging, and treatment. Herein, silica was applied to replace cetyltrimethylammonium bromide (CTAB) molecules on the surface of gold nanorods (GNRs) by the classic Stöber method, thus eliminating their cytotoxicity and improving their biocompatibility. Folic acid molecule was covalently anchored on the surface of GNRs with silane coupling agent. The resultant folic acid-conjugated silica-modified GNRs show highly selective targeting, enhanced radiation therapy

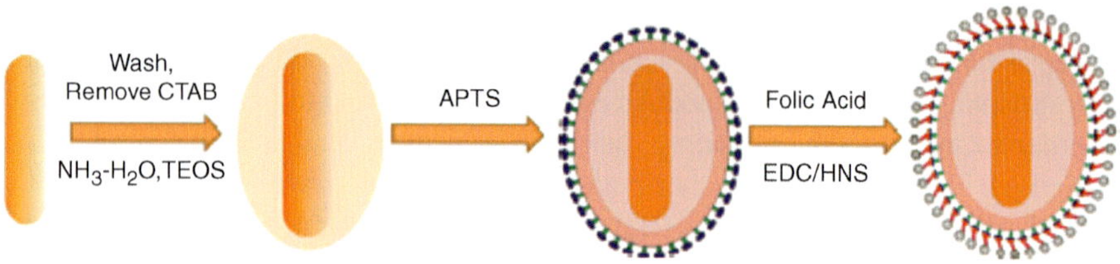


**Fig. 11.15** Synthetic procedure of GNR-SiO2-FA (Permission is from Elsevier Press)

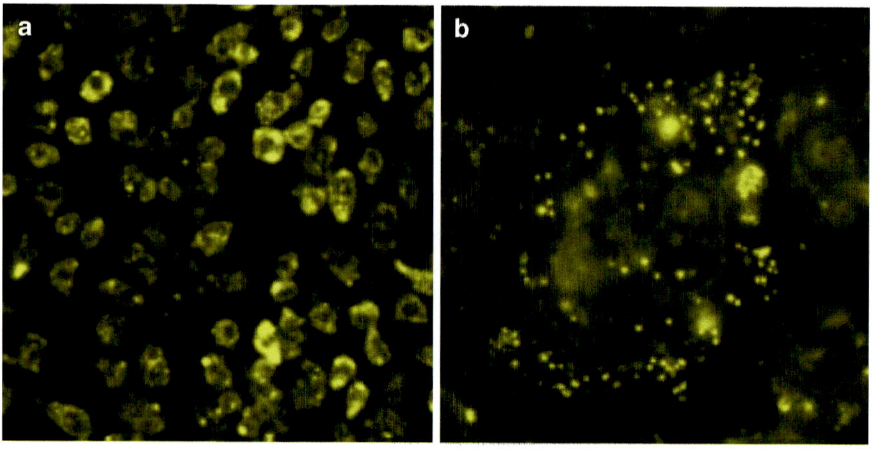


**Fig. 11.16** Dark-field scattering images. (**a**) Low magnification image of targeted MGC803 cells incubated with 50 mM GNR-SiO2-FA for 2 h and (**b**) high magnification image of targeted MGC803 cells incubated with 50 mM GNR-SiO2-FA for 30 min, monitored by dark-field microscopy (Permission is from Elsevier Press)

(RT), and photothermal therapy (PTT) effects on MGC803 gastric cancer cells and also exhibited strong X-ray attenuation for in vivo X-ray and computed tomography (CT) imaging. In conclusion, the as-prepared nanoprobe is a good candidate with excellent imaging and targeting ability for X-ray/CT imaging-guided targeting dual-mode enhanced RT and PTT [44] (Figs. 11.15, 11.16, 11.17, 11.18, and 11.19).

### *11.5.2 Cetuximab-Conjugated Gold Nanoparticles for SERS Imaging and Therapy*

Inspired by the ability of SERS nanoantennas to provide an integrated platform to enhance disease targeting in vivo, we developed a highly sensitive probe for in vivo tumor recognition with the capacity to target specific cancer biomarkers such as epidermal growth factor receptors (EGFR) on human cancer cells and xenograft tumor models. Here, we used ~90 nm gold nanoparticles capped by a Raman reporter, encapsulated and entrapped by larger polymers and an FDA antibody–drug

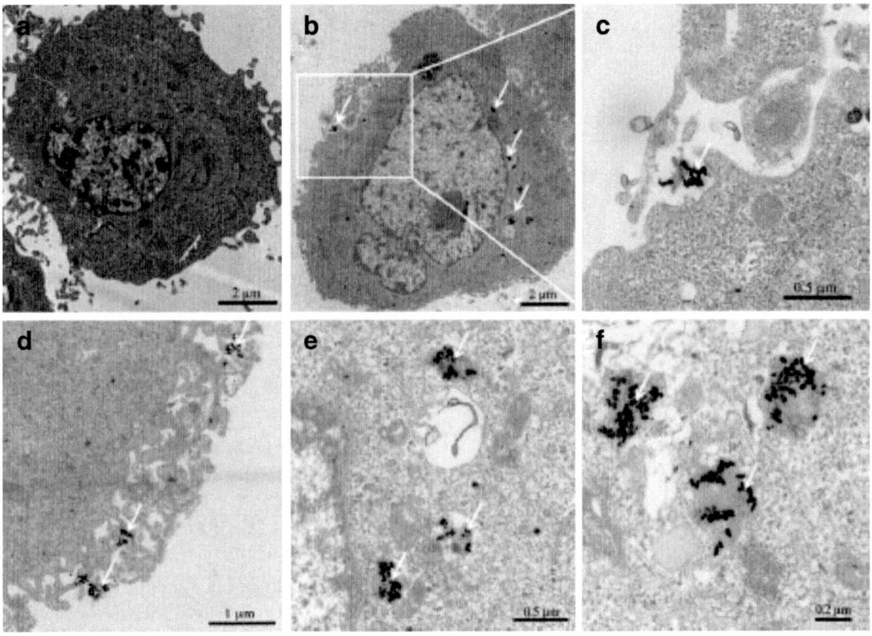


**Fig. 11.17** The thin-section TEM images of MGC803 cells incubated with 50 mM of GNR-SiO2-FA for 2 h. (**a**) Control MGC803 cells, (**b**–**f**) representative ultrastructures of MGC803 cells incubated with GNR-SiO2-FA. *Arrows* denote the GNR-SiO2-FA or their aggregates (Permission is from Elsevier Press)

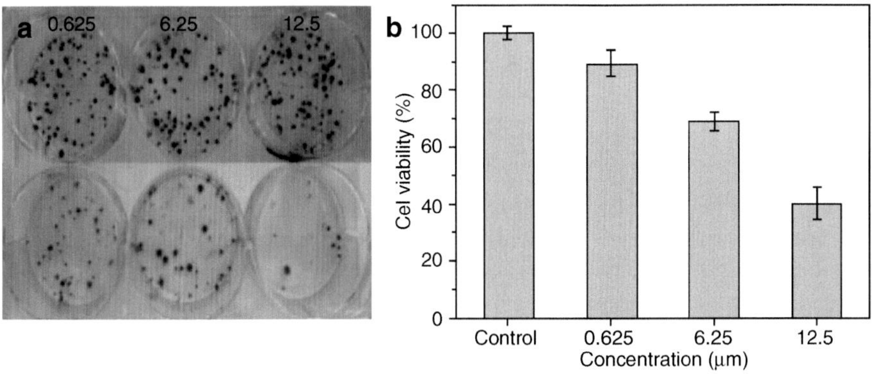


**Fig. 11.18** The proliferation photographs (**a**) and cell viability (**b**) of MGC803 cells incubated with 100 mL of varying concentration of GNR-SiO2-FA for 24 h upon 6 Gy of X-ray irradiation. In the photographs (**a**), the top column of the cells without irradiation, the bottom column of the cells with irradiation (Permission is from Elsevier Press)

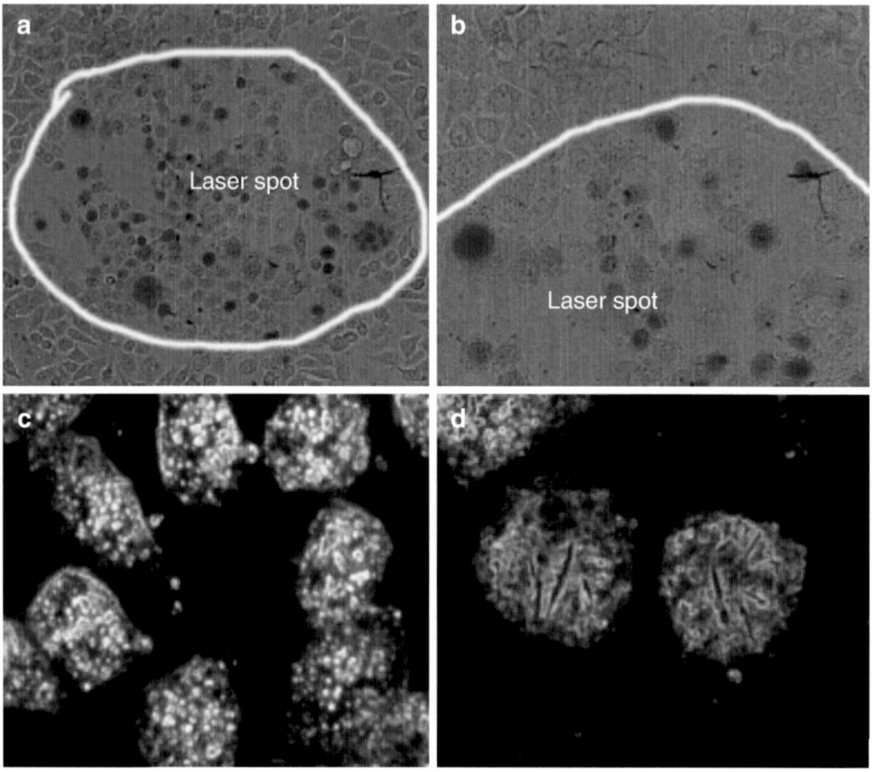


**Fig. 11.19** Photothermal therapy effects on MGC803 cells incubated with 12.5 mM of GNR-SiO2-FA for 24 h at 37 °C in the dark prior to irradiation for 3 min with 808 nm laser. (**a**) MGC803 cells on the laser spot, (**b**) MGC803 cells on the boundary of laser spot, and (**c**, **d**) dark-field images of MGC803 cells after irradiation (Permission is from Elsevier Press)

conjugate – cetuximab (Erbitux®) – that specifically targets EGFR and turns off a main signaling cascade for cancer cells to proliferate and survive. These drug/SERS gold nanoantennas present a high Raman signal both in cancer cells and in mice-bearing xenograft tumors. Moreover, the Raman detection signal is accomplished simultaneously by extensive tumor growth inhibition in mice, making these gold nanoantennas ideal for cancer nanotheranostics, i.e., tumor detection and tumor cell inhibition at the same time [45] (Fig. 11.20).

### *11.5.3 BRCAA1 Antibody-Conjugated Gold Nanoprisms for Targeted Photoacoustic Imaging and Photothermal Therapy*

The development of high-resolution nanosized photoacoustic contrast agents is an exciting technological advance as well as challenging. Herein, two functionalized gold nanoprisms (AuNprs), PEG/RGD-AuNprs and PEG/Brcaa1-AuNprs, were

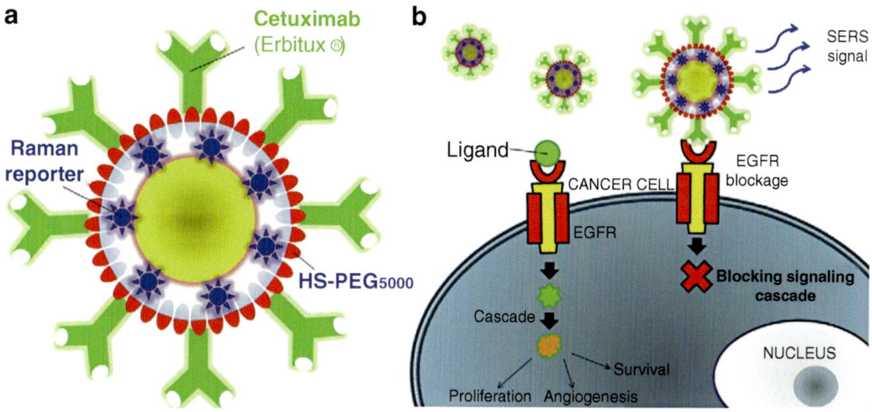


**Fig. 11.20** (**a**) Antibody–drug gold SERS nanoantennas – 90 nm AuNPs surrounded by a Raman reporter, encapsulated and entrapped by a larger polyethylene glycol (PEG) polymer and covered. (**b**) The drug-Raman NPs can easily bind to EGF receptors(EGFR), blocking the EGF protein from reaching the cancer cells and inhibit the signalling cascade. The activation of this cascade will consequently stop proliferation and survival of targeted cells (Permission is from Elsevier Press)

synthesized and used as a combinatorial methodology for in situ photoacoustic imaging, angiography, and local hyperthermia using orthotopic and subcutaneous murine gastric carcinoma models. PEG/RGD-AuNprs are available for tumor angiography, and PEG/Brcaa1-AuNprs are used to target and for in situ imaging of gastric carcinoma orthotopic tumor models. In situ photoacoustic imaging allowed for anatomical and functional imaging at the tumor site. In vivo tumor angiography imaging showed enhancement of the photoacoustic signal in a time-dependent way. Furthermore, photoacoustic imaging showed that tumor vessels were clearly damaged after local hyperthermia. This is the first proof-of-concept using two gold nanoprisms probes as high sensitive contrasts and therapeutic agents for in situ tumor detection and inhibition. These smart antibody/peptide AuNprs can be used as an efficient nanotheranostic platform for in vivo tumor detection with high sensitivity as well as for tumor-targeting therapy, which results in tumor size reduction and increased mice survival after local hyperthermia with a single-dose injection [46] (Fig. 11.21).

### *11.5.4 CD44v6 Antibody-Conjugated Gold Nanostars for Photoacoustic Imaging and Photothermal Therapy of Gastric Cancer Stem Cells*

Developing safe and effective nanoprobes for targeted imaging and selective therapy of gastric cancer stem cells (GCSCs) has become one of the most promising anticancer strategies [47–51]. Herein, gold nanostar-based PEGylated multifunctional nanoprobes were prepared with conjugated CD44v6 monoclonal antibodies (CD44v6-GNS) as the targeting ligands. It was observed that the prepared nanoprobes had high affinity toward GCSC spheroid colonies and destroyed them completely with a low power density upon near-infrared (NIR) laser

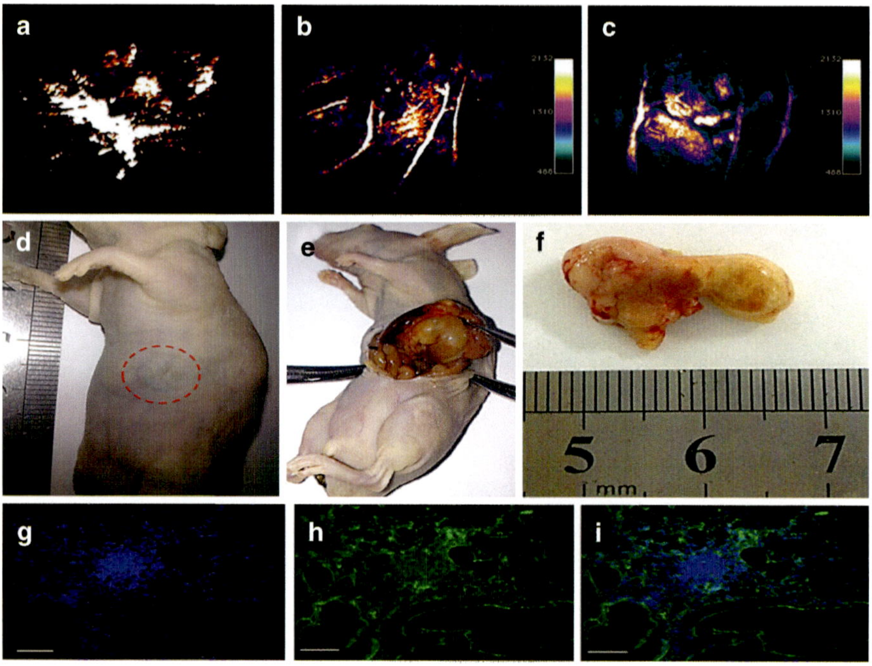


**Fig. 11.21** In vivo photoacoustic imaging study of subcutaneous transplanted tumors. (**a–d**) Sequential PA maximum intensity projection (MIP) frames after tail injected with RGD-AuNprs. (**a**) Before injection with RGD-AuNprs; (**b**) Injection with RGD-AuNprs after 1 h; (**c**) Injection with RGD-AuNprs after 3 h; (**d**) Injection with RGD-AuNprs after 6 h (unpublished data)

treatment (790 nm, 1.5 W/cm$^2$, 5 min) in vitro experiment. Orthotopic and subcutaneous xenografted nude mice models of human gastric cancer were established. Subsequently, biodistribution and photothermal therapeutic effects after being intravenously injected with the prepared nanoprobes were assessed. Photoacoustic imaging revealed that CD44v6-GNS nanoprobes could target the gastric cancer vascular system actively at 4 h postinjection, while the probes inhibited tumor growth remarkably upon NIR laser irradiation, and even extended survivability of the gastric cancer-bearing mice. The CD44v6-GNS nanoprobes exhibited great potential for applications of gastric cancer-targeted imaging and photothermal therapy in the near future [52–68] (Figs. 11.22 and 11.23).

**Fig. 11.23** Infrared microscopic imaging: deionized water (**a**) and GNS (**b**) in a tube upon NIR laser irradiation (790 nm, 0.3 W /cm$^2$, 3 min); subcutaneous tumor of GC without (**c**) and with (**d**) injection of GNS-PEG-CD44v6 upon NIR laser irradiation (790 nm, 1.5 W /cm$^2$, 3 min). The nude mouse of GC subcutaneous xenograft, injected with nanoparticles before (**e**) and after (**f**) laser irradiation treatment (790 nm, 1.5 W/cm$^2$, 3 min); (**g**) tumor growth curves of four groups after treatment with GNS-PEG-CD44v6, GNS-PEG and PBS respectively upon NIR laser irradiation (790 nm, 0.8 W/cm$^2$, 5 min); and the untreated control group. Error bars represent standard deviation ($n$=10/group). *$p$<0.05, **$p$<0.01 (student's $t$ test); (**h**) survival rate of GC tumor-bearing mice within 8 weeks after treatment (Permission is from IVY Press)

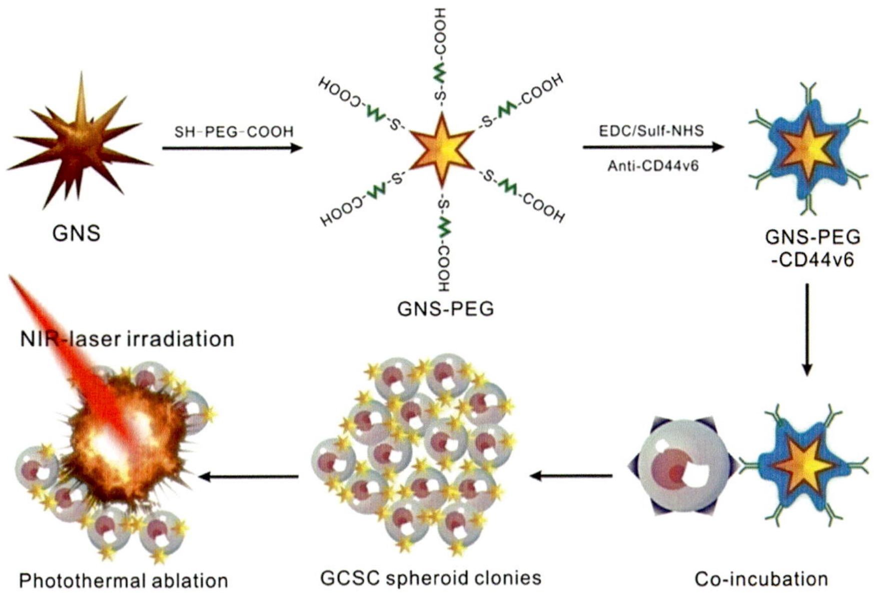


Fig. 11.22 Schematic illustration of the GNS-PEG-CD44v6 nanocomplex synthesis process and the mechanism of GCSC targeted PTT (Permission is from IVY press)

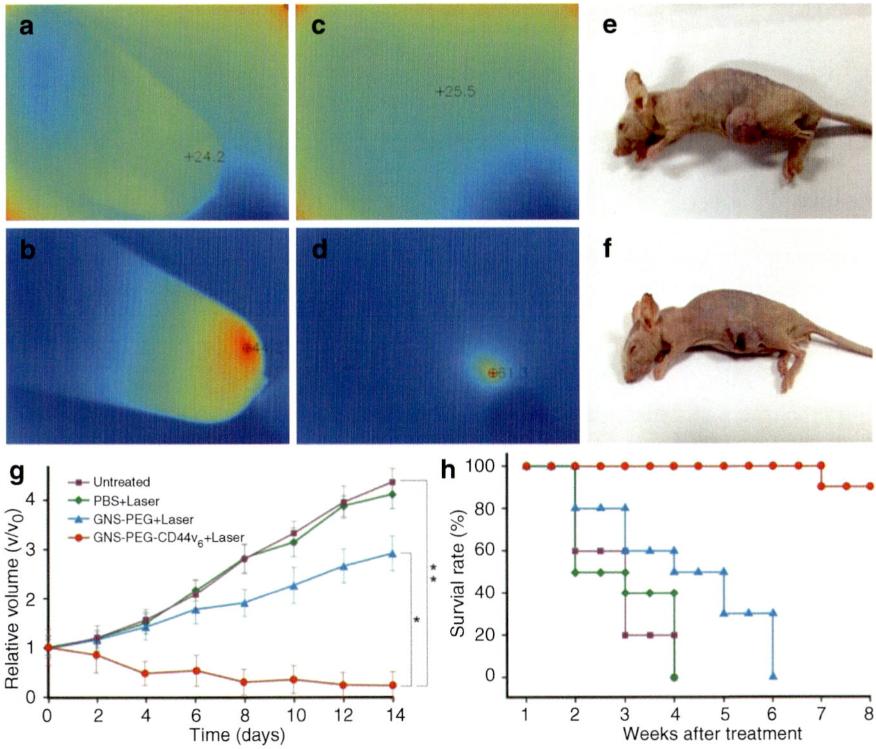

### 11.5.5 Folic Acid-/ce6-Conjugated Gold Nanoclusters for NIR Fluorescent Imaging and Photodynamic Therapy

Gold nanoclusters (GNCs) attract increasing attention due to their potential applications in sensing, catalysis, optoelectronics, and biomedicine. Herein, the formation of highly fluorescent glutathione (GSH)-capped GNCs is achieved through the delicate control of the reduction kinetics and thermodynamic selection of the Au(I)–SG complexes. Furthermore, the GNC-based nanoprobes are developed by the covalent coupling folic acid (FA) and polyethylene glycol (PEG) on the surface of GNCs directly, followed by trapping photosensitizer (chlorin e6, Ce6) within PEG networks and attaching to the GNCs surface. The fabricated nanoprobes (Ce6@GNCs-PEG 2 K -FA) possess a uniform particle size (hydrodynamic diameter $\approx$6.1 ± 1.2 nm), without affecting the yield of singlet oxygen of the trapped Ce6. In vitro studies show the enhanced cellular uptake and satisfactory photodynamic therapy (PDT) effectiveness toward MGC-803 cells when compared with free Ce6. The biodistribution and excretion pathway studies of the nanoprobes in MGC-803 tumor-bearing nude mice reveal their superior penetration and retention behavior in tumors, while the preserved features of renal clearance and stealthy to reticuloendothelial system are mainly attributed to the small hydrodynamic diameters and the FA-capped PEGylated ligands. The enhanced PDT efficacy and the nontoxicity to mice provide an exciting new nano-platform with promising clinical translational potential [69–71] (Figs. 11.24 and 11.25).

### 11.5.6 Gold Nanoparticles as a High Efficient siRNA Delivery System

Gold glyconanoparticles (GlycoNPs) are full of promise in areas like biomedicine, biotechnology, and materials science due to their amazing physical, chemical, and biological properties. Here, siRNA GlycoNPs (AuNP@PEG@Glucose@siRNA) in comparison with PEGylated GlycoNPs (AuNP@PEG@Glucose) were applied in vitro to a luciferase-CMT/167 adenocarcinoma cancer cell line and in vivo via intratracheal instillation directly into the lungs of B6 albino mice grafted with luciferase-CMT/167 adenocarcinoma cells. siRNA GlycoNPs but not PEGylated GlycoNPs induced the expression of proapoptotic proteins such as Fas/CD95 and caspases 3 and 9 in CMT/167 adenocarcinoma cells in a dose-dependent manner, independent of the inflammatory response, evaluated by bronchoalveolar lavage cell counting. Moreover, in vivo pulmonary delivered siRNA GlycoNPs were capable of targeting c-Myc gene expression (a crucial regulator of cell proliferation and apoptosis) via in vivo RNAi in tumor tissue, leading to an $\sim$80 % reduction in tumor size without associated inflammation [72–74] (Fig. 11.26).

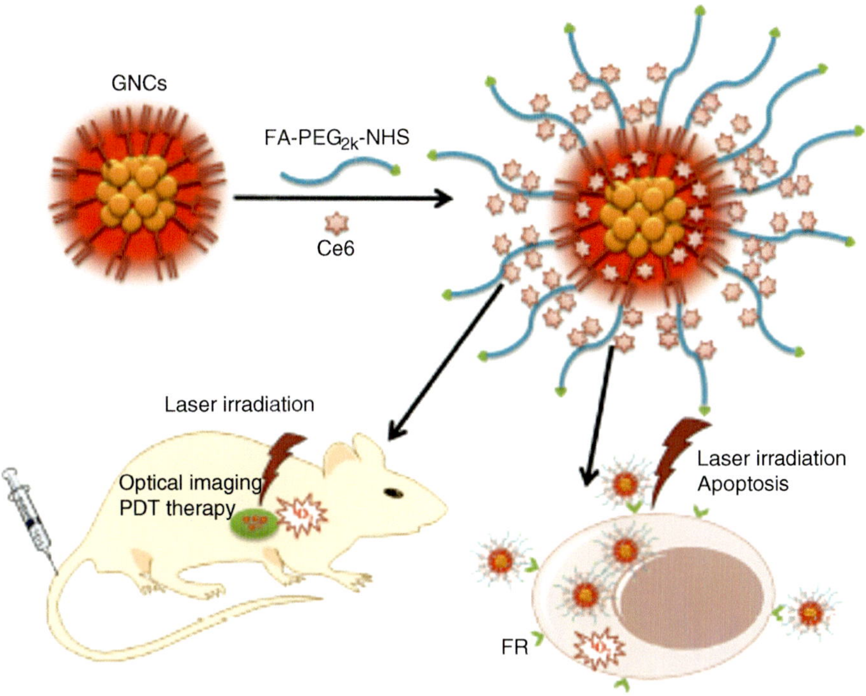


**Fig. 11.24** Schematic illustration of the preparation of the GNC-based nanoprobes and their applications in vitro and in vivo (Permission is from Wiley Press)

Up to now, functionalized gold nanoparticles have been optimized as an effective intracellular in vitro delivery vehicle for siRNAs to interfere with the expression of specific genes by selective targeting and provide protection against nucleases. However, few examples of suchlike in vivo applications have been described so far. We report the use of siRNA/RGD gold nanoparticles capable of targeting tumor cells in a lung cancer syngeneic orthotopic murine model. Therapeutic RGD–nanoparticle treatment resulted in successful targeting evident from significant c-myc oncogene downregulation followed by tumor growth inhibition and prolonged survival of lung tumor-bearing mice, possibly via avb3 integrin interaction. Our results suggest that RGD gold nanoparticles-mediated delivery of siRNA by intratracheal instillation in mice leads to successful suppression of tumor cell proliferation and respective tumor size reduction. These results reiterate the capability of functionalized gold nanoparticles for targeted delivery of siRNA to cancer cells toward effective silencing of the specific target oncogene. What is more, we demonstrate that the gold nanoconjugates trigger a complex inflammatory and immune response that might promote the therapeutic effect of the RNAi to reduce tumor size with low doses of siRNA [75, 76].

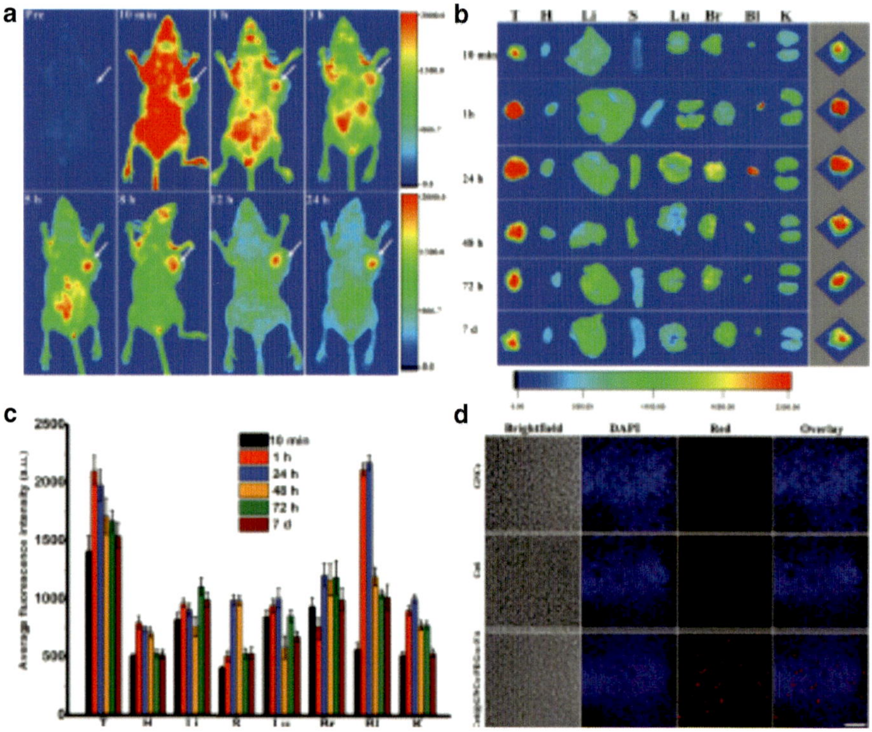


**Fig. 11.25** (**a**) In vivo fluorescence imaging of MGC-803 tumor-bearing mouse after intravenous injection of Ce6@GNCs-PEG 2 K -FA over a period of 24 h. (**b**) Representative ex vivo fluorescence images of major organs and tumors of mice after different postinjection times. Tumor (T), heart (H), liver (Li), spleen (S), lung (Lu), brain (Br), bladder (Bl), kidney (K). The right columns were the corresponding 3D surface plot images of pixel intensities obtained from the tumor regions using NIH Image J software. (**c**) Quantitative bio-distribution of Ce6@GNCs-PEG 2 K -FA in different organs and tumors based on ROIs analysis of the average fluorescence intensity from the tumors and normal tissues (3 mice per group). (**d**) Representative fluorescence images of tumor cryosections (10 μm). Images presented are bright-field, blue fluorescence (DAPI), red fluorescence (>590 nm), and merged (overlap) images. *Scale bar*, 100 μm (Permission is from Wiley Press)

## 11.6 RNA Nanoparticles for Targeted Imaging and siRNA Therapy of Gastric Cancer

In recent years, there have been several new nano-delivery systems with different materials and physic-chemical properties to be developed. However, effective strategies to block tumor progression and prevent metastasis are still very few; there exist several challenges including specific cancer targeting, tissue penetration, intracellular delivery, toxicities, and side effects due to organ accumulation, nonspecific cell entry, particle heterogeneity, aggregation, dissociation due to dilution after systemic injection, and unfavorable pharmacological profiles. In recent years, RNA

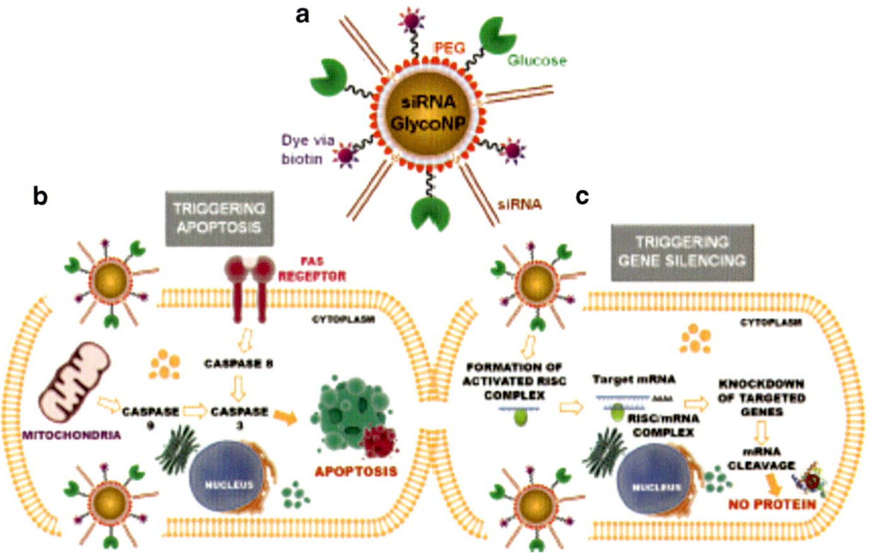


**Fig. 11.26** (**a**) Multifunctional siRNA glyconanoparticles (siRNA GlycoNPs) trigger apoptotic pathways (**b**) with the expression of cell death receptors (Fas) and caspases. The death domain-containing receptor Fas can sense an external signal and activate the apoptosis pathway through the Fas-related death domain. This pathway is mediated by the activation of caspase-8, followed by direct cleavage of downstream effector caspases. The apoptosis pathway can also be initiated in the cytoplasm through activation of intracellular changes resulting in the release of proapoptotic factors from the mitochondria. The release of these factors leads to the activation of caspase-9 and ultimately results in the activation of effector caspases (e.g., caspase 3) and consequently to cell death by apoptosis. (**c**) The siRNA glycoNPs also have the capacity to trigger gene silencing via activation of the RNA interference pathway, by double-stranded RNA (i.e., siRNA), promoting nucleolytic degradation of the target mRNA and/or translational suppression (Permission is from RSC Press)

nanotechnology, as an emerging nano-platform, has shown great prospect in clinical theranostic applications. RNA nanoparticles can be fabricated with precise control of shape, size, and stoichiometry, as demonstrated by the packaging RNA (pRNA) of the bacteriophage phi29 DNA packaging motor, which forms dimmers, trimers, and hexamers, via hand-in-hand interactions of the interlocking loops [49, 50]. The pRNA contains an ultra-stable three-way junction (3WJ) motif, which can be assembled from three short fragments with extremely high affinity. Recently, Prof. Peixuan Guo's group has obtained the crystal structure of the pRNA-3WJ motif, and a variety of therapeutic RNA nanoparticles using the pRNA-3WJ and pRNA-X motifs as scaffolds have been constructed. The pRNA-3WJ nanoparticles display thermodynamically stable properties, including high melting temperature with low free energy, resistance to denaturation in 8 M urea, and resistance to dissociation at very low concentrations in the blood. Boiling-resistant RNA nanoparticles with controllable shapes and defined stoichiometry have recently been reported. Various functional groups, such as imaging agents; targeting ligands, such as receptor binding aptamers; and therapeutic modules, such as siRNA, miRNA, or ribozymes, can be integrated into the 3WJ scaffold without affecting the folding and functionality of the core motif and incorporated functional moieties. Upon 2′-fluoro (2′-F) modifications of uracil (U) and cytosine (C) nucleotides, the RNA nanoparticles become resistant to RNase degradation with enhanced in vivo half-life while retaining authentic functions of the incorporated modules. Furthermore, the pRNA nanoparticles are nontoxic, non-immunogenic, and display favorable biodistribution and pharmacokinetic profiles in mice. These favorable findings prompted the use of this novel platform for the treatment of stomach cancer, which is one of the challenging tasks in clinical oncology.

Such targeted delivery systems need a ligand–receptor pair that is specifically existed in cancer cells. Many, but not all, cancer cells, including the stomach, ovary, lung, breast, kidney, endometrium, colon, and hematopoietic cells, overexpressed folate receptors (FRs) than normal cells for high uptake of folate, since folate is an essential component during DNA replication and methylation in highly proliferating cells. Folic acid (FA), a synthetic oxidized form of folate, has been widely used as a ligand conjugate in various cancer targeting materials. *Breast cancer-associated antigen 1* (*BRCAA1, AF208045*) has been confirmed to exhibit overexpression in breast cancer and gastric cancer and no or lower expression in normal gastric mucosa and normal breast tissues. Our previous studies have demonstrated that gastric cancer MGC803 cells were transfected with constructed plasmids of shRNA-BRCAA1; the cell growth was greatly inhibited, and the rate of cell apoptosis was significantly higher than those of untransfected group and mock plasmid transfected group. We also screened out a new antigen epitope, SSKKQKRSHK, screened out matched two monoclonal antibody cell lines, successfully prepared monoclonal antibody-conjugated fluorescent magnetic nanoparticles, and realized targeted imaging and hyperthermal therapy of in vivo gastric cancer. Therefore, the *BRCAA1* gene is a potential therapeutic target for gastric cancer.

My group cooperated with Prof. Peixuan Guo's group, adopted an innovative RNA nanotechnology approach to overcome some of the aforementioned challenges, and report for the first time a new strategy to target and deliver therapeutic BRCAA1 siRNA to in vivo stomach cancer tissues using FA-conjugated pRNA-3WJ nanoparticles. Our objective is to construct multifunctionally, thermodynamically, and chemically stable RNA nanoparticles harboring sequences that allow specific binding to stomach cancer-specific cell surface antigens or receptors resulting in the internalization of RNA nanoparticles into target cells and delivery of the siRNA, miRNA, and drugs for attaining synergistic effects for the treatment of stomach cancer; we also investigated the effects of prepared RNA nanoparticles on the regression of gastric cancer tissues in vivo, and potential molecular mechanism, with the aim of laying foundation for further clinical application in the near future [77, 78] (Figs. 11.27, 11.28, and 11.29).

RNA interference via the use of small interfering RNA (siRNA) is a powerful and useful tool to block gene function through sequence-specific posttranscriptional gene silencing, playing an important role in the downregulation of gene expression. siRNAs can be transfected into mammalian cells by a variety of methods that influ-

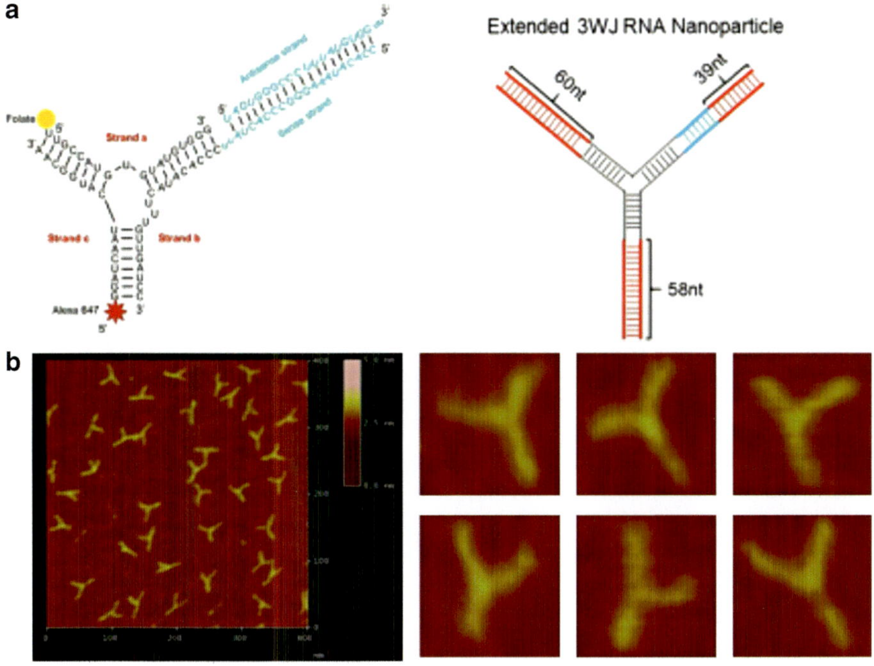


**Fig. 11.27** Global structure of the therapeutic RNA nanoparticles with BRCAA1 siRNA. (**a**) Design of the RNA nanoparticles. Left is the one used in animal trial. Right is the extended one to prepare the AFM images. (**b**) AFM image of extended 3WJ RNA nanoparticles. The RNA complex in left of a is estimated to be around 10 nm. Due to convolution of the tip size (~10 nm in diameter) in AFM images, features smaller than the size of the tip cannot be resolved. To characterize the structure of the RNA constructs, the 3WJ nanoparticles were extended by 39–60 base pairs (in *red* color), which is within the persistence length of dsRNA and will not affect the 3WJ folding as described before 31, to generate the AFM image as shown (Permission is from Nature Press)

ence the strength and duration of the silencing response, which in turn is affected by the amount of siRNA effectively delivered and by the potential of each siRNA to suppress its target. Nevertheless, naked siRNAs show extremely short half-lives due to RNases activity, poor chemical stability, and dissociation from the vector. In fact, the major obstacle to clinical application is the uncertainty about how to deliver siRNA with maximal therapeutic impact [79].

## 11.7 Carbon Dot-Based Nanoprobes for Targeted Imaging and Photodynamic Therapy

Carbon dots (C-dots), a very recent rising star, as a class of zero-dimensional carbon nanomaterials, possess some of the same major advantageous characteristics of quantum dots (QDs), such as high photostability, tunable emission, and large two-photon excitation cross sections. Moreover, C-dots exhibit nonblinking

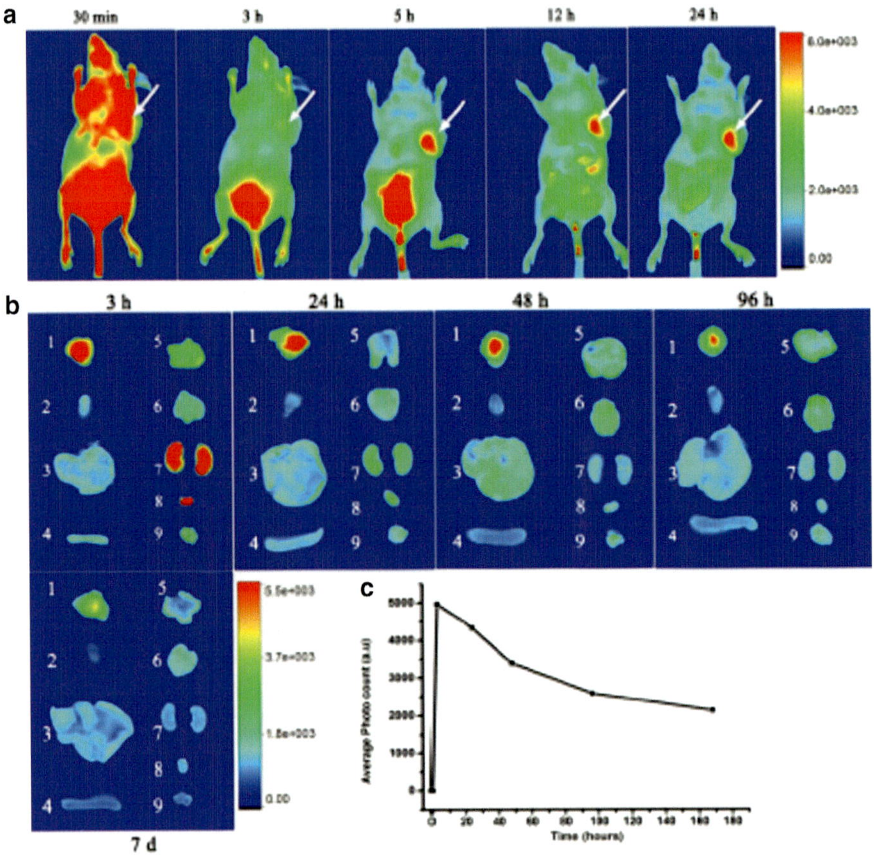


**Fig. 11.28** (**a**) Representative in vivo fluorescence images of MGC803-tumor-bearing mouse after iv-injected with FA-AlexaFluor647-labeled pRNA nanoparticle. The tumor areas are indicated with arrows. (**b**) Representative ex vivo images of tumors and organs. Labels: *1*, tumor; *2*, heart; *3*, liver; *4*, spleen; *5*, lung; *6*, Stomach; *7*, kidneys; *8*, bladder; *9*, muscle. (**c**) The average fluorescence intensities from the tumor areas of postinjection (Permission is from Nature Press)

fluorescence and excellent water solubility and are cheaply produced. Particularly, C-dots without heavy metal content are more environmentally friendly and can be much safer for biological applications. We strategically designed and prepared novel multifunctional chlorin e6-conjugated C-dots (C-dots-Ce6) as the light-triggered theranostics for simultaneous enhanced near-infrared fluorescent imaging and photodynamic therapy (PDT) by FRET mechanism [80] (Fig. 11.30).

Carbon dots exhibit great potential in applications such as molecular imaging and in vivo molecular tracking. However, how to enhance fluorescence intensity of carbon dots has become a great challenge. We report for the first time a new strategy to synthesize fluorescent carbon dots (C-dots) with high quantum yields by using ribonuclease A (RNase A) as a biomolecular templating agent under

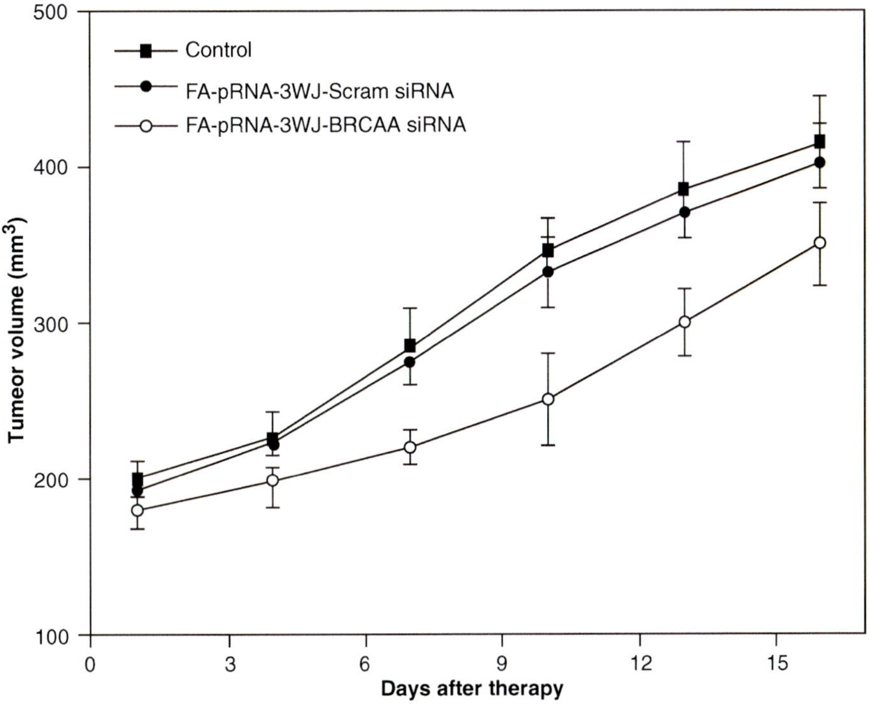


Fig. 11.29 Tumor size curve as the posttreatment time increases (Permission is from Nature Press)

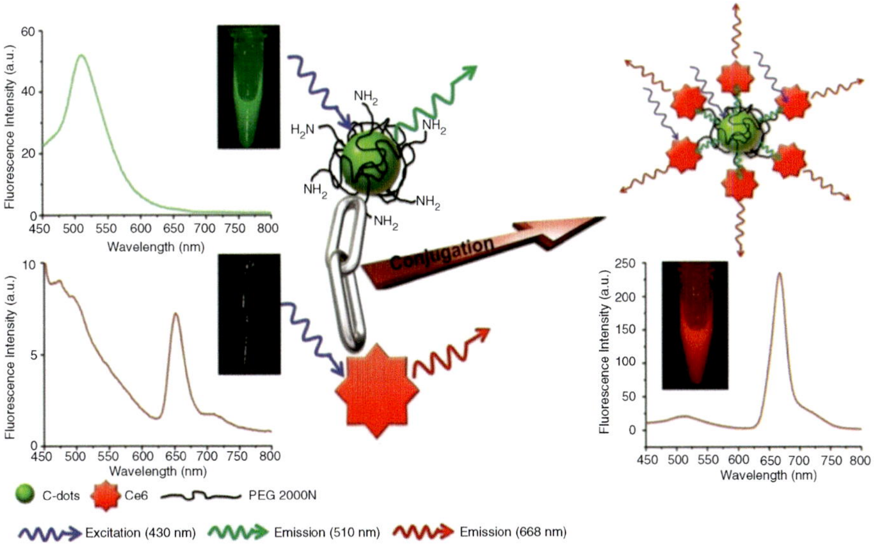


Fig. 11.30 Förster (fluorescence) resonance energy transfer (FRET) process between C-dots and Ce6 (Permission is from Wiley Press)

microwave irradiation. The synthesized RNase A-conjugated carbon dots (RNase A@C-dots) exhibited quantum yields of 24.20 %. The fluorescent color of the RNase A@C-dots can easily be adjusted by varying the microwave reaction time and microwave power. Moreover, the emission wavelength and intensity of RNase A@C-dots displayed a marked excitation wavelength-dependent character. As the excitation wavelength alters from 300 to 500 nm, the photoluminescence (PL) peak exhibits gradually redshifts from 450 to 550 nm, and the intensity reaches its maximum at an excitation wavelength of 380 nm. Its Stokes shift is about 80 nm. Notably, the PL intensity is gradually decreasing as the pH increases, almost linearly dependent, and it reaches the maximum at a pH = 2 condition; the emission peaks also show clearly a redshift, which may be caused by the high activity and perfective dispersion of RNase A in a lower pH solution. In high pH solution, RNase A tends to form RNase A-warped carbon dot nanoclusters. Cell imaging confirmed that the RNase A@C-dots could enter into the cytoplasm through cell endocytosis. 3D confocal imaging and transmission electron microscopy observation confirmed partial RNase A@C-dots located inside the nucleus. MTT and real-time cell electronic sensing (RT-CES) analysis showed that the RNase A@C-dots could effectively inhibit the growth of MGC-803 cells. Intra-tumor injection test of RNase A@C-dots showed that RNase A@C-dots could be used for imaging in vivo gastric cancer cells. The as-prepared RNase A@C-dots are suitable for simultaneous therapy and in vivo fluorescence imaging of nude mice loaded with gastric cancer or other tumors. We have developed a microwave assisted one-step synthesis of C-dots with citric acid as carbon source and tryptophan (Trp) as both nitrogen source and passivation agent. The C-dots with uniform size show superior water solubility, excellent biocompatibility, and high quantum yield. Afterward, the PEI (polyethylenimine)-adsorbed C-dots nanoparticles (C-dots@PEI) were applied to deliver survivin siRNA into human gastric cancer cell line MGC-803. The results have confirmed the nanocarrier exhibited excellent biocompatibility and a significant increase in cellular delivery of siRNA, inducing efficient knockdown for survivin protein to 6.1 %. In addition, PEI@C-dot complexes mediated survivin silencing; the arrested cell cycle progression in G1 phase as well as cell apoptosis was observed [81–83]. Carbon dots may be one kind of good contrast agent and drug delivery system.

## 11.8 Dendritic Cell and Tumor Cell-Fused Vaccine for Targeted Imaging and Enhanced Immunotherapy of Gastric Cancer

Dendritic cells (DCs) have been broadly used for clinical therapy of tumor patients, achieved good therapeutic efficacy, displaying the great promising potential in clinical cancer immunotherapy. How to enhance DCs' immunotherapeutic effect in cancer-targeted immunotherapy and prevention is still a great challenge. My group is for the first time to prepare the allogenic DCs and gastric cancer cell-fused vaccine combined with cytokine induced killing cells (CIKs), which realized targeted imaging and enhanced immunotherapy of gastric cancer (GC). The fused vaccine was prepared by PEG-mediated fusion between the mature DCs and inactive gastric cancer MGC803 cells. The immunotherapeutic and prophylactic potential of the fused cells (FCs) were evaluated in tumor-bearing, post-surgery, and tumor-free mice models. The migration and homing process of near-infrared region quantum dot (NIR-QD)-labeled FCs were investigated by real-time animal imaging system. Results showed that the FCs and FC þ CIKs could trigger the tumor-specific cell toxic lymph cells (CTLs) against GC cells, target the tumor tissue initiatively, enhance the prophylactic effects, suppress the tumor growth remarkably in vivo, and make several tumor tissues in nude model disappeared. This is only very initiative result; further clinical study will continue to carry out. The allogenic DCs and tumor cell-fused vaccine can be used for targeted imaging and enhanced immunotherapy of GC, and the FC þ CIKs strategy owns great potential in clinical applications such as early therapy and prevention of tumor metastasis and relapse in the near future [84–90].

## 11.9 Oral Microcapsule Endoscopy Combined with Nanoprobes for Gastrointestinal Imaging

One kind of oral microcapsule endoscopy was successfully developed by Prof. Guozheng Yan in Shanghai Jiao Tong University as shown in Fig. 11.31a, which is composed of several sensors and power system and realizes real-time ultrasensitive detection of five parameters such as temperature, pH value, pressure, hemoglobin, and imaging in gastrointestinal tract, obtained registration certificate of medical devices in china in 2003, and have been used in almost 100 hospitals [91]. Based on the oral capsule endoscopy system, Prof. Yan designed one kind of a new type of oral fluorescent nanoprobes that obtained clear fluorescent imaging of gastrointestinal lesions as shown in Fig. 11.31c; the control image without

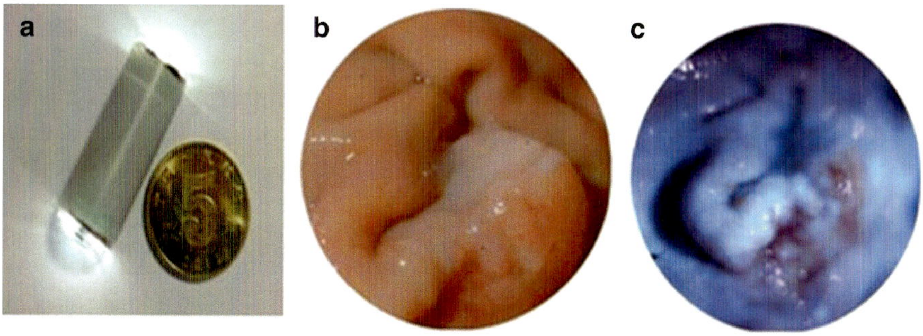


**Fig. 11.31** (**a**) Oral microcapsule; (**b**) normal image of human gastric lesion; (**c**) fluorescent image of human gastric lesion incubated with nanoprobes

fluorescent signal was shown in Fig. 11.31b. Furtherly integrated with oral RGD/BRCAA1 antibody-conjugated gold nanoprism probes, the oral capsule endoscopy system can obtain more clear molecular imaging of gastric intestinal tumors, simultaneously realize photothermal therapy in vivo, which provide a new pathway for quick theranostics of clinical gastrointestinal tumor or precancerous lesions patients (Fig. 11.31).

## 11.10 Clinical Translational Prospects of Multifunctional Nanoprobes

There have been more than several 10,000 publications regarding nanoprobes in the last decade, and very few of the assessed nanoprobes progressed from bench to bedside. According to Drolet and Lorenzi's "continuum" theory, when a new basic science discovery transfers into proposed human application, proven clinical application, and clinical practice and finally raises public health impact, there are three "translation chasm" gaps that must be bridged. These gaps are human (T1), clinical trials (T2), and translation to practice (T3). Although there have been very promising applications of nanoprobes in many areas, they are still at an early stage of development on the way to reaching the bedside; in other words, they are mostly in the T1 stage, although they may come into clinical trials.

Up to date, there are several nanoprobes to enter into clinical trials or obtain FDA permission. For example, in 2011, a clinical trial of insulin-coated AuNPs was approved by Midatech Ltd (UK). In addition, AuroShell® (Au@silica nanoshells) from Nanospectra (Nanospectra Biosciences, TX, USA), which is used for thermotherapy, was approved by the FDA in 2012. This has been proven to be effective at an earlier stage in a murine glioma model and in the ablation of prostate cancer in a canine model. Another example is from McNeil's laboratory; after their reports of the biodistribution, immune response, and interactions of AuNPs with human blood, they used AuNPs as a drug delivery system in order to impair the toxicity of TNF-α. It is confirmed to be safe to inject up to three times the amount that had been lethal with previous versions. The modified drug Aurimune™ (CYT-6091) from CytImmune Sciences (MD, USA) is a TNFα-bound, PEGylated colloidal AuNP. Aurimmune has passed through a Phase I clinical trial and is entering a Phase II trial. It is reported that the systemic TNF toxicity of Aurimune was decreased dramatically compared with TNF alone (e.g., the fever side effect was managed) and the target effect was realized. Another product, AuriTol™ (CYT-21001), also from CytImmune Sciences, has reached the preclinical stage [91–95].

An example of a fast, in vitro, Au-based biomedical application is the FDA-approved Verigene® detector device (Nanosphere, IL, USA), which utilizes oligonucleotide-coated AuNPs in order to capture DNA. The Verigene system enables clinicians to rapidly identify and treat the bacteria and viruses that are responsible for some of the most complicated, costly, and fatal infectious diseases [96].

My group established the new method of RNase A-assisted carbon dot synthesis and finished biosafety evaluation of carbon dots in National Drug Safety Evaluation Center, confirming to be very safe under the dosage of 100 mg/kg body weight; we also prepared hyaluronic acid-/RGD-conjugated carbon dots, realized targeted imaging and operation boundary identification of in situ gastric cancer, and finished 40 specimens of clinical trials, showing the prepared nanoprobes can image the metastasis lymph nodes and display the surgery boundary of gastric cancer.

Although some nanoprobes got through the clinical trials and obtained FDA permission to enter into clinical application, inorganic nanoparticle-based multifunctional nanoprobes exist great disputed for clinical application; the reasons are summarized as being unable to degrade in vivo, long-term leftover in body; the further toxicological mechanism is not clarified well. The development of targeting, safe, high efficient nanoprobes has become a great challenge. Only to solve the biosafety of nanoprobes, multifunctional nanoprobes can be possible to enter into clinical application in the near future.

## 11.11 Conclusion

In summary, gastric cancer is one kind of common tumor worldwide; how to realize theranostics of gastric cancer is a great challenge. Current advances show that gastric cancer therapeutic strategies should focus on early diagnosis and operation therapy, enhanced immunotherapy, and killing gastric cancer stem cells to overcome multidrug resistance (MDR). Facing these strategies, we designed and prepared series of multifunctional nanoprobes for targeted imaging and therapy of gastric cancer, including series of fluorescent magnetic nanoprobes, series of quantum dots nanoprobes and carbon dots, series of gold nanoprobes, series of upconversion nanoprobes, RNA nanoprobes, and nanoprobes for killing gastric cancer stem cells and enhanced immunotherapeutic efficacy. Up to date, carbon dot-based nanoprobes were evaluated to confirm their safety and were used for identifying the boundary of gastric cancer and tracking metastasis lymph nodes, exhibiting clinical application prospect. In the future, the main challenges to the translational research of these nanoprobes will focus on the better understanding of the interactions between nanoparticles and biomacromolecules and the immune system, the homogeneity of the material preparations, the paucity of the knowledge regarding pertinent biomarkers, and the concern regarding biocompatibility. Green synthesis of nanoparticles, multimodality bioimaging (especially PAI), SERS analysis, and theranostics are particular fields of interest. However, designers must possess a systematic view from the beginning as they approach the goal of nanoprobes for translational medical research [97–100].

**Acknowledgment** This work was supported by Chinese Key Basic Research Program (973 Project) (No. 2010CB933901 and 2015CB931802), the National Natural Scientific Foundation of China (Grant No. 81225010, 81327002, and 31170961), and 863 project of China (no. 2012AA022703 and 2014AA020700), Shanghai Science and Technology Fund (No. 13NM1401500).

## References

1. Ferlay J, Shin HR, Bray F, Forman D, Mathers C, Parkin DM. Estimates of worldwide burden of cancer in 2008: GLOBOCAN 2008. Int J Cancer. 2010;127:2893–917.
2. Jemal A, Bray F, Center MM, Ferlay J, Ward E, Forman D. Global cancer statistics. CA Cancer J Clin. 2011;61:69–90.
3. Takahashi T, Saikawa Y, Kitagawa Y. Gastric cancer: current status of diagnosis and treatment. Cancer (Basel). 2013;5:48–63.
4. Dicken BJ, Bigam DL, Cass C, Mackey JR, Joy AA, Hamilton SM. Gastric adenocarcinoma: review and considerations for future directions. Ann Surg. 2005;241:27–39.
5. Uemura N, Okamoto S, Yamamoto S, Matsumura N, Yamaguchi S, Yamakido M, et al. Helicobacter pylori infection and the development of gastric cancer. N Engl J Med. 2001;345:784–9.
6. Comis RL, Carter SK. A review of chemotherapy in gastric cancer. Cancer. 1974;34:1576–86.
7. Kuo CY, Chao Y, Li CP. Update on treatment of gastric cancer. J Chin Med Assoc. 2014;77:345–53.
8. Proserpio I, Rausei S, Barzaghi S, Frattini F, Galli F, Iovino D, et al. Multimodal treatment of gastric cancer. World J Gastrointest Surg. 2014;6:55–8.
9. Zhang D, Fan D. New insights into the mechanisms of gastric cancer multidrug resistance and future perspectives. Future Oncol. 2010;6:527–37.
10. Cui DX, Zhang L, Yan XJ, Zhang LX, Xu JR, Guo YH, et al. A microarray-based gastric carcinoma prewarning system. World J Gastroenterol. 2005;11:1273–82.
11. Zhang YX, Gao G, Liu HJ, Fu HL, Fan J, Wang K, Chen Y, Li BJ, Zhang CL, Zhi X, He L, Cui DX. Identification of volatile biomarkers of gastric cancer cells and ultrasensitive electrochemical detection based on sensing interface of Au-Ag alloy coated MWCNTs. Theranostics. 2014;4:154–62.
12. Wang K, Ruan J, Qian Q, Song H, Bao CC, Kong YF, Zhang CL, Hu GH, Ni J, Cui DX. BRCAA1 monoclonal antibody conjugated fluorescent magnetic nanoparticles for in vivo targeted magnetofluorescent imaging of gastric cancer. J Nanobiotechnol. 2011;9:23.
13. Ruan J, Song H, Qian QR, Li C, Wang K, Bao CC, Cui DX. HER2 monoclonal antibody conjugated RNase-A-associated CdTe quantum dots for targeted imaging and therapy of gastric cancer. Biomaterials. 2012;33:7093–102.
14. He M, Huang P, Zhang CL, Hu HY, Bao CC, Gao G, Chen F, Wang C, Ma JB, He R, Cui DX. Dual phase-controlled synthesis of uniform lanthanide-doped NaGdF4 upconversion nanocrystals via an OA/ionic liquid two-phase system for in vivo dual-modality imaging. Adv Funct Mater. 2011;21:4470–7.
15. Li ZM, Huang P, Zhang XJ, Lin J, Yang S, Liu B, Gao F, Xi P, Ren QS, Cui DX. RGD-conjugated dendrimer-modified gold nanorods for in vivo tumor targeting and photothermal therapy. Mol Pharm. 2010;7:94–104.
16. Huang P, Lin J, Wang XS, Wang Z, Zhang CL, He M, Wang K, Chen F, Li ZM, Shen GX, Cui DX, Chen XY. Light-triggered theranostics based on photosensitizer-conjugated carbon dots for simultaneous enhanced-fluorescence imaging and photodynamic therapy. Adv Mater. 2012;24:5104–10.
17. Zhou ZJ, Zhang CL, Qian QR, Ma JB, He M, Pan LY, Gao G, Fu HL, Wang K, Cui DX. Folic acid-conjugated silica capped gold nanoclusters for targeted fluorescence/X-ray computed tomography imaging. J Nanobiotechnol. 2013;11:17.
18. Zhang CL, Zhou ZJ, Qian QR, Gao G, Li C, Feng LL, Wang Q, Cui DX. Glutathione-capped fluorescent gold nanoclusters for dual-modal fluorescence/X-ray computed tomography imaging. J Mater Chem B. 2013;1:5045–53.
19. Yang W, Raufi A, Klempner SJ. Targeted therapy for gastric cancer: molecular pathways and ongoing investigations. Biochim Biophys Acta. 2014;1846:232–7.

20. Shen M, Huang Y, Han L, Qin J, Fang X, Wang J, et al. Multifunctional drug delivery system for targeting tumor and its acidic microenvironment. J Control Release. 2012;161:884–92.
21. Pan BF, Cui DX, Xu P, Ozkan C, Feng G, Ozkan M, Huang T, Chu BF, Li Q, He R, Hu GH. Synthesis and characterization of polyamidoamine dendrimer-coated multi-walled carbon nanotubes and their application in gene delivery systems. Nanotechnology. 2009;20:125101.
22. Qi L, Wu L, Zheng S, Wang Y, Fu H, Cui DX. Cell-penetrating magnetic nanoparticles for highly efficient delivery and intracellular imaging of siRNA. Biomacromolecules. 2012;13:2723–30.
23. Murphy EA, Majeti BK, Mukthavaram R, Acevedo LM, Barnes LA, Cheresh DA. Targeted nanogels: a versatile platform for drug delivery to tumors. Mol Cancer Ther. 2011;10:972–82.
24. Yu X, Pishko MV. Nanoparticle-based biocompatible and targeted drug delivery: characterization and in vitro studies. Biomacromolecules. 2011;12:3205–12.
25. Zhou J, Shum KT, Burnett JC, Rossi JJ. Nanoparticle-based delivery of RNAi therapeutics: progress and challenges. Pharmaceuticals (Basel, Switzerland). 2013;6:85–107.
26. Guo P. The emerging field of RNA nanotechnology. Nat Nanotechnol. 2010;5:833–42.
27. Guo P, Haque F, Hallahan B, Reif R, Li H. Uniqueness, advantages, challenges, solutions, and perspectives in therapeutics applying RNA nanotechnology. Nucleic Acid Ther. 2012;22:226–45.
28. Guo P, Zhang C, Chen C, Garver K, Trottier M. Inter-RNA interaction of phage phi29 pRNA to form a hexameric complex for viral DNA transportation. Mol Cell. 1998;2:149–55.
29. Shu D, Moll WD, Deng Z, Mao C, Guo P. Bottom-up assembly of RNA arrays and superstructures as potential parts in nanotechnology. Nano Lett. 2004;4:1717–23.
30. Shu Y, Haque F, Shu D, Li W, Zhu Z, Kotb M, et al. Fabrication of 14 different RNA nanoparticles for specific tumor targeting without accumulation in normal organs. RNA (New York, NY). 2013;19:767–77.
31. Shu D, Shu Y, Haque F, Abdelmawla S, Guo P. Thermodynamically stable RNA three-way junction for constructing multifunctional nanoparticles for delivery of therapeutics. Nat Nanotechnol. 2011;6:658–67.
32. Abdelmawla S, Guo S, Zhang L, Pulukuri SM, Patankar P, Conley P, et al. Pharmacological characterization of chemically synthesized monomeric phi29 pRNA nanoparticles for systemic delivery. Mol Ther. 2011;19:1312–22.
33. Zhang H, Endrizzi JA, Shu Y, Haque F, Sauter C, Shlyakhtenko LS, et al. Crystal structure of 3WJ core revealing divalent ion-promoted thermostability and assembly of the Phi29 hexameric motor pRNA. RNA (New York, NY). 2013;19:1226–37.
34. Shu Y, Shu D, Haque F, Guo P. Fabrication of pRNA nanoparticles to deliver therapeutic RNAs and bioactive compounds into tumor cells. Nat Protoc. 2013;8:1635–59.
35. Haque F, Shu D, Shu Y, Shlyakhtenko LS, Rychahou PG, Evers BM, et al. Ultrastable synergistic tetravalent RNA nanoparticles for targeting to cancers. Nano Today. 2012;7:245–57.
36. Zhang F, Braun GB, Pallaoro A, Zhang Y, et al. Mesoporous multifunctional upconversion luminescent and magnetic "Nanorattle" materials for targeted chemotherapy. Nano Letters. 2012;12:61–7.
37. Ma JB, Zhou ZJ, Zhang CL, Gao G, Li C, Cui D. Folic acid-conjugated LaF3:Yb, Tm@SiO2 nanoprobes for targeting dual-modality imaging of upconversion luminescence and X-ray computed tomography. J Phys Chem C. 2012;116:14062–70.
38. Kalli KR, Oberg AL, Keeney GL, Christianson TJ, Low PS, Knutson KL, et al. Folate receptor alpha as a tumor target in epithelial ovarian cancer. Gynecol Oncol. 2008;108:619–26.
39. Teng L, Xie J, Teng L, Lee RJ. Clinical translation of folate receptor-targeted therapeutics. Expert Opin Drug Deliv. 2012;9:901–8.
40. Ly A, Hoyt L, Crowell J, Kim YI. Folate and DNA methylation. Antioxid Redox Signal. 2012;17:302–26.
41. Gao W, Xiang B, Meng TT, Liu F, Qi XR. Chemotherapeutic drug delivery to cancer cells using a combination of folate targeting and tumor microenvironment-sensitive polypeptides. Biomaterials. 2013;34:4137–49.

42. Shi J, Zhang H, Wang L, Li L, Wang H, Wang Z, et al. PEI-derivatized fullerene drug delivery using folate as a homing device targeting to tumor. Biomaterials. 2013;34:251–61.
43. Peng H, Bao L, Chunlei Z, Lin J, Luo T, Yang D, He M, Zhiming L, Gao G, Gao B, Shen F, Daxiang C. Folic acid-conjugated Silica-modified gold nanorods for X-ray/CT imaging-guided dual-mode radiation and photo-thermal therapy. Biomaterials. 2011;32:9796–809.
44. Li ZM, Huang P, He R, Lin J, Yang S, Zhang XJ, Ren QS, Cui DX. Aptamer-conjugated dendrimer-modified quantum dots for cancer cell targeting and imaging. Mat Lett. 2010;64:375–8.
45. Wang Z, Ruan J, Cui DX. Advances and prospect of nanotechnology in stem cells. Nanoscale Res Lett. 2009;4:593–605.
46. Song H, He R, Wang K, Ruan J, Bao CC, Li N, Ji JJ, Cui DX*. Anti-HIF-1 alpha antibody-conjugated pluronic triblock copolymers encapsulated with Paclitaxel for tumor targeting therapy. Biomaterials. 2010;31:2302–12.
47. Liang SJ, Li C, Zhao CL, Chen YS, Xu L, Bao CC, Wang XY, Liu G, Zhang FC, Cui DX. CD44v6 monoclonal antibody-conjugated gold nanostars for targeted photoacoustic imaging and plasmonic photothermal therapy of gastric cancer stem-like cells. Theranostics. 2015;5:879–81.
48. Vinogradov S, Wei X. Cancer stem cells and drug resistance: the potential of nanomedicine. Nanomedicine (Lond). 2012;7:597–615.
49. Zhang D, Fan D. New insights into the mechanisms of gastric cancer multidrug resistance and future perspectives. Future Oncol. 2010;6:527–37.
50. Clarke MF, Dick JE, Dirks PB, Eaves CJ, Jamieson CH, Jones DL, et al. Cancer stem cells-perspectives on current status and future directions: Aacr workshop on cancer stem cells. Cancer Res. 2006;66:9339–44.
51. Gilbertson RJ, Graham TA. Cancer: resolving the stem-cell debate. Nature. 2012;488:462–3.
52. Takaishi S, Okumura T, Tu S, Wang SS, Shibata W, Vigneshwaran R, et al. Identification of gastric cancer stem cells using the cell surface marker cd44. Stem Cells. 2009;27:1006–20.
53. Liu J, Ma L, Xu J, Liu C, Zhang J, Liu J, et al. Spheroid body-forming cells in the human gastric cancer cell line mkn-45 possess cancer stem cell properties. Int J Oncol. 2013;42: 453–9.
54. Li R, Wu X, Wei H, Tian S. Characterization of side population cells isolated from the gastric cancer cell line sgc-7901. Oncol Lett. 2013;5:877–83.
55. Xue Z, Yan H, Li J, Liang S, Cai X, Chen X, et al. Identification of cancer stem cells in vincristine preconditioned sgc7901 gastric cancer cell line. J Cell Biochem. 2012;113:302–12.
56. Brabletz T, Jung A, Spaderna S, Hlubek F, Kirchner T. Opinion: migrating cancer stem cells – an integrated concept of malignant tumour progression. Nat Rev Cancer. 2005;5:744–9.
57. Duan JJ, Qiu W, Xu SL, Wang B, Ye XZ, Ping YF, et al. Strategies for isolating and enriching cancer stem cells: well begun is half done. Stem Cells Dev. 2013;22:2221–39.
58. Marhaba R, Klingbeil P, Nuebel T, Nazarenko I, Buechler MW, Zoeller M. Cd44 and epcam: cancer-initiating cell markers. Curr Mol Med. 2008;8:784–804.
59. Prud'Homme GJ. Cancer stem cells and novel targets for antitumor strategies. Curr Pharm Des. 2012;18:2838–49.
60. Chen T, Yang K, Yu J, Meng W, Yuan D, Bi F, et al. Identification and expansion of cancer stem cells in tumor tissues and peripheral blood derived from gastric adenocarcinoma patients. Cell Res. 2012;22:248–58.
61. Zhang C, Li C, He F, Cai Y, Yang H. Identification of cd44+cd24+ gastric cancer stem cells. J Cancer Res Clin Oncol. 2011;137:1679–86.
62. Chen W, Zhang X, Chu C, Cheung WL, Ng L, Lam S, et al. Identification of cd44+ cancer stem cells in human gastric cancer. Hepatogastroenterology. 2013;60:949–54.
63. Misra S, Heldin P, Hascall VC, Karamanos NK, Skandalis SS, Markwald RR, et al. Hyaluronan-cd44 interactions as potential targets for cancer therapy. FEBS J. 2011;278: 1429–43.
64. Yoshida M, Yasuda T, Hiramitsu T, Ito H, Nakamura T. Induction of apoptosis by anti-cd44 antibody in human chondrosarcoma cell line sw1353. Biomed Res. 2008;29:47–52.

65. Jang BI, Li Y, Graham DY, Cen P. The role of cd44 in the pathogenesis, diagnosis, and therapy of gastric cancer. Gut Liver. 2011;5:397–405.
66. Heider KH, Kuthan H, Stehle G, Munzert G. Cd44v6: a target for antibody-based cancer therapy. Cancer Immunol Immunother. 2004;53:567–79.
67. Chen Y, Huang K, Li X, Lin X, Zhu Z, Wu Y. Generation of a stable anti-human cd44v6 scfv analysis of its cancer-targeting ability in vitro. Cancer Immunol Immunother. 2010;59:933–42.
68. Naor D, Sionov RV, Ish-Shalom D. Cd44: structure, function, and association with the malignant process. Adv Cancer Res. 1997;71:241–319.
69. Zhang CL, et al. Folic acid/ ce6 conjugated gold nanoclusters for NIR fluorescent imaging and photodynamic therapy with enhanced permission and retention. Adv Funct Mater. 2015;28:1314–25.
70. Huang P, Xu C, Lin J, Wang C, Wang X, Zhang C, Zhou X, Guo S, Cui DX. Folic acid-conjugated graphene oxide loaded with photosensitizers for targeting photodynamic therapy. Theranostics. 2011;1:240–50.
71. Chen F, Huang P, Zhu Y, Wu J, Zhang C, Cui DX. The photoluminescence, drug delivery and imaging properties of multifunctional Eu3t/Gd3t dual-doped hydroxyapatite nanorods. Biomaterials. 2011;32:9031–9.
72. Cui D, Jin G, Gao T, Sun T, Tian F, Estrada GG, Gao H. Characterization of BRCAA1 and its novel antigen epitope identification. Cancer Epidemiol. 2004;13:1136–45.
73. Code J, Tian FR, Hemandez Y, Bao CC, Baptisa P, Cui D, Stoeger T, et al. RNAi-based glyconanoparticles trigger apoptotic pathways for in vitro and in vivo enhanced cancer-cell killing. Nanoscale. 2015;7:9083–91.
74. Li C, Yang J, Wang C, Liang S, Zhang C, Chen F, Fu HL, Wang K, Cui D. BRCAA1 antibody- and Her2 antibody-conjugated amphiphilic polymerengineered CdSe/ZnS quantum dots for targeted imaging of gastric cancer. Nanoscale Res Lett. 2014;9:244.
75. Chen L, Zheng J, Zhang Y, Yang L, Wang J, Ni J, Cui D, Yu C, Cai ZL. Tumor-specific expression of MicroRNA-26a suppresses human hepatocellular carcinoma growth via cyclin-dependent and -independent pathways. Mol Ther. 2011;19:1521–8.
76. Fu HL, Ma Y, Lu LG, Hou P, Li BJ, Jin WL, Cui DX. TET1 exerts its tumor suppressor function by interacting with p53-EZH2 pathway in gastric cancer. J Biomed Nanotechnol. 2014;10:1217–30.
77. Khisamutdinov EF, Jasinski DL, Guo P. RNA as a boiling-resistant anionic polymer material to build robust structures with defined shape and stoichiometry. ACS Nano. 2014;8:4771–81.
78. Cui D, Zhang CL, Liu B, Shu Y, Du T, Li C, Pan F, Yang Y, Ni J, Li H, Brand-Saberi B, Guo PX. Regression of gastric cancer by systemic injection of RNA nanoparticles carrying both ligand and siRNA. Sci Rep. 2015;5:10732.
79. Wang X, Yang L, Chen ZG, Shin DM. Application of nanotechnology in cancer therapy and imaging. CA Cancer J Clin. 2008;58:97–110.
80. Huang P, et al. Light-triggered theranostic based on photosensitizer-conjugated carbon dots for simultaneous enhanced-fluorescence imaging and photodynamic therapy. Adv Mater. 2012;24:5104–10.
81. Kim C, Song HM, Cai X, Yao J, Wei A, Wang LV. In vivo photoacoustic mapping of lymphatic systems with Plasmon-resonant nanostars. J Mater Chem. 2011;21:2841–4.
82. Wang S, Huang P, Nie L, Xing R, Liu D, Wang Z, et al. Single continuous wave laser induced photodynamic/plasmonic photothermal therapy using photosensitizer-functionalized gold nanostars. Adv Mater. 2013;25:3055–61; Yuan H, Khoury CG, Hwang H, Wilson CM, Grant GA, Vo-Dinh T. Gold nanostars: surfactant-free synthesis, 3d modelling, and two-photon photoluminescence imaging. Nanotechnology. 2012;23:075102.
83. Chen R, Wang X, Yao X, Zheng X, Wang J, Jiang X. Near-ir-triggered photothermal/photodynamic dual-modality therapy system via chitosan hybrid nanospheres. Biomaterials. 2013;34:8314–22
84. Li C, et al. DC integrated inactive gastric cancer cell fused vaccine for targeted imaging and enhanced immunotherapeutic efficacy of gastric cancer. Biomaterials. 2015;35:177–87.

85. Choi J, Yang J, Bang D, Park J, Suh JS, Huh YM, et al. Targetable gold nanorods for epithelial cancer therapy guided by near-ir absorption imaging. Small. 2012;8:746–53.
86. Yuan H, Khoury CG, Wilson CM, Grant GA, Bennett AJ, Vo-Dinh T. In vivo particle tracking and photothermal ablation using plasmon-resonant gold nanostars. Nanomedicine. 2012;8:1355–63.
87. Van de Broek B, Devoogdt N, D'Hollander A, Gijs HL, Jans K, Lagae L, et al. Specific cell targeting with nanobody conjugated branched gold nanoparticles for photothermal therapy. ACS Nano. 2011;5:4319–28.
88. Park J, Ku M, Kim E, Park Y, Hong Y, Haam S, et al. Cd44-specific supramolecular hydrogels for fluorescence molecular imaging of stem-like gastric cancer cells. Integr Biol. 2013;5:669–72.
89. Jin L, Hope KJ, Zhai Q, Smadja-Joffe F, Dick JE. Targeting of cd44 eradicates human acute myeloid leukemic stem cells. Nat Med. 2006;12:1167–74.
90. Burke AR, Singh RN, Carroll DL, Wood JC, D'Agostino Jr RB, Ajayan PM, et al. The resistance of breast cancer stem cells to conventional hyperthermia and their sensitivity to nanoparticle-mediated photothermal therapy. Biomaterials. 2012;33:2961–70.
91. Lu L, Yan GZ, Zhao K, Xu F. An implantable telemetry platform system with ASIC for in vivo monitoring of gastrointestinal physiological information. IEEE Sens J. 2015;12:3524–34.
92. Weissleder R. A clearer vision for in vivo imaging. Nat Biotechnol. 2001;19:316–7.
93. Pan B, Cui D, Xu P, Ozkan C, Feng G, Ozkan M, et al. Synthesis and characterization of polyamidoamine dendrimer-coated multi-walled carbon nanotubes and their application in gene delivery systems. Nanotechnology. 2009;20:125101.
94. Huang P, Xu C, Lin J, Wang C, Wang X, Zhang C, et al. Folic acid-conjugated graphene oxide loaded with photosensitizers for targeting photodynamic therapy. Theranostics. 2011;1: 240–50.
95. Li Z, Huang P, Zhang X, Lin J, Yang S, Liu B, et al. Rgd-conjugated dendrimer-modified gold nanorods for in vivo tumor targeting and photothermal therapy. Mol Pharm. 2010;7:94–104.
96. Nie L, Chen X. Structural and functional photoacoustic molecular tomography aided by emerging contrast agents. Chem Soc Rev. 2014;43:7132–70.
97. Li W, Sun X, Wang Y, Niu G, Chen X, Qian Z, et al. In vivo quantitative photoacoustic microscopy of gold nanostar kinetics in mouse organs. Biomed Optics Exp. 2014;5:2679–85.
98. Nie L, Huang P, Li W, Yan X, Jin A, Wang Z, et al. Early-stage imaging of nanocarrier-enhanced chemotherapy response in living subjects by scalable photoacoustic microscopy. ACS Nano. 2014;8:12141–50.
99. Chen YS, Frey W, Aglyamov S, Emelianov S. Environment-dependent generation of photoacoustic waves from plasmonic nanoparticles. Small. 2012;8:47–52.
100. Sykes EA, Chen J, Zheng G, Chan WC. Investigating the impact of nanoparticle size on active and passive tumor targeting efficiency. ACS Nano. 2014;8:5696–706.

# Chapter 12
# Upconversion Nanoparticles for Gastric Cancer Targeted Imaging and Therapy

**Yuming Yang and Daxiang Cui**

## 12.1 Introduction

Japan constructed the research association of gastric cancer at the early 1960s. One of the most important achievements for the research of gastric cancer is the proposal of early gastric cancer (EGC) concept in 1962. The proposal of early gastric cancer concept makes the clinical cure and a longer life possible. After half a century of research and diagnosis, the awareness on early gastric cancer is also gradually in depth. The diagnosis of early gastric cancer can be gradually refined, especially in recent years, the guiding ideology for early treatment of gastric cancer, the original pursuit of longer survival period has changed to more requirements in the case of minimally invasive surgery to ensure that patients have a good quality of life, so the treatment is also more personalized. At present, the diagnosis rate of early gastric cancer is still less than 10 %, while Japan is as high as 70–50 %. At present, the most important thing is the changing of diagnosis concept and the improvement of the diagnosis rate of the early gastric cancer.

The field of nanoscience has witnessed a rapid growth in the last decade. Recently, the attention of the nanoscientists has been focusing more and more on biological applications. The development of upconversion nanophosphors over the past decade has facilitated the translation of fluorescence imaging from the microscopic to macroscopic imaging. Although the uses of other conventional fluorescent materials, including organic dyes, fluorescent proteins, metal complexes, or semiconductor quantum dot as based biomarker, have achieved

Y. Yang • D. Cui (✉)
Institute of Nano Biomedicine and Engineering, Shanghai Engineering Research Center for Intelligent Diagnosis and Treatment Instrument, National Center for Translational Medicine, Collaborative Innovational Center for System Biology, Shanghai Jiao Tong University, 800 Dongchuan Road, Shanghai 200240, Peoples Republic China
e-mail: yumingyang@sjtu.edu.cn; dxcui@sjtu.edu.cn

© Springer Science+Business Media B.V. and Shanghai Jiao Tong University Press, Shanghai 2017

D. Cui (ed.), *Gastric Cancer Prewarning and Early Diagnosis System*, Translational Medicine Research, DOI 10.1007/978-94-024-0951-2_12

significant progress in real-time detection and bioimaging, they still have some drawbacks. These fluorescent materials are generally excited by ultraviolet (UV) or visible light, which may induce autofluorescence from the biological tissues and DNA damage and cell death to biological samples, resulting in low signal-to-noise ratio and limited sensitivity. In addition, the broad emission spectra of these fluorescent materials make them unsuitable for multiplex biolabeling and often suffer from low photostability when exposed to external illumination. Quantum dots (QDs) that feature a large molar extinction coefficient, high quantum yield, narrow emission bandwidth, large Stokes' shift, size-dependent tunable emission, and high photostability are attractive as alternative luminescent labels for imaging. However, the potential toxicity of QDs limits their biological applications. In contrast, UCNPs have many good characteristics. The main difference between UCNPs and other luminescent imaging materials is that they can emit visible or near-infrared (NIR) light under NIR irradiation. The NIR irradiation leads to the absence of photodamage to living organisms, very low autofluorescence, high detection sensitivity, and high light penetration depth in biological tissues. In addition, these UCNPs show a sharp emission bandwidth, high photostability, tunable emission, long lifetime, and low cytotoxicity. All these benefits make the UCNPs as probes for intravital imaging which enables the investigation of physiological processes within the context of a living organism and provides a more complete picture of disease pathology and development.

To date, much effort has been put into the design of novel upconversion nanophosphors. Surface-functionalized and multifunctional UCNPs have been synthesized. Great advances on the improvement of probe targeting, pharmacokinetics, biocompatibility, and photophysics and the maturation of multimodal techniques are also made. In this chapter, we will introduce the synthesis, the surface modification method, and the toxicity of the UCNPs. The progress of UCNP applications on biological imaging, cancer, especially gastric cancer, in recent years will also be introduced in detail.

## 12.2 Principles of RE-Based Upconversion Luminescence (UCL)

Upconversion refers to nonlinear optical processes that convert two or more low-energy pump photons to a higher-energy output photon. The upconversion process is basically based on three UCL mechanisms: excited-state absorption (ESA), photon avalanche (PA), and energy transfer upconversion (ETU). ESA involves multi-step excitation by sequentially absorbing one or more photons from the ground state

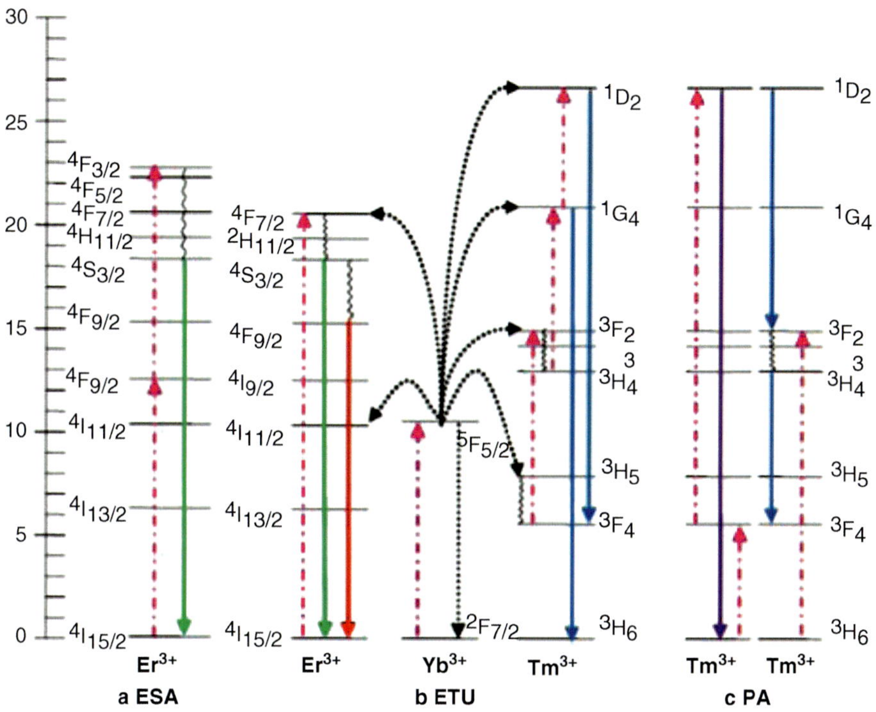


**Fig. 12.1** Schematic illustrations of various upconversion processes: (**a**) excited-state absorption (*ESA*); (**b**) energy transfer upconversion (*ETU*); (**c**) photon avalanche (*PA*)

to intermediate reservoir stage and finally populates at excited state to form the upconversion luminescence, which is illustrated in Fig. 12.1a for the $Er^{3+}$ ion. The PA process basically involves resonant excited-state absorption, efficient cross relaxation, and substantial population of the reservoir level and finally leads to strong upconversion emission. It is an unconventional pumping mechanism because it may lead to strong upconverted emission from level $^1D_2$ without any resonant ground-state absorption (Fig. 12.1c). In the process of ETU, there is resonant non-radiative energy transfer or phonon-assisted nonradiative energy transfer from sensitizer to activator. As shown in Fig. 12.1b, the green emission at 550 nm is generated as a result of two successive resonant energy transfers from $Yb^{3+}$ to $Er^{3+}$ ions, followed by nonradiative decay to the green-emitting level of $^4S_{3/2}$. In the $Yb^{3+}$,$Tm^{3+}$-codoped system, the blue emission from the $^1G_4$ level is accompanied by red emission, which arises from an intermediate transition terminating at the $^3F_4$ level. The ETU is by far the most efficient upconversion process and is suitable for the RE-doped nanophosphors.

## 12.3 Composition of RE-Based Upconversion Nanophosphors

An inorganic UCNP consists of an inorganic host and dopant (activator). The dopant acts as luminescent centers, and the host provides a matrix to bring these centers into optimal position. One of the most substantial ways in which the upconversion nanophosphors differ from ordinary Stokes-type photoluminescent phosphors is that upconversion nanophosphors tend to be more sensitive to vibronic coupling between the host and activator ion.

A large number of suitable hosts doped with actinide [1, 2] and transition metal ions have been reported to show upconversion luminescence, such as $Cm^{3+}$, $U^{3+}$, $Mo^{3+}$, $Os^{4+}$, $Ni^{2+}$, $Ti^{2+}$, and $Re^{4+}$ [3–5], but mainly in the RE elements due to its special $4f^n5d^{0-1}$ inner shell configurations that are well shielded by outer shells and have abundant and unique energy level structures. $Er^{3+}$, $Tm^{3+}$, and $Ho^{3+}$ are currently the most common activators in upconversion phosphors. It is also very critical to choose an appropriate host material, which plays vital roles in the upconversion luminescent process. The choice of the host lattice determines the distance between the dopant ions, the coordination numbers, the relative spatial position, and the type of anions around the dopant, resulting in different optical properties of the UCNPs. Desirable host materials should have adequate transparency within a certain wavelength range, low phonon energy, and high optical damage threshold. In addition, lattice impurities may increase the multiphonon relaxation rates between the metastable states, thereby reducing the overall visible emission intensity. Up to now, various host materials have been studied, including fluoride, oxide, chloride, bromide, iodide, oxysulfide, phosphate, vanadate, etc. The $NaYF_4$ and $LaF_3$ bulky state has been reported to be the most efficient matrix for upconversion nanophosphors [6–9]. Other kinds of fluorides and alkali fluorides, such as $NaGdF_4$, $NaLaF_4$, $LaF_3$, $GdF_3$, $CeO_2$, $LiNaF_4$,$Ca_3$ $(PO_4)_2$,$ZrO_2$, GdOF, etc., have also been considered as excellent host materials in recent years due to their high refractive index and high transparency arising from low-energy phonons [10–15], as low lattice phonon energy could minimize nonradiative losses and increase the luminescence quantum yield. While most of chlorides, bromides, and iodides are sensitive to moisture and thus are not suitable for bioimaging. In addition, there is another kind of impurity substance, such as $Yb^{3+}$, which is usually codoped in the inorganic host as a sensitizer due to the larger absorption cross section of $Yb^{3+}$ in the NIR spectral region to improve UCL efficiency. In the case of $Yb^{3+}$,$Er^{3+}$-codoped system, the energy interval of the ground state of $^2F_{7/2}$ and the $^2F_{5/2}$ excited state of $Yb^{3+}$ matches well with the transition energy between the $^4I_{11/2}$ and $^4I_{15/2}$ states and also the $^4F_{7/2}$ and $^4I_{11/2}$ states of $Er^{3+}$, thus allowing for efficient resonant energy transfer from $Yb^{3+}$ to $Er^{3+}$ ions. $Yb^{3+}$ is not only a common sensitizer for $Er^{3+}$ systems but also for $Tm^{3+}$, $Ho^{3+}$, and $Pr^{3+}$ ions. In order to minimize the cross relaxation energy loss, the content of activators is usually less than 2 mol%.

## 12.4 Synthesis of Rare Earth-Doped UCNPs

Due to the excellent properties and wide application prospect of UCNPs, the development of rare earth has attracted wide attention. At present, different methods have been developed to get upconversion nanophosphors, such as sol–gel [1, 2, 13, 16–21], hydrothermal [6, 22–26], coprecipitation [9, 27–29], thermal decomposition methods [30–43], etc. Nanoparticles synthesized by sol–gel method are usually not suitable for biological applications due to the particle agglomeration in the process of high-temperature treatment. In contrast, hydrothermal method is a good method for the preparation of UCNPs with high crystallinity, homogeneous size, and good dispersion, but the disadvantage is that the process of crystal growth cannot be observed. The thermal decomposition method and coprecipitation method are developed in recent years, which can synthesize UCNPs with small and homogeneous size. At present, many papers have been reported based on both of the two methods, such as Daxiang Cui team of Shanghai Jiaotong University that synthesized $NaGdF_4$:20%Yb, 2%Er based on the thermal decomposition method. The synthesized nanoparticles have good water solubility, and the CT imaging and near-infrared fluorescence imaging are successfully achieved [44] (Fig. 12.2).

In order to obtain the UCNPs with good biocompatibility, "one-step" method was developed in recent years with the help of hydrophilic or binary cooperative ligands, such as polyols [8, 22, 45], EDTA [46–49], citrate [46, 50–54], sodium dodecyl sulfate (SDS) [55], PVP [56], small-molecule binary acid [57, 58], polyethylene glycol (PEG) [58], PVP [59], PAA [60, 61], PEI [60], 3-mercaptopropionic acid (3MA), 6-aminocaproic acid (6AA) [62], etc. This "one-step" method has the advantages of high efficiency and could be operated at mild reaction conditions [63, 64]. Daxiang Cui group reported the UCNPs with different morphologies in the presence of n-octanol and 1-butyl-3-methylimidazolium tetrafluoroborate [65]. The results show that the UCNPs synthesized by this method have good water solubility; 1-butyl-3-methylimidazolium tetrafluoroborate plays a very important role in controlling the size and morphology of the UCNPs (Fig. 12.3).

In addition, Dr. Cui team also synthesized $NaLuF_4$ in the presence of four different surface active agents using the improved one-step method [66]. The TEM images of the upconversion nanorod and upconversion nanoparticles are shown in Figs.12.4 and 12.5.

Cui Daxiang team synthesized different morphologies of UCNPs in the presence of DDBAC, DTAB, and DDAB cationic surfactants. Compared with the other two kinds of surfactants, DDBAC cationic surfactants play very important role in the process of synthesizing $LuF_3:Yb^{3+},Er^{3+}$ and $Na_{1.5}Y_{2.5}F_9:Yb^{3+},Er^{3+}$. In addition, compared with CTAB, the UCNPs synthesized with DDBAC have very good water solubility, which is more conducive to the future in vivo applications [67].

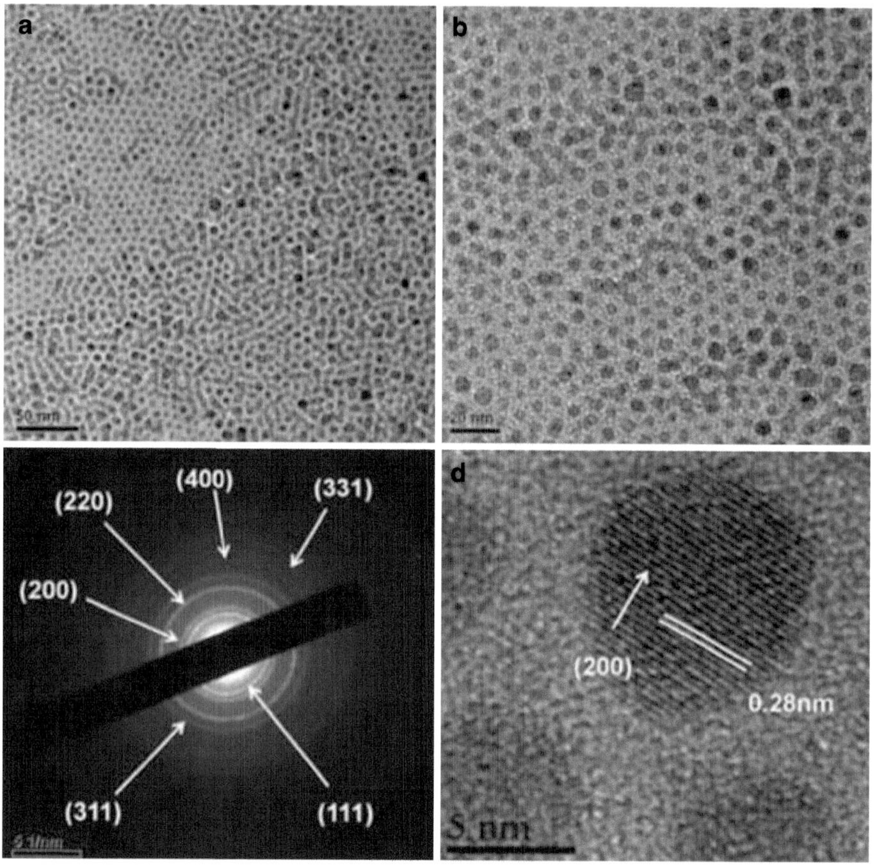


**Fig. 12.2** (**a**, **b**) TEM images of $NaGdF_4$:20%Yb, 2%Er; (**c**) SAED of UCNPs; (**d**) HR-TEM of $NaGdF_4$:20%Yb, 2%Er

## 12.5 Surface Modification of UCNPs

For UCNPs used for bioimaging, surface modification is required to obtain biocompatible nanoparticles with excellent water dispersity, physiologic stability, and functional anchors for further bioconjugation. The methods contain ligand exchange, ligand oxidation [68, 69], ligand attraction [70–73], layer by layer assembly [74–78] and silanization [79–82] etc. Daxiang Cui group has made a detailed summary on this area [83].

### *12.5.1 Ligand Exchange Method*

For the ligand exchange method, the original hydrophobic ligands are displaced by other polymeric molecules, which could provide a hydrophilic surface and anchors for further bioconjugation. Most of the hydrophobic UCNPs are coated

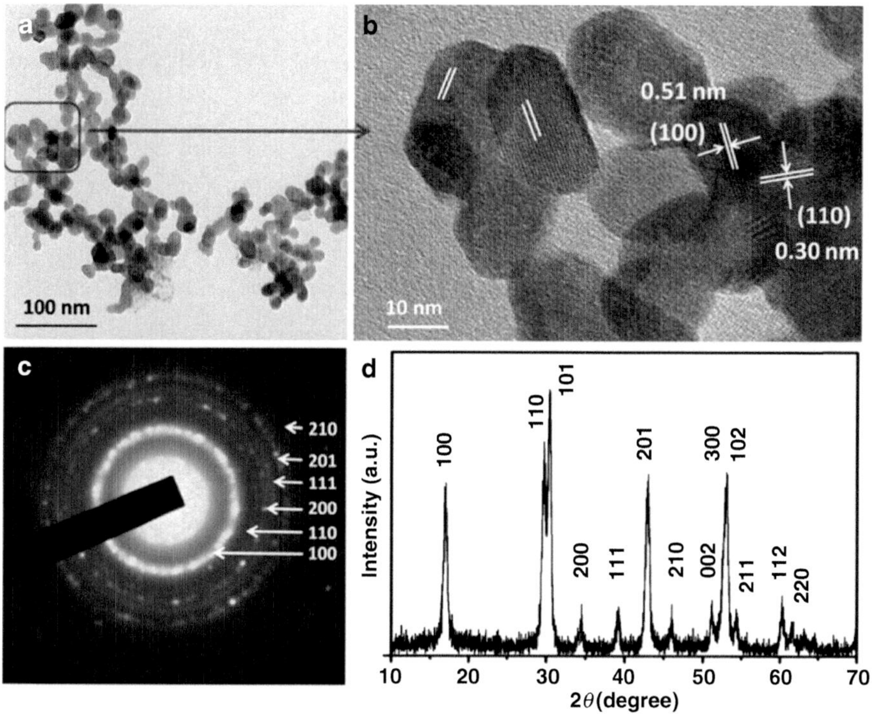


**Fig. 12.3** TEM image of $NaGdF_4$:Yb,Er (n-octanol (10 mL)/Bmim$PF_6$ (15 mL)). (**b**) Magnification of selected area. (**c**) The corresponding SAED pattern. (**d**) The corresponding XRD pattern

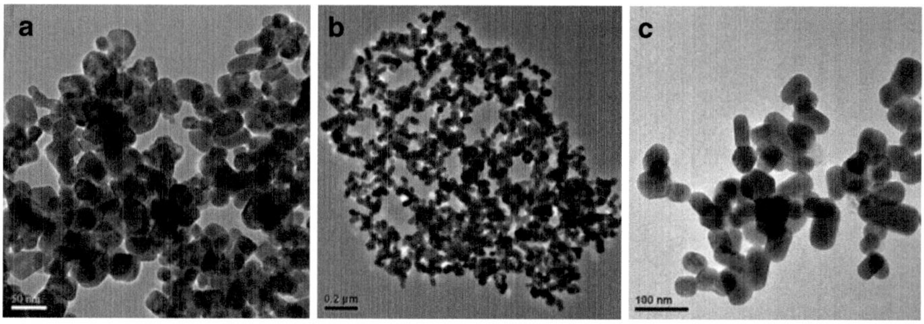


**Fig. 12.4** TEM images of UCNPs synthesized with different surface active agents: (**a**) ILs-UCNPs, (**b, c**) Cit-UCNPs

by OA and oleylamine (OM) surfactants using -COOH and $-NH_2$ as chelating ligands, respectively. The -COOH ligands have stronger coordination ability to the lanthanide ions than the $-NH_2$ ligands. For the UCNPs using -COOH as the chelating ligands, an excess of chelating ligands is required to displace the original -COOH ligands due to its strong interactions with lanthanide ions. A typical example based on the ligand exchange method is reported by Capobianco group in 2009. PAA was chosen as the surface modification ligands. A complete ligand

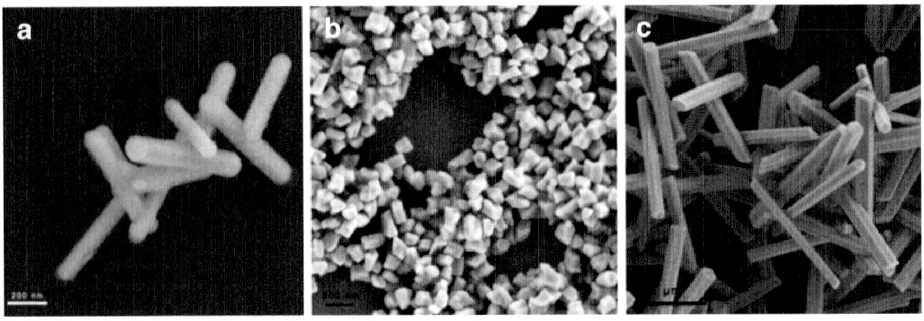


**Fig. 12.5** TEM of UCNPs in the presence of different surface active agents: (**a**) SDS-UCNPs, (**b**) DDBAC-UCNPs, and (**c**) PEG-UCNPs

exchange with PAA was realized and resulted in moderate UCL intensity as well as physical stability of the dispersed nanoparticles in solution for up to 3 months [37]. Van Veggel group also reported a technique for the replacement of oleate with PEG–phosphate ligand as an efficient method for the generation of water-dispersible $NaYF_4$ nanoparticles. Many other kinds of molecules were also chosen as surfactants to displace the OA or OM in the ligand exchange process, such as hexanedioic acid [84, 85], dimercaptosuccinic acid (DMSA) [86], mercaptosuccinic acid (MSA) [87], citrate [7, 88], poly(amidoamine) (PAMAM) [89], PEG diacid (MW 600) [90], PEI [91] and thioglycolic acid (TGA) [92, 93], maleimide–PEG–COOH [94], nitrosonium tetrafluoroborate (NOBF4) [95], etc. Among these surfactants, the PEGylation could shield the UCNPs from quick immune-mediated removal when the UCNPs are to be used intravenously. However, the luminescence intensity of the UCNPs in aqueous environments was found to be severely quenched when compared to the original NPs in organic solvents. This is attributed to an increase in the multiphonon relaxations of the lanthanide excited state in aqueous environments due to high-energy vibrational modes of water molecules.

### *12.5.2 Ligand Oxidation Method*

Ligand oxidation is another method used for the hydrophobic UCNP modifications. This method has no obvious effects on the particle morphology or luminescence properties of the UCNPs, but is only applicable to a few specific kinds of ligands. A specific example is the carbon–carbon double bond of the OA oxidized by Lemieux–von Rudloff reagent [68] or ozone [96] to generate water-soluble carboxylic ligands. The carbon–carbon double bond of the OA was also reported undergoing epoxidation in the first step and further coupling with PEG monomethyl ether (mPEG-OH) to realize the hydrophilic modification [69].

### 12.5.3 Self-Assembly Method

Self-assembly method is based on the electrostatic attraction between the oppositely charged species deposited or between some special host and guest molecules. The major advantage is that it permits the preparation of coated colloids of different shapes and sizes, with uniform layers of diverse composition as well as controllable thickness. Li's group firstly applied this method on the synthesis of hydrophilic UCNPs [77]. The zeta potential of the UCNPs alternated from negative to positive values, along with the alternative absorption of poly(allylamine hydrochloride) PAH and poly(sodium 4-styrenesulfonate) (PSS), resulted in the successful modification of $NaYF_4$:Yb,Er/Tm nanoparticles with rich amino ligands on their surfaces. There are also methods that use excess organic solvents under ultrasonic condition or by changing the pH value of the solution [97, 98] to make ligand-free UCNPs; then hydrophilic and biocompatible molecules could be conjugated on the surface through electrostatic attraction process.

### 12.5.4 Ligand Attraction Method

Ligand attraction approach involves the absorption of an amphiphilic copolymer onto the surface of the UCNPs through the hydrophobic–hydrophobic attraction between the original ligands and the hydrophobic ligands of the surfactant. One typical example is the coating of 25 % octylamine and 40 % isopropylamine modified PAA on the surface of $NaYF_4$:Yb,Er(Tm)/$NaYF_4$ core/shell nanoparticles [99]. The coating of PAA was from the hydrophobic interactions between the octyl and isopropyl groups of PAA and the moctadecyl groups of OM on the UCNPs surface. After coating, the hydrophilic carboxyl groups of PAA extended outward, making the nanoparticles water soluble and allowing further attachment of biomolecules. PEG-block-poly(caprolactone) (PEG-b-PCL), PEG-block-poly(lactic-coglycolic acid) (PEG-b-PLGA), (PEG-block-lactic acid) (PEG-b-PLA) [100, 101], octylamine–PAA–PEG (OA–PAA–PEG) [102], PEG [103, 104], hexadecyltrimethylammonium bromide (CTAB) [105], octylamine-modified PAA (OPA) [88], PEG–phospholipids [106], etc. were also used as the hydrophilic polymer coating on the surface of hydrophobic UCNPs to improve the UCNP water solubility.

### 12.5.5 Surface Silanization Method

Surface silanization is one of the most important methods for the UCNP surface modification. An amorphous silica shell is coating on the surface of the UCNPs by hydrolysis and condensation of siloxane monomers. This method is applicable for coating the silica on both hydrophilic and hydrophobic metals, metal oxides, and QD nanoparticles with controllable thickness of the shell by Stöber method or reverse microemulsion method. In order to make the UCNPs water soluble and able to be conjugated by biological molecules, modified silanes (aminosilanes), such as aminopropyltrimethoxysilane (APTES) and (3-aminopropyl) triethoxysilane (APS), are directly (one step) or indirectly (two steps) reacted with the UCNPs to generate functional amino groups on the surface of the UCNPs [107, 108].

## 12.6 UCNPs Used for Gastric Cancer Imaging

### *12.6.1 UCNPs for Single Model Passive Imaging and Targeted Imaging*

As a new imaging technique, UCNP imaging offers a unique approach for visualizing morphological details in the tissue with subcellular resolution and has come to be a powerful noninvasive tool for visualizing the full range of bio-species from living cells to animals. Exploitation in general internalization studies was performed using bare or modified UCNPs. They have been widely used for in vitro or in vivo imaging to demonstrate their promise in biological in vivo applications [109–112]. Kobayashi et al. demonstrated in vivo multiple color lymphatic imaging using upconverting nanocrystals in 2009 [105]. Multicolor in vivo lymph node mapping UCL imaging was further demonstrated by Liu's group in the following year and found the in vivo detection limit of UCNPs to be at least one order of magnitude lower than that of QDs [102]. Li's group reported a series of articles related to the UCNPs for lymphatic imaging [86, 113, 114]. Niagara et al. applied silica/$NaYF_4$:Yb,Er to dynamically track live myoblast cells in vitro and in a living mouse model of cryo-injured hind limb [115]. In vivo confocal imaging of nanoparticle-loaded cells intravenously injected into a mouse tail vein showed them flowing in the ear blood vessels. Nanoparticle-loaded cells were also unambiguously identified with superior contrast against a negligible background at least 1,300 μm deep in a fully vascularized living tissue upon intramuscular injection.

Tumor-targeted molecular imaging plays a very important role in the diagnosis and prognosis of gastric cancer. It has absorbed a wide range of attention in the use of UCNPs for tumor imaging in vivo. Different kinds of antibodies, aptamers, small molecules, and peptides were conjugated on the surface of the UCNPs for targeted imaging of specific kinds of cells or biological molecules. Zako originally reported the tumor cell-targeted upconversion imaging using UCNPs modified with cyclic RGD peptide (RGD-$Y_2O_3$) [107]. FA is also widely chosen as targeting ligands for specific targeting and imaging of cancer cells. Cui group reported the $NaYbF_4$: 25%Gd, 2%Tm/$SiO_2$ modified by FA, which could be applied on the gastric cancer imaging in vivo [116] (Fig. 12.6).

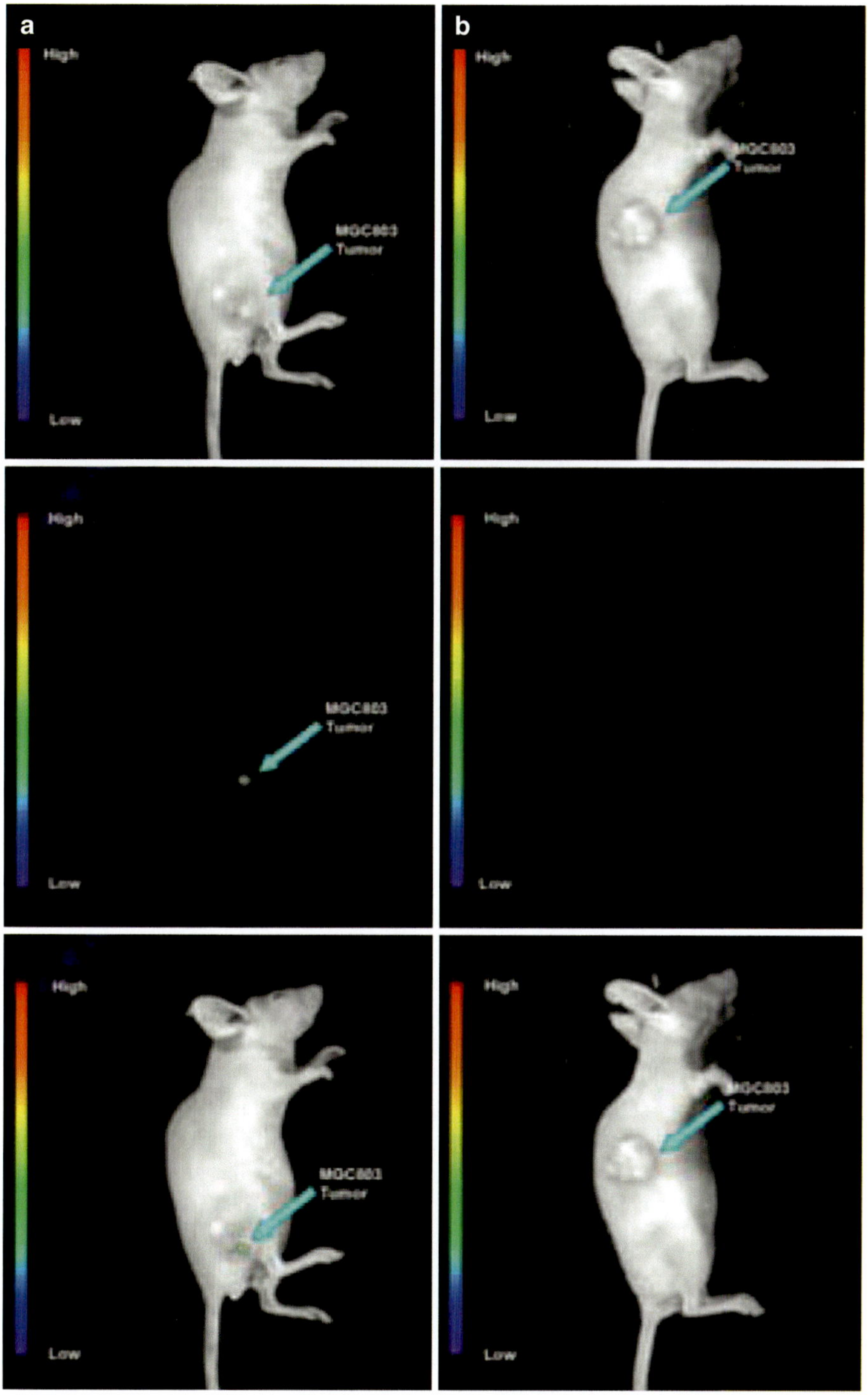


**Fig. 12.6** Targeted in vivo NIR luminescence imaging of subcutaneous MGC803 tumor (indicated by *short arrows*) after intravenous injection of $NaYbF_4$:25%Gd, 2%Tm-FA (**a**) and $NaYbF_4$:25%Gd, 2%Tm@$SiO_2$-$NH_2$. (**b**) Nanoparticles for 4 h

## 12.6.2 UCNPs for Multimodal Bioimaging

In addition to the optical imaging technique, molecular imaging technique also contains positron emission tomography (PET), magnetic resonance imaging (MRI), single-photon emission computed tomography (SPECT), and X-ray computer tomography (CT). These techniques have already become routine clinical examination techniques in hospitals. However, each imaging modality has its own merits and disadvantages, and a single technique does not possess all the required capabilities for comprehensive imaging. Therefore multimodal UCNPs for bioimaging are quickly becoming important tools for biomedical research and clinical diagnostics.

### 12.6.2.1 MRI/UCL and CT/UCL Dual-Modal Imaging

Optical imaging as an inexpensive, robust, and portable method provides the highest sensitivity and spatial resolution for in vitro imaging, but still lacks the full capability to obtain anatomical and physiological details in vivo. MRI and CT techniques have excellent spatial resolution, good depth for in vivo imaging, and exceptional anatomic information, but suffer from limited sensitivity and lack resolution for imaging at the cellular level. Combination of MRI, CT, and optical imaging can lead to the development of new approaches to bridge gaps in the resolution and depth of imaging between these modalities. Potential benefits of combined UCNPs, MRI, and CT have stimulated a development of hybrid nanomaterials for in vitro and in vivo imaging [104, 117–135]. The most commonly used MRI contrast agent is $Gd^{3+}$ chelate complex. The contrast of MRI imaging is improved by changing the exchange rate of the internal and external layers of the water to shorten the longitudinal relaxation time $T_1$. Prasad group is one of the pioneers who originally developed $Gd^{3+}$- and $Er^{3+}/Yb^{3+}/Eu^{3+}$-codoped $NaYF_4$ for the dual modality of optical and MR imaging [117]. In 2011, they further synthesized core/shell $NaYbF_4$:$Tm^{3+}$/$NaGdF_4$ nanocrystals to be used as probes for bimodal NIR-to-NIR UCL and MR imaging. Owing to the large magnetic moment, superparamagnetic $Fe_3O_4$ nanoparticles have been combined with RE-UCNPs together for fabricating magnetic operation, $T_2$-enhanced MR imaging, and UCL imaging [136, 137]. For example, Li's group reported core–shell $NaYF_4$:Yb,Er/Tm@$SiO_2$@$Fe_3O_4$ nanoparticles with very good superparamagnetic and luminescent properties [46]. In particular, Liu's group developed UCL/down-conversion fluorescence/MR imaging nanocomposite for in vivo cancer cell imaging. A chemotherapy drug of DOX is also loaded into the nanocomposite, which enables novel imaging-guided and magnetic-targeted drug delivery [138].

For UCNPs used as CT contrast agents, there are two approaches to improve the X-ray attenuation coefficient. The first method is to increase the lanthanide elements content in a single particle. Liu et al. prepared the PEGylated $Yb_2O_3$:Er nanoparticles with high Yb content in a single particle suitable for both X-ray CT

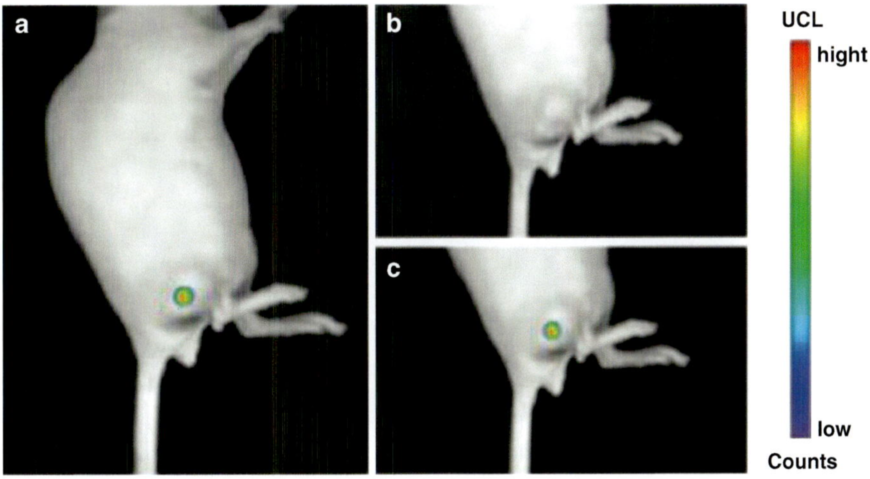


**Fig. 12.7** Real-time in vivo UCL images after intravenous injection of UCNPs@$SiO_2$–FA in nude mice at 2 h

imaging and UCL imaging [139]. The second way is to choose higher atomic number elements among RE elements for the preparation of UCNPs. For the $REF_3$ and $NaREF_4$ fluorides, the La content in $LaF_3$ is 70.9 %, much higher than other RE elements content in $NaREF_4$. Therefore, $REF_3$-based UCNPs can serve as excellent CT contrast agents and ideal building blocks for multimodal imaging agents. Based on this mechanism, FA-conjugated silica-modified $LaF_3$:Yb,Tm UCNPs (UCNPs@$SiO_2$-FA) with high La content in a single particle were strategically designed by Cui group for simultaneously targeted dual-modality imaging of UCL and CT [140]. Figure 12.7 shows the real-time in vivo UCL images after intravenous injection of UCNPs@$SiO_2$-FA in nude mice at 2 h.

$NaGdF_4$ is also a good contrast agent for UCL/CT dual mode imaging [141–145]. Cui team synthesized rare earth ion-doped $NaGdF_4$ micron crystal using one-step synthesis method. This material displays good conversion luminescence, magnetism, and biocompatibility and was successfully used for gastric cancer CT imaging [146], as shown in Fig. 12.8.

#### 12.6.2.2 CT/MRI/UCL, PET/MRI/UCL, and PET/CT/UCL Trimodal Imaging

Compared with CT scan, an MRI is suited for examining the soft tissue in ligament and tendon injury, spinal cord injury, brain tumors, etc., while a CT scan is better suited for bone injuries, lung and chest imaging, and detecting cancers. A combination of UCL, CT, and MRI imaging is no doubt a good way to realize better tissue scans. Up to now, a series of multimodal CT/MRI/UCL imaging nanocomposites have been developed with high $T_1$ enhancement, bright UCL emissions, and

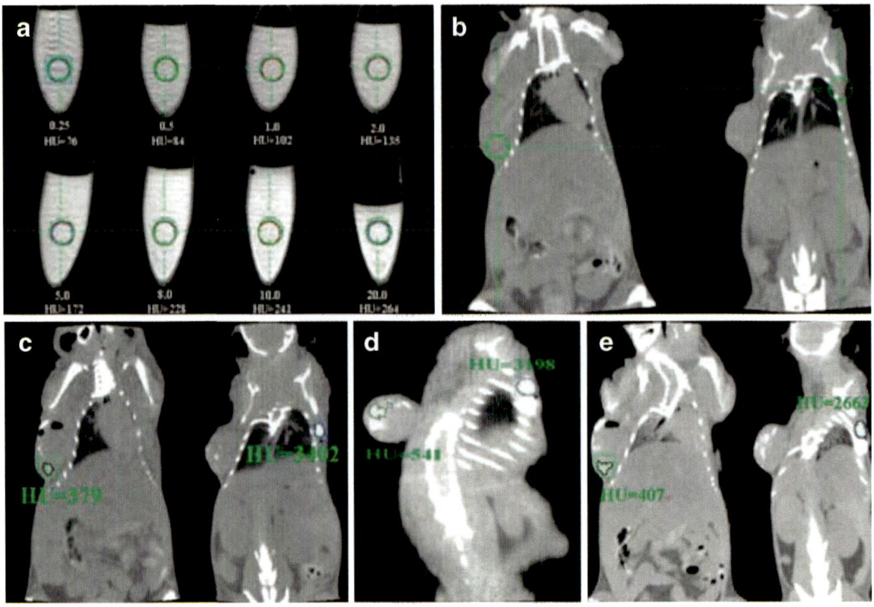


**Fig. 12.8** In vitro CT imaging of $NaGdF_4:Yb^{3+}/Er^{3+}$ submicrocrystals suspended in PBS solution (**a**), in vivo CT imaging of mouse before injection (**b**), after injection (**c**), after 2-h injection, (**d**) and after 4-h injection (**e**). As for the investigated nude mice, tumor sites are located in the left side of the back, and normal tissues are located in the right side of the back. Nude mice loaded with gastric cancer MGC803 cells were selected as the animal model. The tumor was implanted and grown in nude mice to an appropriate size (3–5 mm in diameter)

excellent X-ray absorption coefficient [147–150]. Take the core–shell $Fe_3O_4$@$NaLuF_4$:Yb,Er/Tm nanocomposite which has been developed by Li's group as a typical example; it exhibits superparamagnetic property and $T_2$-enhanced magnetic resonance effect resulting from the $Fe_3O_4$ cores and excellent X-ray attenuation and UCL under excitation at 980 nm. In vivo MR, CT, and UCL images of tumor-bearing mice show that the $Fe_3O_4$@$NaLuF_4$:Yb,Er/Tm nanoparticles can be successfully used in multimodal imaging [151]. Most recently, Liu et al. reported a multifunctional nanoprobe based on PEGylated $Gd_2O_3:Yb^{3+}$, $Er^{3+}$ nanorods for in vivo UCL, $T_1$-enhanced MR, and CT multimodality imaging. The capability of PEG-UCNPs as high-performance contrast agents for UCL/MR/CT imaging is evaluated successfully through small-animal experiments. Additionally, pharmacokinetics, biodistribution, and clearance route are studied after intravenous injection in a mouse model, reflecting their overall safety use for in vivo imaging [152].

PET is a nuclear medical imaging technique that produces a three-dimensional image or picture of functional processes in the body. PET scans are increasingly read alongside CT or MRI scans to provide excellent spatial resolution and high sensitivity, as CT provides exceptional anatomic information, but suffers from limited sensitivity. PET provides a visualization method with high sensitivity, but a low (~mm) spatial resolution. However, CT, PET, and MRI are all unsuitable for visualizing living cells because of low planar resolution, but this can be remedied by combining with UCNPs, which provide the highest spatial resolution and are suitable for imaging living cells.

Li's group has developed a series of nanocomposites for the multimodal PET/MRI/UCL or PET/CT/UCL imaging using 18 F as the radionuclide [42, 153–155]. For example, a simple, rapid, efficient, and general synthesis strategy for $^{18}F$-labeled rare earth nanoparticles through a facile inorganic reaction between rare earth cations and fluoride ions were developed. The $^{18}F$-labeling process based on rare earth elements was chieved efficiently in water at room temperature with an $^{18}F$ labeling field of >90 % and completed within 5 min. The as repared $^{18}F$-labeled rare earth nanoparticles was further valuated by UCL and PET imaging of their in vivo distribution nd application in lymph monitoring [154]. In addition to the method of doping with $Gd^{3+}$ within host material, they also developed PET/MRI/UCL nanoparticles that $Gd^{3+}$ is distributed on the surface of the nanoparticle through cation exchange with $Y^{3+}$. 18 F is introduced on the UCNPs through interaction with the rare earth ions for PET imaging. The versatility of the surface modification approach for incorporating functional molecules and fabricating fluorine-18-labeled magnetic-upconversion nanophosphors as multimodal bioprobes has been demonstrated by targeted cell imaging, in vivo UCL, MR imaging, and PET imaging of whole-body small animals [155].

## 12.7 UCNPs for Biological Detection and Gastric Cancer Therapy

### *12.7.1 UCNPs for Biological Detection*

Apart from being used for cancer cell-related bioimaging, UCNPs were selected as probes for disease-related sensing or monitoring. For example, mesenchymal stem cells have shown great potential in regenerative medicine. Sensitive and reliable methods for stem cell labeling and in vivo tracking are thus of great importance. Liu's group reported the use of oligo-arginine-conjugated UCNPs as an exogenous contrast agent to track mouse mesenchymal stem cells in vivo. As few as ~10 cells labeled with UCNPs are detected in vivo, which highlight the promise of using UCNPs as a new type of ultrasensitive probes for labeling and in vivo tracking of stem cells at nearly the single-cell level [156]. In another case, Liu and coworkers developed $MnO_2$ nanosheet-modified UCNPs for rapid, selective detection of glutathione in aqueous solutions and living cells [157].

Recently, Huang, etc. reported for the first time the application of $LiLuF_4:Ln^{3+}$ core/shell UCNPs as sensitive UCL bioprobes for the detection of an important disease marker β subunit of human chorionic gonadotropin (β-hCG) with a detection limit of 3.8 $ng\,mL^{-1}$. In addition, they applied UCNPs on CT and UCL imaging of cancer cells, revealing the great potential of $LiLuF_4:Ln^{3+}$ UCNPs as efficient nano-bioprobes in disease diagnosis [134].

Tissue hypoxia may induce the resistance of radiotherapy or chemotherapy and tumor metastasis. Liu et al. report a new type of UCNP probe, which can detect the extent of tissue hypoxia in tissue and in vivo, predict the degree of tumor progression, and take effective treatment method [158].

ROS overproduction is closely related with the rheumatoid arthritis inflammation; Chen et al. adopted UCNPs modified with hyaluronic acid chromophore modified for the selective detection of reactive oxygen species and successfully realized the detection of inflammatory cells in vivo [159].

## 12.7.2 UCNPs for Gastric Cancer Targeted Drug Delivery

### 12.7.2.1 UCNPs as Drug Carrier

In recent years, much attention has been paid to developing new drug delivery systems with enhanced bioavailability, greater efficiency, lower toxicity, and controlled release advantages. In general, an efficient drug delivery system should not only deliver the therapeutic drugs to the target cells or tissues but also maintain the optimum concentration and rational toxicity of drugs in precise sites of the organs, which can improve therapeutic efficiency and reduce toxicity.

Up to now, most of the UCNP drug delivery systems were realized by coating PEG or mesoporous silica for the probe attachment or drug loading [160–163]. Liu and coworkers functionalize UCNPs with a PEG-grafted amphiphilic polymer. Then the PEGylated UCNPs are loaded with DOX molecules and conjugated with FA for targeted drug delivery and cell imaging. The loading and releasing of DOX from UCNPs are controlled by varying pH, with an increased drug dissociation rate in acidic environment, favorable for controlled drug release [160]. Cui and coworkers reported a multifunctional "nanorattle" hollow spheres that consist of RE-doped $NaYF_4$ shells with a $SiO_2$-coated $Fe_3O_4$ inner particle. The material emits visible luminescence upon NIR excitation and can be directed by an external magnetic field to a specific target, making it an attractive system for a variety of biological applications. In vivo experiments exhibit encouraging tumor shrinkage with the DOX loading and significantly enhanced tumor targeting in the presence of an applied magnetic field [164].

Lin group reported a series of mesoporous silica-modified UCNP system for drug delivery containing $Gd_2O_3$:$Er^{3+}$/$mSiO_2$ UCNPs [165]; UCNP (β-$NaYF_4$:$Yb^{3+}$, $Er^{3+}$@β-$NaGdF_4$:$Yb^{3+}$)@$mSiO_2$–PEG [166]; $NaYF_4$:$Yb^{3+}$,$Er^{3+}$@silica [167, 168]; $NaYF_4$:$Yb^{3+}$/$Er^{3+}$@$nSiO_2$@$mSiO_2$ [169]; $NaYF_4$:$Yb^{3+}$/$Er^{3+}$@$SiO_2$@P(NIPAM-co-MAA) hybrid microspheres [170]; $NaYF_4$:$Yb^{3+}$,$Er^{3+}$@$SiO_2$ [170]; $Fe_3O_4$@$nSiO_2$@$mSiO_2$@$NaYF_4$:$Yb^{3+}$,$Er^{3+}$/$Tm^{3+}$[171]; etc.

Although the $SiO_2$-modified UCNPs have made a great progress, the core/shell structure needs complicated procedure, and it is difficult to realize batch production. To solve this question, Cui group of Shanghai Jiao Tong University proposed a one-step method for the synthesis of mesoporous structure of UCNPs. Drugs for the

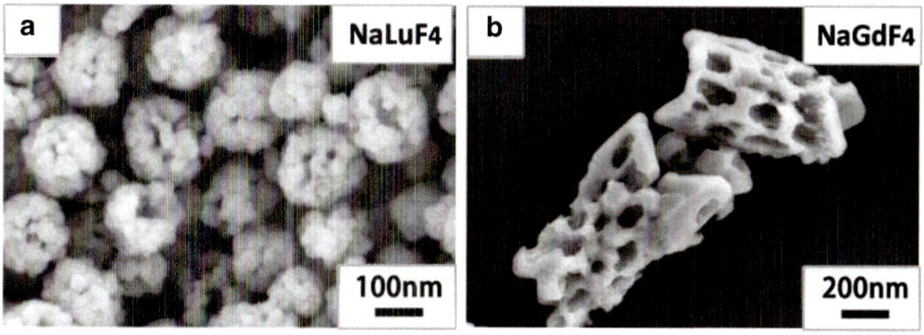


**Fig. 12.9** SEM images of $NaLuF_4$ (**a**) and $NaGdF_4$ (**b**)

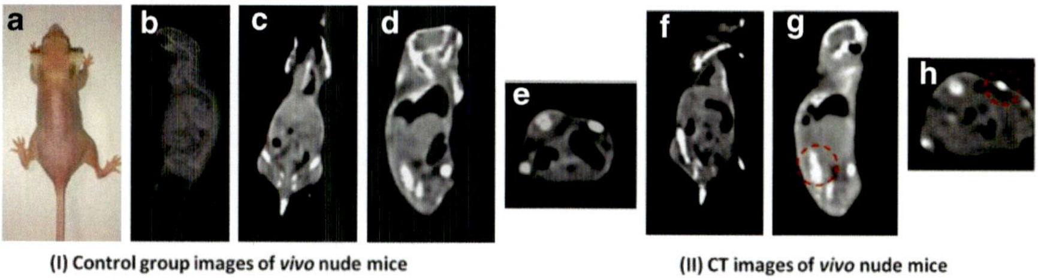


**Fig. 12.10** (**I**) Images of control group before injection: (**a**) the photograph of the nude mouse; (**b**) X-ray image, and (**c**–**e**) CT images of nude mouse model as a control. (**II**) CT images of coronal plane (**f**), sagittal plane (**g**), and transverse plane (**h**) through subcutaneous injection with $NaLuF_4$ upconversion nanoparticles

treatment of gastric cancer can be filled in the mesoporous of UCNPs and then delivered to the tumor site. SEM images of the synthesized mesoporous UCNPs are shown in Fig. 12.9.

In view of magneto-optical imaging features of $NaGdF_4$, the nanoparticles can be successfully applied to CT imaging (Fig. 12.10) and the delivery of antitumor drugs for the treatment of gastric cancer [172].

#### 12.7.2.2 UCNPs for siRNA Delivery

Small interfering RNA (siRNA) has emerged as a gene-based therapy due to their highly desirable roles in RNA interference and gene-silencing effects in biomedical research. The key requirements for an effective siRNA therapy are that sufficient siRNAs need to be introduced into cells or organs and remote control the release of siRNA inside target cells in a highly spatial and temporal precision. Up to now, a few types of UCNPs have been developed for the purpose of effective siRNA delivery. Zhang and coworkers reported a UCNP-based system for targeted delivery of siRNA to cancer cells. The siRNA was attached to anti-Her2 antibody-conjugated UCNPs, and the delivery of these nanoparticles to SK-BR-3 cells was studied, which certified the capability of using UCNPs as a fluorescent probe and delivery system for simultaneous imaging and delivery of biological molecules [173]. Very recently, Yang et al. reported a system of silica-coated UCNPs, which were functionalized with cationic photocaged linkers through covalent bonding. Anionic siRNAs could be effectively absorbed onto the linkers through electrostatic attractions and were easily internalized by living cells. Upon NIR light irradiation, the photocaged linker on the Si-UCNPs surface could be cleaved by the upconverted UV light and thus initiated the intracellular release of the siRNA [174].

### 12.7.3 UCNPs for Photodynamic Therapy (PDT)

PDT is a relatively new clinical therapeutic modality which involves killing of disease cells by excitation of photosensitizer chemicals with high-energy light to generate reactive oxygen species from surrounding dissolved oxygen [175–177]. However, poor tissue penetration of high-energy light and hydrophobic photosensitizers limits the effectiveness to superficial pathologies. UCNPs can be used to activate the photosensitizers in the deep tissue because NIR light can penetrate a few centimeters into the soft tissue due to weak absorption in the optical "transparent window." Four main classes of photosensitizers have been approved by the US Food and Drug Administration for clinical use against cancer cells, containing porphyrin derivatives, chlorins, phthalocyanines, and porphycenes [178, 179]. The second-generation photosensitizers of zinc(II)-phthalocyanine (ZnPc) have been proven highly selective for tumor targets and showed enhanced cytotoxic effects both in vitro and in vivo.

Since the first report on UCNP/merocyanine 540-based PDT drugs in 2007 [180], different methods have been developed for construction of the UCNPs for PDT, containing directly conjugation of photosensitizers through the carboxyl ligands [181] or by coating mesoporous silica [182–184], alpha-cyclodextrin [185], PEI, or PEG onto the surface of UCNPs for the attachment of photosensitizers [186]. Zhang's group reported the attachment of ZnPc photosensitizers to PEI-modified $NaYF_4$ UCNPs and used them as nano-transducers for PDT of viruses [187] or cancer cells [188, 189]. However, the amount of photosensitizers attached is low and the photosensitizers are not stably attached to the UCNPs. In order to solve this problem, Zhang and coworkers further developed mesoporous silica-coated UCNPs for the loading of photosensitizer [190, 191]. The photosensitizer encapsulated in mesoporous silica is protected from degradation in the harsh biological environment. In particular, they use multicolor emission capability of the UCNPs at a single excitation wavelength for simultaneous activation of two photosensitizers for enhanced PDT, which showed a greater PDT efficacy compared with the UCNPs with a single photosensitizer. Moreover, they originally applied photosensitizer-loaded UCNPs as an in vivo-targeted PDT agent. The in vivo studies showed tumor growth inhibition in PDT-treated mice by direct injection of

UCNP into melanoma tumors or intravenous injection of UCNPs conjugated with a tumor-targeting agent into tumor-bearing mice [192].

In order to improve the biocompatibility and make it easier for further chemical modification, an outer layer of PEG on the core of UCNPs is usually used which acts as solubilizing agent and allows the penetration of $O_2$ and diffusion of singlet oxygen [193]. Prud'homme and coworkers reported the preparation of a novel biocompatible poly(ethylene glycol-block-(DL)lactic acid) block copolymer (PEG-b-PLA)-coated β-$NaYF_4$:$Yb^{3+}$, $Er^{3+}$ UCNPs, using meso-tetraphenyl porphine (TPP) as the photosensitizer which is stabilized by PEG-b-PLA. Based on in vitro studies utilizing HeLa cervical cancer cell lines, the composite nanoparticles are shown to exhibit low dark toxicity and efficient cancer cell-killing activity upon NIR excitation [101]. Liu and coworkers load photosensitizer of chlorin e6 (Ce6) on PEG-coated UCNPs, forming a UCNPCe6 supramolecular complex that produces singlet oxygen to kill cancer cells under NIR light. Excellent PDT efficacy is achieved in tumor-bearing mice upon intratumoral injection of UCNPCe6 and then followed by NIR light exposure [103, 194].

Most recently, chitosan derivatives are also used to modify the surface of the hydrophobic UCNPs and efficiently trap the photosensitizers for PDT. In addition to good biocompatibility, chitosan is also biodegradable and provides a hydrophilic environment for solubilizing the nanoconstructs. Xing and coworkers developed a photosensitizer pyropheophorbide a (Ppa) and RGD peptide co-modified chitosan-wrapped $NaYF_4$:Yb/Er upconversion nanoparticle (UCNP-Ppa-RGD) for targeted NIR PDT, which exhibits high phototoxicity against cancer cells upon 980-nm laser irradiation at an appropriate dosage [195]. Gu and coworkers reported the use of FA-modified amphiphilic chitosan (FASOC)-coated UCNPs (FASOC-UCNP) that anchor the ZnPc close to the UCNPs, thereby facilitating resonance energy transfer from UCNPs to ZnPc. The FASOC-UCNPs exhibit higher tumor-targeting ability than none FA-modified UCNPs (SOCUCNP). In vivo PDT treatments for deep-seated tumors demonstrated that NIR light-triggered PDT based on the nanoconstructs possesses remarkable therapeutic efficacy with a tumor inhibition ratio up to 50 % compared with conventional visible light-activated PDT with a noticeable reduced tumor inhibition ratio of 18 % [196].

Methylene blue (MB) is a widely used photosensitizer used to create singlet oxygen when exposed to both oxygen and light [197–200]. Shi et al. developed a new kind of UCNP/MB-based PDT drug, $NaYF_4$:Er/Yb/Gd@$SiO_2$(MB), with a particle diameter less than 50 nm, which provides a potential theranostic nanomedicine for future near-infrared laser-triggered PDT and simultaneous magnetic/optical bimodal imaging [182].

Different from the above methods, the Yan team has developed a new method for the use of upconversion luminescence nanomaterials to absorb near-infrared light and convert it into visible light. By means of a photosensitive o-nitro group, the hybrid structure of polyacrylamide was cross-linked together, and the upconversion luminescent nanoparticles were loaded. Near-infrared light irradiation, the conversion of light into the visible light, can lead to the breaking of the o-nitro group, which is released from the biological molecules adsorbed on the hybrid materials

[201]. This method makes the tumor treatment of PDT not only be used in the surface of the photosensitive material, greatly expanding the application of the light in the field of the conversion luminescence materials.

### 12.7.4 UCNPs for Photothermal Therapy (PTT)

PTT employs photo-absorbers to generate heat from light absorption, leading to thermal ablation of cancer cells. Various nanomaterials with high NIR light absorbance such as gold and silver nanoshells, nanorods, nanocages, etc. have been utilized for PTT treatment of cancer [202]. Song et al. reported the synthesis of core–shell structured hexagonal-phase $NaYF_4$:$Yb^{3+}$,$Er^{3+}$@Ag nanoparticles and their unique biofunctional properties. HepG2 cells from human hepatic cancer and BCap-37 cells from human breast cancer incubated with the composite UCNPs in vitro were found to undergo photothermally induced death on exposure to 980 nm [92]. Liu and coworkers synthesized a new class of multifunctional nanoparticles consisting of a UCNP particle as the core, a layer of ultrasmall iron oxide nanoparticles (IONPs) as the intermediate shell, and a thin layer of gold as the outer shell. The layer of IONPs between UCNPs and the Au shell not only affords the magnetic properties but also significantly reduces the luminescence quenching effects of the gold nanostructure to UCNPs. The UCNP–IONP–Au nanoparticles are then coated with PEG to improve its biocompatibility in physiological solutions. Those multifunctional nanoparticles are used for UCL/MR multimodal imaging as well as in vitro photothermal ablation of cancer cells [203]. In the next year of 2012, they achieved highly efficient in vivo magnetically targeted PTT by using the same multifunctional nanoparticles. In vivo dual-modal optical/MR imaging of mice uncovers that by placing a magnet nearby the tumor; the UCNPs show high tumor accumulation after intravenous injection, which is ~8 folds higher than that of without magnetic nearby. An outstanding PTT therapeutic efficacy with 100 % of tumor elimination in a murine breast cancer model is realized [204].

Most recently, Hilderbrand et al. reported the design and synthesis of $NaYF_4$:$Er^{3+}$,$Yb^{3+}$@$SiO_2$ core–shell nanocomposites with highly absorbing NIR carbocyanine dyes in the outer silica shell for combined NIR imaging and photothermal therapy. Photothermal cell killing under 750-nm excitation light source demonstrated the capability of UCNPs for both diagnostic optical imaging and therapeutic thermal therapy [205].

### 12.7.5 UCNPs for Radiotherapy

Radiotherapy is the controlled use of high-energy X-rays to kill many different types of cancer cells by delivering therapeutic X-rays to tumor regions without surgical risks or systematic toxicity. Inaccurate tumor localization and inherent radioresistance of tumors are the two challenges for the clinical potentials of radiotherapy. Shi and coworkers developed a novel RGD-labeled $BaYbF_5$:2 % $Er^{3+}$ nanocube (UCA-RGD) for the first time to meet these clinical demands. These heavy metal-based nanocubes not only act as CT contrast agents for targeted tumor imaging but also act as irradiation dose enhancers in tumors during radiotherapy, which could greatly enhance therapeutic efficacy and minimize the damage to surrounding tissues [129].

### 12.7.6 *Multimodal Tumor Therapy*

With the development of nanoscience in the field of tumor diagnosis and treatment, the multimodal treatment of cancers based on one kind of nanoparticles has been developed. Recently, Liu group reported a kind of double function UCNPs. Two kinds of material were selected for the surface modification of UCNPs to achieve the dual function of PTT and PDT. Bengal rose is used to absorb the green fluorescence of UCNP to realize the PTT, and the infrared absorption dye of IR825 is used to realize PDT [206].

The Bu team reports a MRI/NIR dual-modal nanoparticle, which can treat the tumor with chemo-/radio-/photodynamic modals [207].

## 12.8 Toxicity of UCNPs

For UCNPs used for bioimaging, fully understanding the toxicity for UCNPs used in vitro and in vivo biological applications is necessary and also very important. Up to now, researches on the toxicity of UCNPs have just begun and the data are rather few and fragmentary. Numerous cytotoxicity tests based on morphology and mitochondrial function assays have suggested the UCNP low cell toxicity, as most of the cell viabilities are more than 85 % when incubated with a certain range of UCNPs for more than 24 h [105, 188, 203, 208–210]. The toxicity of the UCNPs could be improved by modifying with certain kinds of compounds, such as the PEG, PEI, PAA, silica, etc. [72, 79, 211, 212].

Zhang and coworkers reported silica-coated $NaYF_4$ nanocrystals incubated with rat skeletal myoblasts and bone marrow-derived mesenchymal stem cells, and cytotoxicity was assessed by using (3-(4,5-dimethylthiazol-2-yl)-5-(3-carboxymethoxyphenyl)-2-(4-sulfophenyl)-2H-tetrazolium), sodium salts (MTS), and lactate dehydrogenase (LDH) assay. The results from this study revealed that the silica-coated $NaYF_4$ UCNPs displayed good in vitro and in vivo biocompatibility, demonstrating their potential applications in both cellular and animal imaging systems [210]. Yan et al. reported the in vitro and in vivo toxicity assessments of water-soluble $NaYF_4$:Yb,Tm nanoparticles with HeLa cell and *Caenorhabditis elegans* (*C. elegans*) cases. $NaYF_4$:Yb,Tm nanoparticles afforded an efficient NIR

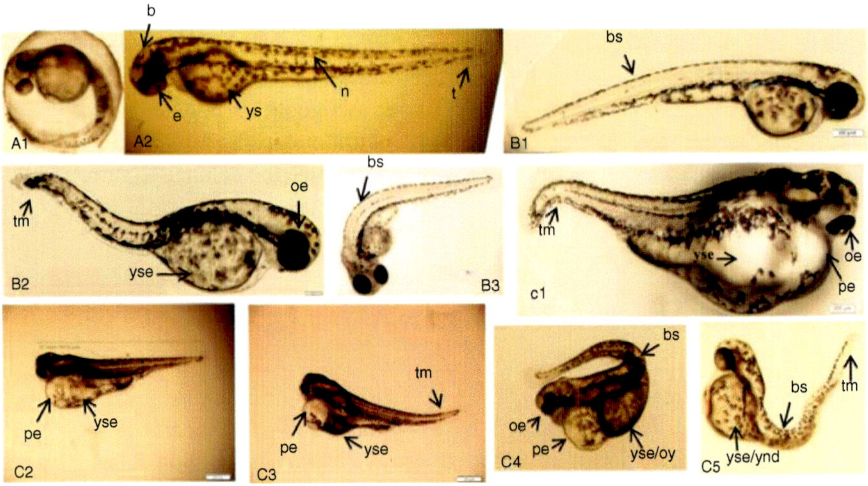


**Fig. 12.11** Phenotypic changes of zebra fish embryos at 48 hpf. (*A-1*, *A-2*) Control group. (*B1-3*) UCNPs<200 μg/mL groups. (*C1-5*) UCNPs 200–400 μg/mL groups. Abbreviations: *b* brain, *e* eye, *n* notochord, *t* tail, *ys* yolk sac, *bs* bent spine, *tm* tail malformation, *oe* ocular edema, *pe* pericardial edema, *oy* opaque yolk, *yes* yolk sac edema, and *ynd* yolk not depleted

image of the HeLa cells with low toxicity. Toxicity studies were further addressed with protein expression, life span, egg production, egg viability, and growth rate of the worms in comparison with those of the intact ones. The feeding of RE fluoride nanoparticles with a dose of 100 μg did not raise obvious toxicity effect from the growth to procreation [213].

Cui and coworkers reported the toxicity effects of UCNPs of $LaF_3$:Yb,Er on zebra fish. Results showed that water-soluble $LaF_3$:Yb,Er did not exhibit obvious toxicity to zebra fish embryos under 100 $\mu g \cdot mL^{-1}$, but exhibited chronic toxicities with 200 $\mu g \cdot mL^{-1}$ in vivo, resulting in malformations and delayed hatching rate and embryonic and larval development. The excretion channels of $LaF_3$:Yb,Er in adult zebra fish were primarily found in the intestine after being injected for 24 h [214] (Fig. 12.11).

Although there have been many reports on the toxicity of the UCNPs, but the biological application of UCNPs is still at the preliminary stage, the data on the toxicology of the UCNPs is still not comprehensive, and there is also a need to carry out a lot of work in this field.

## References

1. Kumar KU, et al. Multicolor upconversion emission and color tunability in Tm3+/Er3+/Yb3+ Tr-doped NaNbO3 nanocrystals. Mater Express. 2012;2(4):294–302.
2. Guo H, et al. Visible upconversion in rare earth ion-doped Gd2O3 nanocrystals. J Phys Chem B. 2004;108(50):19205–9.

3. Deren PJ, et al. Up-conversion in elpasolite crystals doped with U3+. Chem Phys Lett. 2000;332(3):308–12.
4. Stump NA, et al. Stokes and anti-Stokes luminescence from the trihalides of Cm-248. Radiochim Acta. 1993;61(3–4):129–36.
5. Auzel F. Upconversion and anti-Stokes processes with f and d ions in solids. Chem Rev. 2003;104(1):139–74.
6. Yang D, et al. Hollow structured upconversion luminescent NaYF(4):Yb(3)(+), Er(3)(+) nanospheres for cell imaging and targeted anti-cancer drug delivery. Biomaterials. 2013;34(5):1601–12.
7. Cao TY, et al. Water-soluble NaYF4:Yb/Er upconversion nanophosphors: synthesis, characteristics and application in bioimaging. Inorg Chem Commun. 2010;13(3):392–4.
8. Wei Y, et al. Polyol-mediated synthesis of water-soluble LaF3: Yb, Er upconversion fluorescent nanocrystals. Mater Lett. 2007;61(6):1337–40.
9. Yi G-S, Chow G-M. Colloidal LaF3:Yb, Er, LaF3:Yb, Ho and LaF3:Yb, Tm nanocrystals with multicolor upconversion fluorescence. J Mater Chem. 2005;15(41):4460–4.
10. Babu S, et al. Multicolored redox active upconverter cerium oxide nanoparticle for bioimaging and therapeutics. Chem Commun. 2010;46(37):6915–7.
11. Cho JH, et al. Up conversion luminescence of Yb3+–Er3+ codoped CeO2 nanocrystals with imaging applications. J Lumin. 2012;132(3):743–9.
12. Yang DM, et al. Synthesis of Li1-xNaxYF4:Yb3+/Ln(3+) (0<= x<= 0.3, Ln=Er, Tm, Ho) nanocrystals with multicolor up-conversion luminescence properties for in vitro cell imaging. J Mater Chem. 2012;22(38):20618–25.
13. Patra A, et al. Upconversion in Er3+:ZrO2 nanocrystals. J Phys Chem B. 2002;106(8):1909–12.
14. Guo H, et al. Seed-mediated synthesis of NaY F4:Y b, Er/NaGdF4 nanocrystals with improved upconversion fluorescence and MR relaxivity. Nanotechnology. 2010;21(12):125602.
15. Sarakovskis A, et al. Up-conversion processes in NaLaF4:Er3+. Opt Mater. 2009;31(10):1517–24.
16. Patra A, et al. Fluorescence upconversion properties of Er3+–doped TiO2 and BaTiO3 nanocrystallites. Chem Mater. 2003;15(19):3650–5.
17. Wang X, et al. Luminescence spectroscopy and visible upconversion properties of Er3+ in ZnO nanocrystals. J Phys Chem B. 2004;108(48):18408–13.
18. Venkatramu V, et al. Synthesis and luminescence properties of Er3+–doped Lu3Ga5O12 nanocrystals. J Lumin. 2008;128(5–6):811–3.
19. Venkatramu V, et al. Synthesis, structure and luminescence of Er3+–doped Y3Ga5O12 nano-garnets. J Mater Chem. 2012;22(27):13788–99.
20. Rai M, et al. Infrared to visible upconversion in Ho(3)+/Yb(3)+co-doped Y(2) O(3) phosphor: effect of laser input power and external temperature. Spectrochim Acta A Mol Biomol Spectrosc. 2012;97:825–9.
21. Li DY, et al. White upconversion emission in Yb3+/Tm3+/Ho3+ doped SrMoO4 nanocrystals by high excited state energy transfer. J Alloys Compd. 2013;550:509–13.
22. Wei Y, et al. Polyol-mediated synthesis and luminescence of lanthanide-doped NaYF4 nanocrystal upconversion phosphors. J Alloys Compd. 2008;455(1–2):376–84.
23. Zhang F, et al. Uniform nanostructured arrays of sodium rare-earth fluorides for highly efficient multicolor upconversion luminescence. Angew Chem Int Ed. 2007;46(42):7976–9.
24. Jin J, et al. Polymer-coated NaYF(4):Yb(3)(+), Er(3)(+) upconversion nanoparticles for charge-dependent cellular imaging. ACS Nano. 2011;5(10):7838–47.
25. Zhao JW, et al. A facile approach to fabrication of hexagonal-phase NaYF4:Yb3+, Er3+ hollow nanospheres: formation mechanism and upconversion luminescence. Eur J Inorg Chem. 2010;12:1813–9.
26. Liang YJ, et al. Hydrothermal synthesis and upconversion luminescent properties of YVO4:Yb3+, Er3+ nanoparticles. J Alloys Compd. 2013;552:289–93.

27. Heer S, et al. Highly efficient multicolour upconversion emission in transparent colloids of lanthanide-doped NaYF4 nanocrystals. Adv Mater. 2004;16(23–24):2102–5.
28. Heer S, et al. Blue, green, and Red upconversion emission from lanthanide-doped LuPO4 and YbPO4 nanocrystals in a transparent colloidal solution. Angew Chem Int Ed. 2003;42(27):3179–82.
29. Yi G, et al. Synthesis, characterization, and biological application of size-controlled nanocrystalline NaYF4:Yb,Er infrared-to-visible up-conversion phosphors. Nano Lett. 2004;4(11):2191–6.
30. Zhang Y-W, et al. Single-crystalline and monodisperse LaF3 triangular nanoplates from a single-source precursor. J Am Chem Soc. 2005;127(10):3260–1.
31. Yin A, et al. Colloidal synthesis and blue based multicolor upconversion emissions of size and composition controlled monodisperse hexagonal NaYF4: Yb, Tm nanocrystals. Nanoscale. 2010;2(6):953–9.
32. Mai H-X, et al. High-quality sodium rare-earth fluoride nanocrystals: controlled synthesis and optical properties. J Am Chem Soc. 2006;128(19):6426–36.
33. Mai H-X, et al. Highly efficient multicolor Up-conversion emissions and their mechanisms of monodisperse NaYF4:Yb, Er core and core/shell-structured nanocrystals. J Phys Chem C. 2007;111(37):13721–9.
34. Liu Q, et al. Sub-10 nm hexagonal lanthanide-doped NaLuF4 upconversion nanocrystals for sensitive bioimaging in vivo. J Am Chem Soc. 2011;133(43):17122–5.
35. Boyer J-C, et al. Synthesis of colloidal upconverting NaYF4 nanocrystals doped with Er3+, Yb3+ and Tm3+, Yb3+ via thermal decomposition of lanthanide trifluoroacetate precursors. J Am Chem Soc. 2006;128(23):7444–5.
36. Boyer J-C, Cuccia LA, Capobianco JA. Synthesis of colloidal upconverting NaYF4: Er3+/Yb3+ and Tm3+/Yb3+ monodisperse nanocrystals. Nano Lett. 2007;7(3):847–52.
37. Naccache R, et al. Controlled synthesis and water dispersibility of hexagonal phase NaGdF4:Ho3+/Yb3+ nanoparticles. Chem Mater. 2009;21(4):717–23.
38. Vetrone F, Mahalingam V, Capobianco JA. Near-infrared-to-blue upconversion in colloidal BaYF5:Tm3+, Yb3+ nanocrystals. Chem Mater. 2009;21(9):1847–51.
39. Vetrone F, et al. The active-core/active-shell approach: a strategy to enhance the upconversion luminescence in lanthanide-doped nanoparticles. Adv Funct Mater. 2009;19(18):2924–9.
40. Huang WJ, et al. Uniform NaYF4N:Yb, Tm hexagonal submicroplates: controlled synthesis and enhanced UV and blue upconversion luminescence. Mater Res Bull. 2013;48(2):300–4.
41. Huang WJ, et al. Controlled synthesis and upconversion luminescence properties of Yb3+– Tm3+ codoped NaYF4 hexagonal submicroplates. In: Kao JCM, Hou M, Chen R, editors. Frontier of nanoscience and technology II. Stafa-Zurich: Trans Tech Publications Ltd; 2012. p. 117–20.
42. Liu Q, et al. Multifunctional rare-earth self-assembled nanosystem for tri-modal upconversion luminescence/fluorescence/positron emission tomography imaging. Biomaterials. 2011;32(32):8243–53.
43. Pires AM, et al. Low-temperature upconversion spectroscopy of nanosized Y2O3: Er, Yb phosphor. J Appl Phys. 2005;98(6):063529. -063529-7.
44. He M, et al. Dual phase-controlled synthesis of uniform lanthanide-doped NaGdF4 upconversion nanocrystals Via an OA/ionic liquid Two-phase system for in vivo dual-modality imaging. Adv Funct Mater. 2011;21(23):4470–7.
45. Wang J, et al. One-step synthesis of highly water-soluble LaF 3:Ln 3+ nanocrystals in methanol without using any ligands. Nanotechnology. 2007;18(46):465606.
46. Hu D, et al. A facile method to synthesize superparamagnetic and up-conversion luminescent NaYF4:Yb, Er/Tm@SiO2@Fe3O4 nanocomposite particles and their bioapplication. J Mater Chem. 2011;21(30):11276–82.
47. Lu H, et al. Synthesis and characterization of multi-functional nanoparticles possessing magnetic, up-conversion fluorescence and bio-affinity properties. J Mater Chem. 2004;14(8): 1336–41.

48. Zeng JH, et al. Synthesis of complex rare earth fluoride nanocrystal phosphors. Nanotechnology. 2006;17(14):3549–55.
49. Zeng JH, et al. Synthesis and upconversion luminescence of hexagonal-phase NaYF4:Yb, Er3+ phosphors of controlled size and morphology. Adv Mater. 2005;17(17):2119–23.
50. Ghosh P, et al. Influence of surface coating on the upconversion emission properties of LaPO[sub 4]:Yb/Tm core-shell nanorods. J Appl Phys. 2009;105(11):113532–5.
51. Li C, et al. Highly uniform and monodisperse β-NaYF4:Ln3+ (Ln = Eu, Tb, Yb/Er, and Yb/Tm) hexagonal microprism crystals: hydrothermal synthesis and luminescent properties. Inorg Chem. 2007;46(16):6329–37.
52. Ma D-K, et al. Rare-earth-ion-doped hexagonal-phase NaYF4 nanowires: controlled synthesis and luminescent properties. J Phys Chem C. 2009;113(19):8136–42.
53. Wang Z-L, Hao JH, Chan HLW. Down- and up-conversion photoluminescence, cathodoluminescence and paramagnetic properties of NaGdF4: Yb3+, Er3+ submicron disks assembled from primary nanocrystals. J Mater Chem. 2010;20(16):3178–85.
54. Zhao J, et al. Controlled synthesis, formation mechanism, and great enhancement of red upconversion luminescence of NaYF4:Yb3+, Er3+ nanocrystals/submicroplates at low doping level. J Phys Chem B. 2008;112(49):15666–72.
55. Yang D, et al. One-step synthesis of small-sized and water-soluble NaREF(4) upconversion nanoparticles for in vitro cell imaging and drug delivery. Chemistry. 2013;19(8):2685–94.
56. Sikora B, et al. Transport of NaYF4:Er3+, Yb3+ up-converting nanoparticles into HeLa cells. Nanotechnology. 2013;24(23):235702.
57. Yang J, et al. One-step hydrothermal synthesis of carboxyl-functionalized upconversion phosphors for bioapplications. Chemistry. 2012;18(43):13642–50.
58. Wang Z-L, et al. Simultaneous synthesis and functionalization of water-soluble upconversion nanoparticles for in-vitro cell and nude mouse imaging. Nanoscale. 2011;3(5):2175–81.
59. Gao Y, Cao TY, Li FY. Water-soluble upconversion nanophosphors with cooperative ligands for in vivo lymph node imaging. Chin J Inorg Chem. 2012;28(10):2043–8.
60. Wang M, et al. One-step synthesis and characterization of water-soluble NaYF4:Yb, Er/Polymer nanoparticles with efficient up-conversion fluorescence. J Alloys Compd. 2009;485(1–2):L24–7.
61. Wang Z, et al. One-pot synthesis of water-soluble and carboxyl-functionalized beta-NaYF4:Yb, Er(Tm) upconversion nanocrystals and their application for bioimaging. J Mater Chem. 2012;22(24):12186–92.
62. Cao T, et al. High-quality water-soluble and surface-functionalized upconversion nanocrystals as luminescent probes for bioimaging. Biomaterials. 2011;32(11):2959–68.
63. Chen C, et al. Ionic liquid-based route to spherical NaYF4 nanoclusters with the assistance of microwave radiation and their multicolor upconversion luminescence. Langmuir. 2010;26(11):8797–803.
64. Liu X, et al. Ionothermal synthesis of hexagonal-phase NaYF(4):Yb(3+),Er(3+)/Tm(3+) upconversion nanophosphors. Chem Commun (Camb). 2009;43:6628–30.
65. He M, et al. Phase- and size-controllable synthesis of hexagonal upconversion rare-earth fluoride nanocrystals through an oleic acid/ionic liquid two-phase system. Chem-a Eur J. 2012;18(19):5954–69.
66. Zhou N, et al. Shape-controllable synthesis of hydrophilic NaLuF4:Yb, Er nanocrystals by a surfactant-assistant two-phase system. Nanoscale Res Lett. 2013;8:518.
67. Qiu PY, et al. Tuning lanthanide ion-doped upconversion nanocrystals with different shapes via a one-pot cationic surfactant-assisted hydrothermal strategy. CrystEngComm. 2014;16(10):1859–63.
68. Chen Z, et al. Versatile synthesis strategy for carboxylic acid – functionalized upconverting nanophosphors as biological labels. J Am Chem Soc. 2008;130(10):3023–9.
69. Hu H, et al. Facile epoxidation strategy for producing amphiphilic up-converting rare-earth nanophosphors as biological labels. Chem Mater. 2008;20(22):7003–9.

70. Chen K, et al. Fabrication of core/shell structured NaYF4:Yb3+, Er3+/polyphosphazene upconversion nanophosphors functionalized with abundant active amino groups. Mater Lett. 2013;101(15):54–6.
71. Gao X, et al. In vivo cancer targeting and imaging with semiconductor quantum dots. Nat Biotechnol. 2004;22(8):969–76.
72. Jingning Shan JC, Meng J, Collins J, Soboyejo W. Biofunctionalization, cytotoxicity, and cell uptake of lanthanide doped hydrophobically ligated NaYF4 upconversion nanophosphors. J Appl Phys. 2008;104:094308.
73. Luccardini C, et al. Size, charge, and interactions with giant lipid vesicles of quantum dots coated with an amphiphilic macromolecule. Langmuir. 2006;22(5):2304–10.
74. Decher G. Fuzzy nanoassemblies: toward layered polymeric multicomposites. Science. 1997;277(5330):1232–7.
75. Hong X, et al. Fabrication of magnetic luminescent nanocomposites by a layer-by-layer self-assembly approach. Chem Mater. 2004;16(21):4022–7.
76. Wang D, Rogach AL, Caruso F. Semiconductor quantum dot-labeled microsphere bioconjugates prepared by stepwise self-assembly. Nano Lett. 2002;2(8):857–61.
77. Wang L, et al. Fluorescence resonant energy transfer biosensor based on upconversion-luminescent nanoparticles. Angew Chem Int Ed. 2005;44(37):6054–7.
78. Zhang P, et al. Design of a highly sensitive and specific nucleotide sensor based on photon upconverting particles. J Am Chem Soc. 2006;128(38):12410–1.
79. Abdul Jalil R, Zhang Y. Biocompatibility of silica coated NaYF(4) upconversion fluorescent nanocrystals. Biomaterials. 2008;29(30):4122–8.
80. Das GK, Tan TTY. Rare-earth-doped and codoped Y 2O 3 nanomaterials as potential bioimaging probes. J Phys Chem C. 2008;112(30):11211–7.
81. Mader HS, et al. Surface-modified upconverting microparticles and nanoparticles for use in click chemistries. Chemistry. 2010;16(18):5416–24.
82. Sivakumar S, Diamente PR, van Veggel FCJM. Silica-coated Ln3+−doped LaF3 nanoparticles as robust down- and upconverting biolabels. Chemistry (A European Journal). 2006;12(22):5878–84.
83. Qiu PY, et al. Recent advances in lanthanide-doped upconversion nanomaterials: synthesis, nanostructures and surface modification. Nanoscale. 2013;5(23):11512–25.
84. Sisi C, Haiyan C, Yueqing G. Comparison of two strategies for the synthesis of upconverting nanoparticles as biological labels. J Phys Conf Ser. 2011;277(1):012006.
85. Zhang QB, et al. Hexanedioic acid mediated surface-ligand-exchange process for transferring NaYF4:Yb/Er (or Yb/Tm) up-converting nanoparticles from hydrophobic to hydrophilic. J Colloid Interface Sci. 2009;336(1):171–5.
86. Chen Q, et al. Functionalization of upconverted luminescent NaYF4: Yb/Er nanocrystals by folic acid-chitosan conjugates for targeted lung cancer cell imaging. J Mater Chem. 2011;21(21):7661–7.
87. Zhan Q, et al. Using 915 nm laser excited Tm(3)+/Er(3)+/Ho(3)+− doped NaYbF4 upconversion nanoparticles for in vitro and deeper in vivo bioimaging without overheating irradiation. ACS Nano. 2011;5(5):3744–57.
88. Wu S, et al. Non-blinking and photostable upconverted luminescence from single lanthanide-doped nanocrystals. Proc Natl Acad Sci. 2009;106(27):10917–21.
89. Bogdan N, et al. Carbohydrate-coated lanthanide-doped upconverting nanoparticles for lectin recognition. J Mater Chem. 2010;20(35):7543–50.
90. Yi GS, Chow GM. Synthesis of hexagonal-phase NaYF4:Yb, Er and NaYF4:Yb, Tm nanocrystals with efficient up-conversion fluorescence. Adv Funct Mater. 2006;16(18):2324–9.
91. Yi G, Peng Y, Gao Z. Strong red-emitting near-infrared-to-visible upconversion fluorescent nanoparticles. Chem Mater. 2011;23(11):2729–34.
92. Dong B, et al. Multifunctional NaYF4: Yb3+, Er3+@Ag core/shell nanocomposites: integration of upconversion imaging and photothermal therapy. J Mater Chem. 2011;21(17):6193–200.

93. Li D, et al. Influence of the TGA modification on upconversion luminescence of hexagonal-phase NaYF4:Yb3+, Er3+ nanoparticles. J Phys Chem C. 2010;114(18):8219–26.
94. Liebherr RB, et al. Maleimide activation of photon upconverting nanoparticles for bioconjugation. Nanotechnology. 2012;23(48):485103.
95. Dong A, et al. A generalized ligand-exchange strategy enabling sequential surface functionalization of colloidal nanocrystals. J Am Chem Soc. 2011;133(4):998–1006.
96. Zhou H-P, et al. Clean and flexible modification strategy for carboxyl/aldehyde-functionalized upconversion nanoparticles and their optical applications. Adv Funct Mater. 2009;19(24):3892–900.
97. Wang M, et al. Two-phase solvothermal synthesis of rare-earth doped NaYF4 upconversion fluorescent nanocrystals. Mater Lett. 2009;63(2):325–7.
98. Bogdan N, et al. Synthesis of ligand-free colloidally stable water dispersible brightly luminescent lanthanide-doped upconverting nanoparticles. Nano Lett. 2011;11(2):835–40.
99. Yi G-S, Chow G-M. Water-soluble NaYF4:Yb, Er(Tm)/NaYF4/polymer core/shell/shell nanoparticles with significant enhancement of upconversion fluorescence. Chem Mater. 2006;19(3):341–3.
100. Budijono SJ, et al. Synthesis of stable block-copolymer-protected NaYF4:Yb3+, Er3+ upconverting phosphor nanoparticles. Chem Mater. 2009;22(2):311–8.
101. Shan J, et al. Pegylated composite nanoparticles containing upconverting phosphors and meso-tetraphenyl porphine (TPP) for Photodynamic Therapy. Adv Funct Mater. 2011;21(13):2488–95.
102. Cheng L, et al. Highly-sensitive multiplexed in vivo imaging using pegylated upconversion nanoparticles. Nano Res. 2010;3(10):722–32.
103. Wang C, et al. Near-infrared light induced in vivo photodynamic therapy of cancer based on upconversion nanoparticles. Biomaterials. 2011;32(26):6145–54.
104. Park YI, et al. Nonblinking and nonbleaching upconverting nanoparticles as an optical imaging nanoprobe and T1 magnetic resonance imaging contrast agent. Adv Mater. 2009;21(44):4467–71.
105. Kobayashi H, et al. In vivo multiple color lymphatic imaging using upconverting nanocrystals. J Mater Chem. 2009;19(36):6481–4.
106. Nam SH, et al. Long-term real-time tracking of lanthanide ion doped upconverting nanoparticles in living cells. Angew Chem Int Ed. 2011;50(27):6093–7.
107. Zako T, et al. Cyclic RGD peptide-labeled upconversion nanophosphors for tumor cell-targeted imaging. Biochem Biophys Res Commun. 2009;381(1):54–8.
108. Hu H, et al. Multimodal-luminescence core-shell nanocomposites for targeted imaging of tumor cells. Chemistry (A European Journal). 2009;15(14):3577–84.
109. Boyer J-C, et al. Surface modification of upconverting NaYF4 nanoparticles with PEG – phosphate ligands for NIR (800 nm) biolabeling within the biological window. Langmuir. 2009;26(2):1157–64.
110. Wang M, et al. Immunolabeling and NIR-excited fluorescent imaging of HeLa cells by using NaYF4:Yb, Er upconversion nanoparticles. ACS Nano. 2009;3(6):1580–6.
111. Deng M, et al. Monodisperse upconversion NaYF4 nanocrystals: syntheses and bioapplications. Nano Res. 2011;4(7):685–94.
112. Salthouse C, et al. Design and demonstration of a small-animal up-conversion imager. Opt Express. 2008;16(26):21731–7.
113. Li LL, et al. Biomimetic surface engineering of lanthanide-doped upconversion nanoparticles as versatile bioprobes. Angew Chem Int Ed Engl. 2012;51(25):6121–5.
114. Xiong LQ, et al. Synthesis, characterization, and in vivo targeted imaging of amine-functionalized rare-earth up-converting nanophosphors. Biomaterials. 2009;30(29):5592–600.
115. Idris NM, et al. Tracking transplanted cells in live animal using upconversion fluorescent nanoparticles. Biomaterials. 2009;30(28):5104–13.

116. Pan LY, et al. Phase and size controllable synthesis of NaYbF4 nanocrystals in oleic acid/ionic liquid two-phase system for targeted fluorescent imaging of gastric cancer. Theranostics. 2013;3(3):210–22.
117. Kumar R, et al. Combined optical and MR bioimaging using rare earth Ion doped NaYF4 nanocrystals. Adv Funct Mater. 2009;19(6):853–9.
118. Chen D, et al. Lanthanide activator doped NaYb1-xGdxF4 nanocrystals with tunable down-, up-conversion luminescence and paramagnetic properties. J Mater Chem. 2011;21(17):6186–92.
119. Chen F, et al. A "neck-formation" strategy for an antiquenching magnetic/upconversion fluorescent bimodal cancer probe. Chemistry (A European Journal). 2010;16(37):11254–60.
120. Chen H, et al. Synthesis of brightly PEGylated luminescent magnetic upconversion nanophosphors for deep tissue and dual MRI imaging. Small. 2013;10:160–8.
121. Das GK, et al. Gadolinium oxide ultranarrow nanorods as multimodal contrast agents for optical and magnetic resonance imaging. Langmuir. 2010;26(11):8959–65.
122. Debasu ML, et al. (Gd, Yb, Tb)PO4 up-conversion nanocrystals for bimodal luminescence-MR imaging. Nanoscale. 2012;4(16):5154–62.
123. Hou Y, et al. Fe3O4 modified up-conversion luminescent nanocrystals for biological applications. Chin J Chem. 2012;30(12):2774–8.
124. Li FF, et al. Hydrophilic, upconverting, multicolor, lanthanide-doped NaGdF4 nanocrystals as potential multifunctional bioprobes. Chemistry (A European Journal). 2012;18(37):11641–6.
125. Liu C, et al. Magnetic/upconversion fluorescent NaGdF4:Yb,Er nanoparticle-based dual-modal molecular probes for imaging tiny tumors in vivo. ACS Nano. 2013;7:7227–40.
126. Lu Q, et al. A novel contrast agent with rare earth-doped up-conversion luminescence and Gd-DTPA magnetic resonance properties. J Solid State Chem. 2012;192:75–80.
127. Lopez-Mariscal C, et al. Phase dynamics of continuous topological upconversion in vortex beams. Opt Express. 2008;16(15):11411–22.
128. Paik T, et al. Designing tripodal and triangular gadolinium oxide nanoplates and self-assembled nanofibrils as potential multimodal bioimaging probes. ACS Nano. 2013;7(3):2850–9.
129. Wang Y, et al. Upconverting rare-earth nanoparticles with a paramagnetic lanthanide complex shell for upconversion fluorescent and magnetic resonance dual-modality imaging. Nanotechnology. 2013;24(17):175101.
130. Wilhelm S, et al. Magnetic nanosensor particles in luminescence upconversion capability. Angew Chem Int Ed Engl. 2011;50(37):A59–62.
131. Wong HT, et al. Water dispersible ultra-small multifunctional KGdF4:Tm3+, Yb3+ nanoparticles with near-infrared to near-infrared upconversion. J Mater Chem. 2011;21(41):16589–96.
132. Zeng S, et al. Dual-modal fluorescent/magnetic bioprobes based on small sized upconversion nanoparticles of amine-functionalized BaGdF5:Yb/Er. Nanoscale. 2012;4(16):5118–24.
133. Chen H, et al. Synthesis of brightly PEGylated luminescent magnetic upconversion nanophosphors for deep tissue and dual MRI imaging. Small. 2014;10(1):160–8.
134. Huang P, et al. Lanthanide-doped LiLuF(4) upconversion nanoprobes for the detection of disease biomarkers. Angew Chem Int Ed Engl. 2014;53(5):1252–7.
135. Tian G, et al. TPGS-stabilized NaYbF4:Er upconversion nanoparticles for dual-modal fluorescent/CT imaging and anticancer drug delivery to overcome multi-drug resistance. Biomaterials. 2015;40:107–16.
136. Shen J, et al. Superparamagnetic and upconversion emitting Fe3O4/NaYF4:Yb, Er hetero-nanoparticles via a crosslinker anchoring strategy. Chem Commun (Camb). 2010;46(31):5731–3.
137. Zhang L, et al. Magnetic/upconversion luminescent mesoparticles of Fe3O4@LaF3:Yb3+, Er3+ for dual-modal bioimaging. Chem Commun (Camb). 2012;48(91):11238–40.
138. Xu H, et al. Polymer encapsulated upconversion nanoparticle/iron oxide nanocomposites for multimodal imaging and magnetic targeted drug delivery. Biomaterials. 2011;32(35):9364–73.

139. Liu Z, et al. Long-circulating Er3+-doped Yb2O3 up-conversion nanoparticle as an in vivo X-Ray CT imaging contrast agent. Biomaterials. 2012;33(28):6748–57.
140. Ma J, et al. Folic acid-conjugated LaF3:Yb, Tm@SiO2 nanoprobes for targeting dual-modality imaging of upconversion luminescence and X-ray computed tomography. J Phys Chem B. 2012;116(48):14062–70.
141. Xing H, et al. Multifunctional nanoprobes for upconversion fluorescence, MR and CT tri-modal imaging. Biomaterials. 2012;33(4):1079–89.
142. Zeng S, et al. PEG modified BaGdF(5):Yb/Er nanoprobes for multi-modal upconversion fluorescent, in vivo X-ray computed tomography and biomagnetic imaging. Biomaterials. 2012;33(36):9232–8.
143. Xiao Q, et al. Radiopaque fluorescence-transparent TaOx decorated upconversion nanophosphors for in vivo CT/MR/UCL trimodal imaging. Biomaterials. 2012;33(30):7530–9.
144. Xing HY, et al. A NaYbF4: Tm3+ nanoprobe for CT and NIR-to-NIR fluorescent bimodal imaging. Biomaterials. 2012;33(21):5384–93.
145. Zhang G, et al. Dual modal in vivo imaging using upconversion luminescence and enhanced computed tomography properties. Nanoscale. 2011;3(10):4365–71.
146. Gao G, et al. One-pot hydrothermal synthesis of lanthanide ions doped one-dimensional upconversion submicrocrystals and their potential application in vivo CT imaging. Nanoscale. 2013;5(1):351–62.
147. Zhou J, et al. Water-stable NaLuF4-based upconversion nanophosphors with long-term validity for multimodal lymphatic imaging. Biomaterials. 2012;33(26):6201–10.
148. Liu FY, et al. Conjugation of NaGdF4 upconverting nanoparticles on silica nanospheres as contrast agents for multi-modality imaging. Biomaterials. 2013;34(21):5218–25.
149. Xia A, et al. Gd3+ complex-modified NaLuF4-based upconversion nanophosphors for trimodality imaging of NIR-to-NIR upconversion luminescence, X-Ray computed tomography and magnetic resonance. Biomaterials. 2012;33(21):5394–405.
150. Liu F, et al. Facile preparation of doxorubicin-loaded upconversion@polydopamine nanoplatforms for simultaneous in vivo multimodality imaging and chemophotothermal synergistic therapy. Adv Healthc Mater. 2015;4(4):559–68.
151. Zhu X, et al. Core-shell Fe3O4@NaLuF4:Yb, Er/Tm nanostructure for MRI, CT and upconversion luminescence tri-modality imaging. Biomaterials. 2012;33(18):4618–27.
152. Liu Z, et al. Long-circulating Gd2O3:Yb3+, Er3+ up-conversion nanoprobes as high-performance contrast agents for multi-modality imaging. Biomaterials. 2013;34(6):1712–21.
153. Zhou J, et al. Fluorine-18-labeled Gd3+/Yb3+/Er3+ co-doped NaYF4 nanophosphors for multimodality PET/MR/UCL imaging. Biomaterials. 2011;32(4):1148–56.
154. Sun Y, et al. Fluorine-18 labeled rare-earth nanoparticles for positron emission tomography (PET) imaging of sentinel lymph node. Biomaterials. 2011;32(11):2999–3007.
155. Liu Q, et al. 18F-Labeled magnetic-upconversion nanophosphors via rare-Earth cation-assisted ligand assembly. ACS Nano. 2011;5(4):3146–57.
156. Wang C, et al. Towards whole-body imaging at the single cell level using ultra-sensitive stem cell labeling with oligo-arginine modified upconversion nanoparticles. Biomaterials. 2012;33(19):4872–81.
157. Deng R, et al. Intracellular glutathione detection using MnO(2)-nanosheet-modified upconversion nanoparticles. J Am Chem Soc. 2011;133(50):20168–71.
158. Liu J, et al. Ultrasensitive nanosensors based on upconversion nanoparticles for selective hypoxia imaging in vivo upon near-infrared excitation. J Am Chem Soc. 2014;136(27):9701–9.
159. Chen Z, et al. Upconversion nanoprobes for efficiently in vitro imaging reactive oxygen species and in vivo diagnosing rheumatoid arthritis. Biomaterials. 2015;39:15–22.
160. Chao Wang LC, Zhuang L. Drug delivery with upconversion nanoparticles for multi-functional targeted cancer cell imaging and therapy. Biomaterials. 2011;32(4):1110–20.
161. Liu JN, et al. Simultaneous nuclear imaging and intranuclear drug delivery by nuclear-targeted multifunctional upconversion nanoprobes. Biomaterials. 2012;33(29):7282–90.

162. Tian G, et al. Mn2+ dopant-controlled synthesis of NaYF4:Yb/Er upconversion nanoparticles for in vivo imaging and drug delivery. Adv Mater. 2012;24(9):1226–31.
163. Guo H, et al. Upconversion nanoparticles modified with aminosilanes as carriers of DNA vaccine for foot-and-mouth disease. Appl Microbiol Biotechnol. 2012;95(5):1253–63.
164. Zhang F, et al. Mesoporous multifunctional upconversion luminescent and magnetic "nanorattle" materials for targeted chemotherapy. Nano Lett. 2012;12(1):61–7.
165. Xu Z, et al. Facile synthesis of an up-conversion luminescent and mesoporous Gd2O3: Er3+@nSiO2@mSiO2 nanocomposite as a drug carrier. Nanoscale. 2011;3(2):661–7.
166. Li C, et al. Multifunctional upconversion mesoporous silica nanostructures for dual modal imaging and in vivo drug delivery. Small. 2013;9:4150–9.
167. Hou Z, et al. Electrospinning preparation and drug-delivery properties of an up-conversion luminescent porous NaYF4:Yb3+, Er3+@silica fiber nanocomposite. Adv Funct Mater. 2011;21(12):2356–65.
168. Hou Z, et al. Electrospun upconversion composite fibers as dual drugs delivery system with individual release properties. Langmuir. 2013;29(30):9473–82.
169. Kang X, et al. Core–shell structured up-conversion luminescent and mesoporous NaYF4:Yb3+/Er3+@nSiO2@mSiO2 nanospheres as carriers for drug delivery. J Phys Chem C. 2011;115(32):15801–11.
170. Wang M, et al. Synthesis and characterization of NaYF4: Yb, Er upconversion fluorescent nanoparticles via a co-precipitation method. Guang Pu Xue Yu Guang Pu Fen Xi. 2009;29(12):3327–31.
171. Gai S, et al. Synthesis of magnetic, up-conversion luminescent, and mesoporous core–shell-structured nanocomposites as drug carriers. Adv Funct Mater. 2010;20(7):1166–72.
172. Qiu PY, et al. An anion-induced hydrothermal oriented-explosive strategy for the synthesis of porous upconversion nanocrystals. Theranostics. 2015;5(5):456–68.
173. Shan J, et al. NIR-to-visible upconversion nanoparticles for fluorescent labeling and targeted delivery of siRNA. Nanotechnology. 2009;20(15):155101.
174. Yang Y, et al. NIR light controlled photorelease of siRNA and its targeted intracellular delivery based on upconversion nanoparticles. Nanoscale. 2013;5(1):231–8.
175. Koo YE, et al. Photonic explorers based on multifunctional nanoplatforms for biosensing and photodynamic therapy. Appl Opt. 2007;46(10):1924–30.
176. Lee YE, Kopelman R. Polymeric nanoparticles for photodynamic therapy. Methods Mol Biol. 2011;726:151–78.
177. Reddy GR, et al. Vascular targeted nanoparticles for imaging and treatment of brain tumors. Clin Cancer Res. 2006;12(22):6677–86.
178. Gupta A, et al. Multifunctional nanoplatforms for fluorescence imaging and photodynamic therapy developed by post-loading photosensitizer and fluorophore to polyacrylamide nanoparticles. Nanomedicine. 2012;8(6):941–50.
179. Wang S, et al. Novel methods to incorporate photosensitizers into nanocarriers for cancer treatment by photodynamic therapy. Lasers Surg Med. 2011;43(7):686–95.
180. Zhang P, et al. Versatile photosensitizers for photodynamic therapy at infrared excitation. J Am Chem Soc. 2007;129(15):4526–7.
181. Carling C-J, et al. Remote-control photorelease of caged compounds using near-infrared light and upconverting nanoparticles. Angew Chem Int Ed. 2010;49(22):3782–5.
182. Chen F, et al. A uniform sub-50 nm-sized magnetic/upconversion fluorescent bimodal imaging agent capable of generating singlet oxygen by using a 980 nm laser. Chemistry. 2012;18(23):7082–90.
183. Wang FF, et al. Multifunctional up-converting nanocomposites with multimodal imaging and photosensitization at near-infrared excitation. J Mater Chem. 2012;22(47):24597–604.
184. Zhao ZX, et al. Multifunctional core-shell upconverting nanoparticles for imaging and photodynamic therapy of liver cancer cells. Chem Asian J. 2012;7(4):830–7.

185. Tian G, et al. Red-emitting upconverting nanoparticles for photodynamic therapy in cancer cells under near-infrared excitation. Small. 2013;9(11):1929–38.
186. Park YI, et al. Luminescence/magnetic resonance imaging and photodynamic therapy based on upconverting nanoparticles. In: Choi SH et al., editors. Nanosystems in engineering and medicine. Bellingham: SPIE-Int Soc Optical Engineering; 2012.
187. Lim ME, et al. Photodynamic inactivation of viruses using upconversion nanoparticles. Biomaterials. 2012;33(6):1912–20.
188. Chatterjee DK, Rufaihah AJ, Zhang Y. Upconversion fluorescence imaging of cells and small animals using lanthanide doped nanocrystals. Biomaterials. 2008;29(7):937–43.
189. Chatterjee DK, Yong Z. Upconverting nanoparticles as nanotransducers for photodynamic therapy in cancer cells. Nanomedicine. 2008;3(1):73–82.
190. Guo H, et al. Singlet oxygen-induced apoptosis of cancer cells using upconversion fluorescent nanoparticles as a carrier of photosensitizer. Nanomedicine. 2010;6(3):486–95.
191. Qian HS, et al. Mesoporous-silica-coated up-conversion fluorescent nanoparticles for photodynamic therapy. Small. 2009;5(20):2285–90.
192. Idris NM, et al. In vivo photodynamic therapy using upconversion nanoparticles as remote-controlled nanotransducers. Nat Med. 2012;18(10):1580–5.
193. Ungun B, et al. Nanofabricated upconversion nanoparticles for photodynamic therapy. Opt Express. 2009;17(1):80–6.
194. Wang C, et al. Imaging-guided pH-sensitive photodynamic therapy using charge reversible upconversion nanoparticles under near-infrared light. Adv Funct Mater. 2013;23(24):3077–86.
195. Zhou A, et al. Pyropheophorbide A and c(RGDyK) comodified chitosan-wrapped upconversion nanoparticle for targeted near-infrared photodynamic therapy. Mol Pharm. 2012;9(6):1580–9.
196. Cui S, et al. In vivo targeted deep-tissue photodynamic therapy based on near-infrared light triggered upconversion nanoconstruct. ACS Nano. 2013;7(1):676–88.
197. Hah HJ, et al. Methylene blue-conjugated hydrogel nanoparticles and tumor-cell targeted photodynamic therapy. Macromol Biosci. 2011;11(1):90–9.
198. Qin M, et al. Methylene blue covalently loaded polyacrylamide nanoparticles for enhanced tumor-targeted photodynamic therapy. Photochem Photobiol Sci. 2011;10(5):832–41.
199. Tang W, et al. Photodynamic characterization and in vitro application of methylene blue-containing nanoparticle platforms. Photochem Photobiol. 2005;81(2):242–9.
200. Tang W, et al. Encapsulation of methylene blue in polyacrylamide nanoparticle platforms protects its photodynamic effectiveness. Biochem Biophys Res Commun. 2008;369(2):579–83.
201. Yan B, et al. Near infrared light triggered release of biomacromolecules from hydrogels loaded with upconversion nanoparticles. J Am Chem Soc. 2012;134(40):16558–61.
202. Huang X, et al. Cancer cell imaging and photothermal therapy in the near-infrared region by using gold nanorods. J Am Chem Soc. 2006;128(6):2115–20.
203. Cheng L, et al. Facile preparation of multifunctional upconversion nanoprobes for multimodal imaging and dual-targeted photothermal therapy. Angew Chem Int Ed Engl. 2011;50(32):7385–90.
204. Cheng L, et al. Multifunctional nanoparticles for upconversion luminescence/MR multimodal imaging and magnetically targeted photothermal therapy. Biomaterials. 2012;33(7):2215–22.
205. Shan GB, Weissleder R, Hilderbrand SA. Upconverting organic dye doped core-shell nanocomposites for dual-modality NIR imaging and photo-thermal therapy. Theranostics. 2013;3(4):267–74.
206. Chen Q, et al. Protein modified upconversion nanoparticles for imaging-guided combined photothermal and photodynamic therapy. Biomaterials. 2014;35(9):2915–23.
207. Fan W, et al. A smart upconversion-based mesoporous silica nanotheranostic system for synergetic chemo-/radio-/photodynamic therapy and simultaneous MR/UCL imaging. Biomaterials. 2014;35(32):8992–9002.

208. Zhou J, et al. Dual-modality in vivo imaging using rare-earth nanocrystals with near-infrared to near-infrared (NIR-to-NIR) upconversion luminescence and magnetic resonance properties. Biomaterials. 2010;31(12):3287–95.
209. Hilderbrand SA, et al. Upconverting luminescent nanomaterials: application to in vivo bioimaging. Chem Commun (Camb). 2009;0(28):4188–90.
210. Xiong L, et al. Long-term in vivo biodistribution imaging and toxicity of polyacrylic acid-coated upconversion nanophosphors. Biomaterials. 2010;31(27):7078–85.
211. Cheng L, et al. In vivo pharmacokinetics, long-term biodistribution and toxicology study of functionalized upconversion nanoparticles in mice. Nanomedicine (Lond). 2011;6(8):1327–40.
212. Bae YM, et al. Endocytosis, intracellular transport, and exocytosis of lanthanide-doped upconverting nanoparticles in single living cells. Biomaterials. 2012;33(35):9080–6.
213. Zhou JC, et al. Bioimaging and toxicity assessments of near-infrared upconversion luminescent NaYF4:Yb, Tm nanocrystals. Biomaterials. 2011;32(34):9059–67.
214. Wang K, et al. Toxicity assessments of near-infrared upconversion luminescent LaF3:Yb, Er in early development of zebrafish embryos. Theranostics. 2013;3(4):258–66.

# Chapter 13
# Stem Cells and Gastric Cancer

**Meng Yang and Daxiang Cui**

This chapter elaborates the concept of stem cells, summarizes the role of stem cells in tumor occurrence and development, introduces a simple mechanism of stem cell tropism toward tumors, focuses on the application of stem cells as carriers in cancer therapy, and summarizes the security problems of using stem cells in tumor treatment.

## 13.1 The Concept of Stem Cells

Stem cells are undifferentiated cells that can differentiate into specialized cells and can divide to produce more stem cells. They are found in multicellular organisms. Two broad types of stem cells are found in mammals: embryonic stem (ES) cells, which are isolated from the inner cell mass of blastocysts, and adult stem cells, which are found in various tissues. In adult organisms, stem cells and progenitor cells act as a repair system for the body, replenishing tissues.

Stem cells are primitive cells formed in mammalian tissues and organs. Stem cells are usually circular or oval in morphology, small in size with large nuclei, and they have high telomerase activity. Stem cells have the capacity to replicate and differentiate. The fate of stem cells mainly depends on the state of the cells and on microenvironmental factors. The state of stem cells refers to the regulation of cell cycle–dependent kinase and gene transcription factors, these being cell factors that

M. Yang • D. Cui (✉)
Institute of Nano Biomedicine and Engineering, Shanghai Engineering Research Center for Intelligent Diagnosis and Treatment Instrument, National Center for Translational Medicine, Collaborative Innovational Center for System Biology, Shanghai Jiao Tong University, 800 Dongchuan Road, Shanghai 200240, P. R. China
e-mail: yangmeng@sjtu.edu.cn; dxcui@sjtu.edu.cn

© Springer Science+Business Media B.V. and Shanghai Jiao Tong University Press, Shanghai 2017

D. Cui (ed.), *Gastric Cancer Prewarning and Early Diagnosis System*, Translational Medicine Research, DOI 10.1007/978-94-024-0951-2_13

affect asymmetric cell division. Microenvironmental factors include the nature of the surrounding cells, the extracellular matrix, and a variety of soluble factors.

ES cells have totipotency and can differentiate into all adult tissues and organs. Under in vitro culture conditions with leukemia inhibitory factor, ES cells can remain in the undifferentiated state; if ES cells are cultured without leukemia inhibitory factor, the cells rapidly differentiate into cell lines such as muscle cells, blood cells, and nerve cells. Many adult tissues and organs, such as the skin and hematopoietic system, have the capacity for repair and regeneration. Adult stem cells play a key role in this process. Under certain conditions, the adult stem cells produce new stem cells or form cells with new functions to maintain the dynamic balance of tissue or organ growth and recession. Adult stem cells are derived from (1) embryonic cells by differentiation from ES cells; (2) the embryo, by separating the embryonic tissue and submitting it to cell culture; and (3) other tissues, including umbilical cord blood, neonatal placenta, bone marrow, peripheral blood, and fat cells. Hematopoietic stem cells, neural stem cells (NSCs), peripheral blood stem cells, adipose stem cells, and bone marrow mesenchymal stem cells (MSCs) are commonly used adult stem cells. Hematopoietic stem cells are the only source of all the kinds of blood cells in the body; they mainly exist in bone marrow, peripheral blood, and cord blood. The transplantation of hematopoietic stem cells is the most effective treatment for blood system diseases, congenital genetic diseases, and multiple metastatic tumor disease. Compared with the transplantation of bone marrow and peripheral blood stem cells, the advantage of umbilical cord blood stem cell transplantation is that there is no source restriction; the human leukocyte antigen (HLA) matching requirement is not high, and the cells are not easily polluted by virus or tumor. NSC research is still in a primary stage. Some symptoms in patients with Parkinson's disease can be treated by the transplantation of neural stem cells into the brain, producing dopamine. Bone marrow is the most important source of hematopoietic stem cells in the body. Peripheral blood stem cells are derived from bone marrow stem cells (BMSCs). Similar to BMSCs, peripheral blood stem cells are extracted by anesthesia surgery before. Medical research has confirmed that body fat contains large quantities of MSCs. Thus, the fat tissue that used to be discarded after body-shaping surgery can now be looked at as a source of MSCs. These cells have proliferative and multiple differentiation potential in vitro and can be applied to tissue and organ regeneration and repair. BMSCs are an important member of the stem cell family; BMSCs are derived from the early development of mesoderm and ectoderm. BMSCs were originally found in bone marrow, and, because of their multi-directional differentiation potential, BMSCs implanted into support and promote, immune regulation and characteristics of self-replicating and is becoming more and more attention.

Induced pluripotent stem cell (iPSC) technology was pioneered by Shinya Yamanaka's laboratory in Kyoto, Japan; Yamanaka and Takahashi [1] showed, in 2006, that the introduction of four specific gene-encoding transcription factors could convert adult cells into pluripotent stem cells. Yamanaka was awarded the 2012 Nobel Prize in Physiology or Medicine, along with John Gurdon, for the discovery that mature cells can be reprogrammed to become pluripotent. Pluripotent stem cells hold great promise in the field of regenerative medicine. Because they can propagate indefinitely, as well as give rise to every other cell type in the body (such as neurons, and heart, pancreatic, and liver cells), they represent a single source of cells that could be used to replace those cells lost to damage or disease. The most well-known type of pluripotent stem cell is the embryonic stem cell. However, since the generation of embryonic stem cells involves the destruction (or at least the manipulation) of the preimplantation-stage embryo, there has been much controversy surrounding their use. Further, because embryonic stem cells can only be derived from embryos, it has so far not been feasible to create patient-matched embryonic stem cell lines. Since iPSCs can be derived directly from adult tissues, they not only bypass the need for embryos but can be made in a patient-matched manner, which means that each individual could have their own pluripotent stem cell line. These unlimited supplies of autologous cells could be used to generate transplants without the risk of immune rejection. Research using iPSCs has become a hotspot in the stem cell field.

In our previous studies we produced fluorescent magnetic nanoparticles marked human iPSCs by using an efficient nanotechnology method [2]. First we modified magnetic nanoparticles carrying four genes: Oct4, Sox2, LIN28, and Nanog, and then transferred them into 293 T cells. After several days, we collected the supernatant of the 293 T cells and collected virus infected people fibroblasts. After 21 days, a similar kind of embryonic stem cell clone had formed and it was identified as human-induced pluripotent stem cells. Finally, we detected fluorescence and nuclear magnetic signals in iPSc by labeled magnetic fluorescence nanoparticles. Our method is convenient for the in-vivo application of human iPSCs.

## 13.2 The Role of Stem Cells in the Occurrence and Development of Gastric Cancer

It has taken a long time to realize the relationship between chronic inflammation and cancer [3]. An important example is the relationship between infection with *Helicobacter pylori* and gastric cancer [4]. Epidemiological studies have shown that the occurrence and development of gastric cancer are related to many factors, but gastric *H. pylori* infection is the most important cause of chronic inflammation, and about two-thirds of gastric cancer occurrence is related to gastric *H. pylori*. Gastric cancer can be divided into two pathological types: intestinal type and diffuse type. Although gastric *H. pylori* infection is correlated with both types of gastric cancer, it is more characteristic in the intestinal type [5]. The occurrence of the intestinal type of gastric cancer has a typical pathological development process; most begin with chronic gastritis and atrophic gastritis, developing into intestinal metaplasia, then atypical hyperplasia, and finally developing into cancer of the stomach.

Studies have shown that gastric *H. pylori* infection can induce B-cell activation by activation-induced cytidine deaminase (AID) expression. This gene is mainly related to the transformation of immunoglobulin and B-lymphocyte mutation, but studies have also found that the gene shows abnormal expression in cancer tissue; its main function is induced by the point mutation of the tumor suppressor gene P53 [6, 7]. Though a single gene mutation cannot fully explain the occurrence of tumors, this phenomenon could partly explain the relative importance of chronic inflammation in gastric cancer. A chronic inflammatory response provided by the microenvironment is not only beneficial for the occurrence of tumors but it can also be related to stem cell homing. Study results show that bone marrow MSCs can be homed to chronic injury or chronic inflammation [8, 9]. Previous researchers believed that chronic inflammation was not the start of the tumor but only had a role in promoting tumor development. The persistence of chronic inflammation, however, causes DNA damage repair to be incomplete and causes loss of control of the key genes and signaling pathways, and is likely to cause the adult stem cells in the stomach to develop into cancer stem cells. Mesenchymal stem cells that have migrated to the area of chronic inflammation can also lead to the occurrence of gastric cancer [10]. Thus, adult stem cells in the stomach and the homing of stem cells play important roles in the occurrence and development of gastric cancer.

In the nineteenth century, European pathologists put forward the cancer stem cell hypothesis for the first time; they believed that the tumor tissue itself was composed of cells in various differentiation phases and contained different plasmid and normal tissue. In recent years, it has been shown that some cells have self-renewal and differentiation potential. Because these cells can bypass asymmetric division to produce a large number of daughter cells with proliferation ability, they play an important role in tumorigenesis, development, and metastasis, so these cells were called cancer stem cells. Bonnet and Dick [11] found the existence of cancer stem cells for the first time, in human acute myeloid leukemia cells; they showed that the surfaces of CD34+/CD38- leukemia stem cells could form metastases of acute myeloid leukemia in non-obese diabetic/severe combined immunodeficiency (NOD/SCID) mice. In China and abroad, candidate cancer stem cells have been transplanted in NOD/SCID mice and their tumorigenic and differentiation potential has been observed; serum-free culture in vitro has also been performed to determine whether tumorigenicity is the gold standard characteristic of cancer stem cell biology [12]. Numerous studies have reported that cancer stem cells have been separated from breast cancer [13], brain tumor [14], prostate cancer [15], melanoma [16], liver cancer [17], pancreatic cancer [18, 19], colon cancer [20–22], and head and neck cancer cells [23]. These studies also suggest that although cancer stem cells account for only a small number of the tumor cells they show strong tumorigenicity and resistance to radiation and chemotherapy.

In 2008, Takaishi et al. used the reverse transcription polymerase chain reaction (RT-PCR) and fluorescence-activated cell sorting (FACS) technology and found that gastric cancer cell lines could be separated into subgroups of CD44+ and CD44- cells [24]. Gastric cancer was shown in NOD/SCID mice with CD44+ phenotypic cell transplantation in the skin, but gastric cancer did not occur when CD44- phenotypic cells were transplanted, showing that CD44+cells have very strong tumorigenic ability. Further study found that, in vitro, CD44+gastric cancer cells, accounting for 1–10% of the cells in serum-free medium, formed a spherical colony, while CD44- cells did not form a colony, thus proving that the CD44 + cells have self-renewal ability. In 2011, Chaojun Zhang et al., in order to confirm cells with self-renewal ability, separated cells in the human gastric cancer cell line AGS into CD44+, CD24+, CD44-, and CD24- cells by RT-PCR and FACS technology and showed that NOD/SCID mice implanted with CD44+ and CD24+ cells had a high rate of tumor formation; in-vitro experiments also showed that the CD44+ and CD24+ cells formed spherical colonies in serum-free culture [25]. Another study showed an obvious reduction of tumor cells in a tumor when the expression of the CD44 gene on tumor cells was knocked out [26]. The above research proves that CD44 + cells are most likely stomach cancer stem cells. One of the most important transcription factors, Oct-4, is a kind of marker that is related to stem cell self-renewal and the regulation of cell pluripotency [27]. Qin Wei found that the positive expression rate of Oct-4 in the tissues of 63 patients with cancer was 80.95 %, significantly higher than that in the tissue adjacent to the carcinoma, which had a positive expression rate of 5.20 %. Oct-4 expression in gastric cancer tissues promotes gastric cancer formation and maintains the activity of gastric cancer stem cell self-renewal or proliferation [28]. These lines of research provide further evidence of the objective existence of gastric cancer stem cells.

The targeting of CD44+ gastric cancer stem cells plays a key role in the treatment of gastric cancer. Our group has conjugated the CD44 antibody to a gold nanoparticle star (gold nanostar; GNS) [29]; in this way, photoacoustic imaging and heat treatment for gastric cancer stem cells can be realized. In an in-vitro experiment, GNS-conjugated PEG (polyethylene glycol) was conjugated with a CD44v6 (CD44v6-GNS) monoclonal antibody. The CD44v6-GNS complex was incubated with a gastric cancer stem-cell ball, as shown in Fig. 13.1.

The results showed that CD44v6-GNS had very high affinity to gastric cancer stem cells, and the goal of completely damaging the cells was achieved by using low-energy near-infrared light (790 nm, 1.5 W/cm$^2$, 5 min), as shown in Fig. 13.2.

In vivo experiments confirmed that, compared with findings in a control group, CD44v6-GNS can reach the tumor vasculature 4 h after injection. The treatment can restrain the growth of the tumor and prolong the survival time of experimental animals, as shown in Fig. 13.3.

At present there is no definite conclusion as to the origin of gastric cancer stem cells. Researchers generally believe that cancer stem cells may be derived from normal adult stem cells or from directional progenitor cells and the transformation of differentiated cells [30]. Other researchers think that cancer stem cells could be the result of the fusion of adult stem cells with other cells [31]. After gastric mucosal cells are damaged, cell turnover occurs once every 2–7 days under normal circumstances; this suggests that a certain proportion of pluripotent stem cells is present in stomach tissue [32]. Adult gastric stem cells are a type of very primitive cells, so they lack specific markers. Studies have shown that gastric progenitor cells are located at the base of the gastric mucosa, while adult gastric stem cells are located mainly in the neck, and these adult stem cells can differentiate into different types of epithelial cells [33]. Under normal conditions, stem cells maintain organization and stability in the body by the control of proliferation in their own

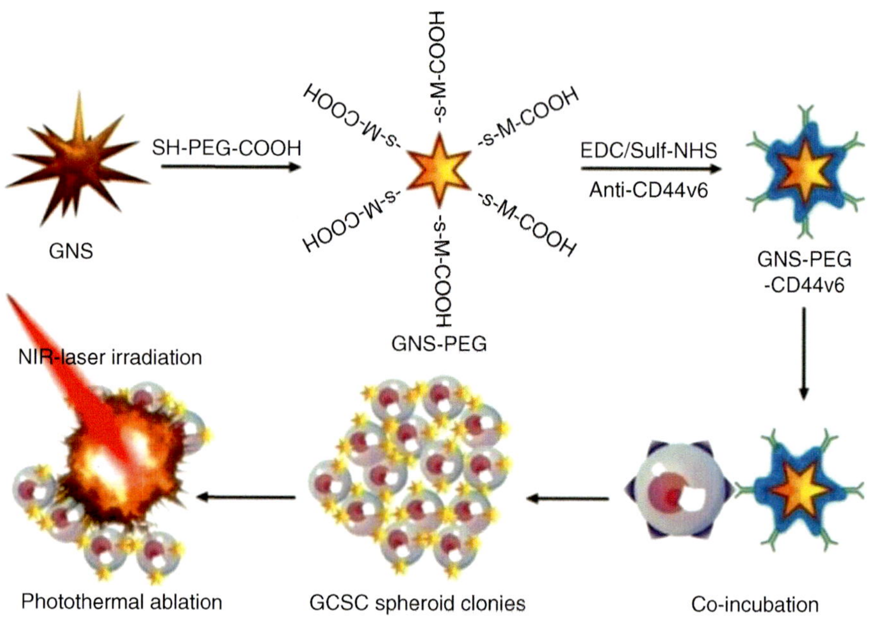


**Fig. 13.1** Schematic illustration of the GNS-PEG-CD44v6 nanocomplex synthesis process and the mechanism of GCSC targeted PTT. *GNS* Gold nonostar; *GCSC* Gastric cancer stem cell; *NIR* Near infrared; *CXCR* C-X-C chemokine receptor type; *SDF*-1 stromal cell-derived factor-1; *mES* mouse embryonic stem cell; *CXCL*12 C-X-C motif chemokine 12; *MFC* mouse fore-stomach carcinoma cell; *MSC* Mesenchymal stem cell; *FITC* Fluorescein isothiocyanate; *PE* Phycoerythrin; *FACS* Fluorescence-activated cell sorting; *FMNP* Fluorescent magnetic nanoparticle; *CCL* Chemokine (C-C motif) ligand

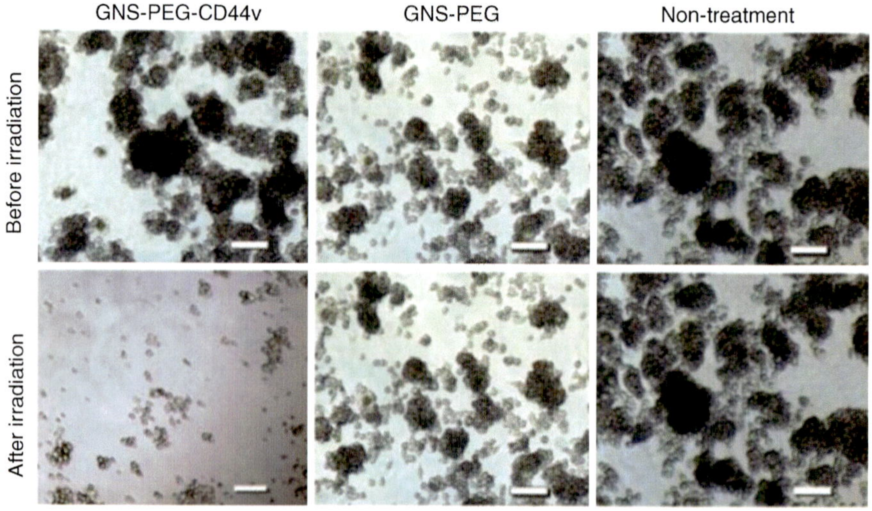


**Fig. 13.2** Microscopy images of spheroid GCSC colonies treated with GNS-PEG and GNS-PEG-CD44v6 for 24 h after NIR laser irradiation (790 nm, 1.5 W/cm 2, 5 min). All scale bars are 100 μm

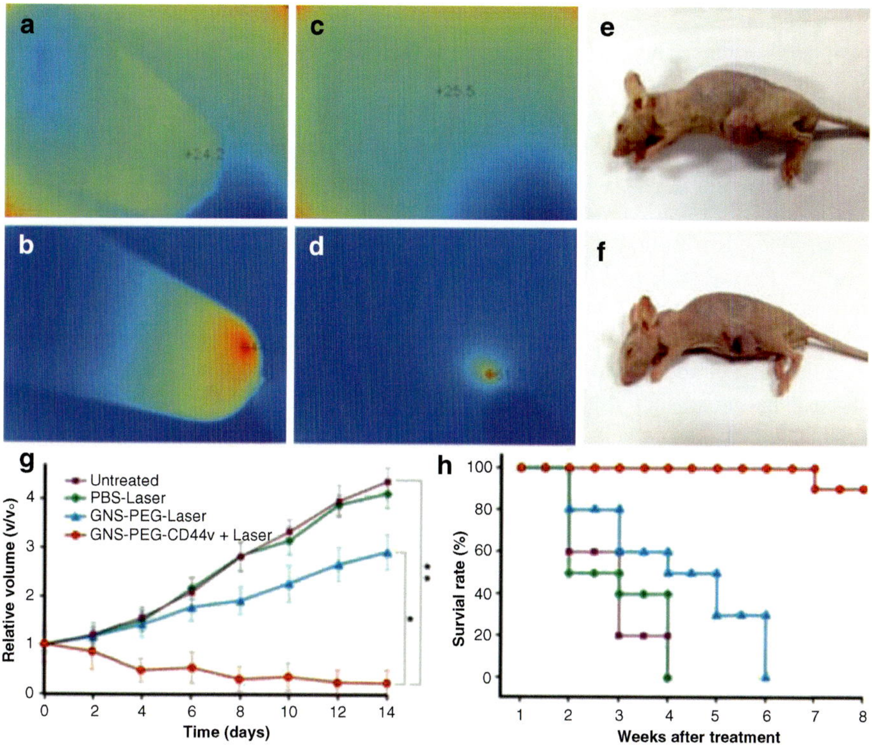


**Fig. 13.3** Infrared microscopic imaging: deionized water (**a**) and GNS (**b**) in a tube upon NIR laser irradiation (790 nm, 0.3 W/cm 2, 3 min); subcutaneous GC tumor without (**c**) and with (**d**) injection of GNS-PEG-CD44v6 upon NIR laser irradiation (790 nm, 1.5 W/cm 2, 3 min). Nude mouse model of subcutaneous GC xenograft, injected with nanoparticles before (**e**) and after (**f**) laser irradiation treatment (790 nm, 1.5 W/cm 2, 3 min); (**g**) tumor growth curves of four groups, after treatment with GNS-PEG-CD44v6, GNS-PEG, and phosphate-buffered saline (PBS), respectively, upon NIR laser irradiation (790 nm, 0.8 W/cm 2, 5 min); and the untreated control group. Error bars represent standard deviation ($n = 10$/group). *$p < 0.05$,**$p < 0.01$, (Student's *t*-test); (**h**) Survival rate of GC tumor-bearing mice within 8 weeks after treatment

environment, achieved with a variety of factors [34]. When the microenvironment of stem cells changes or when internal mutations in the genes for stem cell proliferation and differentiation are no longer subject to regulation, the stem cells transform into cancer stem cells. Thus, in the development of cancer of the stomach, the adult gastric stem cells could be transformed into gastric cancer stem cells. Although the adult gastric stem cells or progenitor cells may be the source of the gastric cancer stem cells, there is no specific experimental basis to confirm this [35]. Bone marrow-derived cells are considered to be the most primitive directional differentiation of adult stem cells; these cells and bone marrow MSCs have a wide range of plasticity and can migrate to inflammation areas or tissue injury sites in peripheral organs. The differentiation patterns and growth control of bone marrow-derived cells and bone marrow MSCs depend largely on the migration sites that provide the corresponding signal [9]. In 2004, Houghton et al. injected *H. pylori* into C57BL/6 mice; this led to chronic stomach inflammation; the mice were damaged by their own bone marrow cells [10]. Then genetically engineered bone marrow cells with a marker were transplanted into the mice. Eight to ten weeks after the *H. pylori* injection, the investigators started testing to determine whether bone marrow-derived cells had migrated to and participated in the repair process of the chronic inflammation of gastric mucosa. Twenty weeks after the *H. pylori* injection, gastric epithelial cells in these mice began to show the marker, and these cells with the marker showed some abnormal biological behavior. Over time, the gastric mucosal apoptosis increase and consist with this beta-galactosidase-positive glands appeared. Fifty-two weeks after the *H. pylori* injection, all of the C57BL/6 mice had developed gastric cancer, and the markers were present in the gastric cancer cells, thus demonstrating that bone marrow stem cells may be the origin of gastric cancer stem cells. Although Houghton and colleagues established an animal model of gastric cancer cells and showed that cancer stem cells may come from bone marrow-derived cells, few experiments have been performed to investigate whether bone marrow-derived cells are directly transformed into gastric cancer stem cells, or whether the bone marrow-derived cells exhibit fusion with gastric cancer cells. Also, there is a lack of specific markers to confirm that human gastric cancer is derived from bone marrow-derived cells.

In conclusion, gastric cancer stem cells exist in gastric cancer tissue. Although whether gastric cancer stem cells are derived from adult gastric stem cells or by the targeting of other stem cells in gastric cancer tissue does is not certain, it is clear that gastric cancer stem cells play an important role in the occurrence and development of stomach cancer. The exact role played by stem cells in the occurrence and development of gastric cancer needs further study; however, there many studies using modified stem cells that home to the tumor site have shown that this process can treat the tumor.

## 13.3 The Mechanism of Tumor Tropism

Many experiments have confirmed that stem cells can be targeted to a tumor site, but the molecular mechanism of this targeting to the tumor site is still not clear. In research to determine the mechanism of stem cell targeting to the tumor site, the most studied pathway is that of stromal cell-derived factor 1 (SDF-1) and its receptor CXCR4. The role of SDF-1 and CXCR4 axis cells in cell migration has been confirmed in adult stem cells [36–39], embryonic stem cells [40], and induced

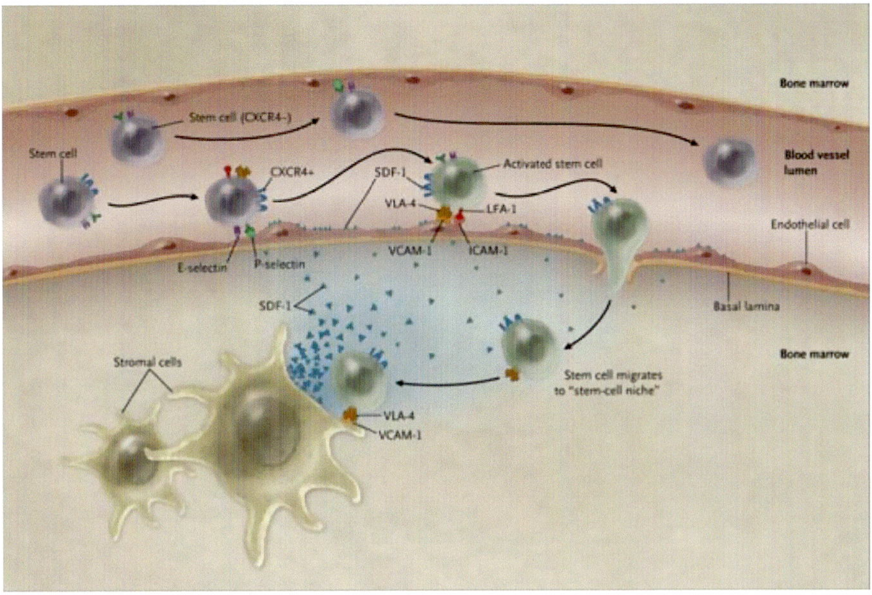


**Fig. 13.4** CXCR4-positive stem cells can migrate to the site of cells secreting SDF1

pluripotent stem cells [41], as shown in Fig. 13.4. Other signaling pathways that affect stem cell migration include PI3K [42], vascular endothelial growth factor receptor 2 [43], and 1-protease activated receptor 1 matrix metalloproteinase [44].

Research results from our team have shown that DIR (1,1′-dioctadecyl-3,3,3′,3′-tetramethyl indotricarbocyanine Iodide) labeled embryonic stem cells can target to a tumor site, and that SDF1/CXCR4 plays a role in this process [45]. DIR is a kind of near-infrared dye; after the DIR labeling, the signal in small living animals can be detected in the imager. Results showed that DIR-labeled embryonic stem cells ($5 \times 10^6$) were detected at the tumor site 10 min after the intravenous injection, and the signal peak occurred at 24 h in the gastric tumor-burdened mice, as shown in Fig. 13.5.

Immunohistochemistry and Western blot results also confirmed that the stem cells at the tumor site showed specificity for SSEA-1 (stage-specific embryonic antigen) expression, as shown in Fig. 13.6.

In-vitro cell migration results showed that SDF1/CXCR4 played a role in this process, as shown in Fig. 13.7.

The extent of stem cell migration to a tumor site is influenced by many factors, such as the nature of the stem cell itself, the culture conditions, the expression of migration factors, and the characteristics of the tumor microenvironment (such as oxygen, blood vessel growth, and immunity). Better understanding of the factors that influence the process of stem cell migration will be conducive to improving the therapeutic effects of stem cells.

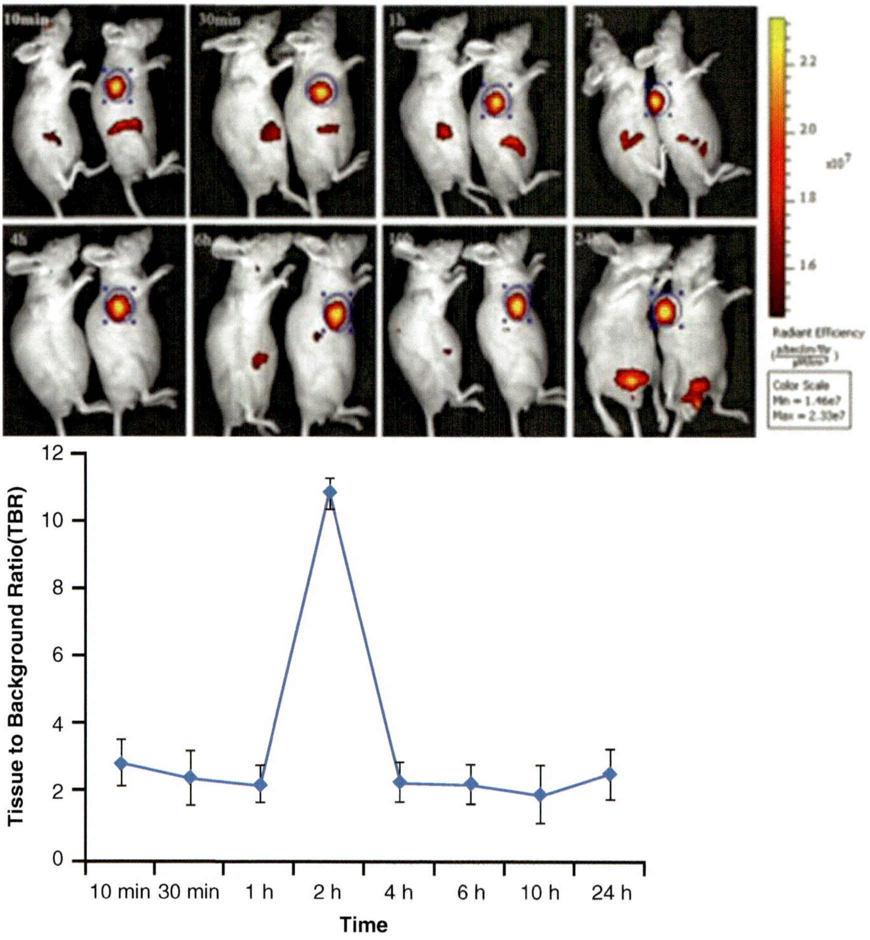


**Fig. 13.5** In-vivo fluorescence images of gastric cancer mouse models at 10 min, 30 min, 1 h, 2 h, 4 h, 6 h, 10 h, and 24 h post-injection of DiR-mES cells in the lateral position. The images show that the tumor tissues had strong fluorescence signals after the injection of DiR-mES cells, as shown in *blue circles*. The tissue-to-background ratio (TBR) values show that DiR-mES cells accumulated in the tumor tissues and reached a peak value at 2 h post-injection

## 13.4 The Use of Stem Cells in Gastric Cancer Treatment

Conventional treatments of gastric cancer patients include surgery, radiotherapy, and chemotherapy. Surgery is the main treatment for gastric cancer and is the only possible cure for advanced gastric cancer. The 5-year survival rate was 90 % after the eradicative resection of early gastric cancer. However, early gastric cancer lacks specific signs, and the majority of patients (>70 %) present with stage III or IV disease. Although radical excision can remove the tumor, there may be subclinical metastases causing relapse or metastasis, so the overall prognosis is poor.

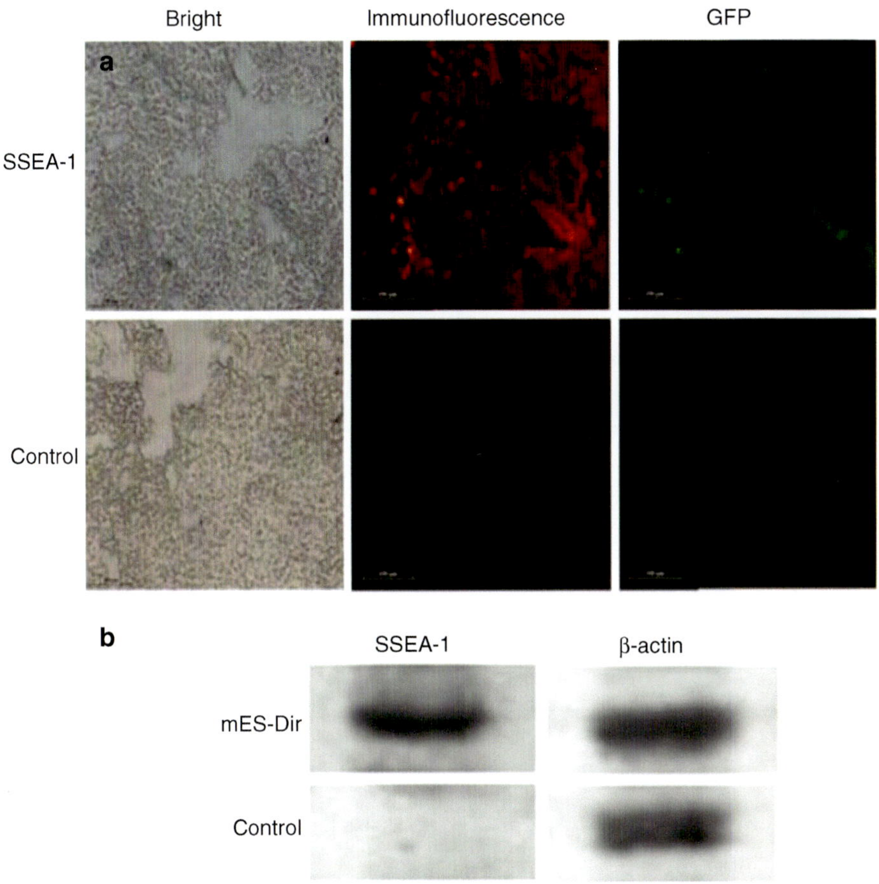


**Fig. 13.6** (**a**) Immunofluorescence analysis of tumor tissues. (**b**) Western blot analysis of tumor tissues

The choice of chemotherapy drugs mainly include alkalating agents such as phosphoric acid amides, antimetabolite drug, plant derived agents and antitumor antibiotics. In addition to traditional chemotherapy alone, new methods of using chemotherapy have been developed, as listed below.

Surgical adjuvant chemotherapy or radiation therapy: Adjuvant chemotherapy has gained attention because the view of tumor development over time is obviously different from that in the past. In the past, it was thought that the tumor began as a localized disease process and then invaded surrounding tissues, first via the lymphatic system, and finally the principal body transfer. So the key for early treatment of the tumor was radical tumor resection. But in recent years, it has been realized that, after the occurrence of the tumor; tumor cells migrate into the blood circulation. Although most of these cells can be annihilated by the body's immune defense mechanisms, a few tumor cells may remain, and this will be the root cause of recurrence and metastasis; therefore, when clinical tumor surgery is performed, the

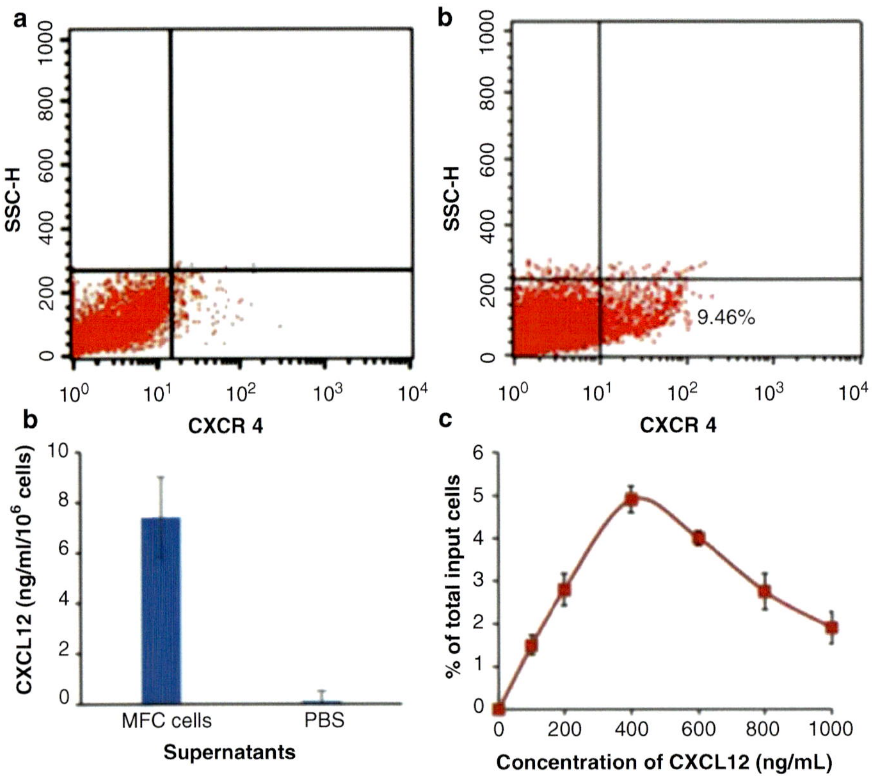


**Fig. 13.7** Analysis of chemotaxis. (**a**) Expression of CXCR4 in isotype control (*a*) and mES cells (*b*) was determined by flow cytometry. (**b**) Secretion of CXCL12 in MFC cells was measured by enzyme-linked immunosorbent assay. (**c**) Cell and chemoattractant dose–response curves

majority of patients will already have distant metastases. So early systemic chemotherapyshould be performed after the surgery, taking advantage of the fact that most of the tumor has been removed, to eliminate tiny lesions in a timely fashion.

Neoadjuvant chemotherapy: Neoadjuvant chemotherapy is adjuvant chemotherapy given before surgery. The period of adjuvant chemotherapy before surgery cannot be too long; usually about three courses of treatment are given. The mechanism of this therapy may be different from that of the 6 to 12 courses of adjuvant chemotherapy given after surgery, so it is not known as preoperative adjuvant chemotherapy, but is referred to as neoadjuvant chemotherapy or induction chemotherapy. The earlier that chemotherapy is started, the less the chance of resistance, so, in recent years, many tumors, such as breast cancer, have been treated with neoadjuvant chemotherapy. The advantages of neoadjuvant chemotherapy are (1) the use of chemotherapy can turn a tumor from untreatable by surgery to treatable by shrinking the volume (2) less extensive surgery (3) lower morbidity. However, not everyone is suitable for therapy in this way because it can be extremely toxic. Some patients may react severely that further treatments, especially surgery, are precluded because the patient is rendered unfit for anesthetic.

Intraperitoneal chemotherapy: At present, radical cure can be obtained for gastrointestinal cancer, and the postoperative survival rate has increased to a certain extent; however, because of the late presentation of most cases, there is a high chance of recurrence, so intraperitoneal chemotherapy is given in order to reduce recurrence in the abdominal cavity. At a certain stage of tumor development, in lesions involving the serosal surface of cancer cells, the cells may become detached and they are then intraperitoneal free cancer cells. Intraoperative or early postoperative intraperitoneal chemotherapy should be started, the in vivo tumor load, minimum corresponding acceleration tumor cell proliferation, sensitive to chemotherapy. If the delay treatment, tumor load is big, poor chemotherapy effect, another surgery intraperitoneal adhesion solution, and the new adhesive has not yet formed, drug is easy to reach all parts of the abdominal cavity. After intraperitoneal chemotherapy is mainly used in ovarian cancer resection of tiny residual lesions, gastrointestinal carcinoma postoperative residual, or have a high risk of recurrence and metastasis, peritoneal mesothelioma, etc. For intraperitoneal chemotherapy, the dosing method has a single point of targeted drug delivery, such as an indwelling catheter.

Arterial infusion chemotherapy: Arterial infusion chemotherapy, compared with systemic intravenous chemotherapy, has the following characteristics: (1) The local tumor drug concentration is increased significantly, and the whole-body systemic drug concentration is decreased. (2) There is a significantly lower incidence of systemic side effects and drug interactions and local organs are relatively heavy. (3) Local perfusion can greatly improve the effective dose. (4) The curative effect is improved. The main methods for administering arterial infusion chemotherapy are to insert an arterial catheter, and via the catheter, to perfuse the chemotherapy drugs. Arterial infusion chemotherapy is now mainly used for the treatment of liver cancer. The artery intubation method can be open intubation or femoral artery intubation. In recent years, the application of subcutaneous infusion pumps has greatly simplified arterial perfusion. Complications of arterial perfusion chemotherapy mainly include infection at the catheter site, catheter blockage, catheter shedding, and the chemotherapy itself causing liver damage and bone marrow suppression.

The ultimate goal of cancer therapy is to develop antitumor agents that robustly target cancerous cells while sparing normal cells. Currently, the major drawback of using conventional treatments is their lack of selectivity, often resulting in considerable loss of healthy tissue. Many adult stem cells show intrinsic tumor-tropic properties, making them attractive candidates for the targeted delivery of anticancer biological agents (Table 13.1) [46]. The strategy is two-fold: Stem cells can disseminate in solid tumors and migrate toward micrometastatic lesions, enabling

**Table 13.1** Stem cell sources

| Name | Sources | Advantages | Disadvantages |
|---|---|---|---|
| Embryonic SCs | Inner cell mass of the mammalian blastocyst | Pluripotent—enabling them to form derivatives of all three germ layers | Results in destruction of the embryo, making them an ethically controversial source<br>Form teratomas when transplanted in vivo, limiting their current clinical value |
| Hematopoietic SCs | Bone marrow, umbilical cord blood, and peripheral blood<br>Embryonic SCs<br>Induced pluripotent SCs | Multipotent—can form lymphoid and myeloid blood cells<br>Extracted in high yields<br>Readily cryopreserved<br>Intrinsic tumor-tropic properties | Limited differentiation potential |
| Mesenchymal SCs | Fetal origins (Wharton's jelly and cord blood)<br>Developing tooth bud of the mandibular third molar<br>Adult tissues such as bone marrow and adipose tissue<br>Embryonic SCs<br>Induced pluripotent SCs | Differentiate into mesenchymal lineages that make up bone, cartilage, fat, and muscle<br>Readily cryopreserved<br>Many types are tumor-tropic<br>Readily genetically manipulated | Limited differentiation potential<br>Limited yield depending on source |
| Neural SCs | Brain, spinal cord, and retina<br>Embryonic SCs<br>Induced pluripotent SCs | Multipotent—can give rise to neurons, astrocytes, and oligodendrocytes<br>Tumor-tropic properties<br>Readily genetically manipulated | Limited differentiation potential<br>Difficult to source |
| Endothelial SCs | Bone marrow<br>Embryonic SCs<br>Induced pluripotent SCs | Multipotent—give rise to endothelial precursor cells, which form blood and lymphatic vessels<br>Readily cryopreserved<br>Readily genetically manipulated | Limited differentiation potential |
| Induced SCs | Derived from somatic cells using reprogramming technologies | Can be driven into many different cell types<br>Creation of patient-specific cell types<br>Can be made using various viral and nonviral methods | Tumorigenicity poses a considerable clinical hurdle<br>Heterogeneity in final induced SC population<br>Technically challenging<br>Low efficiency of conversion |

site-specific delivery. Furthermore, stem cells can be modified to stably express or release various anticancer agents, thereby circumventing the short half-lives that many chemotherapeutic agents exhibit.

### 13.4.1 Genetic Modification of Stem Cells to Secrete Anticancer Proteins

Stem cell secretion of therapeutic proteins can be divided into two broad categories depending on whether these agents act directly on malignant cells or on supporting cells in the tumor environment, such as blood vessel cells and stromal vessel cells. Stem cells are typically modified by viral transduction to express transgenes encoding secretable effector proteins. Direct effectors include the proapoptotic protein, tumor necrosis factor-related apoptosis-inducing ligand (TRAIL), which binds to death receptor 4 (DR4; also known as TRAILR1). TRAIL is a very promising antitumor molecule; it can cause apoptosis in many types of tumor cells and does not damage normal cells [47]. The main difficulties in applying TRAIL to clinical tumor treatment are that its action time is short, and that it cannot be effectively delivered to the tumor site. These difficulties can be effectively solved through the use of stem cells as a carrier. Table 13.2 shows a summary of the applications of TRAIL with stem cells in tumor treatment.

Using proteins that can outcompete or statically block the binding of endogenous ligands to their cognate receptors is another strategy that results in the inhibition of proliferation pathways in the cancer and associated cells. For example, the stem cell expression of biological agents that bind to epidermal growth factor receptor (EGFR) [60] or its tumor-specific variant EGFRvIII [61] and cytokines, such as interferon-β (IFN-β) [62–65] and IFN-α [66], has been shown to negatively regulate tumor growth in various preclinical cancer models.

Indirect effectors include agents that inhibit the formation of tumor-associated vasculature, such as antiangiogenic thrombospondin-1 [67] (TSP1; also known as THBS1) and PEX (cross-linked polyethylene) [68] (a fragment of MMP2 (metalloproteinases 2)); these agents effectively limit the growth of the tumor mass by creating a hostile, nonpermissive microenvironment.

### 13.4.2 Genetic Modification of Stem Cells to Induce Cancer Cell Death

Stem cell-mediated suicide gene therapy describes a strategy whereby stem cells are engineered to express an enzyme that converts a separately administered nontoxic prodrug into a cytotoxic drug that can efficiently kill surrounding cells by the bystander effect. An added benefit of stem cell-mediated suicide therapy is that the stem cell is eliminated after exerting its therapeutic effect, thereby abolishing any

**Table 13.2** Stem cell delivery of TRAIL

| Stem cell type | Form of TRAIL | Tumor type | Combination therapy | Principal finding | Reference |
|---|---|---|---|---|---|
| Human UCB-MSCs | Adenoviral infection of secretable trimeric TRAIL | U87-MG | Irradiation of tumor | Irradiation enhances tumor tropism and therapeutic potential of SCs | [48] |
| Human BM-MSCs | Lentiviral infection of secretable TRAIL | Primary and established human GBM | – | Use of real-time imaging to follow migration and therapeutic effect of MSCs on tumor volume in vivo | [49] |
| Human A-MSCs | Non-viral nucleofection of TRAIL | Rat glioma model | – | Reduction of tumor volume and significant survival benefit in vivo | [50] |
| Rat BM-MSCs | Adenoviral transduction of dodecameric TRAIL | Renal cell carcinoma (RCC) | Coexpression of herpes simplex virus thymidine kinase | Complete elimination of established RCC in vivo | [51] |
| Human MSC/ mouse MSC | Secretable TRAIL | Established GBMs | Stem cells express herpes simplex virus thymidine kinase | Stem cells are eliminated as the therapeutic effect of the addition of the prodrug ganciclovir | [52] |
| MSCs | Secretable TRAIL introduced using nonviral PEI(600)-Cyd | Lung metastasis | – | Lung tumor homing and reduction in metastasis | [53] |
| Mouse NSCs | Lentiviral infection of secretable TRAIL | GBM6/8/12 in vitro, Gli36-EGFRVIII in vivo | PI-103 (PI3-kinase/ Mtor inhibitor) | PI-103 augments in-vivo response of gliomas to TRAIL | [54] |
| Mouse NPCs | Lentiviral infection of secretable TRAIL | Human U87 established glioma model | Locked nucleic acid (LNA) anti-miR-21 oligonucleotides | Synergism with TRAIL resulting in eradication of tumor in vivo | [55] |

**Table 13.2** (continued)

| Stem cell type | Form of TRAIL | Tumor type | Combination therapy | Principal finding | Reference |
|---|---|---|---|---|---|
| Human MSCs | LV-TRAIL under tet promoter | Pulmonary metastasis model | Conditional expression of TRAIL using DOX | Cleared metastatic disease in lung cancer model | [56] |
| Human A-MSCs | Secretable TRAIL | Malignant fibrous histiocytoma | – | Decrease in metastasis | [57] |
| Mouse NSC/ human MSC | Secretable TRAIL | Established and primary GBMs | Stem cells encapsulated in sECM | Increased retention of stem cells within resection cavity | [58] |
| Human MSC | Inducible TRAIL | Breast and bone metastasis model | Stem cells encapsulated in silk scaffold | Halts breast cancer growth and decreases degree of bone and lung metastasis | [59] |
| Mouse NSC | LV-EGFR-nanobody TRAIL (ENb2-TRAIL) | Established GBMs | – | Targets EGFR and TRAIL signaling pathways simultaneously | [60] |

*UCB-MSCs* umbilical cord blood-derived mesenchymal stem cells, *BM-MSCs* bone marrow-derived mesenchymal stem cells, *A-MSCs* adipose-derived mesenchymal stem cells, *NSCs* neural stem cells, *NPCs* neural progenitor cells, *sECM* synthetic extracellular matrix, *EGFR* epidermal growth factor receptor, *U87-MG* human malignant glioma cells, *GBM* Glioblastoma multiforme, *PEI*(600)-*Cyd* PEI molecules with MW of 600 Da to form a CyD-containing polymer with MW of 61 kDa, *Gli*36-EGFRVIII human glioma cells, *PI* 3,8-Diamino-5-[3-(diethylmethylammonio) propyl]-6-phenylphenanthridinium diiodide. *mTOR* The mammalian TOR, *miR*: MicroRNA, *LV* Lentivirus, *DOX* Doxorubicin, *sECM* extracellular matrix, *ENb* EGFR- specific nanobodies

concern over its long-term fate. Three major suicide gene systems are currently used. Cytosine deaminase (CD) converts 5-fluorocytosine (5-FC) to the toxic anti-metabolite 5-fluorouracil. The herpes simplex virus thymidine kinase (HSV-tk) converts ganciclovir (GCV) to GCV-monophosphate, which is further phosphorylated to GCV-triphosphate, which potently blocks DNA synthesis. In addition, carboxylesterase (CE) converts the prodrug irinotecan to the potent topoisomerase inhibitor SN-38. The CD/5-FC system has been used in modified MSCs and NSCs and applied in mouse models to treat tumors of the brain, such as glioblastoma [69–72] and medulloblastoma [73], often resulting in regression of the tumor mass and prolonged survival [74]. The HSV-tk system, which relies on the formation of gap junctions between the stem cells and surrounding target cells for an efficient bystander effect, has shown efficacy in several animal models of cancer, including glioblastoma [52, 75] and breast [76] and prostate cancer [77]. Human NSCs harnessing the CE–irinotecan system have proved to be effective in preclinical models of ovarian [78] and lung cancers [79], as well as in medulloblastoma [80].

### 13.4.3 Stem Cells as Nanoparticle Carriers

Nanoscience and nanotechnology operate on the scale of 1–100 nm and the material characters are different from large scale materials. The development of nanoscience and nanotechnology in the medical field has brought about profound changes, especially in aspects such as the treatment of cancer and cardiovascular disease. The size of nanomaterials commonly used in the medical field is 10–100 nm. Nanodrugs, compared with conventional drugs, have the following characteristics: small particles, large specific surface area, high surface reactivity, an active center, and strong adsorption ability. Therefore, nanodrugs may have the following features: (1) They are slow-release drugs, changing the half-life of the drug in the body, prolonging action time. (2) They reduce or eliminate side effects, and, as a result of reduced drug dosage, they reduce tolerance and improve the curative effect. (3) They change the membrane transport mechanism, increasing biological membrane permeability, and this is advantageous for transdermal drug absorption and efficacy of action in the cell. (4) They show increased solubility, thus improving the drug preparation. (5) Because of their local retention, contact with the intestinal wall is increased in both time and area, improving the degree of oral drug absorption. At present, the use of nanodrugs is aimed at intractable diseases, including tumors. Due to the effect of nanodrugs in enhancing tumor tissue permeability and drug retention, tumor-targeting is more effective, and nanodrugs have largely overcome or compensated for the drawbacks of traditional cancer treatments, such radiotherapy and chemotherapy which candidate drugs in clinical oncology therapy. Although nanomaterials applied in medical research show many advantages, there are also some drawbacks, such as their short retention in the body, lack of effective diffusion in solid tumors, and inability to target small metastatic lesions [81]. A strategy to solve these problems is to use stem cells as a carrier of the nanomaterials [82].

Nanomaterials loaded with stem cells can target a tumor site in the body and release nanomaterials in the tumor itself, so as to achieve efficient treatment of the tumor [83]. When stem cells are used as a carrier, technical issues to consider include whether or not the stem cells provide an effective loading material, and control of the nanomaterial release so as to ensure long-term effective tumor treatment. Recently, a new method has been developed. This involves making a drug "sandwich" between silica nanoparticles on MSCs. MSCs from the tumor cell secretion of cytokines can attract active tracking to tumor cells, can send drug-loaded nanoparticles to various parts of the tumor tissue, and the released drugs lead to tumor cell apoptosis. This new targeted method employs chemotaxis between MSCs as a "targeted carrier", which, like a "Trojan horse", go inside the tumor cells and completely kill them; compared with the traditional targeting method it is more proactive and shows greater targeting properties [84].

Generally, stem cells loading in stem cells via passive transport or assist with stem cell membrane protein can target to tumor. The tumor therapy results mainly depend on nanomaterial such as convert light to thermal energy thus cause a high temperature in tumor [85–87].

Our group's research results in gastric cancer have shown that magnetic fluorescent nanoparticle-labeled MSCs can be used effectively for hyperthermia treatment [88]. The synthesis of fluorescent magnetic nanoparticles and their co-culture with mouse bone marrow MSCs, in order to achieve cell labeling, are shown in Fig. 13.8.

On transmission electron microscopy, nuclear magnetic and fluorescence microscopic mirror results confirmed that the magnetic fluorescent nanoparticles were markers of gastric cancer cells, as shown in Fig. 13.9.

In vivo experimental results have demonstrated that the labeled bone marrow MSCs could target to tumor site after tail vein injection and stable in tumor site 14 days detected by the fluorescent and magnetic signal as shown in Fig. 13.10. The tumor growth could be inhibited by thermal treatment by alternating a magnetic field as shown in Fig. 13.11.

The research results show that CCL19/CCR7 and CXCL12/CXCR4 played a role in the mechanism of tumor tropism, as shown in Fig. 13.12.

### *13.4.4 Stem Cells Loaded with Oncolytic Virus*

Oncolytic viruses have the ability to selectively replicate in and kill tumor cells while sparing healthy cells. In 1991, Martuza and others [89] published an article in the journal *Science,* demonstrating that a thymidine kinase-negative mutant of herpes simplex virus-1 has a certain effect in malignant glioma therapy; using HSV for soluble virus tumor treatment has since attracted increasing attention. The principle of this treatment is based on the fact that the weak form of virus used to create special soluble tumor virus proliferation in tumor cells and then the large number of replication virus destroy cancer cells [90]. In recent decades, although the use of soluble tumor viruses has caused widespread public concern over the treatment, related research has made great progress.

However, the therapeutic efficacy of soluble tumor viruses is confounded by clearance of the virus by host defense mechanisms following systemic administration, and insufficient viral spread following intratumoral administration. In an effort to improve the delivery of oncolytic virus therapeutics and circumvent antiviral immunity, several studies have explored the possibility of using stem cells as delivery vehicles [91–94]. Different types of stem cells have been used as carriers of soluble tumor viruses that are released at the tumor site in ovarian cancer [92], breast cancer [95], and glioblastoma [96–98]. In animal experiments, the use of stem cell-loaded viruses significantly reduced tumor size and extended the life of the animal [96–100]. Figure 13.13 is a summary of the applications of stem cells as carriers in tumor treatment [46].

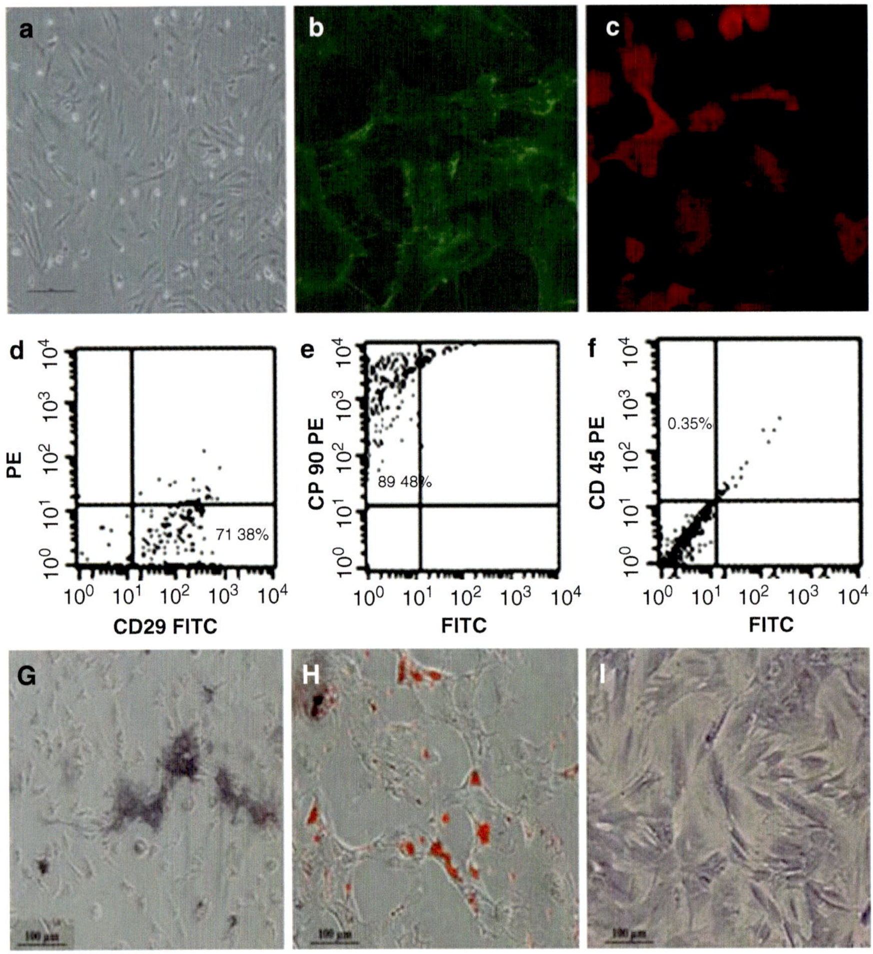


**Fig. 13.8** Morphology and immunophenotypic characterization of mesenchymal stem cells (MSCs). (**a**) The fibroblastic morphology of passage-three MSCs (magnification = ×100); (**b**) MSCs stained with FITC-conjugated CD29 antibody (×200); (**c**) MSCs stained with PE-conjugated CD90 antibody (×200); (**d–f**) MSCs analyzed by FACS for the positive expression of CD29 (**d**) and CD90 (**e**) and negative expression of CD45 (**f**). (**g**) Differentiated osteoblasts tested with alkaline phosphatase staining (×100); (**h**) differentiated adipocytes characterized by oil red O staining (×100); (**i**) differentiated chondrocytes verified by toluidine blue staining (×100)

## 13.5 The Safety of Stem Cells

Stem cells have two important properties: the ability to self-renew and differentiate into specialized cell types and the capacity to home toward pathology sites and malignant lesions. Although these features are very important from a regenerative standpoint, they present potential safety concerns when the cells are introduced into

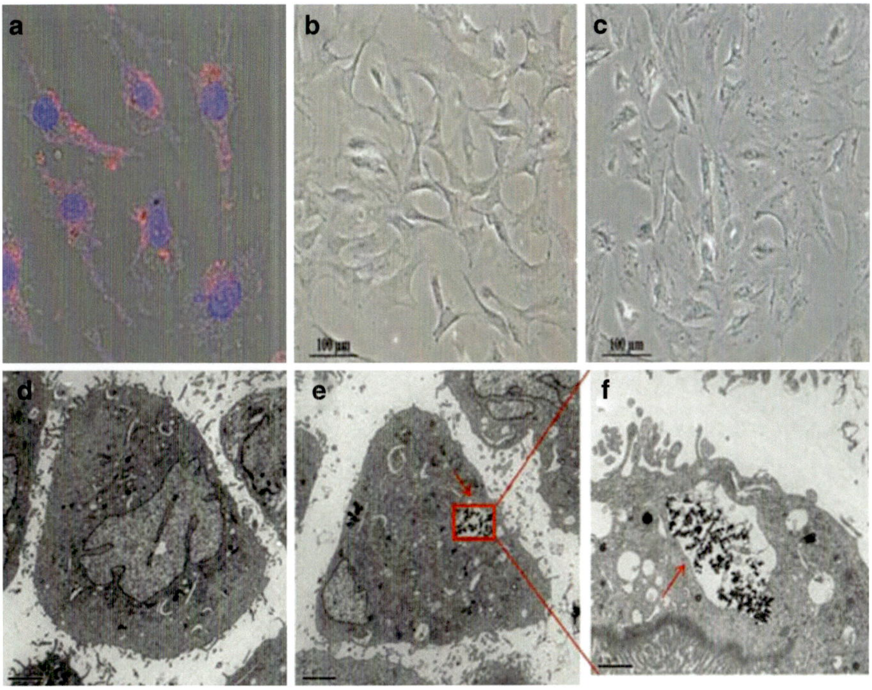


**Fig. 13.9** Evaluation of FMNP-labeled MSCs. (**a**) Cell nuclei and FMNPs were visible in the MSCs by fluorescence microscope observation (×200); (**b**) FMNP-labeled MSCs were detected by using Prussian blue staining (×100); (**c**) unlabeled MSCs were detected by using Prussian blue staining (×100); (**d**) transmission electron microscope (TEM) image of unlabeled MSCs. (**e**) The TEM image of FMNP-labeled MSCs illustrates that amino-modified FMNPs were randomly distributed in the cytoplasm of MSCs (×6,000). (**f**) Magnifiied image of FMNPs distributed in the cytoplasm of MSCs (×12,000)

a recipient. Of particular concern is whether stem cells promote the growth of certain tumors [101–103] or indeed form tumors themselves [104, 105]. Table 13.3 shows a summary of the issues that will be faced when stem cells are used in the clinical setting [46].

Non-immortalized adult stem cells (such as MSCs, NSCs, hematopoietic stem cells, and endothelial stem cells) cause fewer safety concerns than their immortalized counterparts (such as embryonic stem cells and induced pluripotent stem cells), especially when they are autologous and delivered into a niche similar to the one from which they were derived [46]. Stem cells that are engineered to express antitumor agents need rigorous testing to ensure that the stem cells have not been rendered tumorigenic. The incorporation of suicide genes into therapeutic stem cells enables their controlled eradication, thereby providing a safety mechanism to alleviate this concern. Furthermore, many stem cell-based therapies, including the use of oncolytic viruses and the delivery of nanoparticles, result in the death of the stem cells upon release of the therapeutic agent, thereby

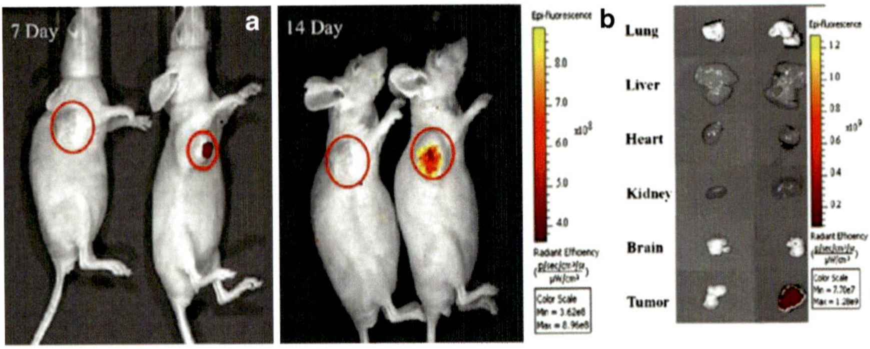


**Fig. 13.10** Fluorescence imaging of FMNP-labeled MSCs targeting gastric cancer cells in vivo. (**a**) The in-vivo fluorescent images show that tumor sites of the mice in the test group had fluorescent signals at 7 and 14 days after the injection of FMNP-labeled MSCs (*right*), while the tumor sites of the mice in the control group had no fluorescent signal at 7 and 14 days after the injection of FMNPs (*left*). (**b**) Fluorescence imaging of major organs shows that no signal was detected in the tumor or organs of the control group (*left*), while obviously fluorescent signals were detected in the tumor tissues of the test group (*right*)

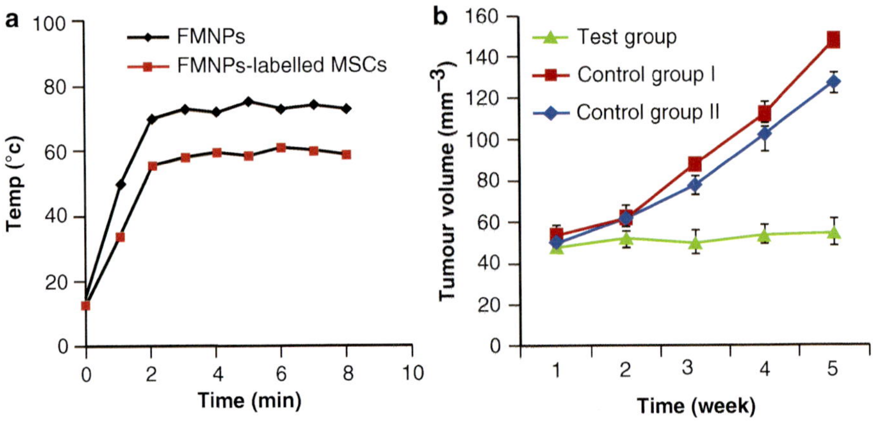


**Fig. 13.11** Effects of alternating magnetic field on gastric tumor size. (**a**) Time-temperature curve of FMNP-labeled MSCs under an alternating magnetic field with 63 kHz and 7 kA/m; (**b**) the growth of tumors was inhibited with hyperthermia therapy

effectively abolishing the possibility of any tumor formation or errant differentiation. The advent of reprogramming technologies — the ability to convert a somatic cell into a pluripotent or multipotent stem cell — provides additional avenues for creating therapeutic patient-derived cells. Despite their huge potential, these cells have been shown to form teratomas in mice, and this is a considerable clinical hurdle that must be overcome before these cells can be transplanted into patients; overcoming this challenge is an area of intense research [106]. This challenge has already been overcome in the case of induced NSCs that were created from human fetal fibroblasts and were shown not to aberrantly proliferate when implanted into

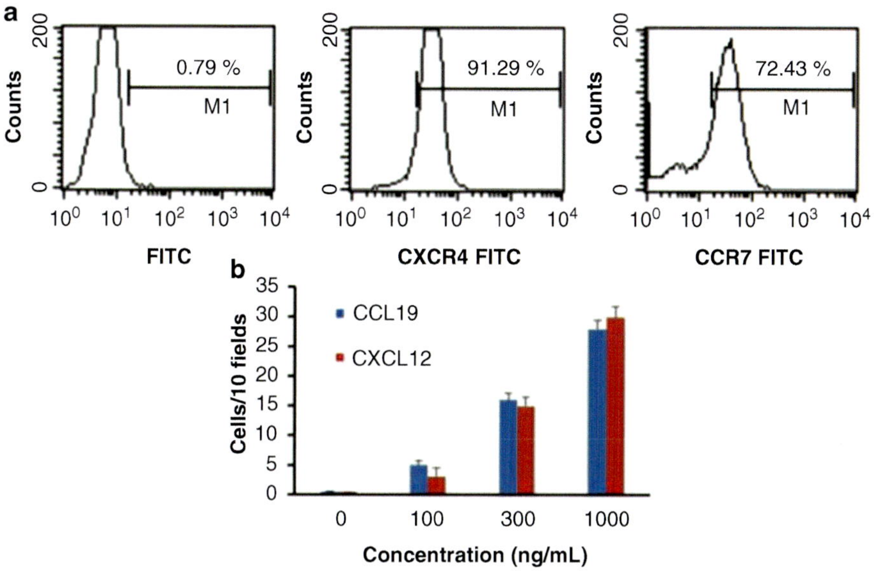


**Fig. 13.12** Analysis of chemokine receptors and chemotaxis. (**a**) The expressions of CXCR4 and CCR7 on MSCs were determined by FACS. (**b**) The migration of MSCs to cells with different concentrations of CXCL12 and CCL19 was evaluated by chemotaxis assay

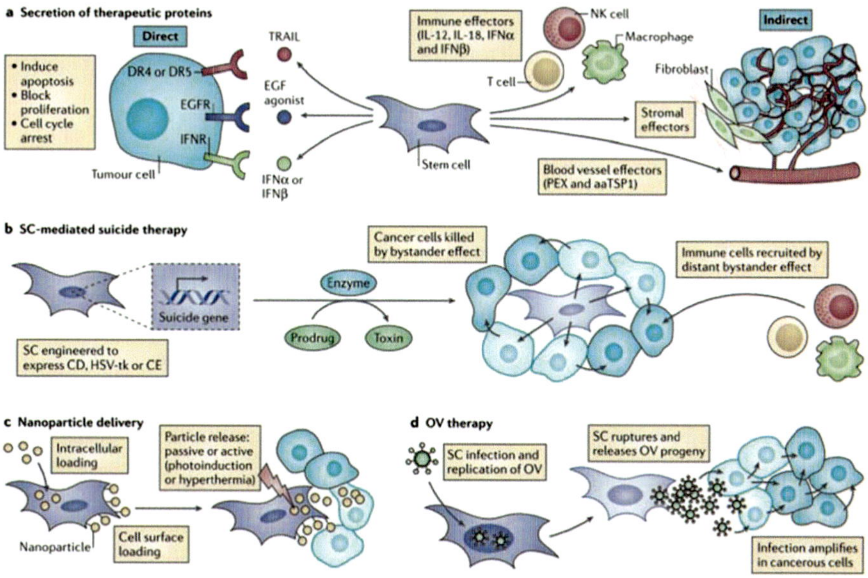


**Fig. 13.13** Stem cells promote tumor cell death

**Table 13.3** Hurdles on the way to the clinic

| Consideration | Reason | Finding a solution |
|---|---|---|
| Optimal SC type | SCs vary greatly, even within the same class of SC | Thorough molecular and functional characterization is required |
| Quality control | Minimize variability to ensure homogeneity of SC population | Good manufacturing practice-compliant isolation and culture |
| Route of administration or cell dosing | Maximize in vivo effectiveness | Rigorous preclinical characterization and appropriate scaling from animal to human required |
| Tumorigenicity or cell fate | Avoid the occurrence of therapeutic SCs forming tumors or aberrantly differentiating in host | Test tumorigenicity or differentiation potential in preclinical models<br>Ensure efficacy of suicide gene therapy to eliminate SCs post therapy |
| Poor engraftment or survival | Efficacy of therapy will be impaired if SCs do not survive or are poorly engrafted in recipient | Use autologous SCs that are less readily rejected<br>Use SCs derived from a similar niche to that in which the cancer resides<br>Engineer SCs with enhanced survival capacity |
| Funding | Stopping a clinical trial midway could jeopardize patient health and hinder medical benefit | Make sure costs are carefully considered and the solvency of funding sources are verified |
| Ensure modifications to SCs to maintain their characteristics | Therapeutic modifications might adversely affect performance and/or safety of SCs | Test modified SCs for inappropriate genomic modifications and verify functionality in a preclinical context |
| Therapeutic SCs or secreted proteins breaking host tolerance to self-antigens | Evoking an immune response might impair effectiveness of therapy and provoke additional complications in the patient | Consider using autologous SCs<br>Engineer SCs to exhibit an immunoevasive phenotype<br>Engineer secreted proteins that are hypoimmunogenic |

mice [107]. Clearly, these safety concerns need to be addressed to limit undue harm to the patient and to prevent a situation whereby the therapeutic stem cell exacerbates cancer progression. Careful selection of the stem cell type, maintaining the purity of the stem cell population, and determination of the effects of any modifications on the function of the stem cells should be established in preclinical studies and facilitated by a thorough understanding of stem cell biology.

## References

1. Takahashi K, Yamanaka S. Induction of pluripotent stem cells from mouse embryonic and adult fibroblast cultures by defined factors. Cell. 2006;126:663–76.
2. Ruan J, Shen J, Wang Z, Ji J, Song H, Wang K, et al. Efficient preparation and labeling of human induced pluripotent stem cells by nanotechnology. Int J Nanomedicine. 2011;6:425–35.
3. Balkwill F, Mantovani A. Inflammation and cancer: back to Virchow? Lancet. 2001;357:539–45.
4. Wang TC, Goldenring JR, Dangler C, Ito S, Mueller A, Jeon WOOKYU, et al. Mice lacking secretory phospholipase A 2 show altered apoptosis and differentiation with Helicobacter felis infection. Gastroenterology. 1998;114:675–89.
5. Dongyan C, Dong H. Advances of gastric cancer stem cells. Int J Tumor. 2011;38:51–3.
6. Matsumoto Y, Marusawa H, Kinoshita K, Endo Y, Kou T, Morisawa T, et al. Helicobacter pylori infection triggers aberrant expression of activation-induced cytidine deaminase in gastric epithelium. Nat Med. 2007;13:470–6. doi:10.1038/nm1566.
7. Takaishi S, Wang TC. Providing AID to p53 mutagenesis. Nat Med. 2007;13:404–6. doi:10.1038/nm0407-404.
8. Jiang Y, Jahagirdar BN, Reinhardt RL, Schwartz RE, Keene CD, Ortiz-Gonzalez XR, et al. Pluripotency of mesenchymal stem cells derived from adult marrow. Nature. 2002;418:41–9. doi:10.1038/nature05812.
9. Krause DS, Theise ND, Collector MI, Henegariu O, Hwang S, Gardner R, et al. Multi-organ, multi-lineage engraftment by a single bone marrow-derived stem cell. Cell. 2001;105:369–77. doi:10.1016/S0092-8674(01)00328-2.
10. Houghton J, Stoicov C, Nomura S, Rogers AB, Carlson J, Li H, et al. Gastric cancer originating from bone marrow – derived cells. Science. 2004;306:1568–71.
11. Bonnet D, Dick JE. Human acute myeloid leukemia is organized as a hierarchy that originates from a primitive hematopoietic cell. Nat Med. 1997;3:730–7.
12. Lobo NA, Shimono Y, Qian D, Clarke MF. The biology of cancer stem cells. Annu Rev Cell Dev Biol. 2007;23:675–99.
13. Lee J, Olofsson BA, Mwidau A, Escudero M, Flemington E, Azizkhan J, et al. Prospective identification of tumorigenic breast cancer cells. Proc Natl Acad Sci U S A. 2003;100:3983–8.
14. Singh SK. Identification of a cancer stem cell in human brain tumors. Cancer Res. 2003;63:5821–8.
15. Collins AT, Berry PA, Hyde C, Stower MJ, Maitland NJ. Prospective identification of tumorigenic prostate cancer stem cells. Cancer Res. 2005;65:10946–51.
16. Fang D, Nguyen TK, Leishear K, Finko R, Kulp AN, Hotz S, et al. A tumorigenic subpopulation with stem cell properties in melanomas. Cancer Res. 2005;65:9328–37.
17. Ma S, Chan KW, Hu L, Lee TKW, Wo JYH, Ng IOL, et al. Identification and characterization of tumorigenic liver cancer stem/progenitor cells. Gastroenterology. 2007;132:2542–56.
18. Hermann PC, Huber SL, Herrler T, Aicher A, Ellwart JW, Guba M, et al. Distinct populations of cancer stem cells determine tumor growth and metastatic activity in human pancreatic cancer. Cell Stem Cell. 2007;1:313–23.
19. Li C, Heidt DG, Dalerba P, Burant CF, Zhang L, Adsay V, et al. Identification of pancreatic cancer stem cells. Cancer Res. 2007;67:1030–7.
20. Ricci-Vitiani L, Lombardi DG, Pilozzi E, Biffoni M, Todaro M, Peschle C, et al. Identification and expansion of human colon-cancer-initiating cells. Nature. 2007;445:111–5.
21. Dalerba P, Dylla SJ, Park I-K, Liu R, Wang X, Cho RW, et al. Phenotypic characterization of human colorectal cancer stem cells. Proc Natl Acad Sci U S A. 2007;104:10158–63.

22. O'Brien CA, Pollett A, Gallinger S, Dick JE. A human colon cancer cell capable of initiating tumour growth in immunodeficient mice. Nature. 2007;445:106–10.
23. Prince ME, Sivanandan R, Kaczorowski A, Wolf GT, Kaplan MJ, Dalerba P, et al. Identification of a subpopulation of cells with cancer stem cell properties in head and neck squamous cell carcinoma. Proc Natl Acad Sci U S A. 2007;104:973–8.
24. Takaishi S, Okumura T, Tu S, Wang SSW, Shibata W, Vigneshwaran R, Gordon SAK, Shimada Y, Wang TC. Identification of gastric cancer stem cells using the cell surface marker CD44. Stem Cells. 2009;27:1006–20.
25. Zhang C, Li C, He F, Cai Y, Yang H. Identification of CD44+CD24+ gastric cancer stem cells. J Cancer Res Clin Oncol. 2011;137:1679–86.
26. Zeilstra J, Joosten SPJ, Dokter M, Verwiel E, Spaargaren M, Pals ST. Deletion of the WNT target and cancer stem cell marker CD44 in Apc(Min/+) mice attenuates intestinal tumorigenesis. Cancer Res. 2008;68:3655–61.
27. Trosko JE. From adult stem cells to cancer stem cells Oct-4 gene, cell-cell communication, and hormones during tumor promotion. Ann N Y Acad Sci. 2006;1089:36–58.
28. Wei Q, Xiaowei Z, Li Z, Fengchun Z. Expression of Oct-4 in gastric carcinoma and its clinical significance. J Shanghai Jiao Tong University (Medical science). 2009;29:733–6.
29. Liang S, Li C, Zhang C, Chen Y, Xu L, Bao C, et al. CD44v6 monoclonal antibody-conjugated gold nanostars for targeted photoacoustic imaging and plasmonic photothermal therapy of gastric cancer stem-like cells. Theranostics. 2015;5:970–84.
30. Tokar EJ, Diwan BA, Waalkes MP. Arsenic exposure transforms human epithelial stem/progenitor cells into a cancer stem-like phenotype. Environ Health Perspect. 2010;118:108–15.
31. Bomken S, Fiser K, Heidenreich O, Vormoor J. Understanding the cancer stem cell. Br J Cancer. 2010;103:439–45.
32. Lever E, Sheer D. Gastrointestinal stem cells in development and cancer. J Pathol. 2010;220:114–25.
33. Sciences H, Arab U, Biology C. Defining epithelial cell progenitors in the human oxyntic mucosa. Stem Cells. 2003;21:322–36.
34. Spradling A, Drummond-Barbosa D, Kai T. Stem cells find their niche. Nature. 2001;414:98–104.
35. Liu G, Zhou C, Xie C, Yang Z, Lv N, Liu G, et al. Recent advances in research of gastric cancer stem cells. World Chinese J Digestol. 2012;20:574–9.
36. Suárez-Álvarez B, López-Vázquez A, López-Larrea C. Mobilization and homing of hematopoietic stem cells. Adv Exp Med Biol. 2012;741:152–70.
37. Shi M, Li J, Liao L, Chen B, Li B, Chen L, et al. Regulation of CXCR4 expression in human mesenchymal stem cells by cytokine treatment: role in homing efficiency in NOD/SCID mice. Haematologica. 2007;92:897–904.
38. Wynn RF, Hart CA, Corradi-Perini C, O'Neill L, Evans CA, Wraith JE, et al. A small proportion of mesenchymal stem cells strongly expresses functionally active CXCR4 receptor capable of promoting migration to bone marrow. Blood. 2004;104:2643–5.
39. Nie F, Yu X-L, Wang X-G, Tang Y-F, Wang L-L, Ma L. Down-regulation of CacyBP is associated with poor prognosis and the effects on COX-2 expression in breast cancer. Int J Oncol. 2010;37:1261–9.
40. Guo Y, Hangoc G, Bian H, Pelus LM, Broxmeyer HE. SDF-1/CXCL12 enhances survival and chemotaxis of murine embryonic stem cells and production of primitive and definitive hematopoietic progenitor cells. Stem Cells. 2005;23:1324–32.
41. Koizumi S, Gu C, Amano S, Yamamoto S, Ihara H, Tokuyama T, et al. Migration of mouse-induced pluripotent stem cells to glioma-conditioned medium is mediated by tumor-associated specific growth factors. Oncol Lett. 2011;2:283–8. doi:10.3892/ol.2011.234.
42. Kendall SE, Najbauer J, Johnston HF, Metz MZ, Li S, Bowers M, et al. Neural stem cell targeting of glioma is dependent on phosphoinositide 3-kinase signaling. Stem Cells. 2008;26:1575–86.

43. Schmidt NO, Przylecki W, Yang W, Ziu M, Teng Y, Kim SU, et al. Brain tumor tropism of transplanted human neural stem cells is induced by vascular endothelial growth factor. Neoplasia (New York, NY). 2005;7:623–9.
44. Ho IAW, Chan KYW, Ng W-H, Guo CM, Hui KM, Cheang P, et al. Matrix metalloproteinase 1 is necessary for the migration of human bone marrow-derived mesenchymal stem cells toward human glioma. Stem Cells. 2009;27:1366–75.
45. Ruan J, Song H, Li C, Bao C, Fu H, Wang K, et al. DiR-labeled embryonic stem cells for targeted imaging of in vivo gastric cancer cells. Theranostics. 2012;2:618–28.
46. Stuckey DW, Shah K. Stem cell-based therapies for cancer treatment: separating hope from hype. Nat Rev Cancer. 2014;14(10):683–91.
47. Stuckey DW, Shah K. TRAIL on trial: preclinical advances in cancer therapy. Trends Mol Med. 2013;19:685–94.
48. Kim SM, Oh JH, Park SA, Ryu CH, Lim JY, Kim D-S, et al. Irradiation enhances the tumor tropism and therapeutic potential of tumor necrosis factor-related apoptosis-inducing ligand-secreting human umbilical cord blood-derived mesenchymal stem cells in glioma therapy. Stem Cells. 2010;28:2217–28.
49. Sasportas LS, Kasmieh R, Wakimoto H, Hingtgen S, van de Water JAJM, Mohapatra G, et al. Assessment of therapeutic efficacy and fate of engineered human mesenchymal stem cells for cancer therapy. Proc Natl Acad Sci U S A. 2009;106:4822–7.
50. Choi SA, Hwang S-K, Wang K-C, Cho B-K, Phi JH, Lee JY, et al. Therapeutic efficacy and safety of TRAIL-producing human adipose tissue-derived mesenchymal stem cells against experimental brainstem glioma. Neuro Oncol. 2011;13:61–9.
51. Kim SW, Kim SJ, Park SH, Yang HG, Kang MC, Choi YW, et al. Complete regression of metastatic renal cell carcinoma by multiple injections of engineered mesenchymal stem cells expressing dodecameric TRAIL and HSV-TK. Clin Cancer Res. 2013;19:415–27.
52. Martinez-Quintanilla J, Choi SH, Bhere D, Heidari P, He D, Mahmood U, et al. Therapeutic efficacy and fate of bimodal engineered stem cells in mice models of malignant brain tumors. Stem Cells. 2013;15:1706–14.
53. Hu YL, Huang B, Zhang TY, Miao PH, Tang GP, Tabata Y, et al. Mesenchymal stem cells as a novel carrier for targeted delivery of gene in cancer therapy based on nonviral transfection. Mol Pharm. 2012;9:2698–709. doi:10.1021/mp300254s.
54. Bagci-Onder T, Wakimoto H, Anderegg M, Cameron C, Shah K. A dual PI3K/mTOR inhibitor, PI-103, cooperates with stem cell-delivered TRAIL in experimental glioma models. Cancer Res. 2011;71:154–63.
55. Corsten MF, Miranda R, Kasmieh R, Krichevsky AM, Weissleder R, Shah K. MicroRNA-21 knockdown disrupts glioma growth in vivo and displays synergistic cytotoxicity with neural precursor cell-delivered S-TRAIL in human gliomas. Cancer Res. 2007;67:8994–9000.
56. Loebinger MR, Eddaoudi A, Davies D, Janes SM. Mesenchymal stem cell delivery of TRAIL can eliminate metastatic cancer. Cancer Res. 2009;69:4134–42.
57. Lee HJ, Yang H-M, Choi Y-S, Park S-H, Moon S-H, Lee Y-S, et al. A therapeutic strategy for metastatic malignant fibrous histiocytoma through mesenchymal stromal cell-mediated TRAIL production. Ann Surg. 2012;257:1.
58. Kauer TM, Figueiredo J-L, Hingtgen S, Shah K. Encapsulated therapeutic stem cells implanted in the tumor resection cavity induce cell death in gliomas. Nat Neurosci. 2011;15:197–204.
59. Reagan MR, Seib FPP, McMillin DW, Sage EKW, Mitsiades CS, Janes SM, et al. Stem cell implants for cancer therapy: TRAIL-expressing mesenchymal stem cells target cancer cells in situ. J Breast Cancer. 2012;15:273–82.
60. Van De Water JAJM, Bagci-onder T, Agarwal AS, Wakimoto H, Roovers RC. Therapeutic stem cells expressing variants of EGFR-specific nanobodies have antitumor effects. Proc Natl Acad Sci U S A. 2012;109(41):16642–7.

61. Balyasnikova IV, Ferguson SD, Sengupta S, Han Y, Lesniak MS. Mesenchymal stem cells modified with a single-chain antibody against EGFRvIII successfully inhibit the growth of human xenograft malignant glioma. PLoS One. 2010;5:e9750.
62. Dembinski JL, Wilson SM, Spaeth EL, Studeny M, Samudio I, Roby K, et al. Tumor stroma engraftment of gene-modified mesenchymal stem cells as anti-tumor therapy against ovarian cancer. Cytotherapy. 2014;15:20–32. doi:10.1016/j.jcyt.2012.10.003.Tumor.
63. Ito S, Natsume A, Shimato S, Ohno M, Kato T, Chansakul P, et al. Human neural stem cells transduced with IFN-beta and cytosine deaminase genes intensify bystander effect in experimental glioma. Cancer Gene Ther. 2010;17:299–306.
64. Studeny M, Marini FC, Champlin RE, Zompetta C, Fidler IJ, Andreeff M. Bone marrow-derived mesenchymal stem cells as vehicles for interferon-beta delivery into tumors. Cancer Res. 2002;62:3603–8.
65. Ponnazhagan S. Cancer gene therapy using mesenchymal stem cells expressing interferon-β in a mouse prostate cancer lung metastasis model. Gene Ther. 2009;15:1446–53.
66. Ponnazhagan S. Therapeutic potential of mesenchymal stem cells producing IFN-α in a mouse melanoma lung metastasis model. Stem Cells. 2008;26:2332–8.
67. Manuscript A. Human stem cells expressing novel TSP-1 variant have anti-angiogenic effect on brain tumors. Changes. 2012;29:997–1003.
68. Kim S-K, Cargioli TG, Machluf M, Yang W, Sun Y, Al-Hashem R, et al. PEX-producing human neural stem cells inhibit tumor growth in a mouse glioma model. Clin Cancer Res Off J Am Assoc Cancer Res. 2005;11:5965–70.
69. Kosaka H, Ichikawa T, Kurozumi K, Kambara H, Inoue S, Maruo T, et al. Therapeutic effect of suicide gene-transferred mesenchymal stem cells in a rat model of glioma. Cancer Gene Ther. 2012;19:572–8.
70. Zhao Y, Lam DH, Yang J, Lin J, Tham CK, Ng WH, et al. Targeted suicide gene therapy for glioma using human embryonic stem cell-derived neural stem cells genetically modified by baculoviral vectors. Gene Ther. 2012;19:189–200.
71. Aboody KS, Najbauer J, Metz MZ, Apuzzo MD, Annala AJ, Synold TW, et al. Neural stem cell – mediated enzyme/prodrug therapy for glioma : preclinical studies. Sci Transl Med. 2013;5:1–11. doi:10.1126/scitranslmed.3005365.
72. Altaner C, Altanerova V, Cihova M, Ondicova K, Rychly B, Baciak L, et al. Complete regression of glioblastoma by mesenchymal stem cells mediated prodrug gene therapy simulating clinical therapeutic scenario. Int J Cancer. 2014;134:1458–65.
73. Kim SK, Kim SU, Park IH, Bang JH, Aboody KS, Wang KC, et al. Human neural stem cells target experimental intracranial medulloblastoma and deliver a therapeutic gene leading to tumor regression. Clin Cancer Res. 2006;12:5550–6.
74. You M-H, Kim W-J, Shim W, Lee S-R, Lee G, Choi S, et al. Cytosine deaminase-producing human mesenchymal stem cells mediate an antitumor effect in a mouse xenograft model. J Gastroenterol Hepatol. 2009;24:1393–400. doi:10.1111/j.1440-1746.2009.05862.x.
75. Ryu CH, Park KY, Kim SM, Jeong CH, Woo JS, Hou Y, et al. Valproic acid enhances anti-tumor effect of mesenchymal stem cell mediated HSV-TK gene therapy in intracranial glioma. Biochem Biophys Res Commun. 2012;421:585–90.
76. Yang J, Lam D, Goh S, Lee E, Zhao Y. Tumor tropism intravenously injected human-induced pluripotent stem cell-derived neural stem cells and their gene therapy application in a metastatic breast cancer. Stem Cells. 2012;30:1021–9. doi:10.1002/22.
77. Lee WYW, Zhang T, Lau CPY, Wang CC, Chan K-M, Li G. Immortalized human fetal bone marrow-derived mesenchymal stromal cell expressing suicide gene for anti-tumor therapy in vitro and in vivo. Cytotherapy. 2013;15:1484–97.
78. Kim KY, Kim SU, Leung PCK, Jeung EB, Choi KC. Influence of the prodrugs 5-fluorocytosine and CPT-11 on ovarian cancer cells using genetically engineered stem cells: tumor-tropic potential and inhibition of ovarian cancer cell growth. Cancer Sci. 2010;101:955–62.

79. Hong SH, Lee HJ, An J, Lim I, Borlongan C, Aboody KS, et al. Human neural stem cells expressing carboxyl esterase target and inhibit tumor growth of lung cancer brain metastases. Cancer Gene Ther. 2013;20:678–82.
80. Gutova M, Shackleford GM, Khankaldyyan V, Herrmann KA, Shi X-H, Mittelholtz K, et al. Neural stem cell-mediated CE/CPT-11 enzyme/prodrug therapy in transgenic mouse model of intracerebellar medulloblastoma. Gene Ther. 2013;20:143–50.
81. Bertrand N, Wu J, Xu X, Kamaly N, Farokhzad OC. Cancer nanotechnology: the impact of passive and active targeting in the era of modern cancer biology. Adv Drug Deliv Rev. 2014;66:2–25.
82. Roger M, Clavreul A, Venier-Julienne MC, Passirani C, Sindji L, Schiller P, et al. Mesenchymal stem cells as cellular vehicles for delivery of nanoparticles to brain tumors. Biomaterials. 2010;31:8393–401.
83. Auffinger B, Morshed R, Tobias A, Cheng Y, Ahmed AU, Lesniak MS. Drug-loaded nanoparticle systems and adult stem cells: a potential marriage for the treatment of malignant glioma? Oncotarget. 2013;4:378–96.
84. Li L, Guan Y, Liu H, Hao N, Liu T, Meng X, et al. Silica nanorattle-doxorubicin- anchored mesenchymal stem cells for tumor-tropic therapy. ACS Nano. 2011;5:7462–70.
85. Duchi S, Sotgiu G, Lucarelli E, Ballestri M, Dozza B, Santi S, et al. Mesenchymal stem cells as delivery vehicle of porphyrin loaded nanoparticles: effective photoinduced in vitro killing of osteosarcoma. J Control Release. 2013;168:225–37.
86. Schnarr K, Mooney R, Weng Y, Zhao D, Garcia E, Armstrong B, et al. Gold nanoparticle-loaded neural stem cells for photothermal ablation of cancer. Adv Healthc Mater. 2013;2:976–82.
87. Rachakatla RS, Balivada S, Seo GM, Myers CB, Wang H, Samarakoon TN, et al. Attenuation of mouse melanoma by A/C magnetic field after delivery of bi-magnetic nanoparticles by neural progenitor cells. ACS Nano. 2010;4:7093–104.
88. Ruan J, Ji J, Song H, Qian Q, Wang K, Wang C, et al. Fluorescent magnetic nanoparticle-labeled mesenchymal stem cells for targeted imaging and hyperthermia therapy of gastric cancer. Nanoscale Res Lett. 2012;7:309.
89. Martuza RL, Malick A, Markert JM, Ruffner KL, Coen DM. Experimental therapy of human glioma by means of a genetically engineered virus mutant. Science. 1991;252:854–6.
90. Aghi M, Martuza RL. Oncolytic viral therapies - the clinical experience. Oncogene. 2005;24:7802–16.
91. García-Castro J, Alemany R, Cascalló M, Martínez-Quintanilla J, Arriero MDM, Lassaletta A, et al. Treatment of metastatic neuroblastoma with systemic oncolytic virotherapy delivered by autologous mesenchymal stem cells: an exploratory study. Cancer Gene Ther. 2010;17:476–83.
92. Komarova S, Kawakami Y, Stoff-Khalili MA, Curiel DT, Pereboeva L. Mesenchymal progenitor cells as cellular vehicles for delivery of oncolytic adenoviruses. Mol Cancer Ther. 2006;5:755–66.
93. Kranzler J, Tyler MA, Sonabend AM, Ulasov IV, Lesniak MS. Stem cells as delivery vehicles for oncolytic adenoviral virotherapy. Curr Gene Ther. 2009;9:389–95.
94. Ahmed AU, Rolle CE, Tyler MA, Han Y, Sengupta S, Wainwright DA, et al. Bone marrow mesenchymal stem cells loaded with an oncolytic adenovirus suppress the anti-adenoviral immune response in the cotton rat model. Mol Ther J Am Soc Gene Ther. 2010;18:1846–56.
95. Stoff-Khalili MA, Rivera AA, Mathis JM, Banerjee NS, Moon AS, Hess A, et al. Mesenchymal stem cells as a vehicle for targeted delivery of CRAds to lung metastases of breast carcinoma. Breast Cancer Res Treat. 2007;105:157–67.
96. Sonabend AM, Ulasov IV, Tyler MA, Rivera AA, Mathis JM, Lesniak MS. Mesenchymal stem cells effectively deliver an oncolytic adenovirus to intracranial glioma. Stem Cells. 2008;26:831–41.

97. Manuscript A, Cytokines P, Transplantation P. Human bone marrow-derived mesenchymal stem cells for intravascular delivery of oncolytic adenovirus Delta24-RGD to human gliomas. Cancer Res. 2010;9:1–14.
98. Ahmed AU, Tyler MA, Thaci B, Alexiades NG, Han Y, Ulasov IV. A comparative study of neural and mesenchymal stem cell- based carriers for oncolytic adenovirus in a model of malignant glioma. Mol Pharm. 2012;8:1559–72.
99. Mader EK, Butler G, Dowdy SC, Mariani A, Knutson KL, Federspiel MJ, et al. Optimizing patient derived mesenchymal stem cells as virus carriers for a phase I clinical trial in ovarian cancer. J Transl Med. 2013;11:20. doi:10.1186/1479-5876-11-20.
100. Ong HT, Federspiel MJ, Guo CM, Ooi LL, Russell SJ, Peng KW, et al. Systemically delivered measles virus-infected mesenchymal stem cells can evade host immunity to inhibit liver cancer growth. J Hepatol. 2013;59:999–1006.
101. Karnoub AE, Dash AB, Vo AP, Sullivan A, Brooks MW, Bell GW, et al. Mesenchymal stem cells within tumour stroma promote breast cancer metastasis. Nature. 2007;449:557–63. doi:10.1038/nature06188.
102. Yang T, Zhang X, Wang M, Zhang J, Huang F, Cai J, et al. Activation of mesenchymal stem cells by macrophages prompts human gastric cancer growth through NF-kB pathway. PLoS One. 2014;9:1–11.
103. Rowan BG, Gimble JM, Sheng M, Anbalagan M, Jones RK, Frazier TP, et al. Human adipose tissue-derived stromal/stem cells promote migration and early metastasis of triple negative breast cancer xenografts. PLoS One. 2014;9:e89595.
104. Røsland GV, Svendsen A, Torsvik A, Sobala E, McCormack E, Immervoll H, et al. Long-term cultures of bone marrow-derived human mesenchymal stem cells frequently undergo spontaneous malignant transformation. Cancer Res. 2009;69:5331–9.
105. Amariglio N, Hirshberg A, Scheithauer BW, Cohen Y, Loewenthal R, Trakhtenbrot L, et al. Donor-derived brain tumor following neural stem cell transplantation in an ataxia telangiectasia patient. PLoS Med. 2009;6:0221–31.
106. Lee AS, Tang C, Rao MS, Weissman IL, Wu JC. Tumorigenicity as a clinical hurdle for pluripotent stem cell therapies. Nat Med. 2013;19:998–1004. doi:10.1038/nm.3267.
107. Li G, Walker D, Zhang WR, Kreitzer AC. Direct reprogramming of mouse and human fibroblasts into multipotent neural stem cells with a single factor. Cell Stem Cell. 2013;11: 100–9.

# Chapter 14
# Safety Assessment of Nanoprobes

**Yanlei Liu, Yuxia Wang, and Daxiang Cui**

## 14.1 Introduction

The application of nanotechnology in different parts of our life is expanding significantly with the rapid development of nanoscience and nanotechnology. The nanomaterials exhibit excellent properties in many fields. From the time that the concept of nanotechnology was firstly proposed to now, a period of more than 50 years, nanotechnology and nanomaterials have been applied in many a fields such as medicine, some food, cosmetics and dye industries, as well as the place to treat environmental pollution. Nanotechnology and nanomaterials are significant to the development of economy, the improvement of technology. Moreover, it is domestic to find that the nanotechnology research and related investments are favorable. However, the biological safety of the new materials and new technologies may still be problematic. Furthermore, the scientists begin to realize that there is a new problem: nanotechnology is similar to the industrial revolution, which brings us huge benefits and changes the quality of life. However, at the same time, it also causes environmental pollution, ecological destruction, and many kinds of diseases.

Y. Liu (✉) • Y. Wang • D. Cui
Institute of Nano Biomedicine and Engineering, Shanghai Engineering Research Center for Intelligent Diagnosis and Treatment Instrument, National Center for Translational Medicine, Collaborative Innovational Center for System Biology, Shanghai Jiao Tong University, 800 Dongchuan Road, Shanghai 200240, P. R. China
e-mail: dxcui@sjtu.edu.cn

© Springer Science+Business Media B.V. and
Shanghai Jiao Tong University Press, Shanghai 2017

D. Cui (ed.), *Gastric Cancer Prewarning and Early Diagnosis System*,
Translational Medicine Research, DOI 10.1007/978-94-024-0951-2_14

## 14.2 International Opinions and Measurement on the Safety of Nanotechnology

### *14.2.1 International Organizations*

In August 2003, an article by the Greenpeace organization was published in the *New Scientist*: "Technology in the future, today's choice". The report warned of problems that may arise with nanotechnology, and a hazard analysis was performed. The authors noted that nanomaterials may contain new pollutants that scientists do not yet fully understand; they also noted the emphasis on the current development and application of nanotechnology, whereas studies of its safety are seriously lagging behind. Further calls were made for governments and nanoindustries to invest more funds for the more comprehensive and in-depth assessments of the safety of nanotechnology. And the authors warned the nanotechnology industry of serious negative issues arising from nanotechnology.

### *14.2.2 Other Organizations*

In 2003, the United States federal government allocated 5 million to the US Center for Nanotechnology for an investigation of the application of nanotechnology and of ethical issues associated with the expansion of the investigation, and the National Science Foundation allocated 2 million for sociologists to examine the social consequences of nanotechnology research. In June 2003, the British Government, the Royal Academy of Engineering, and the Royal Society reviewed the ethical and social significance of nanotechnology.

### *14.2.3 The Organizations of China*

In China, nanomaterials have caused security issues and some initial results have been achieved with regard to the safety of these materials. The theme of a conference of experts from the fields of nanoscience, biology, chemistry, medicine, physics, and environment, held in Xiangshan in 2004, was "The biological effects of nanoscale materials," and more than 40 experts from these fields called for strengthening the biological safety of nanomaterials and nanotechnology. A number of organizations around the world have shown great responses and close attention to the biosafety issues of nanomaterials and they are striving to clarify the mechanisms underlying the biological effects of nanomaterisls before nanotechnology matures to an industrial scale.

This chapter explains the applications of nanomaterials in medicine and related fields, and evaluates the safety of nanomaterials. These materials may have a potential impact on organisms, and thus safety evaluation is a necessity.

## 14.3 Nanotechnology and Nanomaterials in Biomedicine Field [1–5]

Nanomaterials have extremely widespread applications in the biomedical field and play an important role in clinical preventive medicine and in the field of biomedical detection and diagnosis.

### *14.3.1 Medical Diagnosis*

With the development of nanoscience and nanotechnology, medical laboratory diagnosis has seen unprecedented development. In medical laboratory diagnosis, nanomaterials are used in three main ways: (1) to label specific elements in vivo, allowing a target element to be traced in order to assess the accumulation and excretion of that element in the body; (2) to label blood cells with specific target nanomaterials and use in vitro diagnostic tests to detect the labeled nanomaterials in blood; and (3) for genetic diagnosis in the clinical field, molecular hybridization and fluorescence detection of target DNA fragments are used; this particular application has great prospects. Gold nanoparticles were hybridized with DNA fragments, and then into the chromosome of the body nucleus and nucleus-specific combination, because of its high specificity, this method can solve the problem currently facing genetic diagnosis.

### *14.3.2 Diagnostic Imaging*

Particles of different sizes have different features with regard to selective enrichment in the cells, tissues, and organs of organisms. To achieve passive targeting, biocompatible nanoparticles are needed, and they can be enriched in certain cells, tissues, or organs in vivo. Alternatively, nanomaterials can be functionally modified, and their surfaces can react with antibody molecules by chemical bonding, Such nanoparticles may be active in vivo for targeting molecules to a relevant antigen or antigen expression in the target cells. In magnetic resonance imaging diagnosis, with the intravenous injection of magnetic nanoparticle contrast agents, these agents can be taken to various organs in the body. However, magnetic nanoparticles can be captured only by reticular endothelial cells in the liver and spleen, and as malignant tissue contains only very small numbers of such cells, normal tissue and cancer tissue can be clearly distinguished. The specificity of the diagnostic agent is extremely important, especially for the early diagnosis of liver cancer.

### 14.3.3 Drug Carriers

In the pharmaceutical field, nanomaterials are widely used as drug carriers. In normal tissue, endothelial cells are closely packed, and the distance between the cells is not more than 2 nm; in tumor tissue, the distance between the vascular endothelial cells is greater than 100 nm, due to the rapid growth of the cancer cells, and there are new blood vessels and structural defects. The tumor vascular tissue is incomplete, and the permeability of the vessel wall is greater than that in nontumorous vascular tissue; the small size of the nanoparticles allows them to accumulate at the tumor site after passing through the vessel wall, and a low drug dosage can achieve a high therapeutic effect. Nanoparticles show the following advantages as drug carriers: (1) Targeted positioning; after nanomaterials are combined with the drug, the nanoparticles enter the body in two ways, through active targeting (nanomaterials link with a target cell-specific ligand) and passive targeting (Enhance permeability and retention (EPR) effect). (2) Controlled release of the drug; with nanoparticles, the drug treatment window is maintained for a longer time, thus reducing the frequency of administration. For ready degradation of many traditional drugs when they went via liver or other organs, nanomaterials can also protect drug adsorption, after arrival in specific parts of the nanoparticles, the drug release, thereby enhancing the efficacy of drugs. (3) Increased drug absorption rates; nanomaterials can alter the membrane transport mechanism of drugs, and can transport drugs through the three biological barriers (blood–brain barrier, air – blood barrier, and placental barrier) into tissues and organs.

### 14.3.4 Photothermal Therapy

Some time ago, scientists discovered that high temperatures could induce normal cell death; when the temperature rises to 42 C or more, the heat begins to damage cells and cause cell death. Thermal killing of cells depends on the exposure level (the amount of cancer cell was infiltrated in nanodrugs) and exposure time at high temperatures, and this heat-killing phenomenon is more obvious in tumor cells. Currently used thermal treatments such as ultrasound, microwave, and radiofrequency do not achieve exclusive heat transfer to tumor cells. In the heat treatment of tumors, thermal damage also occurs in normal tissue around the tumor site, and this greatly restricts the use of heat treatment. Nanomaterials such as gold star, gold rod, and gold corner nanoparticles have certain physical and chemical properties that lend themselves to use in photothermal therapy. In the near infrared range, these nanoparticles have a strong surface plasmon resonance absorption effect, and carbon quantum dots are efficiently converted into heat. In the body, gold nanoparticles can accumulate selectively at a tumor site, and when the tumor site is irradiated with a near-infrared laser, the gold nanoparticles absorb near-infrared light and efficiently convert this into heat, so that a high temperature is produced at the tumor site, thus killing the tumor cells. Because few nanoparticles are distributed in normal tissues and organs, and these tissues do not receive near-infrared light irradiation, these tissues are not subjected to excessive temperatures or injury; thus, this use of nanoparticles significantly improves the efficacy and safety of thermal treatment.

### *14.3.5 Artificial Bone*

Artificial nanomaterials are used for bone repair and to promote the reconstruction of damaged tissue; these materials have a variety of excellent physical and chemical properties and biological characteristics that simulate the basic physiological composition of natural bone. These nanomaterials have good biocompatibility; they are self-curing; they are of appropriate mechanical strength, and they are degraded and absorbed by the body. They also induce the growth of bone cells and blood vessels, and the materials have no obvious side effects. Artificial bone has a porous bone structure similar to that of natural bone; the porous structure is conducive to vessel formation and some osteoblast invasion, and induces the growth of bone cells and blood vessels, promoting the formation of new bone. The body shows no obvious rejection response to artificial bone, due to its biological activity, and the artificial bone causes no immune activity. Currently the main nanomaterials used in nanobones are: nano-hydroxyapatite, tricalcium phosphate, alumina/zirconia, and certain polymers. These nanomaterials have demonstrated excellent biological performance and have broad application prospects in biological tissue engineering research.

Nanomaterials show great application prospects in the biomedical field; however, the side effects of nanoparticles are, more and more, a focus of people's attention. As nanomaterials and nanotechnology are becoming more closely connected with human relations, it is urgent to resolve the following problems. How many ways can nanomaterials enter the body? How are nanoparticles distributed in and excreted from the body? What will lead to adverse consequences? Multidisciplinary crossover studies are needed to solve these problems.

## 14.4 Methods of Nanoparticles Exposure and Distribution

With the development of the biomedical applications of nanotechnology and nanoparticles, the chance of nanoscale materials entering the body is increasing significantly. People will be exposed to different types of nanomaterials in their daily lives, in such products as cosmetics, textiles, and decoration supplies, while in outdoor activities, people will be exposed to automobile exhausts, which contain large quantities of nanoparticles. These nanomaterials enter the human body by passive exposure routes, the main routes of exposure being the respiratory system and skin. In the biomedical applications of nanomaterials, the designed exposure pathways are primarily the intravenous and gastrointestinal routes. Because nanomaterials are still little used clinically these pathways are less frequent than the passive routes of exposure, although these biomedical exposure approaches claim the most attention in experimental studies. In daily life, as stated above, nanomaterials enter the body mainly through the respiratory system and exposed skin, and production and research and development staff are more likely than the general population to have access to nanomaterials through both these exposure pathways. The chemical and physical properties of nanoparticles, such as their small particle size, allows them to pass through biological barriers and enter cells. Because of its large surface area, strong absorption capacity and high chemical activity, even though same composition, the character and intensity of biological effects on body may change; The same physical and chemical properties of nanomaterials, after surface modification of different functional groups, the physiological distribution and metabolic dynamics of the nanoparticles are significantly different in the body's. In order to more fully utilize the advantages of nanoparticles and use them safely, much research is needed. Currently in-vivo nanoparticle research is carried out primarily on the four exposure pathways described below.

### *14.4.1 Inhalation Exposure*

According to the prediction model of the International Commission on Radiological Protection (ICPR; 1994) (Fig. 14.1), different sizes of nanoparticles (10 nm ~ 10,000 nm) entering the respiratory system may show different proportions after deposition in the nasal cavity, trachea, bronchi, and alveoli; more than 80 % of the inhaled nanoparticles (100 nm) can be deposited in the respiratory tract. The smaller-size particles tend to be deposited in the upper respiratory tract, e.g., 90 % of 1-nm particles are deposited in the nasopharynx, 10 % are deposited in the tracheobronchial area, and there is almost no deposition of this particulate matter in the alveolar region; the particulate matters of 5-nm size are totally deposited over 30% in upper respiratory tract; up to 50 % of 20-nm particles are deposited in the alveolar region; and only 15 % of these particles are deposited in the nasopharynx and tracheobronchial area; the smaller nanoparticles prior to entering into the tracheobronchial and alveolar deposition, due to strong dispersion effects in the upper airway of the nasopharynx. The particles distributed in the alveoli, in airway epithelial cells, can be transported to the interstitial lung tissue and across the air – blood barrier directly into the blood circulation, or they may be distributed in the body through the lymphatic system. Oberdorster et al. [6] found that, after inhalation, the particles can pass through the olfactory mucosa to the olfactory nerve and into the central nervous system, and this is also another confirmed pathway [7]; magnetic iron oxide nanoparticles can pass through the olfactory nerve into the brain before crossing the blood–brain barrier. Experimental studies have reported that inhaled carbon nanotubes can

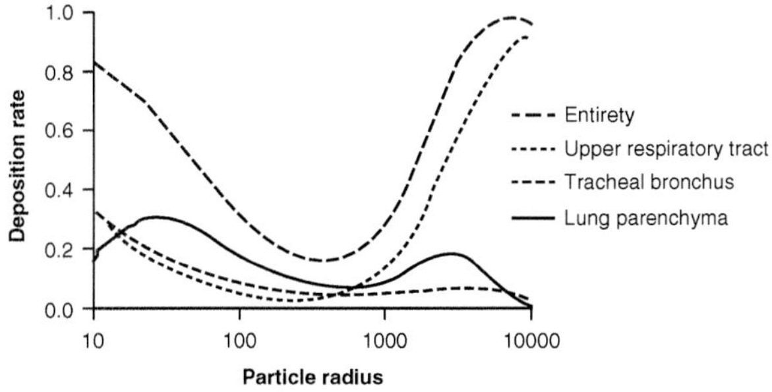


**Fig. 14.1** Nanoparticles deposition in the respiration system

penetrate the trachea and pulmonary blood vessels, and then be distributed to various body organs.

### *14.4.2 Intravenous Injection*

Intravenous exposure is an aggressive pathway. After nanoparticles enter the blood circulation through intravenous administration, they may be rapidly distributed to various organs; the distribution of particulate matter in the body is affected by many factors, such as the size of the particles, their shape, and their surface chemical groups, as well as blood flow and other factors [8]. Most current research focuses on the impact of the particle size on the distribution of particles in the body; the liver is the main organ to which the particles are distributed, followed by the spleen [9–11]. The mononuclear phagocyte system and monocytes/macrophages play a crucial role in the distribution of nanoparticles in a living body; these cells capture the particles and engulf them into the cell interior [12]. It should be noted that there other routes for the distribution of nanoparticles in the body after their distribution in the blood circulation.

### *14.4.3 Gastrointestinal Exposure*

Gastrointestinal exposure is another important way that nanoparticles enter the body, either when swallowed in food contaminated with nanoparticles or when nanoparticles, as carriers of drugs, enter the digestive system after oral administration. However, the transport and distribution of nanoparticles after oral administration is rarely studied, although some studies show that nanoparticles can also be absorbed through the gastrointestinal tract. As for oral radioisotope-labeled rat functional group of the C60 fullerene (polyethylene glycol and albumin dissolution), within 48 h of administration, 98 % of the nanoparticles were excreted in feces and 2 % in urine, indicating that nanoparticles may be absorbed via the gastrointestinal tract into the blood circulation [13]. Another study found that, after oral administration, titanium dioxide particles of 150–200 nm were absorbed from the gastrointestinal tract and distributed in the liver [14].

### 14.4.4 Skin Exposure

Skin exposure to nanoparticles is an important route of exposure in the environment. Because nanoparticles in the air can be precipitated on the skin surface, and nanotechnology is increasingly being used in clothing and cosmetics, this greatly increases the chances of skin being exposed to nanoparticles. Studies of ultraviolet absorption have examined the nanoscale titanium dioxide content of sunscreen to test its ability to penetrate the skin; the results show that the nanoparticles do not penetrate the stratum corneum layer [15, 16]. But research on quantum dot skin penetration shows that the dots can penetrate the skin, although this ability is limited; some kinds of quantum dots can penetrate the stratum corneum layer, in which even can penetrate skin reach leather gathered [17, 18]. There is no study showing that nanoparticles can enter the body through completely healthy skin; they are only distributed, to a limited depth, in the cortex.

### 14.4.5 Distribution in Vivo

A large amount of literature has reported on the distribution of different nanomaterials. Using fluorescent signals or fluorescent material itself as a beacon, dynamic observations of the in-vivo distribution of nanomaterials have been performed, and these have been combined with in-vitro phase and slice observations for the qualitative analysis of the in-vivo distribution results. Inductively coupled plasma mass spectrometry (ICP – MS) is used to quantitatively determine the content of nanomaterials in various organs. The research results show that many factors can influence the distribution of nanomaterials in the body; mainly the size of the nanomaterial, its geometry, and surface modifications of functional groups. A study by Tan et al. compared intravenous drug delivery with nanomaterials of two different geometric shapes: nanorods and nanoballs; the nanoballs showed faster spread in blood than the nanorods [8]. And the smaller the particle size of the nanomaterials or material surface modification polyethylene glycol (PEG), the longer was the blood circulation time, and the greater the body distribution. The longer the circulation time of the nanomaterials, the more material accumulated in the liver, spleen, kidneys, lungs, and other organs, and nanomaterials even accumulated in bone marrow and lymph nodes. In our study, in order to determine the distribution kinetics of nanoclusters in the blood, we designed the following experiment, with Sprague Dawley (SD) rats as the test animals. The dosage of nanoclusters and exposure manner were: 10 mg/kg, iv, 0.25 ml/100 g

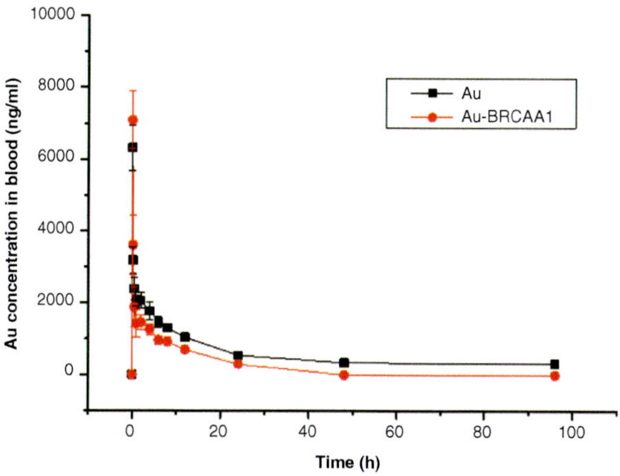


**Fig. 14.2** The distribution of gold nanoclusters in the blood

(0.5 ml/200 g), by slow injection. The two experimental groups (*n*=6 for each) were a gold cluster group and a gold nanocluster -BRCAA1 group. Blood collection time points were: pre-dose (0) and 2 min, 10 min, 0.5 h, 1 h, 2 h, 4 h, 6 h, 8 h, 12 h, and 24 h after injection. Finally, the content of gold nanoparticles in the sample was determined with the ICP-MS assay; the experimental results are shown in Fig. 14.2. The two types of gold particles were injected intravenously into the mice, and the particles were rapidly distributed by the blood to various body organs within in 24 h after injection; after this time the concentration of gold particles in the blood dropped quickly, and the blood concentration of the remaining gold material did not change.

Nanomaterials in the body enter organs with the blood flow. In Hardman's study, 133 days after the intravenous injection of quantum dots, quantum dots were found in the bone marrow and lymph nodes [19]. Many studies have reported on the injection of nanomaterials of various particle sizes modified with a variety of functional groups; after the tail vein injection of nanomaterials, the liver showed the highest aggregation [10, 11, 20, 21]. In our experiment, fluorescent magnetic nanoparticles were injected intravenously into mice and these aggregated in the lungs, the liver, and spleen (Fig. 14.3). The result showed that, within 6 h after injection, the greatest distribution was in the blood, and then nanomaterials aggregated in various organs, the largest amount being in the liver, a finding that is consistent with the experimental results of other groups. After these nanomaterials enter the blood circulation, they can enter other organs, they can cross the blood–brain barrier into the brain, andt hey can cross the placental barrier and be distributed from mother to fetus. It is not known what kind of impact these nanomaterials would have on the brain, and we need to carry out further studies to determine the biological effects overall, as well as the effects at the organ, cellular, and molecular levels.

In the study of nanomaterials, nano-preparations of liposomes have recently been investigated; these particles can enter the blood and undergo metabolic degradation and be excreted in the urine and feces. Nanoparticles made of syn-

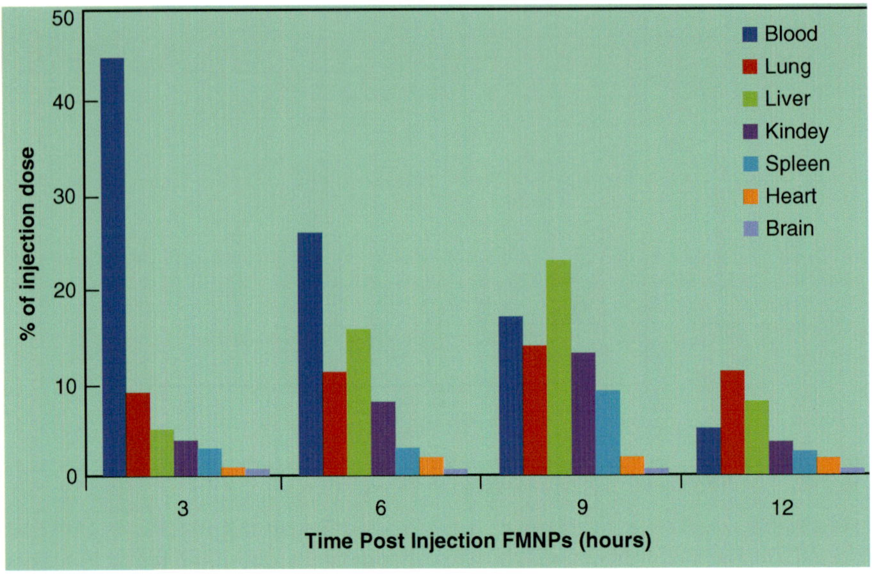


**Fig. 14.3** The distribution of fluorescent magnetic nanoparticles in organs

thetic material or containing a metal element cannot be degraded in a biological environment, and to a large extent, the distribution of nanoparticles depends on the nanomaterial's morphology, surface chemical structure, and size [22, 23]. Hirn et al. [24] prepared gold nanoparticles of five different sizes (1.4, 5, 18, 80, and 200 nm) and different chemical surface modifications, and injected these preparations into the tail veins of mice. Quantitative determination of the distribution and accumulation of the gold nanoparticles in body organs was performed by gamma spectrometry. The research results showed that the distribution of gold nanoparticles in the body was determined by the gold particle surface charge and size; with an increase in the gold nanoparticle size, its concentration in the liver increased, the range from 50% enrichment of 1.4-nm to 99% enrichment of 200-nm in liver, but there are small amounts accumulated in other organs. In another experiment, 2.8-nm gold nanoparticles were prepared with positive or negative charges, to study the influence of the different charges on the distribution of the particles in various organs. With the different charges, the distribution of gold nanoparticles in various organs was quite different. In the liver, the accumulation of negatively charged gold nanoparticles was 81 %, significantly higher than that of the positively charged gold nanoparticles (72 %), and in the spleen the accumulation of positively charged gold nanoparticles (11.4 %) was significantly higher than that of the negatively charged gold nanoparticles (8.6 %).

These different physiological effects can lead to effects that are opposite to the biological effects of conventional materials. A positive biological effect of nanomaterials is their application in the field of nanomedicine; for example, to carry drugs through biological barriers. However, there can be negative biological effects of nanomaterials being carried through biological barriers, as they may and show biological toxicity. From the viewpoint of the healthy development of nanomaterials and nanotechnology, research on the biological toxicity of nanomaterials is urgent.

## 14.5 Nanoparticle Metabolism (Secretion)

In many studies, nanoparticles that contain a metal element are used, and these nanoparticles cannot be metabolized by the living body, and can only be cleared from the body by the excretory system. From the perspective of pharmacology and toxicology, if nanomaterials cannot be effectively metabolized by the body, in vivo accumulation may reach a toxic threshold, resulting in toxicity. Therefore, the study of the metabolic (secretory) pathways by which nanomaterials are cleared from the body is essential for their safe use. Recent research work on the metabolism of nanomaterials and nanoparticles has shown that they are cleared from the body in various ways – from the kidneys, liver, and lungs [25, 26]; however, these studies focused on the clearance mechanisms from a single organ, rather than the overall metabolism of nanomaterials [27]. Current research results show that nanomaterials are cleared from the body in two main ways: by the liver – fecal route and the kidney – urine pathway.

### *14.5.1 Liver – Fecal Route*

Nanoparticles of larger particle size are mainly secreted by the liver – fecal secretory pathway [28, 29]. In the liver, it was shown that nanomaterials were not immediately caught Khufu cells, but there are holes through capillary walls into the Disse space, then after cellular uptake by the liver into the bile duct, the materials finally pass through the bile duct into the duodenum [30]; this study also found that the liver TPPMS (sodium triphenylphosphine meta sulfonate) encapsulated nanoparticles secretion rates in the range of 2.8–200 nm, and the the particle size was in inverse linear correlation with carbon quantum dots (Fig. 14.4). The smaller the particle size, the more easily the particles passed through the liver, where they were secreted with the bile into the small intestine, but when the particle size was 1.4 nm or more, the liver clearance no longer followed the above linear relationship, and the clearance rate was higher than 4.6 % [24].

In our above-mentioned experiment, magnetic fluorescent nanoparticles were injected intravenously into mice and we found that these nanoparticles aggregated in the lungs, the liver, and the spleen (Fig. 14.3). The fluorescent magnetic nanoparticles in the liver increased with time, and the largest amount was seen at the 12-h time point, after which the liver aggregate amount of the nanoparticles began to decline, indicating that the nanoparticles were cleared through the liver – fecal exreretory pathway.

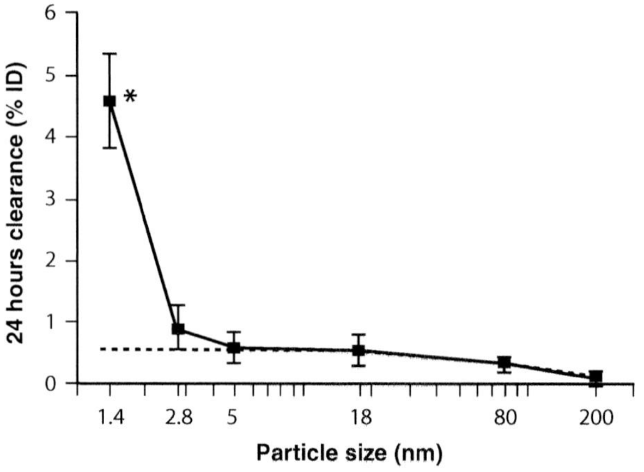


**Fig. 14.4** The rate of removal of nanoparticles from the lungs

### *14.5.2 Kidney – Urine Pathway*

Smaller particle size nanomaterials, such as gold nanoclusters, fullerenes, and quantum points are mainly cleared by the kidney – through secretion in urine [25, 31, 32]. In the normal physiological state, most nanomaterials of molecular weight less than 50 KD and hydration radius of less than 5.5 nm are cleared through the kidney – urine secretory pathway. In one study, polyethylene glycol-wrapped quantum dots administered intravenously into mice accumulated in the kidney 2 h after the injection. Histological analysis of the quantum dots is closely related to the aggregation of mesangial cells, which can make the quantum dots can also phagocytosis by nanomaterials secretion from the kidney is closely associated with the kinds of cells [33]. In our study, we prepared gold clusters with a hydration radius of less than 5.5 nm, and used SD rats as test animals; dose and mode of administration were: 10 mg/kg, intravenous injection, 0.25 ml/100 g (0.5 ml/200 g); the rats were divided into two experimental groups: gold cluster group ($n$=12) and gold nanocluster -BRCAA1 group ($n$=12), with two animals in a metabolic cage for the collection of urine samples. Urine collection times were: pre-dose (0 h), and 0–2 h, 2–4 h, 4–8 h, 8–12 h, 12–24 h, 24–48 h, and 48–96 h after injection, i.e., eight time periods. Finally, by determining the gold content in the urine we were able to determine the passage of the gold nanoclusters through the kidney – urine excretion pathway. The experimental results are shown in Fig. 14.5; 2 h after the injection, the urine showed the highest concentration of gold nanoclusters, as the concentration of gold nanoclusters gradually declined, the gold nanoclusters could readily reach kidney in vivo, and then passed through glomerular cell gap, finally they were eliminated with urine.

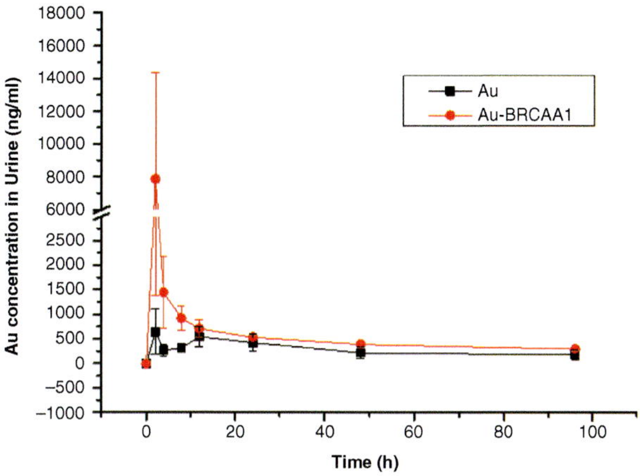


**Fig. 14.5** The distribution of gold nanoclusters in urine

## *14.5.3 Other Pathways*

Current in-vivo studies have shown that there may be metabolic pathways for nanomaterials other than those noted above. Souris et al. found that, after intravenous injections of high concentrations of silica nanospheres (50–100 nm), the intestinal wall and fecal concentrations of the silica nanospheres were higher than the liver concentration [32], and the nanospheres had accumulated in the intestinal wall cells. Meanwhile, another study found that unmodified nanoparticles could be cleared from the zebrafish gut, and the clearance of 50-nm nanoparticles was higher than that of 500-nm particles, and a particle size of 500 nm was far beyond the liver's clearance capacity [34]. Wang et al. found that intestinal goblet cells ingested nanoparticles [35]. The above study [35] showed that organisms secrete gastrointestinal system may exist through another one that is not known nano-particles. At present there are many unknowns about nanomaterial metabolic pathways in the body, and knowledge of these pathways is a necessary condition for the safe use of nanomaterials. Existing research results show that the secretion of nanomaterials from the body depends not only on their material composition and chemistry, but also on their size and shape. It is very difficult to determine the "blood" concentration of nanomaterials via these different exposure pathways in the organism. Because the nanoparticles easily aggregate, particle size may change with changes in the biological microenvironment, and the toxicity of nanomaterials is closely related to their concentration and particle size, and organisms are very complex. After nanomaterials enter the body, "power distribution channels" intersect (Fig. 14.6), and the toxicity of the nanomaterials is closely related to their concentrations in various organs. Therefore, in order to accurately assess the safety of nanomaterials, we need to solve these problems.

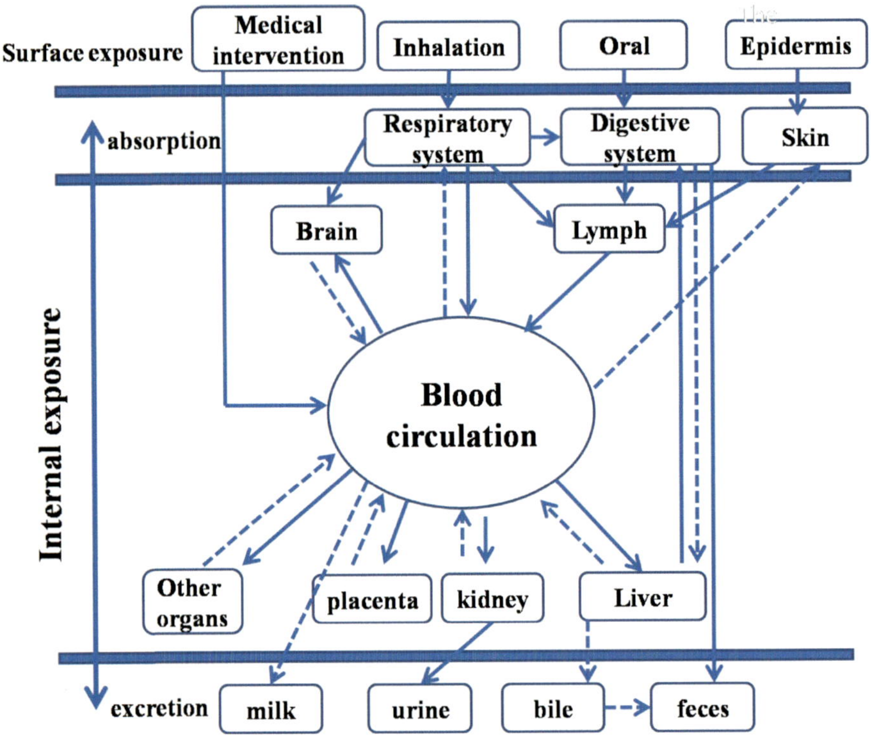


**Fig. 14.6** Diagram of nanoparticle dynamics in the body

## 14.6 Nanoparticle Toxicology

When materials from the "macro-world" are scaled down to the nanometer scale, even if they have the same chemical composition, there are fundamental changes in their physical and chemical properties. These changes will make the nanomaterials exhibit characteristics different from those of macroscopic matter, and also from bizarre physical and chemical properties of a single atom, so when nanomaterials are in a living body, their physiological behavior is very likely to be different from that of conventional materials. For example, they may penetrate biofilm, and, whereas metal ions such as sodium and potassium, and nonmetallic ions such as chloride cannot move freely in and out of the cell membrane, many nanoparticles, such as gold nanorods and single-walled carbon nanotubes can easily move across the cell membrane into the cell. This phenomenon, however, cannot explain the existence of a gap in the biofilm. This is because the molecules after nano materials, the surface charge can be changed, in particular reduced surface charge can increase its fat-soluble, and thus can more easily pass through the biological membrane into the cell. In the large-scale synthesis of nanomaterials, considerable numbers of these materials showed toxicity in animal studies. Many substances at the micron

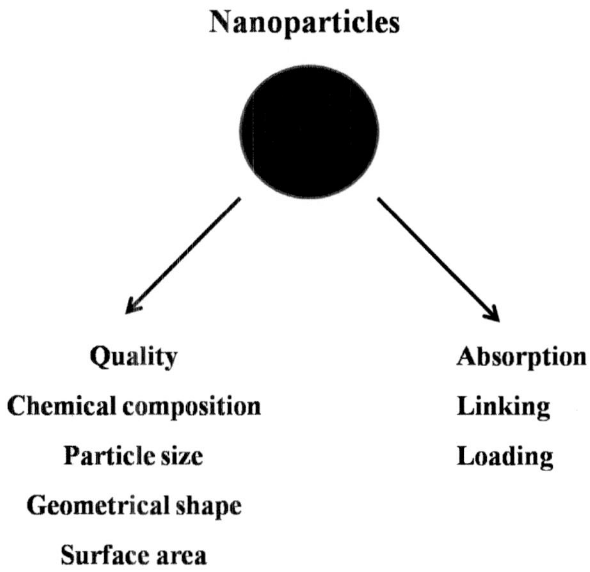


**Fig. 14.7** Properties of nanomaterials and possible related biological effects

scale showed no toxicity, but at the nanoscale they exhibited significant toxicity; this was especially notable with inhaled nanomaterials, which caused inflammation of the lungs, and even death. In a study, the rats were exposed to 20-nm particle size polytetrafluoroethylene for 14 minutes, most of which died within 4 hours, while the rats exposed to a particle size of 120 nm of polytetrafluoroethylene did not show significant toxicity. Therefore, the safety evaluation of macroscopic matter may not apply to nanoscale materials, so when performing safety evaluation of nanomaterials, in reference to traditional toxicology, we must also look into nanotoxicology, The toxicity of nanomaterials is closely related to their morphology, particle size, potential and surface-modified nanomaterials (Fig. 14.7). Nanomaterials used in a living organism must have good biological compatibility, which is a necessary condition for their safety. However, nanomaterials are exogenous substances, i.e., "foreign bodies", and, in general, foreign bodies can cause injury to organisms. For nanoparticles used in living organisms, we need to study the morphology, particle size, potential and surface modification and other aspects of the nanomaterials in relation to the molecules, cells, organs, systems, and biological effects in vivo. We need to identify the relationship between the particle size of nanomaterials, their geometry, surface functional groups, and chemical composition and their toxic properties. Possible mechanisms of the toxic effects of nanomaterials may be free radical mechanisms, molecular mechanisms, and immune mechanisms.

The safety of nanomaterials has been studied in many laboratories, but the results obtained so far are inconclusive.

Braydich-Stolle et al. used original stem cells in mice as a cell model to assess the safety and toxicity of nanomaterials; the nanomaterials showed significant

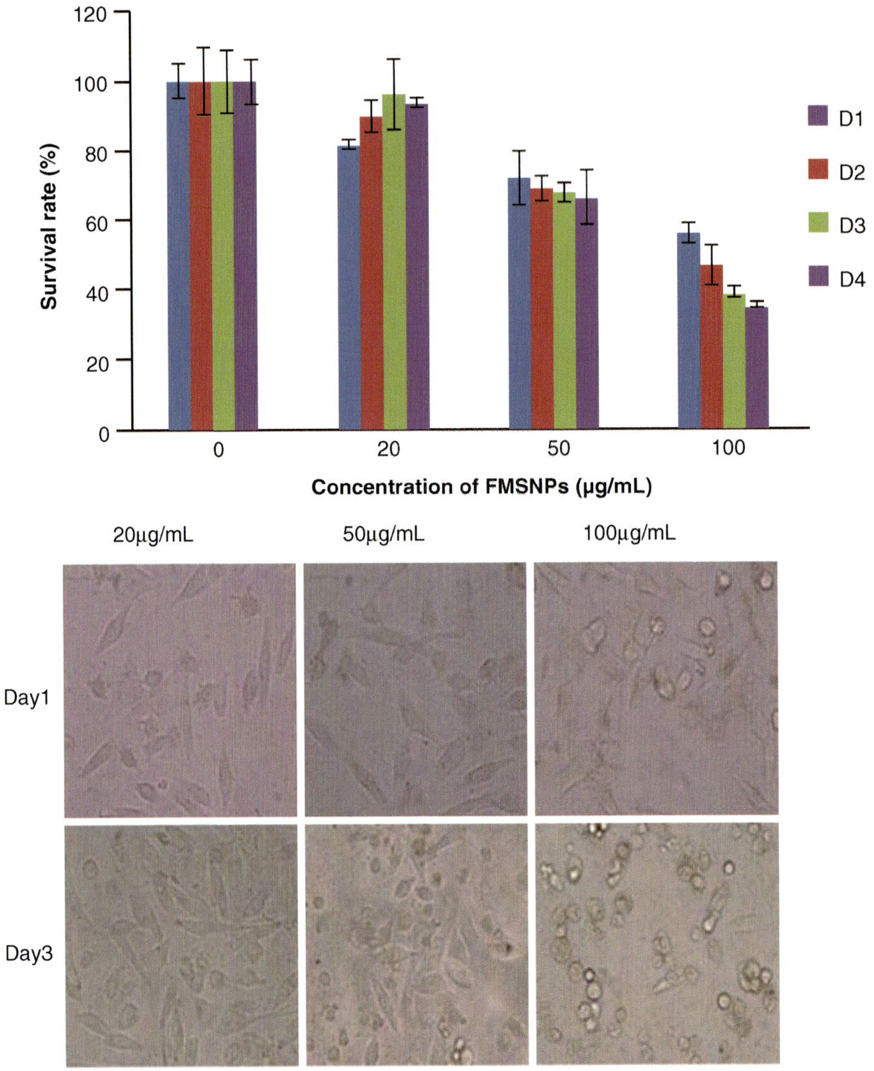


**Fig. 14.8** The effect of magnetic nanoparticles on cell

toxicity, and the toxicity was dose-dependent, with increasing doses showing greater toxicity [36]. Similarly, in our experiment, as shown in Fig. 14.8, with an increased concentration of fluorescent magnetic nanoparticles, the cell survival rate was lower.

Goodman et al. showed that positively charged gold particles had obvious cytotoxicity, while negatively charged gold particles were relatively non-toxic, and there was also a clear dose relationship: high concentration of nanoparticle had greater toxicity, as well as the nanoparticles with different charges had different

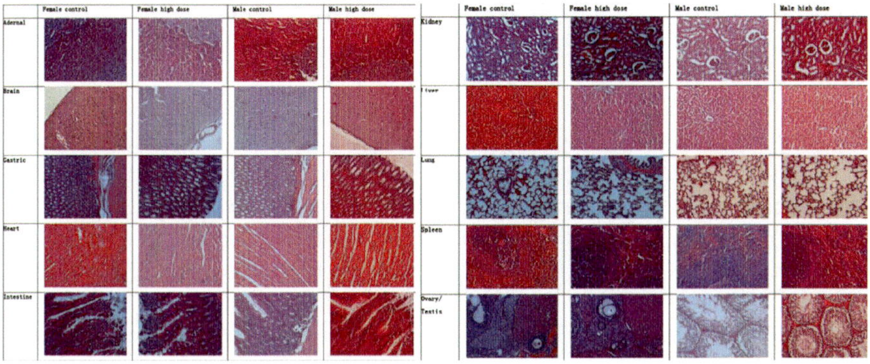


**Fig. 14.9** Pathology samples of the main organs of mice (× 100)

cell toxicity, which determined the amounts of nanomaterials when they went into cells, so that it shown different toxicity [37]. Derfus et al. Used primary hepatocytes as an experimental model for the toxicity testing of nuclear CdSe quantum dots, and the results showed that, under certain conditions, nanomaterials exhibited acute toxicity in primary stem cells. But the toxicity was significantly reduced with the functionalized modified surface of quantum dots, even little toxicity [38]. Currently, most nanomaterial toxicity testing has been at the cellular level, with less testing at the tissue and organ levels. We carried out animal toxicity tests, to observe the toxic effects of gold nanorods on major tissues and organs in mice after the intravenous injection of gold nanorods (experimental design: high-dose group (0.500 mg/kg), medium-dose group (0.050 mg/kg), and low-dose group (0.005 mg/kg); 28 days after the injection, organs and tissues in the control group (saline) and high-dose group were submitted to pathological examination; the results are shown in Fig. 14.9. Regarding the heart, the myocardial fibers showed no degeneration, necrosis, atrophy, or hypertrophy in either test group; no obvious hyperplasia, inflammatory cell infiltration or interstitial or connective tissue proliferation was observed. Regarding the brain, the organizational structure of the brain was normal in both test groups, with no significant pathological changes. Regarding the lung, lung tissue structure was normal in the two test groups, but two individual animals showed mild inflammatory changes, manifested as alveolar wall vascular congestion and inflammatory cell accumulation, with no pathological significance. Regarding the liver, we observed two normal lobules and no liver cell degeneration; periportally, there was no bile duct or connective tissue proliferation. In the spleen, the capsule was intact in both test groups; red, white pulp clear, the spleen showed normal trabecular structure, with no significant proliferation of lymphoid nodules. In the kidney the renal capsule was intact; glomeruli were evenly distributed in the cortex, which was structurally normal, and there was no inner tubular, interstitial, or connective tissue proliferation. The adrenal capsule was intact in both test groups, and the cortex and medulla showed normal structure. In the stomach, the gastric mucosa was normal both test groups, with no abnormal differentiation. There was no bleeding or ulceration. In the duodenum there were two ulcers and duodenal mucosa showed no erosion. In regard to the gonads, ovarian follicular development was visible at each stage, and no bleeding or cysts were observed. Testicular seminiferous tubules and spermatogenic cells were normal.

The difference of these security measurement may be due to the difference in the conditions of lab experiments and the test of nanomaterials and cell lines. Not yet in a unified under the condition of experiment was carried out on the many kinds of nanometer materials more cell lines safety assessment, so the safety of nano-materials is temporarily still couldn't get a clear conclusion, so now it is can't clearly identified that nanomaterials is safe or dangerous, can only get a degree of evaluation, in order to get the exact conclusion also need a large number of experimental studies, the unified evaluation to the safety of nanomaterials

### 14.6.1 Nanomaterial Toxicity

The chemical composition of the surface, the size, the shape, the physical and chemical structure, the solubility, and the state of aggregation of nanomaterials can greatly affect their biological effects. Before safety assessment is carried out, we must first determine the type of nanomaterial; from the biomedical perspective these materials can be divided into organic, inorganic, and hybrid organic/inorganic categories. In the organic category are nanoliposomes, while silica balls are in the inorganic class, and gold nanorods with surfactants or surface modification are categorized as hybrid organic/inorganic. These features of nanomaterials directly influence the ways they enter cells, their transport, and organelles, and protein, the distribution, gathered in the body, and generation of drainage, causing a specific biological response. In the safety assessment of nanomaterials, the physical and chemical properties of the material itself must be considered, as nanomaterials with the same chemical composition may have completely different biological effects because of different morphological structures. In experimental studies nanoclusters and gold nanorods showed significantly different effects although both were toxic to cells [39, 40]. In the preparation of nanomaterials such as carbon nanotubes, metal is needed as a catalyst, and upon completion of the preparation, the metal element cannot be completely removed from the carbon nanotubes, and the metal element is the source of toxicity. Warheit et al. examined pulmonary perfusion in experimental animals to evaluate acute toxicity, using single-walled carbon nanotubes (SWCNTs); they found that the SWCNTs caused inflammatory reactions and multiple granulomas [41]. Kagan et al. found that SWCNTs without purification of metal ions caused an oxidative stress reaction, in which oxygen free radicals were exchanged for hydroxyl radicals [42]. Therefore, when safety assessment is carried out on nanomaterials, we must fully consider the chemical elements used in preparing the nanomaterials, as these elements may have toxic effects. The nanomaterial itself can also cause acute or subacute toxicity in the body; we chose

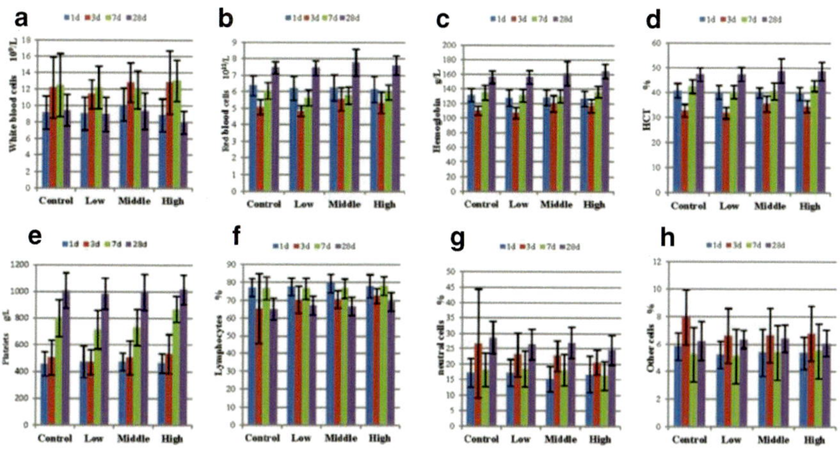


**Fig. 14.10** Routine blood and biochemical indexes in mice. (**a**) White blood cells. (**b**) Red blood cells. (**c**) Hemoglobin. (**d**) HCT. (**e**) Platelets. (**f**) Lymphocytes. (**g**) Neural cells. (**h**) Other cells

to investigate the effects of gold nanorods and gold nanoclusters as two kinds of nanomaterials.

To summarize, carbon point design: acute toxicity test of nanorods in Balb/C mice: high-dose group (1.0 mg/kg), low-dose group (0.1 mg/kg), control group (0 mg/kg), with 20 in each group, 10 each male and female, 1 mg/kg body weight (BW) was given to both male and female animals, by tail vein injection; the control group received intravenous saline in the tail. Within the scope of the subjects dose, no statistic difference were found in rat weekly weight, weekly intake, weekly food utilization, dynamiting, total food intake, food utilization, main viscera wet weight and dirty/body ratio. Compared with negative control group, the difference in some index of routine blood and biochemical has statistical significance but no biological significance. Difference in other index of routine blood and biochemical was not found (Fig. 14.10). There were no difference in gross anatomy and histological examination (Fig. 14.11). In our experiment, we found that no carbon points had acute toxic effects on the mice. We also carried out acute toxicity tests in mice injected with gold nanoclusters; the results showed that the gold nanoclusters did not cause acute toxic effects in the mice.

### *14.6.2 Immune Response*

The first problem with the biocompatibility of nanomaterials is the immunoreactivity of these materials; when nanoparticles enter an organism, the innate immune response is quickly activated, often before toxic symptoms occur. But the domestic and international study on immunotoxicology is barely, the current research is mainly based on the change of inflammatory factors and the influence

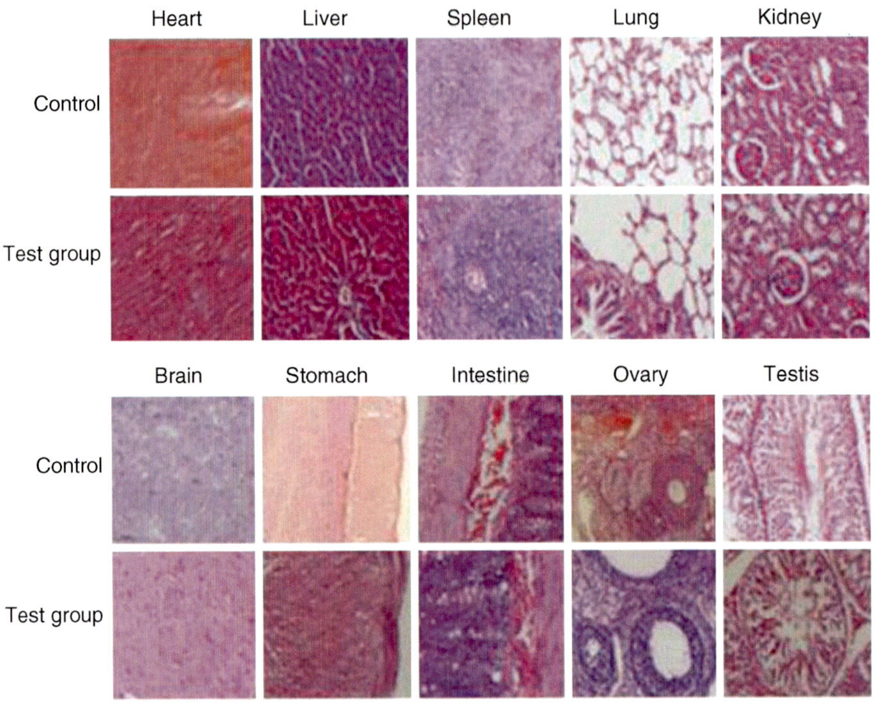


**Fig. 14.11** Pathology observations in mice

phagocyte cells related to nanoparticles. Different interactions between nanoparticles and the immune system can cause two very different reactions, either immune suppression or immune enhancement. Nanoparticles can be used to treat autoimmune disease [43], and it has been reported that the transmission of interleukin (IL)-10 coding DNA nanoparticles as a drug carrier effectively suppressed autoimmune diabetes in mice [44]. And some studies have found that nanoparticles may enhance immunity by activating the complement system of tumor locus and stimulating tumor-associated immune cells convert to tumor helper immune cells, promoting the progress of cancer cells. But by subcutaneous injection of nanometer particles, and through the complement activation in order to enhance the effectiveness of the vaccine, such as nano liposomes and functionalization of carbon nanotubes can activate the complement system appears [45]. To maximize the excellent properties of nanomaterials, we should fully understand these materials and their relationship to the immune system, as well as the disadvantages of their use. We studied the effects of carbon on immune function in mice [46]. Our team carried out an in-depth study to determine whether carbon causes immune function changes in mice. We used Balb/C mice, divided into three treatment groups: 2, 10, and 50 mg/kg of carbon quantum dots was administered. Spleen lymphocyte proliferation experiments showed that, compared with the

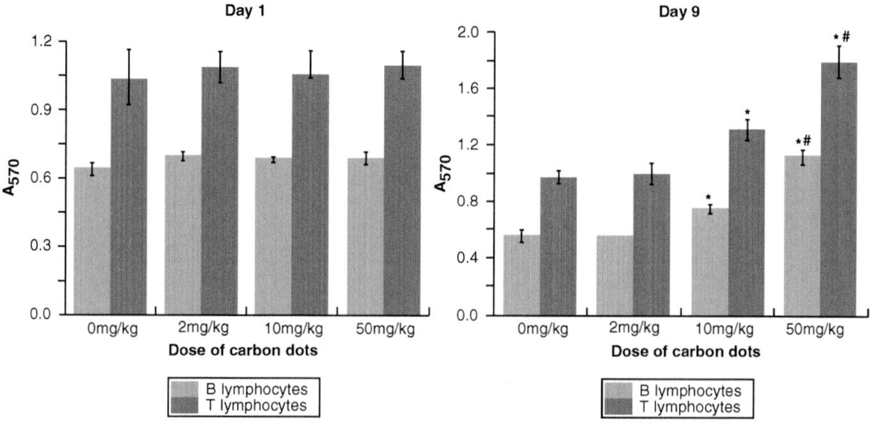


**Fig. 14.12** The influence of carbon dots on B- and T-lymphocytes cells

saline control group, 1 day after the treatment, B- and T-lymphocytes in all treatment groups showed no significant differences, while 9 days after the treatment, B- and T-lymphocyte proliferation activity had increased significantly, and there was a clear concentration-response relationship (Fig. 14.12).

One day after the administration of carbon quantum dots, compared with the control group, the high-dose group (50 mg/kg) showed that the CD3 + CD3 +/CD19 + ratio was significantly increased. T-cell subsets showed that the CD8 + percentage in the 2 mg/kg and 50 mg/kg administration groups was significantly decreased, while the CD4 + percentage in the 10 mg/kg group was significantly decreased; the CD4 +/CD8 + ratios in the three dosing groups showed a significant reduction, but there was no dose- dependent effect. Nine days after the administration of carbon quantum dots, compared with the control group, the CD3 +, CD19 + percentage and the CD3 +/CD19 + ratio were significantly higher in most treatment groups; T-cell subsets showed that the CD4 +, CD8 + percentage was significantly increased in majority administration group, but the CD4 +/CD8 + ratio did not change significantly (Fig. 14.13). Compared the 1 day, 9 days after administration with the control group, there is no obvious changes in each group of mice's thymus and spleen organ coefficients after being administered. The organ coefficient of the three dose groups of mice existed no significant difference. And, no significant pathological changes were found in the liver, spleen, or thymus in any administered group.

### *14.6.3 Mechanical Injury*

Nanomaterials differ from pure chemical particles in relation to mechanical interaction with biological tissues. After nanomaterials enter the body in various ways they may persist in the blood vessels; these materials can be of monomeric form of nano-free, or may be present in the form of aggregates. Aggregated

| Groups | CD3+ (%) | CD19+ (%) | CD3+/CD19+ |
|---|---|---|---|
| Saline | 36.01 ± 1.62 | 58.01 ± 1.55 | 0.62 ± 0.04 |
| Carbon dots | | | |
| 2 mg/kg | 37.44 ± 0.32 | 57.44 ± 0.55 | 0.65 ± 0.01 |
| 10 mg/kg | 35.12 ± 0.39 | 58.09 ± 0.32 | 0.60 ± 0.01 |
| 50 mg/kg | 36.97 ± 1.81**↑ | 55.81 ± 0.73*↓ | 0.70 ± 0.02**↑ |

The effects were recorded 1 day after administration. The data are presented as mean ± standard deviations, $n = 5$. $^{*}P < 0.05$ and $^{**}P < 0.01$ compared with the saline group (control). Significant difference was calculated by one-way ANOVA using SPSS19.0.

| Groups | CD3+ (%) | CD19+ (%) | CD3+/CD19+ |
|---|---|---|---|
| Saline | 18.00 ± 1.40 | 28.74 ± 1.14 | 0.63 ± 0.02 |
| Carbon dots | | | |
| 2 mg/kg | 26.48 ± 0.52**↑ | 33.88 ± 0.56**↑ | 0.78 ± 0.02**↑ |
| 10 mg/kg | 25.50 ± 0.36**↑ | 35.95 ± 0.94**↑ | 0.71 ± 0.03*↑ |
| 50 mg/kg | 26.68 ± 0.57**↑ | 29.87 ± 1.07 | 0.89 ± 0.05**↑ |

The effects were recorded 9 days after administration. Data are presented as mean ± standard deviations, $n = 5$. $^{*}P < 0.05$ and $^{**}P < 0.01$ compared with the saline group (control). Significant difference was calculated by one-way ANOVA using SPSS19.0.

| Groups | CD4+ (%) | CD8+ (%) | CD4+/CD8+ |
|---|---|---|---|
| Saline | 25.97 ± 0.65 | 9.94 ± 1.01 | 2.63 ± 0.21 |
| Carbon dots | | | |
| 2 mg/kg | 24.95 ± 0.20 | 12.54 ± 0.26**↑ | 1.99 ± 0.04**↓ |
| 10 mg/kg | 24.31 ± 0.41**↓ | 11.00 ± 0.14 | 2.21 ± 0.05**↓ |
| 50 mg/kg | 26.51 ± 0.44 | 12.75 ± 0.12**↑ | 2.08 ± 0.04**↓ |

The effects were recorded 1 day after administration. Data are presented as mean ± standard deviations, $n = 5$. $^{**}P < 0.01$ compared with the saline group (control). Significant difference was calculated by one-way ANOVA using SPSS19.0.

| Groups | CD4+ (%) | CD8+ (%) | CD4+/CD8+ |
|---|---|---|---|
| Saline | 10.85 ± 1.15 | 5.47 ± 0.62 | 1.99 ± 0.17 |
| Carbon dots | | | |
| 2 mg/kg | 16.05 ± 0.24**↑ | 9.89 ± 0.40**↑ | 1.63 ± 0.09*↓ |
| 10 mg/kg | 15.77 ± 0.59**↑ | 9.16 ± 0.28**↑ | 1.73 ± 0.12 |
| 50 mg/kg | 16.56 ± 0.28**↑ | 9.65 ± 0.44**↑ | 1.72 ± 0.05 |

The effects were recorded 9 days after administration. The data are presented as mean ± standard deviations, $n = 5$. $^{*}P < 0.05$ and $^{**}P < 0.01$ compared with the saline group (control). Significant difference was calculated by one-way ANOVA using SPSS19.0.

**Fig. 14.13** Effect of carbon dots on cytokines in mice

nanomaterials reaching to a certain concentration may induce mechanical vascular blockage or devastating vascular disease [47]. Nanomaterials deposit in the organization, across the cytomembrane into the cells, and this process may injury cytomembrane, lipid and protein [48], resulting to the change in the molecular structure of cytomembrane. Adsorption may lead to the restructuring of the lipid molecules, and the topological structure of the particle surface may stimulate the extension of actin in the biofilm. Moller et al. studied the impact of several types of nanoparticles on the macrophage cytoskeleton, and they found that high concentrations of the nanoparticles caused loss of the macrophage cytoskeleton, leading to increased cell stiffness, reducing the phagocytic activity of these cells [49]. Nanomaterials may aggregate in organs, causing serious mechanical damage. For example, inhaled nanoparticles deposited in the lungs may induce pulmonary fibrosis, leading to chronic lung disease, the final result of such mechanical respiratory disorders being exhaustion and death. These nanomaterials caused mechanical injury, mainly at the cell membrane by depositing the adhesion, or to enter into binding on the cell organelles, mechanically limit the normal physiological activity of cells.

### 14.6.4 Chemical Injury

The chemical composition of the nanoparticle surface determines their biochemical properties upon contact with organisms. Current research reports on nanomaterials can be broadly divided into two types: organic and inorganic; however, to further improve the biocompatibility of nanomaterials, some organic materials, such as

| Group | | No. of mice | Mean ± S.D. | |
|---|---|---|---|---|
| | | | ‰MNPCE | PCE/NCE |
| Negative control (PBS) | | 5 | 0.80 ± 0.84 | 1.01 ± 0.46 |
| CdSe/ZnS dots | 100nM | 5 | 1.20 ± 0.84 | 1.22 ± 0.21 |
| | 200nM | 5 | 1.00 ± 00.71 | 1.22 ± 0.26 |
| | 400nM | 5 | 2.20 ± 1.92 | 1.18 ± 0.18 |
| Positive control (Cyclophosphamide) | | 5 | 33.60 ± 5.59* | 0.81 ± 0.18 |

**Fig. 14.14** Bone marrow micronucleus test

nanogold bars and PEG-modified carbon nanotubes, may be added to inorganic nanomaterials in the preparation process. The reason of nanomaterials chemical toxicity is itself component or carried by the chemical composition caused, and the concentration of the chemical toxicity of these nanomaterials and material positive correlation, There is a significant toxic dose-effect relationship; for example, 0.2 mg/L of C60 can cause zebrafish embryos to be deformed and show increased mortality. Nanomaterials producing chemical damage may also cause DNA damage, causing genetic toxicity. To examine the genotoxicity of nanomaterials, we designed experiments to show quantum dot chromosomal damage in lung cancer cells and the effects of quantum dots in relation to micronucleus formation, as shown in Figs. 14.14 and 14.15. Within dose tested CdSe/ZnS quantum dots can cause structural aberrations in Chinese hamster lung cell line (CHL) cells staining occurs, no significant dose-effect relationship, but also within the test dose of CdSe/ZnS quantum dots can cause micronuclei in mouse bone marrow cells mutation. Cytotoxic function of the quantum dots produced with its own lattice core release of heavy metals, Cd2+release of reactive oxygen species (ROS)can cause oxidative stress leads to increased cytotoxicity [50, 51]. Beer et al. reported similar findings; the 3-(4,5-Dimethylthiazol-2-yl)-2,5-diphenyltetrazolium bromide (MTT) assay was used for the measurement of Ag nanoparticle suspensions containing different concentrations of Ag ions, to test the influence of Ag ions on A549 cell proliferation, and the results showed that with the increase of Ag ion concentration in the suspension, A549 cell proliferation decreased [52]. These findings remind us that for the safety assessment of nanomaterials, we need to take into account the whole nanomaterial preparation, i.e., the material itself and any free molecules (ions) when testing for biological toxicity.

### *14.6.5 Oxidative Stress*

Many studies have reported that, in chemical toxicity tests in vivo, nanomaterials such as C60, SWCNTs, and quantum dots can cause increases in ROS [53–55]. Other toxicity tests showed that the toxicity of nanomaterials led to the release of

| Dose | S9 mix | Expose time | No. of cells | % Structure aberrations | | % Polyploid |
|---|---|---|---|---|---|---|
| | | | | + gap | − gap | |
| 0 | − | 4 | 200 | 3 | 0 | 1 |
| 25 nM[a] | − | 4 | 200 | 6 | 4 | 3 |
| 50 nM[a] | − | 4 | 200 | 4 | 3 | 5 |
| 100 nM[a] | − | 4 | 200 | 3 | 9 | 6 |
| MMC | − | 4 | 200 | 13 | 1 | 8 |
| 0 | − | 24 | 200 | 3 | 0 | 2 |
| 25 nM[b] | − | 24 | 200 | 7 | 7 | 6 |
| 50 nM[b] | − | 24 | 200 | 3 | 3 | 9 |
| 100 nM[b] | − | 24 | 200 | 1 | 3 | 4 |
| MMC | − | 24 | 200 | 22 | 0 | 3 |
| 0 | + | 4 | 200 | 3 | 0 | 0 |
| 25 nM[c] | + | 4 | 200 | 2 | 3 | 4 |
| 50 nM[c] | + | 4 | 200 | 2 | 5 | 4 |
| 100 nM[c] | + | 4 | 200 | 2 | 5 | 5 |
| CP | + | 4 | 200 | 19 | 1 | 15 |
| CP | − | 4 | 200 | 3 | 0 | 1 |

**Fig. 14.15** Chromosome aberration test

free radicals and ROS, with increased oxidative stress causing multiple organ damage. The physical form and chemical composition of nanomaterials had an impact on mitochondrial metabolism. The mitochondria are the main source of ROS production; under normal physiological conditions, there is some single-electron leakage from the electron transport chain, and biologically important radicals are formed, including superoxide anion radicals, hydroxyl radicals, lipid alkyl oxy radicals and lipid peroxide radicals. However, after mitochondria are subjected to the external effects of toxic substances, the single-electron leakage from the electron transport chain increases, with the formation of an excess of free radicals, causing cellular oxidative stress. Our quantum safety assessment tests also confirmed this concept; ZnS quantum dots in human lung cancer cells were incubated with different concentrations of carboxylated CdSe; after incubation with increasing concentrations of quantum dots, ROS generation increased, indicating that quantum dots can cause an increase in intracellular ROS content (Fig. 14.16).

Under the action of nanometer materials, which may lead to mitochondrial function, electron transfer efficiency, leakage of a single electron number increases, the free radicals increases, the excess free radicals by means of nucleic acid, protein and lipid polyunsaturated against oxidative damage caused cells (Fig. 14.17). The mechanism of the oxidative stress injury caused by nanoparticles has been the subject of much research, but many problems are still not completely resolved.

**ROS in HUVECs treated with CdSe/ZnS**

| Dosage | DCF Mean ± SD |
|---|---|
| Control | 10.47 ± 0.54 |
| 25nM | 12.98 ± 1.35 |
| 50nM | 37.62 ± 0.58*↑ |
| 100nM | 231.33 ± 25.18**↑ |

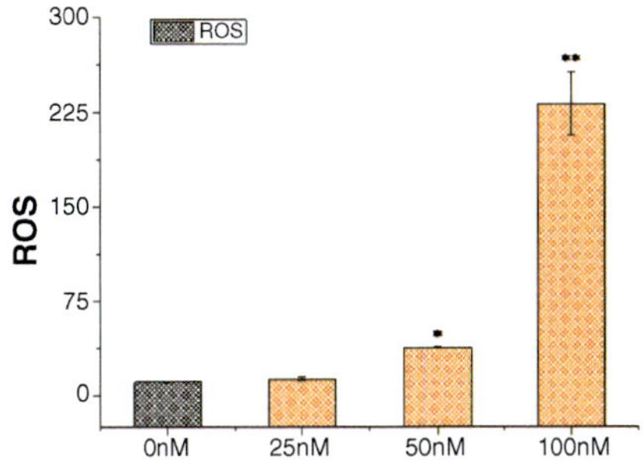


**Fig. 14.16** The generation of reactive oxygen species by CdSe/ZnS quantum dosts

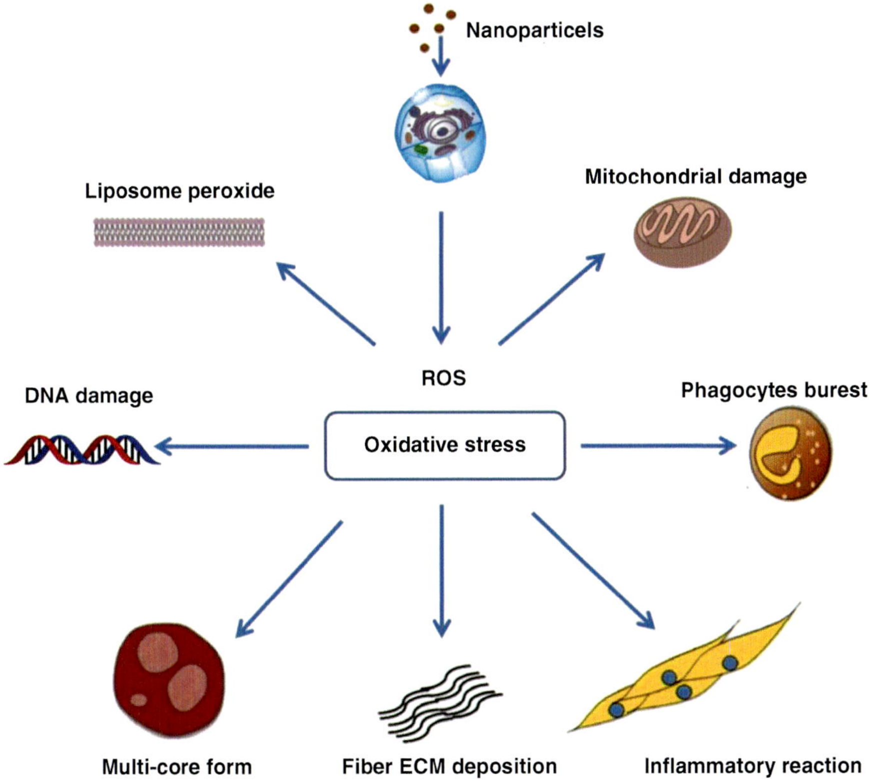


**Fig. 14.17** Oxidative stress damage caused by nanomaterials

## *14.6.6 Organ Damage*

When nanomaterials remain in the body for a long time, particular organs will have high accumulation of the materials, and this may cause organ damage. In addition to the nanomaterial size, geometry, and the physical and chemical properties of the

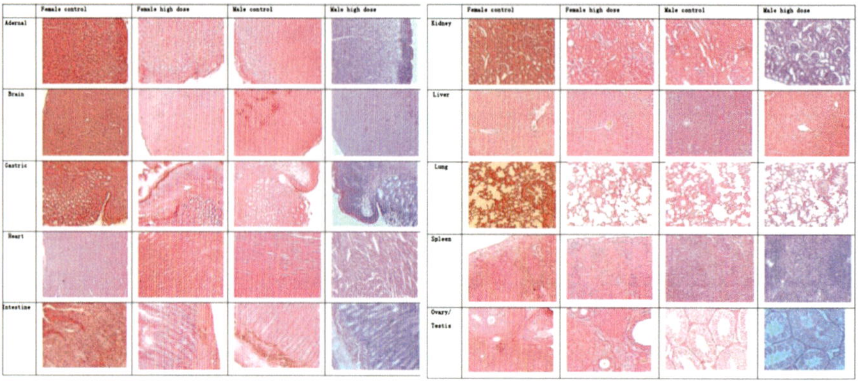


**Fig. 14.18** Tissue pathology of mice

material itself, the exposure dose is another important factor in causing organ damage. Biological factors are also involved in this damage; physiological characteristics, such as rich blood flow, may lead to the accumulation of nanomaterials in organs such as the liver, spleen, and lungs. In our evaluation of the safety of gold nanoclusters, we found that in test (high dose group of 10 mg/kg) injection, using liver: negative control, gold nano clusters – BRCAA1Ab high dose group of female and male animal liver tissue structure and no abnormal liver cell differentiation. Each animal hepatic lobule and portal area needs occasional individual scattered focal point, is given priority to with lymphatic, monocytes, inflammatory cells infiltration, slight degree, each animal portal area did not see small bile duct and connective tissue hyperplasia. Spleen: negative control, gold nano clusters – BRCAA1Ab high dose group of male and female animals spleen is membrane integrity, red, white pulp is clear, appropriate proportion, spleen trabecular structure is normal, no obvious hyperplasia, lymphoid nodule atrophy lesions. Lung: negative control, gold nano clusters – BRCAA1Ab normal high dose group of male and female animals lung structure, bronchial wall structural integrity, at all levels not seen abnormal epithelial cell differentiation, no epithelial cell degeneration, necrosis, hyperplasia and hypertrophy change; No inflammatory exudation in the alveolar space and hemorrhage, alveolar walls are not seen congestion and inflammatory lesion (Fig. 14.18).

As stated earlier, other exposure pathways can also cause damage to different organs; for example, inhalation of nanomaterials can cause an inflammatory pulmonary reaction, leading to chronic lung disease [56, 57], finally leading to respiratory function failure. After intravenous administration, nanomaterials accumulate in the liver, causing oxidative stress and liver toxicity, [58, 59], and the same nanoparticles can cause kidney injury [60]. Another characteristic of nanomaterials is that they can cross biological barriers, although it is unclear whether nanomaterials cross the blood-brain barrier, and whether nanomaterials cause brain toxicity is not clear.

## 14.7 Functionalized Modified Nanomaterials

In view of the problem of the increasing toxicity of nanomaterials, a new research direction has emerged in the field of nanoscience, the functionalization of nanoscale materials. And the functional resorcinarene purpose of nanomaterials largely in order to achieve a certain functional maximize the toxicity of its at the same time. Cetyltrimethyl ammonium bromide (CTAB) is needed for the preparation of gold nanorods, but as CTAB has strong cytotoxicity, in subsequent experiments other surfactants with good biocompatibility were used instead [61]. Currently, the main methods for the surface modification of nanomaterials are: (1) a physical modification method and (2) a chemical modification method. (1) The physical method involves modification of the surface by physical means such as adsorption, coating and coating of nano material surface physical modification: Improving the ability of the nanomaterials by the adsorption, coating and capping methods. The directly surface physical modification will introduce the organic/inorganic materials, inorganic materials usually include ferroferric oxide and silicon dioxide.And the organic materials mainly contain the silk protein, bovine serum albumin (BSA) and so on, or polyvinyl pyrrolidone (PVP), PEG or some other surfactants. (2) the surface chemical modification method: is modified by nano surface atoms and molecules produces chemical reaction, such as esterification, coupling agent, surface graft modification method. Except these traditional modification method, have spirit of deposition method, the rapid expansion of supercritical fluid method, precipitation method, heterogeneous nucleation method and so on. The aim of this technology is to reduce or eliminate the biological toxicity of nanomaterials, but the nano material after surface modification, may significantly alter the physical and chemical, such as Zhang in the experiments, through make orthosilicate reduction in alkaline environment for silica, the parcel evenly on the surface of gold nanorods, can be controlled by adjusting the quantity of orthosilicate packages of the thickness of the silicon dioxide, at the same time of changing the size of the nanoparticles, and the introduction of hydroxy, gold nanorods is changed from positive to negative surface charge of [62–64]. In the experiment of Zhang [65], a layer of silicon dioxide firstly coated on the surface of gold nanorods, and modified a 70 nm thickness hydrogels as the drugs carriers. In his experiment, he mainly consider the cytotoxicity of the introduced of hydrogels to the cells after the laser of the near infrared light. The result indicated that there is no significant cytotoxicity of the modified hydrogels. After fully considering the introduction of hydrogels, into, after near infrared nanorods toxic effect, through the determination of cytotoxicity, no significant cytotoxicity. Because after various functionalized modifications, nanometer materials into the body does change caused by the biological effect, so in the safety assessment, the need to fully consider the surface functionalization modified after its brings the change of biological characteristics. After functionalization of the biological effects of nanomaterials caused by the change also caused the attention of the researchers, as in many research reports, quantum dots, gold clusters and gold nanorods metal nano particles are often modified with specific functional groups, their surface physical and chemical properties, great changes have taken place in chengdu on the changes of nanomaterials hydrophilic or lipotropy, geometry, surface charge properties, such as the nature of the changes will greatly affect the distribution and metabolism within the biological organism, Especially the change of surface charge directly affects the ability of the nanometer materials into the cell, and biological effects of nanomaterials also occur correspondingly changed greatly. Nanomaterials before be being designed so, we should give full consideration to the changes caused by potential new biological effect.

## 14.8 Models and Methods of Safety Evaluation

Preliminary design of safety assessment model and methods for the safety assessment of nanomaterials.

Establishment of in-vivo experimental model:

1. Zebrafish animal model, determination of quantum dots (PEG modification and undecorated), conversion on nano material, activated carbon, carbon nanotubes, nanomaterial genotoxicity, developmental toxicity, and in-vivo distribution and excretion.
2. Acute toxicity in mice, immune function in mice; subacute toxicity experiments in rats to determine the toxicity of nanomaterials and perform pathological analysis.

Four types of in-vitro experimental models:

1. Double cell absorption and curl rat intestinal organ model used for measuring the gastrointestinal absorption of different nanomaterials, determination of whether they cross the blood–brain barrier, and measurement of absorption and efflux rate.
2. Atomic force microscopy analysis technology, to observe the surface features of gastric cancer cells and the features of nanomaterial quantum dots, such as aggregation and the morphological characteristics of the cell surface.
3. Nanoparticle and protein interaction, interaction of fullerene derivatives Gd @ C82 (OH) x and protein and the influence of the structure and function of the protein.
4. Experimental determination of toxic effects of different nanomaterials (growth inhibition, apoptosis, and DNA damage).

Main experiment content:

1. Research on the immunological effect of nanoactivated carbon adsorption of mitomycin C; gastrointestinal absorption of carbon nanotubes, toxicology, target vessel organelles, toxicity mechanism and toxicity studies; with 2-methoxyestradiol antitumor effect of single-walled carbon nanotubes.
2. Acute toxicity of magnetic nanoparticles in mice, in-vivo metabolic profile in mice.
3. Acute toxicity of carbon dots in mice, subacute toxicity and immunotoxicity study in rats.
4. Studies of acute toxicity of gold nanorods in mice.

5. Studies of acute toxicity of gold nanoclusters in mice.
6. Studies of distribution and excretion of fluorescent quantum dots in zebrafish.

For the safety assessment of nanomaterials, our laboratory selected and prepared six nanomaterials: nanocarbon, magnetic nanoparticles, gold bars, gold clusters, quantum dots, and transformed nanomaterials upconversion nanoparticles (UCNPs). And a variety of stomach cancer. Two nanoprobes were used: gold nanocluster -BRCAA1-Ab, and point carbon -BRCAA1-Ab. These nanomaterials and nanoprobes were submitted to safety assessment by the safety evaluation model.

In our experiment, we mainly carry out the test through above-mentioned the safety evaluation model and method of the nanomaterials. First of all, to prepare and characterize the nanomaterials. Then, to select the animal model and establish the last of nanomaterials for cell and animal toxicity tests. This part mainly focuses on the application of the zebrafish model in the nanomaterials safety assessment. And in the experiment, we choose the UPCNs as the model. Owing to the fish's self-advantages including large amount of spawning, the shortly fertilized egg hatching time, the sensitivity to toxicity, it is a good animal model of nanomaterials safety assessment. The following content uses zebrafishes the model to simulate in vivo toxicity studies of UCNPs, [66]. Firstly, 2–20 nm UCNPs was prepared. Then zebrafish model was used for UCNP safety assessment. Safety assessment contents: UCNPs toxicity to zebrafish fish; The influence of UCNPs on zebrafish hatching rate and survival rate; The influence of UCNPs on zebrafish hatching rate and survival rate; The influence of UCNPs on the larval form; the elimination of the UCNPs in fish.

Experimental results: when the UCNPs nanomaterials has the developmental toxicity to the zebrafish embryo. The nanomaterials has effect on the selenium protein expression and the development of bone and muscle. And the UCNPs greatly influence the embryonic development, the voluntary movement frequency (Fig. 14.19). Moreover, we also observed the effect of UCNPs on the zebrafish eggs hatching rate. And there is a dose – effect relationship of UCNPs on hatchability, the higher the concentration is, the slower incubation speed and the lower hatchability will be. With the concentration of UNCPs raised, the zebrafish hatching rate and survival rate dropped, however there is not an obvious effect on low concentration area. And there is obvious influence in the range from 200 to 400 μg/ml (Fig. 14.20). The morphology of zebrafish embryos 48 hpf found that there is no obvious effect of UNCPs with low concentration, but there is a great influence in the range from 200–400 μg/mL, the zebrafish embryos will appear on different degrees of deformity (Fig. 14.21), the length of young fish will shorten and become malformed, etc. The zebrafish in thin slice of experimental groups showed that the injected UCNPs has a distribution in zebrafish lumen, suggesting that UCNPs may be excreted from the zebrafish body (Fig. 14.22). Above-mentioned experimental results indicated that the hatching rate and survival rate of zebrafish has significant drop at high dose of UCNPs.

We also investigated the excretion, genotoxicity and the in-vivo detailed distribution of quantum dots, and then studied that eye targeting of quantum dots. We injected the quantum dots into zebrafish, and observed that the quantum dots can be specifically distributed in zebrafish eye (Fig. 14.23). Microarray analysis, in situ

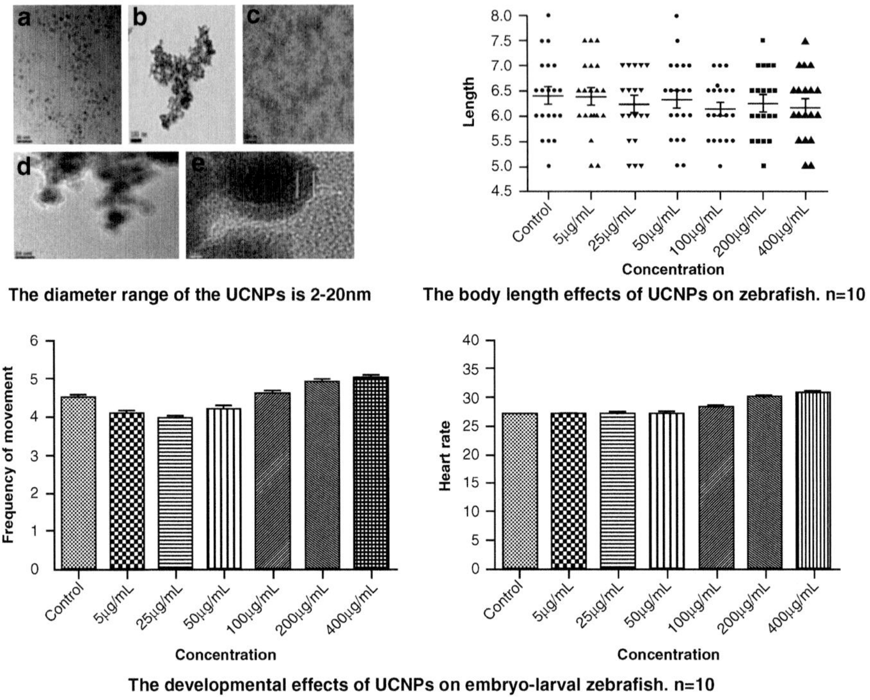


**Fig. 14.19** The effects of UCNPs on zebrafish larvae

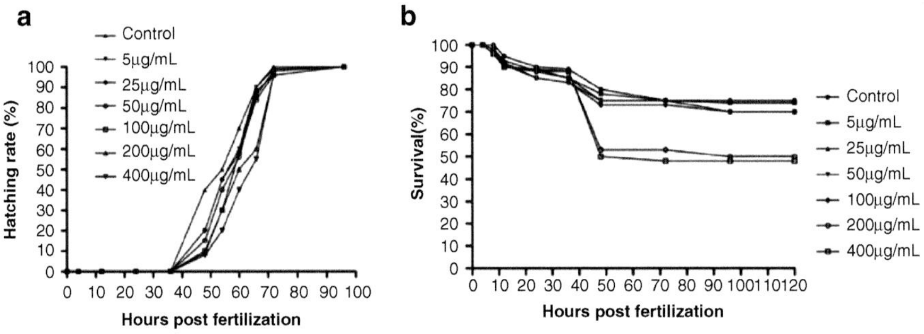


**Fig. 14.20** The effects of UCNPs on hatching rate and survival rate of zebrafish

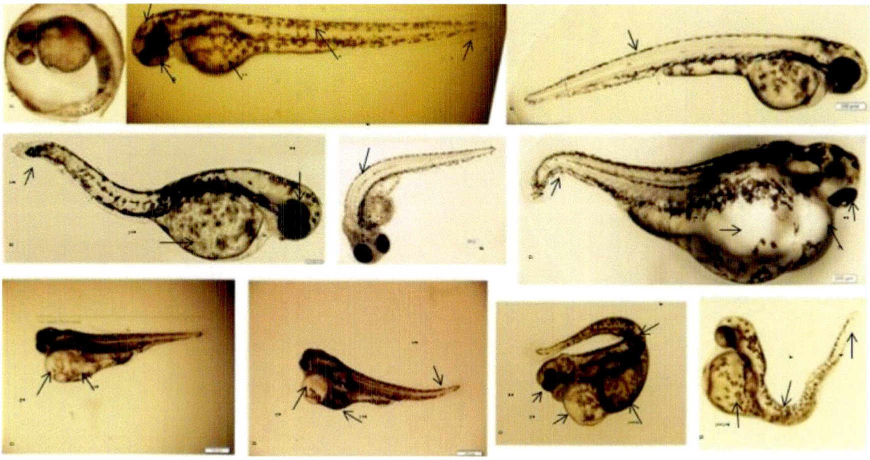

Fig. 14.21 Morphology of zebrafish embryo

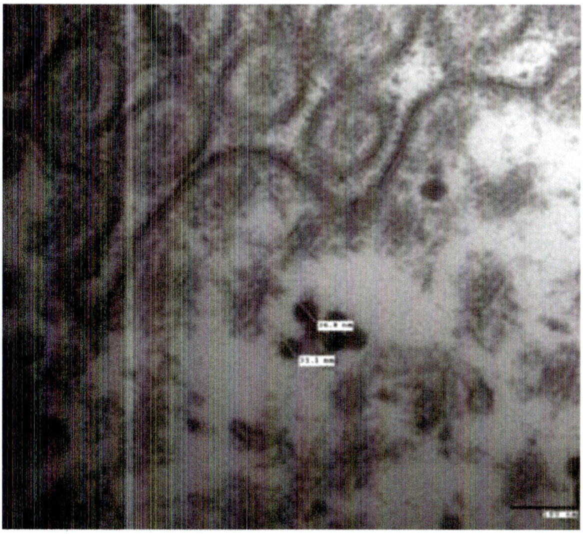

Fig. 14.22 The distribution of UCNPs in the lumen of zebrafish

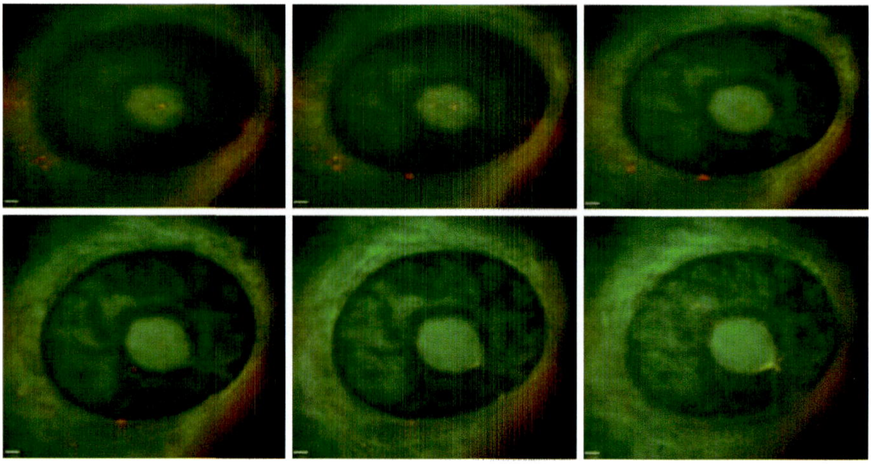

**Fig. 14.23** Target aggregation of quantum dots in zebrafish eye

**Fig. 14.24** Expression of specific genes in eye of zebrafish

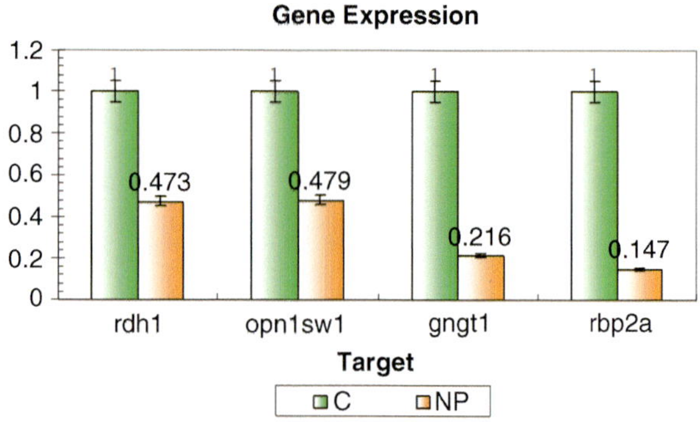


hybridization of tissue-specific gene found 13, including four for the eye-specific expression analysis by Graphene Oxide (GO), larger more affected biological processes and molecular functions associated with the eye (Fig. 14.24). The mechanism of zebrafish eye targeted of quantum dots, need for further research.

At present, both in China and abroad, although there have been many experimental studies of nanomaterials and nanotechnology, tests of nanomaterial toxicity have been confined to single toxicity tests for specific organs, and there is a lack of systematic assessments of the safety of these materials. Conventional toxicology and pathology studies have paid little attention to the mechanism of action of nanoscale particles in major diseases, and therefore there have been no findings on the relationship between nanoscale particles and these diseases. Although there are many problems to be solved, scientists have been aware of the importance of these scientific issues, and realize that nanoscience offers tremendous benefits to humankind. However, we cannot ignore the effects of nanomaterials on human health and the harmful effects on the environment. In this regard, we need to focus on biosafety issues with nanomaterials, and maximize the positive use of nanotechnology, while trying to avoid its hazards. We need to understand and solve these problems, and also to promote and safeguard the necessary conditions for the sustainable development of nanoscience. Based on the current status of the safety assessment of nanomaterials, it can be seen that much research of these materials is still required. This is just the start of a new multidisciplinary area, and we need to use research tools and knowledge in many fields –nanotechnology, biology, medicine, chemistry, physics, and others –to realize the great opportunities now presented for scientific innovation and the fulfillment of human needs at the forefront of science.

## References


1. Boisselier E, Astruc D. Gold nanoparticles in nanomedicine: preparations, imaging, diagnostics, therapies and toxicity. Chem Soc Rev. 2009;38:1759–82.
2. Chen HY, Zhang X, Dai SH, et al. Multifunctional gold nanostar conjugates for tumor imaging and combined photothermal and chemo-therapy. Theranostics. 2013;3(9):633–49.
3. Wang C, Li ZM, Liu B, et al. Dendrimer modified SWCNTs for high efficient delivery and intracellular imaging of surviving siRNA. Nano Biomed Eng. 2013;5(3):125–30.
4. Homan KA, Souza M, Truby R, et al. Silver nanoplate contrast agents for in vivo molecular photoacoustic imaging. ACS Nano. 2012;6(1):641–50.
5. Li W, Zhao Z, Xiong J, et al. The modification experimental study in vivo of nano-bone gelatin. Artif Cells Nanomed Biotechnol. 2013;6:1–7.
6. Oberdorster G, Sharp Z, Atudorei V, et al. Traslocation of inhaled ultrafine particles to the brain. Inhal Toxicol. 2004;16:437–45.
7. Elder A, Gelein R, Silva V, et al. Translocation of inhaled ultrafine manganese oxide particles to the central nervous system. Environ Health Perspect. 2006;8:1172–8.
8. Tan JF, Shah S, Thomas A, et al. The influence of size, shape and vessel geometry on nanoparticle distribution. Micofluid Nanofluidics. 2013;14(1–2):77–87.
9. Crayton SH, Elias DR, Al Zaki A, et al. ICP-MS analysis of lanthanide-doped nanoparticles as a non-radiative, multiplex approach to quantify biodistribution and blood clearance. Biomaterials. 2012;33(5):1509–19.
10. Balasubramanian SK, Jittiwat J, Manikandan J, et al. Biodistribution of gold nanoparticles and gene expression changes in the liver and spleen after intravenous administration in rats. Biomaterials. 2010;31:2034–42.
11. De Jong WH, Hagens WI, Krystek P, et al. Particles size-dependent organ distribution of gold nanoparticles after intravenous administration. Biomaterials. 2008;29:1912–9.
12. Fraga S, Ana B, Maria ES, et al. Short- and long-term distribution and toxicity of gold nanoparticles in the rat after a single-dose intravenous administration. Nanomed : Nanotechnol, Biol Med. 2014;10(8):1757–66.
13. Yamago S, Tokayama H, Nakamura E, et al. In vivo biological behavior of a water-miscible fullerene:14C labeling, absorption, distribution, excretion and acute toxicity. Chem Biol. 1995;2:385–9.
14. Jani PU, Mccarthy DE, Florence AT. Titanium dioxide (rutile) particle uptake from the rat GI tract and translocation to systemic organs after oral administration. Intl J Pharmaceut. 1994;105:157–68.
15. Gamer AO, Leibold E, van Ravenzwaay B. The in vitro absorption of microfine zinc oxide and titanium dioxide through porcine skin. Toxicol In Vitro. 2006;20:301–7.
16. Lademann J, Weigmann H, Rickmeyer C, et al. Penetration of titanium dioxide microparticles in a sunscreen formulation into the horny layer and the follicular orifice. Skin Pharmacol Appl Skin Physiol. 1999;12:247–56.
17. Ryman-Rasmussen JP, Riviere JE, Monteiro-Riviere NA. Penetration of intact skin by quantum dots with diverse physicochemical properties. Toxicol Sci. 2006;91:159–65.

18. Baroli B, Ennas MG, Loffredo F, et al. Pentration of metallic nanoparticles in human full-thickness skin. J Invest Dermatol. 2007;127:1701–12.
19. Hardman R. A toxicologic review of quantum dots: toxicity depends on physicochemical and environmental factors. Environ Health Perspect. 2006;114:165–72.
20. Terentyuk GS, Maslyakova AV, Suleymanova LV, et al. Circulation and distribution of gold nanoparticles and induced alterations of tissue morphology at intravenous particle delivery. J Biophotonics. 2009;2:292–302.
21. Elder A, Gelein R, Finkelstein JN, et al. Effects of subchronically inhaled carbon black in three species. I. Retention kinetics, lung inflammation, and histopathology. Toxicol Sci. 2005;2:614–29.
22. Bar-llan O, Albrecht RM, Fako VE, et al. Toxicity assessments of multisized gold and silver nanoparticles in zebrafish embryos. Small. 2009;5:1897–910.
23. Chithrani BD, Ghazani AA, Chan WC, et al. Determining the size and shape dependence of gold nanoparticles uptake into mammalian cells. Nano Lett. 2006;6:662–8.
24. Hirn S, Manuela SB, Caesten S, et al. Particle size-dependent and surface charge-dependent biodistribution of gold nanoparticles after intravenous administration. Eur J Pharm Biopharm. 2011;77:407–16.
25. Singh R, Pantarotto D, Lacerda L, et al. Tissue biodistribution and blood clearance rates of intravenously administered carbon nanotube radiotracers. Proc Natl Acad Sci U S A. 2006;103(9):3357–62.
26. Feng Y, Zong Y, Ke T, et al. Pharmacokinetics, biodistribution and contrast enhanced MR blood pool imaging of Gd-DTPA cystine copolymers and Gd-DPTA cystine diethy1 ester co-polymers in a rat model. Pharm Res. 2006;23:1736–42.
27. Hagens WI, Oomen AG, de Jong WH, et al. What do we (need to) know about the kinetic properties of nanoparticles in the body? Regul Toxicol Pharmacol. 2007;49:217–29.
28. Renaud G, Hamilton RL, Haval RJ. Hepatic metabolism of colloidal gold-low-density lipoprotein complexes in the rat: evidence for bulk excretion of lysosomal contents into bile. Hepatology. 1989;9:380–92.
29. Souris JS, Lee CH, Cheng CT, et al. Surface charge-mediated rapid hepatobiliary excretion of mesoporous silica nanoparticles. Biomaterials. 2012;31:5564–74.
30. Johnston HJ, Hutchison G, Christensen FM, et al. A review of the in vivo and in vitro toxicity of silver and gold particulates: particle attributes and biological mechanisms responsible for the observed toxicity. Crit Rev Toxicol. 2010;40:328–46.
31. Choi HS, Liu W, Misra P, et al. Renal clearance of quantum dots. Natl Biotecnol. 2007;25:1165–70.
32. Choi CH, Zuckerman JE, Webster P, et al. Targeting kidney mesangium by nanoparticles of defined size. Proc Natl Acad Sci U S A. 2011;108(16):6656–61.
33. Pollinger K, Hennig R, Bauer S, et al. Biodistribution of quantum dots in the kidney after intravenous injection. J Nanosci Nanotechnol. 2014;14:3313–9.
34. Manabe M, Tatarazako N, Kinoshita M. Uptake, excretion and toxicity of nano-sized latex particles on medaka (Oryzias latipes) embryos and larvae. Aquat Toxicol. 2011;105:576–81.
35. Wang YY, Lai SK, Suk JS, et al. Addressing the PEG mucoadhesivity paradox to engineer nanoparticles that "slip" through the human mucus barrier. Angew Chem. 2008;47:9575–781.
36. Braydich-Stolle L, Hussain S, Schlager JJ, et al. In vitro cytotoxicity of nanoparticles in mammalian stem cells. Toxicol Sci. 2005;88:412–9.
37. Goodman CM, Mccusker CD, Yilmaz T, et al. Toxicity of gold nanoparticles functionalized with cationic and anionic side chains. Bioconjug Chem. 2004;15:897–900.
38. Derfus AM, Chen AA, Min DH, et al. Targeted quantum dot conjugates for siRNA delivery. Bioconjug Chem. 2007;18:1391–6.
39. Zhang CL, Zhou ZJ, Zhi X, et al. Insights into the distinguishing stress-induced cytotoxicity of chiral gold nanoclusters and the relationship with GSTPI. Theranostics. 2015;5(2):134–49.
40. Huang P, Bao L, Zhang CL, et al. Folic acid-conjugated silica-modified gold nanorods for X-ray/CT imaging-guided dual-mode radiation and photo-thermal therapy. Biomaterials. 2011;32:9797–809.
41. Warheit DB, Laurence BR, Reed KL, et al. Comparative pulmonary toxicity assessment of single-wall carbon nanotubes in rats. Toxicol Sci. 2004;77:9.

42. Kagan VE, Tyurina YY, Tyurina VA, et al. Direct and indirect effects of single walled carbon nanotubes on RAW 264.7 macrophages: role of iron. Toxicol Lett. 2006;165:88–100.
43. Basarkar A, Singh J. Poly(lactide-co-glycolide)-poly-methacrylate nanoparticles for intra muscular delivery of plasmid encoding interleukin-10 to prevent auto immune diabetes in mice. Pharm Res. 2009;26:72–81.
44. Wu W, Lee WK, Ryoo JW, et al. Suppression of collagen-induced arthritis by single administration of poly(lactic-co-glycolic acid) nanoparticles entrapping type II collagen: a novel treatment strategy for induction of oral tolerance. A rthritis Rheum. 2002;46:1109–20.
45. Salvador M, Flahaut E. Complement activation and protein adsorption by carbon nanotubes. Mol Immunol. 2006;43:193–201.
46. Gao ZC, Shen GX, Zhao XN, et al. Carbon dots, a safety nano-scale substance to immunologic system of mice. Nanoscale Res Lett. 2013;8:276.
47. Nehler MR, Tayior LM, Porter JM. Iatrogenic vascular trauma. Semin Vasc Surg. 1998;11:283–93.
48. Christie MS, John DF, Wenh G. The differential cytotoxicity of water soluble fullerenes. Nano Lett. 2004;4:1881.
49. Moller W, Hofer T, Ziesenis A, et al. Ultrafine particles cause cytoskeletal dysfunction in macrophages. Toxicol Appl Pharm. 2002;182:197.
50. Hamilton SJ. Review of selenium toxicity in the aquatic food chain. Sci Total Environ. 2004;326:1–31.
51. Henson MC, Chedrese PJ. Endocrine disruption by cadmium, a common environmental toxicant with paradoxical effects on reproduction. Exp Biol Med. 2004;229:383–92.
52. Christiane B, Rasmus F, Yuya H, et al. Toxicity of silver nanoparticles—Nanoparticle or silver ion? Toxicity Lett. 2012;3:286–92.
53. Christie MS, Andre MG, Kevin DA. Nano C60 cytotoxicity is due to lipid peroxidation action. Biomaterials. 2005;26:7587.
54. Li N, Sioutas C, Cho A, et al. Ultrafine particlate pollutants induce oxidative stress and mitochondrial damage. Environ Health Perspect. 2003;3:455.
55. Derfus AM, Chan WC, Bhatia SN. Probing the cytotoxicity of semiconductor quantum dots. Nano Lett. 2004;4:11.
56. Ahamed M. Silica nanoparticles-induced cytotoxicity, oxidative stress and apoptosis in cultured A431 and A549 cells. Hum Exp Toxicol. 2013;32:186–95.
57. Gonzalez L, Thomassen LC, Plas G, et al. Exploring the aneugenic and clastogenic potential in the nanosize range: A549 human lung carcinoma cells and amorphous monodisperse silica nanoparticles as models. Nanotoxicology. 2010;4:382–95.
58. Sankar P, Telang AG, Kalaivanan R, et al. Effects of nanoparticle-encapsulated curcumin on arsenic-induced liver toxicity in rats. Enviromental Toxicol. 2013;2:1–10.
59. Paget V, Sergent J A, Grall R, et al. Carboxylated nanodiamonds are neither cytotoxic nor genotoxic on liver, kidney, intestine and lung human cell lines[J]. Nanotoxicology. 2014;8(1): 46–56.
60. Abdelhalim MA, Moussa SA. The gold nanoparticle size and exposure duration effect on the liver and kidney function of rats: in vivo. Saudi J Biol Sci. 2013;20:177–81.
61. Wang YC, Black KC, Luehmann H, et al. A comparison study of gold nanohexapods, nanorods, and nanocages for photothermal cancer treatment. ACS Nano. 2013;7(3):2068–77.
62. Zhang ZJ, Wang LM, Wang J, et al. Mesoporous silica-coated gold nanorods as a light-mediated multifunctional theranostic platform for cancer treatment. Adv Mater. 2012;24: 1418–23.
63. Li YY, Wen T, Zhao RF, et al. Localized electric field of plasmonic nanoplatform enhanced photodynamic tumor therapy. ACS Nano. 2014;8(11):11529–42.
64. Nima ZA, Mahmood M, Xu Y, et al. Circulating tumor cell identification by functionalized silver-gold nanorods with multicolor, super-enhanced SERS and photothermal resonances. Sci Rep. 2014;4:4752.
65. Zhang ZJ, Wang J, Nie X, et al. Near infrared laser-induced targeted cancer therapy using thermoresponsive polymer encapsulated gold nanorods. J Am Chem Soc. 2014;136:7317–26.
66. Wang K, Ma J, He M, et al. Toxicity assessments of near-infrared upconversion iuminescent LaF3:Yb,Er in early development of zebrafish embryos. Theranostics. 2013;3:258–66.

# Chapter 15
# Gastric Cancer Prewarning Database and Bioinformatics Analysis

**Cheng Shangli and Daxiang Cui**

Gastric cancer is the most common tumor, development of gastric cancer prewarning, and early diagnosis system needs a database; this chapter mainly summarized the database including genes and proteins associated with gastric cancer.

## 15.1 Gastric Prewarning Databases

The database is an organized data warehouse, which, according to the data structure, store and manage data warehouses. The database was created with the development of information technology and the market in the 1950s of the last century, and it was greatly and wildly used after the 1990s. Now, a database is no longer just a memory and management data warehouse. It has many new functions, and the converted database can satisfy various needs of the users [1]. There are many types of databases. For example, a simple database could be a data table, which stores various data. And a systematic database can be able to carry out mass data, which can be wildly used in various aspects.

With the development of bioinformatics, more and more data was used in the biology and medicine [2]. The term of bioinformatics was first defined in the information theory seminars in Gatlinburg. With the development of gene sequencing

---

**Electronic supplementary material** The online version of this chapter (doi:10.1007/978-94-024-0951-2_15) contains supplementary material, which is available to authorized users.

C. Shangli (✉) • D. Cui
Institute of Nano Biomedicine and Engineering, Shanghai Engineering Research Center for Intelligent Diagnosis and Treatment Instrument, National Center for Translational Medicine, Collaborative Innovational Center for System Biology, Shanghai Jiao Tong University, 800 Dongchuan Road, Shanghai 200240, P. R. China
e-mail: chengshali@sjtu.edu.cn; dxcui@sjtu.edu.cn

© Springer Science+Business Media B.V. and Shanghai Jiao Tong University Press, Shanghai 2017
D. Cui (ed.), *Gastric Cancer Prewarning and Early Diagnosis System*, Translational Medicine Research, DOI 10.1007/978-94-024-0951-2_15

technology, the data, obtained from molecular and biological experiments, has dramatically grown. The data need to be processed. Under this circumstance, bioinformatics is used to solve the problems [3]. Bioinformatics is an interdisciplinary science, which is based on mathematics, computer science, and life sciences. It mainly included data extraction, data analysis, data process, data storage, and data usage. In order to efficiently use the growing data, the researchers around the world shall share previous research results. Thus, bioinformatics becomes more and more important in the processing and storage of biological data [4].

### *15.1.1 Introduction of Biological Databases*

With the development of bioinformatics, the number of bioinformatics database is constantly increasing. The internal structure of the database has also become complicated and has a more refined feature. Depending on the type of data, the database can be divided into the nucleic acid sequence databases, protein sequence database, and three-dimensional molecular structure of the database.

The nucleic acid sequence databases include common nucleic acid sequence databases such as GenBank and the European Molecular Biology Laboratory (EMBL) (the European Bioinformatics Institute) nucleic acid databases, DNA Data Bank of Japan (DDBJ), and so on. GenBank database [5, 6] contains all known nucleic acid and protein sequences and their associated biological literature of the books and notes. It is built and maintained by the US National Center for Biotechnology Information (NCBI). Every day, GenBank will be synchronized with the European Molecular Biology Laboratory (EMBL), DNA Data Bank of Japan (DDBJ), and other databases. People can use GenBank by downloading it for free from the NCBI Ftp server complete library, or download new data accumulation. NCBI also provides a wide range of data query, sequence similarity search, and other analytical services. Users can find these services from the NCBI's home page. The EMBL Nucleotide Sequence Database [7, 8] is maintained and constituted with the nucleic acid sequence data by the European Bioinformatics Institute (EBI). Since the data exchange with the GenBank and DDBJ cooperation, the EMBL Nucleotide Sequence Database is also a comprehensive database of nucleic acid sequences. The database is maintained by the Oracle database management system, and the work of string query can be done through the sequence extraction systems on the Internet.

With the deepening of the human genome project and the advances in sequencing technology, the protein sequence information has also grown exponentially. The protein sequence database is mainly these sequences and structures. The protein sequence database mainly contains these sequences and structure information, including protein primary structure and quaternary structure. Currently, large-scale comprehensive protein sequence databases include the SWISS-PROT database and protein structure database of PDB. The SWISS-PROT database [9, 10] was manually annotated protein sequence information, and the database was maintained by the European Bioinformatics Institute (EBI). The SWISS-PROT database is constituted by the protein sequence entries. For each entry, it includes protein sequence information, citation information, classified information, comments, protein annotation features, posttranslational modifications, special sites and regions, secondary structure, quaternary structure, similarity to other sequences, sequence incomplete relationship with the disease, sequence variants, and conflict information. The SWISS-PROT database has reduced the redundant sequences and established a cross-reference with other databases, including nucleic acid sequence databases, protein sequence databases, protein structure databases, and so on. By sequential extraction system (SRS), it can be an easy retrieval for SWISS-PROT and other EBI databases. The protein structure database (PDB) [11, 12] was created in 1971 by the US Brookhaven National Laboratory in New York. The PDB database is the most important collection of biological macromolecules, including the database of three-dimensional structure of proteins, nucleic acids, and glucose. The three-dimensional data is determined by means of X-ray diffraction experiments, nuclear magnetic resonance, and electron diffraction. With the continuous improvement of crystal diffraction techniques, the structure determination of the speed and accuracy has gradually improved. Since the 1990s, with the maturity of the solution conformation of the multidimensional NMR measurement method, it becomes possible for the hard crystal structure determination of protein molecules. The amount of protein molecule structure in the database has a rapid rise. According to the statistics in 2015, the PDB database has been stored for more than 100,000 sets of atomic coordinates, which mostly are proteins, including peptides and viruses [13]. In addition, the PDB database also contains nucleic acid, protein and nucleic acid complex, and a small amount of polysaccharide molecules.

## 15.1.2 Primary Gastric Cancer Database

Currently, the ArrayExpress databases of the National Center for Biotechnology Information (NCBI) [14], the Gene Expression Omnibus (GEO) [15], and the EBI are publicly provided data query and data submitting. In addition, the common databases also include Harvard University's chip Express DB, Stanford SMD, Yale University YMD, and the University of Pennsylvania's EPODB [16–19].

### 15.1.2.1 Gene Expression Omnibus

The database of Gene Expression Omnibus (GEO) [20], as the first gene expression database (Fig 15.1), is a high-throughput gene expression and hybridization array database developed by NCBI in 2000. It contains the memory chip data submitted by scientific institutions, second-generation sequencing data, and high-throughput functional genomic data. Now, the GEO database contains more than 13,000 laboratories provided by more than 32,000 public series [21]. The GEO database presents the data into three levels of entity types: platform (platform), sample (sample), and series (series).

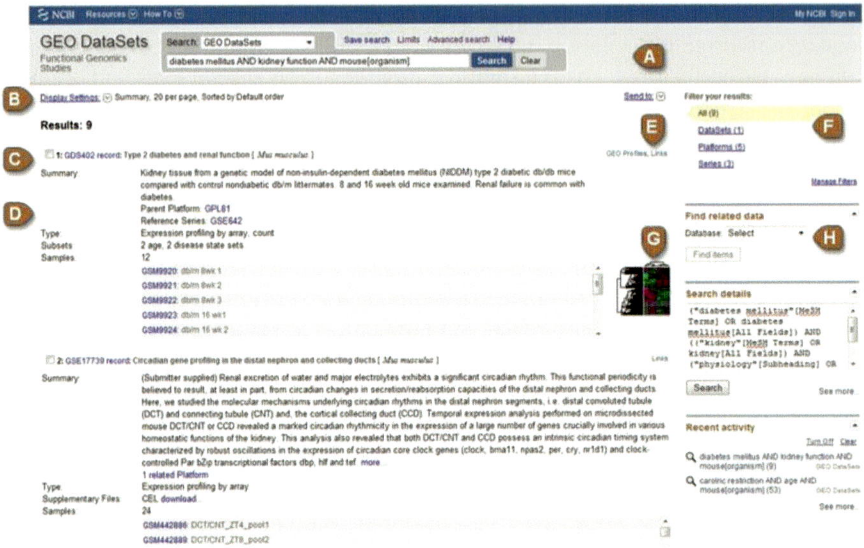


**Fig. 15.1** GEO database preview image (http://www.ncbi.nlm.nih.gov/geo/info/datasets.html)

**Table 15.1** The gastric data platform in GEO database

| Platform number | Platform title |
|---|---|
| 1 | [HG-U133_Plus_2] Affymetrix Human Genome U133 Plus 2.0 Array |
| 2 | [Mouse430_2] Affymetrix Mouse Genome 430 2.0 Array |
| 3 | [MG_U74Av2] Affymetrix Murine Genome U74A Version 2 Array |
| 4 | [HuGene-1_0-st] Affymetrix Human Gene 1.0 ST Array [transcript (gene) version] |
| 5 | [HG-U133A_2] Affymetrix Human Genome U133A 2.0 Array |
| 6 | [Hu6800] Affymetrix Human Full Length HuGeneFL Array |
| 7 | Center for Functional Analysis of Human Genome Human KUGI 14 K (stomach cancer) |
| 8 | PeterMac 8 k cDNA chip |

In the GEO database, the data of gastric cancer contains a total of eight platforms (Table 15.1). It reported a total of nine sets of data, including gastric cancer cell line AZ-521 and MKN28, gastric adenocarcinoma, and gastric tubular adenocarcinoma.

### 15.1.2.2 TCGA Database

The Cancer Genome Atlas (TCGA) database is set up on the basis of cancer and tumor gene mapping program in 2005 sponsored by the US government. Through the application of genome analysis technologies, especially the use of large-scale

**Table 15.2** Gastric data in TCGA

| Gastric cancer | Exon | SNP | Methylation | mRNA | miRNA |
|---|---|---|---|---|---|
| Number of samples | 441 | 443 | 443 | 414 | 436 |

genome sequencing and systems analysis, TCGA database is used to find out the entire human cancer genome variation map. The aim of the database is to find small variations of carcinogenic and tumor suppressor genes, to understand cancer cells and the mechanism of development, and to establish new diagnostic and treatment methods [22, 23]. TCGA is supervised by the National Human Genome Research Institute and the US National Cancer Institute and funded by the US government. The project of TCGA is started in 2006. The main focuses of the three characteristics of human cancers are glioblastoma multiforme tumors, lung cancer, and ovarian cancer. Until 2009, it expanded to phase II, in which it planned to complete genomic features in 20–25 different tumor types and sequence analysis by 2014. The project plans to use 500 patient samples for the comparisons of genomic research and different techniques to analyze the samples. Furthermore, the plan includes gene expression profiling technology, copy number variation analysis, SNP genotyping, genome-wide DNA methylation profiles, microRNA expression profiling, as well as sequencing of exons with at least 1,200 genes. TCGA studies of the tumors are whole-genome sequencing, and they include at least 6000 candidate genes and microRNA sequences. The targeted sequencing is carried out by three sequencing centers using hybrid capture technology. In the second phase, TCGA is performing 80 % of the entire exon and 80 % of the cases. The whole-genome research project is used for sample sequencing [24].

The "standard of care" of the United States is targeting tumor types in TCGA by the 25 major types of cancer. It focuses on the removal of the standard of care before secondary treatment. The sample's availability also plays a vital role in determining the type of cancer research before the beginning of the cancer project. In TCGA, it records the more common tumors. These common tumor types, such as colon, lung, and breast cancer, have become the first to enter the project of tumor types. In the rare tumor types, TCGA targetedly recorded the tumor types, including lung squamous cell carcinoma, renal papillary carcinoma, renal cell carcinoma, breast cancer, diffuse large B-cell lymphoma, renal cell cancer, cervical cancer (squamous cell carcinoma), colon cancer, colorectal cancer, stomach cancer, liver cancer, astrocytoma, head and neck squamous cell carcinoma (oral), thyroid cancer, bladder urothelial carcinoma, non-papillary renal cell carcinoma, endometrial (endometrial cancer), invasive bladder urothelial carcinoma, pancreatic adenocarcinoma, acute myeloid leukemia, prostate cancer, lung cancer, skin melanoma, breast lobular cancer, and multiple myeloma. For gastric cancer, in TCGA database, there are 443 sample data (Table 15.2 below), containing exons, information SNP, methylation, mRNA, miRNA, and clinical data.

#### 15.1.2.3 ONCOMINE

The ONCOMINE database [25] is currently the world's largest cancer microarray database and integrated data-mining platform, which is used for digging cancer gene-related information. So far, the database has collected 715 gene expression datasets, and the sample data contains more than 80,000 cancer and normal tissues. In the ONCOMINE database, the samples expressed as microarray data, and its source is measured for mRNA or DNA copy number in different levels in primary tumor cell lines and a transplant body, which is usually extracted from the studies that have been published. This data includes metadata and sample analysis results, which is based on the data and other sources, including cancer and normal control results, etc. In the ONCOMINE database, it allows users to search for any of the following keywords or indicators, including: genes, cancer type, type of analysis, data collection, definition by name and by type of research, concepts according to the name or type of sample properties, and other information (Fig. 15.2).

### *15.1.3 Other Databases*

#### 15.1.3.1 Human Gene Mutation Database

The Human Gene Mutation Database (HGMD) is a specialized database [26, 27], which is the mutation of human genetic diseases with the gold standard integrated resource data. The database is an essential tool to analyze thousands of samples in

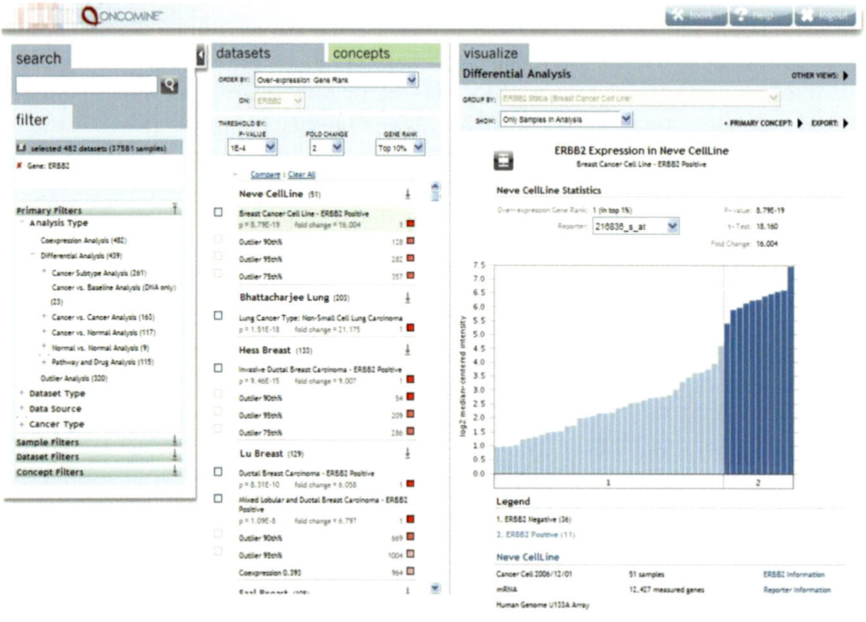

**Fig. 15.2** Preview of ONCOMINE database (https://www.oncomine.org/content/org/help/OncomineHelpFile.htm)

the large genomic data. Today, the human gene mutation has been widely used in the study of genetics and genomics. In this database, users can use the online search interface to quickly find individual mutations, as well as to determine the gene mutation in the name of all data published. Also, users can use advanced search application functions to find data related with a particular gene or disease mutations, which includes a part of the donor or acceptor splice, so that a particular amino acid mutation or change could be found out by more translation results. Moreover, the unique combination of the databases coupled with content helps to quickly identify and distinguish the variants and contains entries of biological evidence and disease.

#### 15.1.3.2 Human Tumor Gene Database

The Tumor Gene Database contains information of the tumor gene family, which is a target gene information of oncogenic mutations. It is comprised of proto-oncogenes and tumor suppressor genes. The goal of the database is to provide a standard set of facts, such as protein size, biochemical activity, and chromosomal localization, which is related genes of all known tumor. Currently, the database contains more than 2600 facts and more than 300 genes. The purpose of these databases is to provide data services for biomedical researchers and human cancer genes. Historically, besides the unexpected tumor genetic data, the Tumor Gene Database, as a more professional database, includes the data of breast cancer gene and the data of oral cancer gene.

#### 15.1.3.3 Genetic Association Database

The purpose of the Genetic Association Database is to collect and standardize genetic association studies in data archiving, making it easy to be used by researchers [28]. The database consists of three main components: a network interface, GAD Perl modules, and databases. The database is in the form of Oracle RDBMS. The database has three levels, in which the genetic and disease data is organized into a large fact table. It also contained a three-dimensional view of the middle layer and the top layer. The GDA database records 429 cancer-related genes.

#### 15.1.3.4 Cancer Genome Anatomy Project

The database [29] is the Cancer Genome Anatomy Project established by NCI. The aim of the database is trying to determine the normal gene, gene expression profiling, gene expression profiles of precancerous lesions, and cancer gene expression profiles. This information is used to make improvements in detection, diagnosis, and treatment of cancer for cancer patients. The goal of CGAP is to create a variety of different types of cancer in different stages of development, including normal gene expression, gene expression before cancer, and its database. It is for the future cancer research to depict complete molecular characteristics of tumor cells. On the basis, the establishment of a complete set of genes and their variation directory is not only conducive to evaluate the degree of risk of cancer but also can determine the prevention or treatment strategies based on genetic changes, ultimately achieving the purpose of treatment based on molecular characteristics. The database contains the expressed sequence tag (EST) and cDNA libraries [30]. By the database, users can obtain the expression of genes in cancer and normal tissues. The tissues are of 51 kinds, including the breast, colon, adrenal gland, liver, eyes, brain, and heart. Gene expression is recorded in the database, including in normal tissue and cancer tissue, gene only expressed in normal tissues, and gene only expressed in cancer tissues and expressed in normal and cancerous tissues.

#### 15.1.3.5 MethyCancer Database

The MethyCancer database studied the interaction of DNA methylation, gene expression, and tumor. The database contains the methylation sites associated with gastric cancer over 7000 pieces. It is shown in the following table [31]. The database is highly integrated with data from other databases, including DNA methylation, gene mutations associated with cancer, and annotation information for cancer genes. This includes the public resource information in other databases as well as the large-scale sequencing information of CpG Island (CGI). The database also analyzes the data contained in the database and the correlation between different data types and puts forward a new methylation site with cancer. Database access can be performed using search tools and the graphical interface software MethyView, through which the user can access all data in the database including the relationship between DNA methylation and cancer in the perspective of genomics and genetics (Table 15.3).

## 15.2 Bioinformatics Analysis of Gastric Cancer

The development of bioinformatics has mainly experienced three stages: the pre-genomic era, the era of the genome, and the post-genomic era. The main work of the former genome era includes: DNA sequence analysis, the creation of biological

**Table 15.3** MethyCancer database information

| Data content | Data statistics |
|---|---|
| Cancer-associated gene | 7075 |
| Annotated gene | 485 |
| Methylation data | 437 |
| Methylation data | 5598 |
| Experimental verification data | 3698 |
| CGI forecast data | 1900 |
| Experiment support | 6847 |
| CGI prediction support | 4714 |

database, the development of search tools, etc. The main work of the genome era includes: gene identification and discovery, the creation of network database system, the development of interactive tools, etc. The main work of the post-genome era includes: data integration, data analysis, large-scale genome analysis, and proteome analysis. With the completion of the human genome sequencing project and the rapid development of molecular biology, more and more gene sequences have been determined. The data of gene expression are growing at an unprecedented rate and are becoming more and more important in clinical diagnosis. Gene chip technology is a great challenge to the existing data processing and analysis methods, which brings the huge and complex gene expression data.

## 15.2.1 Bioinformatics on Gastric Cancer Information Science

The incidence of gastric cancer is regional. The study found that Asian regions, such as China, Japan, and South Korea, have a high incidence of disease. Secondly, the cause of gastric cancer is more complex; at present, the research showed that the cause of gastric cancer is related to the daily dietary factors, gastric ulcer disease, gastric polyps, and *Helicobacter pylori* infection. Through continuous research, it is found that improving the diagnosis effect of gastric cancer can improve the 5-year survival rate of gastric cancer patients [32]. Existing diagnostic methods of gastric cancer include: barium meal radiography diagnosis method, method of electronic gastroscope diagnosis, pathological biopsy diagnosis method, and gastric cancer marker for diagnosis method of screening. The methods of barium meal radiography and gastroscope and pathological examination have a high accuracy for early gastric cancer diagnosis. But due to the complexity of the process of these detection methods, it brought damages to the patients in physiology and the cost is higher. At the same time, domestic and foreign scholars have carried out a more thorough study on gastric cancer markers. By comparing the differences between patients with gastric cancer and healthy group, the markers of gastric cancer were found to be used in clinical diagnosis. These markers are usually expressed in group level, protein level, and metabolic level. However, the detection of these markers is complex and requires large instruments, and the diagnostic accuracy is still at a low level. The popularization rate is relatively low in China. The comprehensive analysis shows that the screening rate of early gastric cancer in China is still low. The markers of gastric cancer are influenced by individual differences, living habits, living environment, and other factors of the body. Usually, gastric cancer marker not only exists in the cancer but also exists in some normal tissues and benign lesions.

In 1978, the tumor marker (TM) was first proposed by the American scholar Herberman for the first time in the National Cancer Institute. Tumor markers are expressed in the tumor cells and their specific properties in the process of malignant tumor cells. Or the patient has a variety of physiological or pathological responses to the tumor, which is secreted into the tissue or body fluids, such as certain proteins, small molecules, and so on. These are collectively referred to as tumor markers. Tumor markers can be used to detect tumor and play a supporting role in the diagnosis [20]. According to the principle of its production, tumor markers are divided into two categories: one is the tumor itself, and the other is the interaction between the tumor and the body. We can reflect the characteristics of phenotype and genotype in all stages of malignant evolution and the characteristics of the material, which are called tumor markers.

## *15.2.2 Analysis of Gastric Cancer on Genomic Level*

### 15.2.2.1 Gastric Cancer Gene Analysis

The human genome is made up of 23 pairs of chromosomes. One chromosome is composed of the DNA molecules, A, G, T, and C. The gene is a fragment of the genetic effect on the DNA molecule. It is the structure and function of the genetic information. Gene is the chemical carrier of human genetic information, which decides the similarity and un-similarity between one person and his predecessors. In the "work" of the gene, our bodies can develop normally and function properly. If a gene is not normal, even if a very small piece of the gene is not normal, it can cause developmental abnormalities, diseases, and even death. The health of the body depends on its constant metabolic actions, ensuring normal protein quantity and quality. These proteins cooperate with each other to ensure the normal execution of various functions of the body. Each protein is a product of the corresponding gene. Genes can change, and some changes do not cause changes in the quantity or quality of the protein. This change in the gene is called a gene mutation. The change in quantity or quality of the protein may cause abnormal physical function of the body. Modern medical studies have shown that, with the exception of the injury, almost all diseases are related to genes. Like blood in different blood groups, the normal human gene in the human body is also divided into different genotypes, that is, gene polymorphism. Different genotypes of the environmental factors, the sensitivity of the genotype in the role of environmental factors, can cause disease.

For gene identification, gene chips are usually used. Gene chip, also known as DNA array, microarray chip, is a chip containing thousands of probes and can complete the detection of multiple genomes in a single experiment. Genes associated with gastric cancer, NCBI, have more than 1000 genes (shown in Supplementary table 1: Gastric cancer gene).

### 15.2.2.2 Gene Copy Number Analysis of Gastric Cancer

In the study of gastric cancer, one of the important indicators is copy number variation, that is, CNV (gene copy number variations) [33]. CNV is the abnormality caused by the structural variation of the genome, which makes the copy number of the genes changed. It mainly shows that the DNA copy number of the gene is amplified, gained, lost, and deleted. In the development of gastric cancer, the accumulation of gene copy number variation may result in the transformation of cells, which is usually the change of apoptosis and related to the apoptosis of gastric cancer. More and more evidence shows that the gene of CNV changes can be taken as a biomarker of gastric cancer and the target of treatment [34]. Twenty genes with CNV associated with gastric cancer are reported in the literature (shown in Supplementary table 2: Gastric cancer gene associated with CNV).

#### 15.2.2.3 SNP Analysis of Gastric Cancer

Single nucleotide polymorphisms (SNPs) are DNA sequence variations in populations (about 1 % of the populations) in a single nucleotide of A, T, C, or G (or other shared sequence). For example, two of the DNA fragments were sequenced from different individuals, AAGCCTA to AAGCTTA, and contained within a single nucleotide difference. The distribution of SNP in the genome is not uniform: in the non-encoding zone, SNP occurs more frequently than that in the encoding zone. Generally speaking, the role of the natural selection and the "fixed" allele (excluding other variants) are the most favorable genetic adaptation of SNP.

Biotechnology has been developed, along with the development of a class of high throughput, and high degree of automation SNP detection methods are created, such as direct sequencing, DNA chips, denaturing high performance liquid chromatography (DHPLC), mass spectrometry detection technology, high-resolution melting (HRM) curve, and so on. Direct sequencing of SNP was the most direct and reliable method. The detection rate was 100 %, which represents the sequencing technology, micro-sequencing technology, etc. The method is to detect SNP by comparing the results of PCR amplification products of the same gene or gene fragment of different samples or sequencing analysis of the sequence tags and the expression sequence tags. PCR product can be purified and recovered by connecting to the carrier for sequencing, and it can also be directly sequenced. By means of sequence alignment, the type and location of SNP mutation can be detected accurately. By this technique, 47,172 SNPs were detected in the human genome. According to the known SNP sites, two kinds of probes and the DNA were designed. The probe was immobilized on a special carrier, and the DNA was hybridized with the fixed probe. The sequence column of single mismatch was not hybridized with the probe. After a literature search, a total of 49 gastric cancer-related SNPs were found, as shown in Supplementary table 3: Gastric cancer SNP.

#### 15.2.2.4 mRNA Analysis of Gastric Cancer

Messenger RNA is a kind of single-strand RNA, which can guide the protein synthesis by DNA and is used as a template to carry the genetic information [35].

Multiple genes (or even the whole genome) are obtained by gene chips. A set of data in different physiological processes is called the data of large-scale gene expression. In the reverse transcription process, the test sample and reference sample RNA were labeled with different red and green fluorescent dyes and then hybridized with the probe sequence on the microarray. Then, the laser scanner was used to scan the chip, and the intensity of each point was obtained by using special image analysis software. The expression level of the gene in the experimental sample was obtained by the ratio of the gene [36]. Usually, in order to reflect the multiple relationships between the level of a gene expression in the experimental sample and reference sample, the ratio of the transformation is used to characterize the expression level of the gene [37]. In the NCBI database, mRNAs associated with gastric cancer have a total of more than 2000 (shown in Supplementary table 4: Gastric cancer mRNA).

#### 15.2.2.5 Analysis of Gastric Cancer in the Level of Proteomics

Protein is the material basis of life, which is the organic macromolecule, the basic organic matter, and the main carrier of life activities. The content of the protein includes the structure of the carrier, the carrier of the transport, antibody, immune, catalytic, and hormonal regulation. In 1994, Wilkins Marc of the University of Australia first proposed the concept of proteomics, which are all proteins expressed in the genome. By the analysis of comparing the proteome, the proteins identified, quantified, and characterized by differential expression are detected, and the new markers of disease are used as a target for the treatment and drug development [38].

With the application of proteomics in gastric cancer, the expression of specific proteins in gastric cancer was analyzed. Through these analysis results, the diagnosis of gastric cancer can be realized, and it can provide the related solutions to the treatment of gastric cancer [39].

#### 15.2.2.6 Proteomics of Gastric Cancer in Blood

In the human body, the blood is easy to obtain and use as the clinical specimen. Thus, blood can be used as an important source of finding and screening for biomarkers of gastric cancer. The methods for the study of proteomics in the blood are commonly matrix-assisted laser desorption ionization, time-of-flight mass spectrometry (MALDI-TOF-MS), and liquid chromatography tandem mass spectrometry [40, 41]. In the method of proteomics research, the combination with multiple different techniques can improve the sensitivity and specificity of the search for potential biomarkers. In the recent study, for example, the expression of SELDI-TOF-MS and protein chip was used to analyze the expression of proteins in the serum of patients with gastric cancer [42]. All the potential biomarkers in the blood of gastric cancer can be used as a potential marker for the diagnosis of gastric cancer, such as apoC-III and apoC-I [43]. Using thrombin light chain A [44], as a marker of gastric cancer, the diagnostic sensitivity of gastric cancer can be more than 89.9 %, while the specificity of the diagnosis more than 71 %.

#### 15.2.2.7 Proteomics of Gastric Cancer in Gastric Juice

Gastric juice can also be used as a source of protein markers of gastric cancer. The composition of gastric juice includes water, hydrochloric acid, lipase, pepsinogen, rennet, mucus, and other inorganic ions. The patient's stomach fluid can be obtained by gastroscopy. Through the detection of gastric liquid protein in gastric cancer patients and healthy people, it was discovered that there is a variety of gastric cancer-related proteins, including PGC, pepsin A [45], GKN1, and trefoil factor 1 [46].

#### 15.2.2.8 Proteomics of Gastric Cancer in Tissues

For the study of protein markers of gastric cancer in the tissue, the analysis by gastric cancer tissues and adjacent tissues of gastric cancer is usually used to analyze the correlation of the protein markers. Matrix-assisted laser desorption ionization, time-of-flight mass spectrometry, or liquid chromatography tandem mass spectrometry are commonly used methods [47, 48]. In the published studies, the potential protein markers in the tissue can be used to detect gastric cancer. For example, by HER2 for gastric cancer detection, the sensitivity can reach 65 %, while the specificity of detection is 92 % [49]. In other studies, the researchers found that there were significant differences between human neutrophil peptide 1–3 and MIF in gastric cancer tissues and healthy tissues [50].

#### 15.2.2.9 Proteins Associated with Gastric Cancer

In the NCBI database, a total of more than 380 proteins associated with gastric cancer, as shown in Supplementary table 5: Proteins associated with gastric cancer. With the continuous development of the research level of the protein group, there is continuous improvement of technical means, such as protein separation and identification technology and continuous improvement of information technology, to explore the pathogenesis, diagnosis, and treatment of gastric cancer and provide important theoretical basis for the development of new drugs.

### *15.2.3 Metabolic Analysis of Gastric Cancer*

Metabolomics was proposed by Professor Nicholson Jeremy at Imperial College of Science and Engineering in the UK in 1999. The study of metabolic components, which is based on the analysis of group index and the high throughput and data processing, is a branch of system biology. It is the science of the changes of all metabolites produced by external stimuli. It is a study on the metabolic changes of gastric cancer, which can be found in gastric cancer and is related to gastric cancer. The small molecular metabolite of gastric cancer is beneficial to the early detection and treatment of gastric cancer, so as to improve the quality of life of patients with gastric cancer.

#### 15.2.3.1 Metabolism in Gastric Cancer Serum and Plasma

Serum is the light yellow transparent liquid after the blood coagulation precipitation. The collection process is done by taking serum out from the inside of the vessel and into test tubes without anticoagulant. In this case, the coagulation reaction is activated, blood rapidly solidified into a jelly. The light yellow transparent liquid is the serum. The plasma is a pale yellow liquid, is the whole blood after anticoagulant treatment, through the centrifugal precipitation, the obtained the liquid of the cell component. In the study of gastric cancer, the proportion of plasma and serum was about 1:1. In the course of the metabolism of small molecules in the serum or plasma and normal population of patients with gastric cancer, the methods that are usually used include gas chromatography [52, 53] and liquid chromatography mass spectrometry [54]. By gas chromatography mass spectrometry, amino acid, fatty acid, cholesterol, carbohydrate nitriles, and small molecule compounds can be detected for gastric cancer diagnosis.

#### 15.2.3.2 Metabolism in Gastric Cancer Tissues

For the study of protein markers of gastric cancer in the tissue, the analysis of gastric cancer tissues and adjacent tissues of gastric cancer is usually used to analyze the correlation of the protein markers. The present study shows that there are abnormalities [55, 56] in the tissues of gastric cancer. Commonly used techniques include gas chromatography mass spectrometry and capillary electrophoresis. By comparing the detection of small molecules in gastric cancer tissues and healthy tissues, a variety of small molecules, such as amino acids, fatty acids, alcohols, acids, and bases, have been up- or downregulated [51, 57].

#### 15.2.3.3 Metabolism in Gastric Cancer Urine

In the course of the study of gastric cancer, the researchers analyzed the small molecular metabolites in urine. In the course of the study, sample selection was included in the model rats and patients with gastric cancer [58, 59]. By liquid chromatography coupled with mass spectrometry, the researchers found that there was a metabolic change in the metabolism of alanine, succinate, citric acid, uric acid, taurine, and so on.

#### 15.2.3.4 Small Molecule Associated with Gastric Cancer

Based on the published literature of gastric cancer metabolism, at present, there are many small molecules that are different in the metabolism of gastric cancer (shown in Table 15.4).

**Table 15.4** The molecules associated with gastric cancer

| Experiment object | Sample source | Differential expression | |
|---|---|---|---|
| | | Upregulation | Downregulation |
| Rat [58] | Urine | Lactic acid, malic acid, uric acid, glycerol | |
| Rat [57] | Urine | Citrate, 2-oxoglutarate, 3-indole sulfate | Taurine, trimethylamine, oxaloacetate, trimethylamine oxide, hippuric |
| Human [60] | Tissue | Lactic acid, malic acid, fumaric acid, guanosine monophosphate | Glucose, citric acid |
| Human [61] | Tissue | Fumaric acid, ketoglutaric acid, xylose acid, stearic acid, phenylpropionic acid, 1-phenanthrene carboxylic acid | n-Hexadecanoic acid, cis-octadecenoic acid, arachidonic acid, 9-oleoylethanolamide, squalene, 3-hydroxybutyric acid |
| Human [62] | Tissue | Lactic acid, fructose, glycerol, pyruvate, isocitrate | Fumaric acid |
| Human [63] | Tissue and plasma | Lactic acid, citric acid, malic acid, fumaric acid, cysteine docosahexaenoic acid, heptanoic acid, 9-z- palmitoleic | Glucose, fructose 6-phosphate glycerol, maltose, ribose, 2,3-diphosphonic acid, cholesterol, uridine, inositol, ribitol, β-d-methyl glucoside |
| Human [64] | Tissue | Furan-galactosidase, valine, isoleucine, serine, glutamine, pimelic acid, propionic acid, phenylalanine | Altrose, mannose, ribofuranose, phosphoserine, inositol |
| Human [54] | Serum | Serine, glutamic acid, ornithine, proline | Aspartic acid, valine, methionine, tyrosine, histidine, tryptophan, phenylalanine, leucine |
| Human [52] | Serum | Sarcosine, valine, cholesta-3,5-diene, pentafluoropropionic acid, cholesterol, cholesterol, cholest-5-ene | Fumaric acid, 2-O-mesyl arabinose, glutamine, adipic acid, 9,12-octadecadienoic acid, oleic acid |
| Human [53] | Serum | 3-Hydroxypropionic acid | Pyruvic acid, 3-hydroxy acid, phosphoric acid |

# References


1. Ponniah P. Chapter 1: the database approach. In: Database design and development: an essential guide for IT professionals, Vol. 1; 2005: p. 1–35.
2. Hogeweg P. The roots of bioinformatics in theoretical biology. PLoS Comput Biol. 2011;7(3):e1002021.
3. Oliver GR, Hart SN, Klee EW. Bioinformatics for clinical next generation sequencing. Clin Chem. 2015;61(1):124–35.
4. Abbott A. Bioinformatics institute plans public database for gene expression data. Nature. 1999;398(6729):646.
5. Benson DA, Cavanaugh M, Clark K, Karsch-Mizrachi I, Lipman DJ, Ostell J, Sayers EW. GenBank. Nucleic Acids Res. 2013;41(Database issue):D36–42.
6. Benson DA, Clark K, Karsch-Mizrachi I, Lipman DJ, Ostell J, Sayers EW. GenBank. Nucleic Acids Res. 2015;43(Database issue):D30–5.

7. Li W, Cowley A, Uludag M, Gur T, McWilliam H, Squizzato S, Park YM, Buso N, Lopez R. The EMBL-EBI bioinformatics web and programmatic tools framework. Nucleic Acids Res. 2015;43(W1):W580–4.
8. Kulikova T, Aldebert P, Althorpe N, Baker W, Bates K, Browne P, van den Broek A, Cochrane G, Duggan K, Eberhardt R, Faruque N, Garcia-Pastor M, Harte N, Kanz C, Leinonen R, Lin Q, Lombard V, Lopez R, Mancuso R, McHale M, Nardone F, Silventoinen V, Stoehr P, Stoesser G, Tuli MA, Tzouvara K, Vaughan R, Wu D, Zhu W, Apweiler R. The EMBL nucleotide sequence database. Nucleic Acids Res. 2004;32(Database issue):D27–30.
9. Boutet E, Lieberherr D, Tognolli M, Schneider M, Bairoch A. UniProtKB/Swiss-Prot. Methods Mol Biol. 2007;406:89–112.
10. Boeckmann B, Bairoch A, Apweiler R, Blatter MC, Estreicher A, Gasteiger E, Martin MJ, Michoud K, O'Donovan C, Phan I, Pilbout S, Schneider M. The SWISS-PROT protein knowledgebase and its supplement TrEMBL in 2003. Nucleic Acids Res. 2003;31(1):365–70.
11. Bhat TN, Bourne P, Feng Z, Gilliland G, Jain S, Ravichandran V, Schneider B, Schneider K, Thanki N, Weissig H, Westbrook J, Berman HM. The PDB data uniformity project. Nucleic Acids Res. 2001;29(1):214–8.
12. Laskowski RA. PDBsum: summaries and analyses of PDB structures. Nucleic Acids Res. 2001;29(1):221–2.
13. Touw WG, Baakman C, Black J, te Beek TA, Krieger E, Joosten RP, Vriend G. A series of PDB-related databanks for everyday needs. Nucleic Acids Res. 2015;43(Database issue): D364–8.
14. Jenuth JP. The NCBI. Publicly available tools and resources on the Web. Methods Mol Biol. 2000;132:301–12.
15. Edgar R, Domrachev M, Lash AE. Gene Expression Omnibus: NCBI gene expression and hybridization array data repository. Nucleic Acids Res. 2002;30(1):207–10.
16. Agarwala R, Barrett T, Beck J, Benson DA, Bollin C, Bolton E, Bourexis D, Brister JR, Bryant SH, Canese K, Clark K, DiCuccio M, Dondoshansky I, Federhen S, Feolo M, Funk K, Geer LY, Gorelenkov V, Hoeppner M, Holmes B, Johnson M, Khotomlianski VE, Kimchi A, Kimelman M, Kitts P, Klimke W, Krasnov S, Kuznetsov A, Landrum MJ, Landsman D, Lee JM, Lipman DJ, Lu ZY, Madden TL, Madej T, Marchler-Bauer A, Karsch-Mizrachi I, Murphy T, Orris R, Ostell J, O'Sullivan C, Panchenko A, Phan L, Preuss D, Pruitt KD, Rubinstein W, Sayers EW, Schneider V, Schuler GD, Sherry ST, Sirotkin K, Siyan K, Slotta D, Soboleva A, Soussov V, Starchenko G, Tatusova TA, Trawick BW, Vakatov D, Wang YL, Ward M, Wilbur WJ, Yaschenko E, Zbicz K, Coordinators NR. Database resources of the National Center for Biotechnology Information. Nucleic Acids Res. 2015;43(D1):D6–17.
17. Sayers EW, Barrett T, Benson DA, Bolton E, Bryant SH, Canese K, Chetvernin V, Church DM, DiCuccio M, Federhen S, Feolo M, Fingerman IM, Geer LY, Helmberg W, Kapustin Y, Krasnov S, Landsman D, Lipman DJ, Lu ZY, Madden TL, Madej T, Maglott DR, Marchler-Bauer A, Miller V, Karsch-Mizrachi I, Ostell J, Panchenko A, Phan L, Pruitt KD, Schuler GD, Sequeira E, Sherry ST, Shumway M, Sirotkin K, Slotta D, Souvorov A, Starchenko G, Tatusova TA, Wagner L, Wang YL, Wilbur WJ, Yaschenko E, Ye J. Database resources of the National Center for Biotechnology Information. Nucleic Acids Res. 2012;40(D1):D13–25.
18. Wheeler DL, Chappey C, Lash AE, Leipe DD, Madden TL, Schuler GD, Tatusova TA, Rapp BA. Database resources of the National Center for Biotechnology Information. Nucleic Acids Res. 2000;28(1):10–4.
19. Goujon M, McWilliam H, Li WZ, Valentin F, Squizzato S, Paern J, Lopez R. A new bioinformatics analysis tools framework at EMBL-EBI. Nucleic Acids Res. 2010;38:W695–9.
20. Wilhite SE, Barrett T. Strategies to explore functional genomics data sets in NCBI's GEO database. Methods Mol Biol. 2012;802:41–53.
21. Barrett T, Wilhite SE, Ledoux P, Evangelista C, Kim IF, Tomashevsky M, Marshall KA, Phillippy KH, Sherman PM, Holko M, Yefanov A, Lee H, Zhang NG, Robertson CL, Serova N, Davis S, Soboleva A. NCBI GEO: archive for functional genomics data sets-update. Nucleic Acids Res. 2013;41(D1):D991–5.

22. Cancer Genome Atlas Research Network. Comprehensive genomic characterization defines human glioblastoma genes and core pathways. Nature. 2008;455(7216):1061–8.
23. Network TC. Corrigendum: comprehensive genomic characterization defines human glioblastoma genes and core pathways. Nature. 2013;494(7438):506.
24. Zhu Y, Qiu P, Ji Y. TCGA-assembler: open-source software for retrieving and processing TCGA data. Nat Methods. 2014;11(6):599–600.
25. Rhodes DR, Yu J, Shanker K, Deshpande N, Varambally R, Ghosh D, Barrette T, Pandey A, Chinnaiyan AM. ONCOMINE: a cancer microarray database and integrated data-mining platform. Neoplasia. 2004;6(1):1–6.
26. Stenson PD, Ball EV, Mort M, Phillips AD, Shaw K, Cooper DN. The Human Gene Mutation Database (HGMD) and its exploitation in the fields of personalized genomics and molecular evolution. Curr Protoc Bioinformatics. 2012;Chapter 1:Unit1 13.
27. Cooper DN, Stenson PD, Chuzhanova NA. The Human Gene Mutation Database (HGMD) and its exploitation in the study of mutational mechanisms. Curr Protoc Bioinformatics. 2006;Chapter 1:Unit 1 13.
28. Becker KG, Barnes KC, Bright TJ, Wang SA. The genetic association database. Nat Genet. 2004;36(5):431–2.
29. Cheng J, Zeng X, Ren G, Liu Z. CGAP: a new comprehensive platform for the comparative analysis of chloroplast genomes. BMC Bioinformatics. 2013;14:95.
30. Chen Y, Hong J, Cui W, Zaneveld J, Wang W, Gibbs R, Xiao Y, Chen R. CGAP-align: a high performance DNA short read alignment tool. PLoS One. 2013;8(4):e61033.
31. He X, Chang S, Zhang J, Zhao Q, Xiang H, Kusonmano K, Yang L, Sun ZS, Yang H, Wang J. MethyCancer: the database of human DNA methylation and cancer. Nucleic Acids Res. 2008;36(Database issue):D836–41.
32. Werner S, Chen H, Tao S, Brenner H. Systematic review: serum autoantibodies in the early detection of gastric cancer. Int J Cancer. 2015;136(10):2243–52.
33. Wheeler E, Huang N, Bochukova EG, Keogh JM, Lindsay S, Garg S, Henning E, Blackburn H, Loos RJ, Wareham NJ, O'Rahilly S, Hurles ME, Barroso I, Farooqi IS. Genome-wide SNP and CNV analysis identifies common and low-frequency variants associated with severe early-onset obesity. Nat Genet. 2013;45(5):513–7.
34. Liang L, Fang JY, Xu J. Gastric cancer and gene copy number variation: emerging cancer drivers for targeted therapy. Oncogene. 2015;35(12):1475–82.
35. Miller BL. mRNA regulation: a patch for a splice. Nat Chem Biol. 2015;11(7):454–5.
36. Shao H, Chung J, Lee K, Balaj L, Min C, Carter BS, Hochberg FH, Breakefield XO, Lee H, Weissleder R. Chip-based analysis of exosomal mRNA mediating drug resistance in glioblastoma. Nat Commun. 2015;6:6999.
37. Tabata N, Sakuma Y, Honda Y, Doi N, Takashima H, Miyamoto-Sato E, Yanagawa H. Rapid antibody selection by mRNA display on a microfluidic chip. Nucleic Acids Res. 2009;37(8):e64.
38. Tomlinson AJ, Hincapie M, Morris GE, Chicz RM. Global proteome analysis of a human gastric carcinoma. Electrophoresis. 2002;23(18):3233–40.
39. Lin LL, Huang HC, Juan HF. Discovery of biomarkers for gastric cancer: a proteomics approach. J Proteomics. 2012;75(11):3081–97.
40. Liu W, Liu B, Xin L, Zhang Y, Chen X, Zhu Z, Lin Y. Down-regulated expression of complement factor I: a potential suppressive protein for gastric cancer identified by serum proteome analysis. Clin Chim Acta. 2007;377(1–2):119–26.
41. Chong PK, Lee H, Loh MC, Choong LY, Lin Q, So JB, Lim KH, Soo RA, Yong WP, Chan SP, Smoot DT, Ashktorab H, Yeoh KG, Lim YP. Upregulation of plasma C9 protein in gastric cancer patients. Proteomics. 2010;10(18):3210–21.
42. Ebert MP, Meuer J, Wiemer JC, Schulz HU, Reymond MA, Traugott U, Malfertheiner P, Rocken C. Identification of gastric cancer patients by serum protein profiling. J Proteome Res. 2004;3(6):1261–6.
43. Cohen M, Yossef R, Erez T, Kugel A, Welt M, Karpasas MM, Bones J, Rudd PM, Taieb J, Boissin H, Harats D, Noy K, Tekoah Y, Lichtenstein RG, Rubin E, Porgador A. Serum

apolipoproteins C-I and C-III are reduced in stomach cancer patients: results from MALDI-based peptidome and immuno-based clinical assays. PLoS One. 2011;6(1):e14540.
44. Ebert MP, Lamer S, Meuer J, Malfertheiner P, Reymond M, Buschmann T, Rocken C, Seibert V. Identification of the thrombin light chain a as the single best mass for differentiation of gastric cancer patients from individuals with dyspepsia by proteome analysis. J Proteome Res. 2005;4(2):586–90.
45. Kon OL, Yip TT, Ho MF, Chan WH, Wong WK, Tan SY, Ng WH, Kam SY, Eng A, Ho P, Viner R, Ong HS, Kumarasinghe MP. The distinctive gastric fluid proteome in gastric cancer reveals a multi-biomarker diagnostic profile. BMC Med Genomics. 2008;1:54.
46. Kam SY, Hennessy T, Chua SC, Gan CS, Philp R, Hon KK, Lai L, Chan WH, Ong HS, Wong WK, Lim KH, Ling KL, Tan HS, Tan MM, Ho M, Kon OL. Characterization of the human gastric fluid proteome reveals distinct pH-dependent protein profiles: implications for biomarker studies. J Proteome Res. 2011;10(10):4535–46.
47. Chen J, Kahne T, Rocken C, Gotze T, Yu J, Sung JJ, Chen M, Hu P, Malfertheiner P, Ebert MP. Proteome analysis of gastric cancer metastasis by two-dimensional gel electrophoresis and matrix assisted laser desorption/ionization-mass spectrometry for identification of metastasis-related proteins. J Proteome Res. 2004;3(5):1009–16.
48. Kikuta K, Kubota D, Saito T, Orita H, Yoshida A, Tsuda H, Suehara Y, Katai H, Shimada Y, Toyama Y, Sato K, Yao T, Kaneko K, Beppu Y, Murakami Y, Kawai A, Kondo T. Clinical proteomics identified ATP-dependent RNA helicase DDX39 as a novel biomarker to predict poor prognosis of patients with gastrointestinal stromal tumor. J Proteomics. 2012;75(4):1089–98.
49. Balluff B, Elsner M, Kowarsch A, Rauser S, Meding S, Schuhmacher C, Feith M, Herrmann K, Rocken C, Schmid RM, Hofler H, Walch A, Ebert MP. Classification of HER2/neu status in gastric cancer using a breast-cancer derived proteome classifier. J Proteome Res. 2010;9(12): 6317–22.
50. Mohri Y, Mohri T, Wei W, Qi YJ, Martin A, Miki C, Kusunoki M, Ward DG, Johnson PJ. Identification of macrophage migration inhibitory factor and human neutrophil peptides 1–3 as potential biomarkers for gastric cancer. Br J Cancer. 2009;101(2):295–302.
51. Jayavelu ND, Bar NS. Metabolomic studies of human gastric cancer: review. World J Gastroenterol. 2014;20(25):8092–101.
52. Song H, Peng JS, Yao DS, Yang ZL, Liu HL, Zeng YK, Shi XP, Lu BY. Serum metabolic profiling of human gastric cancer based on gas chromatography/mass spectrometry. Braz J Med Biol Res. 2012;45(1):78–85.
53. Ikeda A, Nishiumi S, Shinohara M, Yoshie T, Hatano N, Okuno T, Bamba T, Fukusaki E, Takenawa T, Azuma T, Yoshida M. Serum metabolomics as a novel diagnostic approach for gastrointestinal cancer. Biomed Chromatogr. 2012;26(5):548–58.
54. Miyagi Y, Higashiyama M, Gochi A, Akaike M, Ishikawa T, Miura T, Saruki N, Bando E, Kimura H, Imamura F, Moriyama M, Ikeda I, Chiba A, Oshita F, Imaizumi A, Yamamoto H, Miyano H, Horimoto K, Tochikubo O, Mitsushima T, Yamakado M, Okamoto N. Plasma free amino acid profiling of five types of cancer patients and its application for early detection. PLoS One. 2011;6(9):e24143.
55. Sabatini DM. mTOR and cancer: insights into a complex relationship. Nat Rev Cancer. 2006;6(9):729–34.
56. Kuhajda FP. Fatty-acid synthase and human cancer: new perspectives on its role in tumor biology. Nutrition. 2000;16(3):202–8.
57. Chan AW, Gill RS, Schiller D, Sawyer MB. Potential role of metabolomics in diagnosis and surveillance of gastric cancer. World J Gastroenterol. 2014;20(36):12874–82.
58. Hu JD, Tang HQ, Zhang Q, Fan J, Hong J, Gu JZ, Chen JL. Prediction of gastric cancer metastasis through urinary metabolomic investigation using GC/MS. World J Gastroenterol. 2011;17(6):727–34.
59. Liang Q, Wang C, Li B. Metabolomic analysis using liquid chromatography/mass spectrometry for gastric cancer. Appl Biochem Biotechnol. 2015;176(8):2170–84.
60. Kim KB, Yang JY, Kwack SJ, Park KL, Kim HS, Ryu DH, Kim YJ, Hwang GS, Lee BM. Toxicometabolomics of urinary biomarkers for human gastric cancer in a mouse model. J Toxicol Environ Health A. 2010;73(21–22):1420–30.

61. Hirayama A, Kami K, Sugimoto M, Sugawara M, Toki N, Onozuka H, Kinoshita T, Saito N, Ochiai A, Tomita M, Esumi H, Soga T. Quantitative metabolome profiling of colon and stomach cancer microenvironment by capillary electrophoresis time-of-flight mass spectrometry. Cancer Res. 2009;69(11):4918–25.
62. Cai Z, Zhao JS, Li JJ, Peng DN, Wang XY, Chen TL, Qiu YP, Chen PP, Li WJ, Xu LY, Li EM, Tam JPM, Qi RZ, Jia W, Xie D. A combined proteomics and metabolomics profiling of gastric cardia cancer reveals characteristic dysregulations in glucose metabolism. Mol Cell Proteomics. 2010;9(12):2617–28.
63. Aa JY, Yu LZ, Sun M, Liu LS, Li MJ, Cao B, Shi J, Xu J, Cheng LP, Zhou J, Zheng T, Wang XW, Zhao CY, Gu RR, Zhang FY, Shi RH, Wang GJ. Metabolic features of the tumor microenvironment of gastric cancer and the link to the systemic macroenvironment. Metabolomics. 2012;8(1):164–73.
64. Wu H, Xue RY, Tang ZQ, Deng CH, Liu TT, Zeng HZ, Sun YH, Shen XZ. Metabolomic investigation of gastric cancer tissue using gas chromatography/mass spectrometry. Anal Bioanal Chem. 2010;396(4):1385–95.

GPSR Compliance

*The European Union's (EU) General Product Safety Regulation (GPSR) is a set of rules that requires consumer products to be safe and our obligations to ensure this.*

*If you have any concerns about our products, you can contact us on ProductSafety@springernature.com*

In case Publisher is established outside the EU, the EU authorized representative is:

Springer Nature Customer Service Center GmbH
Europaplatz 3
69115 Heidelberg, Germany

**Batch number: 10756376**

Printed by Printforce, the Netherlands